The Etymological Dictionary of Earth Science

Matthew Horrigan
and
David Mustart

Department of Geosciences
San Francisco State University

Kendall Hunt
publishing company

Cover image © NASA.

Kendall Hunt
publishing company

www.kendallhunt.com
Send all inquiries to:
4050 Westmark Drive
Dubuque, IA 52004-1840

Copyright © 2009 by Matthew Horrigan

ISBN 978-0-7575-6711-7

Printed in the United States of America
10 9 8 7 6 5 4

Contents

User's Guide to the Dictionary

    Users of reference books generally resort to dictionaries when seeking an immediate clarification of meaning, confirmation of spelling, or choice of synonym. A dictionary, therefore, should be openable, useable, and understandable without the need to refer to explanatory notes, appendices, and lists of abbreviations. Ease of use has been a constant consideration while making this dictionary, the following user guidelines are brief. Reading this section will make this dictionary easier to use, and increase your understanding of its contents.

    I. Entry Term: The main entry term appears in boldface with a colon at the end. It is flush left to the margin of the column following a line space after the previous term.

    II. Etymology: The etymology is contained in brackets immediately after the main entry. It is a brief linguistic history of the term, showing which language or languages the term derives from, what the root words mean, and what, if any, special etymological forms apply. With the exception of Indo-European (abbreviated IE), all source languages are entirely spelled out. The root words from each source language are written in italic letters. An 'equals sign' (=) and the translated meaning of the root word follow the root word, provided it has a unique meaning compared to older words, it may be derived from. When there is a string of root words leading back in time through various languages, a less than sign (<) follows each root word of a different language, and signifies that it is derived from the next word in the string. When a term is composed of more than one fundamental root, those discrete root strings are separated by a plus sign (+). For example:

        endothermic{Greek *endon* = within + *therme* = heat < *thermos* = hot.]

    III. Part of Speech: When the term is a noun, adjective, or verb, an abbreviation (n., adj., or v.) identifying it as one of these parts of speech immediately follows the etymology. Prefixes, suffixes, acronyms, and the like are not assigned a part of speech indicator. For example:

        acicular: [Latin *acus* = needle.] adj.

IV. Definition: The definition is an explanation of the contemporary meaning of the term (often the term has multiple meanings in various disciplines). It may also include aspects of: usage, local variation, readily identifiable features, and physical characteristics that support the basic definition. For example:

> **lacustrine:** [Latin <i>lacus</i> = lake, pond, basin, tub] adj.
> In general usage, pertaining to lakes. In limnology, said of
> sediment and rock strata, which originated by deposition at the
> bottom of lakes, and of the flora and fauna resident in a lake.

V. Historical Connection: This feature of the dictionary explains how the modern meaning of the term relates back to its etymological roots. Because it is logically connected to the definition, the historical connection is part of the definition entry. It explains the transition from the current meaning to the term's past history.

VI. Related Terms: Related terms are often included in the historical connection, and are noted as 'Related terms'. They are words that share a common linguistic root with the main entry. They are chosen either because they are geological terms, or because they shed light on the historical connection.

VII. Indo-European Derivatives: Because so many different contemporary languages have arisen from a common Indo-European ancestral language, there are often numerous cognate words in English that have evolved via very different etymological routes. They therefore frequently have significantly different sounds and nuances of meaning from the main entry. Nevertheless, they also can appreciably enhance our understanding of the term's linguistic history, as well as provide insight into the evolution of human communication. For these reasons, Indo-European derivatives are often included as part of the historical connection. They are referred to by the abbreviated phrase, 'IE derivatives.'

VIII. Cited Usage: Within the body of the main entry a cited usage often follows the definition and historical connection section. It starts on a new line under the separate heading, '*Cited usage.*' It is a quote from a verified source that uses the main entry in context. Whenever possible, the passage relating to the original coining of the term is given. When the first usage in context has not yet been found, either an early usage of the term or a more recent quote that sheds light on some aspect of the term's use is given. Bibliographic references for the quote are given in brackets after each cited usage.

IX. Alternate Spellings: Alternate spellings are included under the separate heading, '*Alt. spelling.*' They comprise those variations of the term that sound significantly alike to be considered one and the same. Rare or irregular singular and plural forms are also included in this section.

X. Synonyms: Synonyms are those words that can be used interchangeably with the main entry. For all general purposes they have the same meaning. When synonyms exist, they are listed after their own italicized heading, '*Synonyms.*'

XI. Similar Terms: Because this dictionary adheres to a strict definition of what constitutes a synonym, there is also a section called, '*Similar terms.*' These are terms that have similar meanings to the main entry, but differ in at least one significant way.

XII. See Also: This section refers the reader to other terms in the dictionary that are in some way relevant to the main entry. It is in cross-referenced format, such that terms referred to in this section usually refer back to the main entry in their '*See also*' section.

XIII. Appendix I: Following the main body of the dictionary (A-Z), there is an appendix of linguistic terms that may help the reader understand the meaning of certain etymological forms, such as: back-formations, portmanteau words, and metanalysis.

XIV. Appendix II: Appendix II is a list of abbreviations used in the dictionary. Most of them are found in the cited usages.

The potential pitfalls and hidden hazards of creating a scientific document that purports to be true and correct, as well as the amount of time required to create such a work, makes us appreciate all the more the anonymous comment that, "working in academia is like committing suicide by throwing oneself in front of a moving glacier." Extending the metaphor then, writing this Earth science dictionary has been like walking through a skauk blindfolded. The following statements by some of history's premier lexicographers have helped me maintain perspective.

Lexicographer: a writer of dictionaries,
a harmless drudge, that busies himself in tracing the original,
and detailing the signification of words.
       -Dr. Samuel Johnson, *Dictionary of the English Language (1755),*

Down to the time of the first lexicographer
(Heaven forgive him!)
no author ever had used a word that *was* in the dictionary.
       -Ambrose Bierce, *The Devil's Dictionary (1906)*

Lexicographers...are always playing God.
Or if not God, then at least Moses,
descending from Sinai with the tablets of the law.
       -Jonathon Green, *Chasing the Sun (2002)*

If you steal from one author, it's plagiarism;
But if you steal from many, it's research.
       -Wilson Mizner

aa: [Hawaiian *ah'ah* = reduplication of *'ah* = stony, to burn.] adj. A clinkered, rough, and broken-surfaced lava flow comprising one of the two chief forms of lava emitted from shield volcanoes of the Hawaiian type. Folk etymology relates that the Hawaiian pronunciation of an *a'a* lava flow is *ahhhh!*, and refers to the sound one makes when walking across its surface barefoot.
Cited usage: The lavas (are) called by the natives "a-a," and its contrast with pahoehoe is about the greatest imaginable. It consists mainly of clinkers sometimes detached, sometimes partially agglutinated together with a bristling array of sharp, jagged, angular fragments [1883, C. E. Dutton, U.S.G.S. 4th Annual Report, 95].
See also: pahoehoe

abapical: [Latin *ab* = directed away from + *apex* = tip, top, pointed end.] adj. Directed away from the apex, especially when referring to gastropod or cephalopod shells, or the apical system of an echinoderm.
See also: adapical, adaxial, apex

abaxial: [Latin *ab* = away + *axis* = axle.] adj. Away from the center line, as in synforms, antiforms, or the ventral axis of an organism.
See also: abapical, adapical, adaxial, apex

ablation: [Latin *ablatus* = carried away < *ab* = away + *latus* = carried.] n. In glaciology, the loss of glacial ice by melting and evaporation (sublimation). Used also as an adjective in compound terms like ablation moraine, which is formed by rock debris that slowly accumulates at the toe of an ablating glacier.
Cited usage: Ablation till is lowered rather gently to the ground by wasting ice and is generally loose [1974, H. F. Garner, Origin of Landscapes viii. 488/2].

abrasion platform: [Latin *ab* = away + *radere* = to scrape + French *plate-forme* = diagram; < Latin *plattus* = flat + *forma* = form.] n. A flat surface at the edge of the sea produced by the abrading action of the waves. Etymologically related terms include: erode, corrode, rodent, and erase.
Cited usage: Lastly we have the more extensive, nearly horizontal plane produced by long-continued wave erosion, and commonly called the abrasion platform [1922, Flattely & Walton, Biology Sea-Shore ii. 36].

abyssal: [Greek *abyssos* = the deep < *a* = without + *byssos* = bottom.] adj. Referring to those features occurring at great depth, usually in reference to ocean bottom features, such as an abyssal plain or abyssal hill. Specifically, relating to the region of the deep ocean between the bathyal and hadal zones, from depths of approximately 2,000 to 6,000 meters. Figuratively in literature, the great deep, the primal chaos, the bowels of the earth, the cavity of the lower world, and the infernal pit. The Greek reference to 'the deep' stems from their ancient cosmogony that the abyssos was a subterranean reservoir of water lying beneath the Earth as a bottomless pit leading to the infernal regions. This concept led to numerous biblical and poetic references of the *abyssos* as: the great deep, the primal chaos, the bowels of the

earth, the cavity of the lower world, and the infernal pit.
Cited usage: O conscience, into what abyss of fears and horrors hast thou driven me, out of which I find no way, from deep to deeper plunged [1667, John Milton, Paradise Lost bk. X, line 842].
See also: bathyal, hado-

abysso-: [Greek *abyssos* = 'bottomless', when used as an adjective, or 'the deep' when used as a noun < *a* = without + *byssos* = bottom.] In marine science, a combining prefix meaning deep pelagic water. Specifically, a depth between two and six-thousand meters, corresponding to the continental rise; deeper than bathy-, but shallower than hado-.
Cited usage: The name of this zone (abysso-pelagic) comes from the Greek meaning "no bottom", and refers to the ancient belief that the open ocean was bottomless [2003, Anne Stewart, http://oceanlink.island.net].
See also: bathy-, hado-

acanthite: [Greek *akantha* = thorn.] n. A black, monoclinic mineral, silver sulfide (Ag_2S). It is dimorphic with argentite and is stable below 170°C. Acanthite commonly forms as branching or reticulated groups, and the name is an allusion to this habit. It is an important ore of silver and occurs in epithermal veins with native silver and the ruby silvers.

accretion: [Latin *ad* = to + *crescere* = to grow.] n. The process of growth of an inorganic body by external addition of fresh material, as in planetary accretion, continental accretion, accretionary prism, or accretionary lapilli. The Latin prefix *ad*-takes on a consonance to agree with what follows, so *ad* + *clamat*, meaning acclimate; and *ad* + *rivo*, meaning arrive.
Cited usage: As for stones, seeing they are made by the accretion of many hard particles within the earth [1656, Hobbes's Elements of Philosophy 479].

accurate: [Latin *accurare* = to do with care < *ac* = to + *curare* = to care for < *cura* = care.] adj. Conforming exactly to fact; errorless. Deviating only slightly or within acceptable limits from a standard. There is a notable distinction between accuracy and precision. Although its etymology means careful, accuracy implies correctness, whereas precision derives from to *caedere*, meaning to cut, and implies scrupulosity in measurement.
See also: precise

acicular: [Latin *acicula* = hairpin (diminutive of) *acus* = needle; < IE *ak* = sharp.] adj. A slender needlelike body, such as the spines of a porcupine, or the needlelike crystals of certain minerals, actinolite for example. IE derivatives include: acrid, acerbic, acid, acute, and vinegar.
Cited usage: The phosphate of lime forms small acicular prisms [1836, Todd's Cyclopaedia of Anatomy & Physiology II. 234/2].

acid rock: [Latin *acidus* < *acere* = to be sour; < IE *ak* = sharp + French *roche* < Old French *roque* < Latin *rocca* = rock.] n. Obsolete term relating to the supposed existence of silicic acid (H4SiO4) in a molten rock. Term was applied to igneous rocks having greater than 66% silica content. Often loosely and incorrectly equated with felsic. In one silica-content classification scheme, rocks with more than 66% silica are termed acid. In general, the gradation from acid to basic corresponds to an increase in color index (percent of dark minerals), which has led to the development of a parallel classification system from felsic to mafic (i.e., light to dark).
Synonyms: silicic
See also: felsic, leucocratic

acmite: [Greek *acme* = point; IE *ak* = sharp.] n. A brownish, brittle mineral belonging to the pyroxene group of silicates. First named for the shape of its crystals.

Cited usage: The most common indications of peralkaline rocks are the sodium pyroxene aegerine (acmite) and the sodium amphibole riebeckite [2003, IUGS Rock Names].

acre: [Sanskrit *ajras* = plain, untenanted land, open country, forest.] n. A unit of area used in the U.S. and Britain, employed in land and sea floor measurement, and equal to 160 square rods, or 1/640 square mile. Originally, as much as a yoke of oxen could plough in a day, consisting of thirty-two furrows of the plough, each a furlong in length. Adopted by medieval Latin speakers as *acra*, hence the spelling *acre* as opposed to the regular *aker*.

acritarch: [Greek *akritos* = uncertain < *a* = not + *kritos* = to decide + *arch* = origin.] n. Any of a group of unicellular, planktonic microfossils of various uncertain taxonomic affinities having hollow, organic tests; held to represent the earliest known eukaryotic organisms.
Cited usage: Left behind by this transfer is a 'residue' of forms of unknown affinities for which the name Hystrichosphaerida is no longer appropriate. It is for this 'residue' that I propose the name acritarchs.... The name chosen implies no affinity with any other organisms and is not derived from the name of any taxon included in the group [1963, W. R. Evitt, Proc. Nat. Acad. Sci. XLIX. 158].

acronym: [Greek *akros* = extreme, tip, peak, summit + *onuma* = name; < IE *ak* = sharp.] n. A word formed by combining the first letters or parts of a series of words in a descriptive name. Prefix chosen to emphasize 'beginning'. Etymologically related terms include: acid, oxygen, acropolis, acrobat, and acrophobia.
Cited usage: Words made up of the initial letters or syllables of other words I have seen called by the name acronym [1943, American Notes & Queries Feb. 167].

actinic light: [Greek *aktin* = ray + Middle English *liht*; < Old English *leoht*; < IE *leuk* = light.] n. Photochemically active radiation, as of the sun.
Cited usage: A beam of solar light is made up of three distinct sets of rays—the luminous, the calorific, and the chemical or actinic rays [1845, Penny Cyclopedia I. 167/2].

actinolite: [Greek *aktin* = ray + *lithos* = rock, stone.] n. A greenish variety of amphibole, usually in aggregated, long prismatic crystals. Named for its common radiating habit.
Cited usage: Actinolites exhibit a tendency to radiation in the mineral kingdom [1835, Wm. Kirby, The History, Habits & Instincts of Animals I. vi. 193].
See also: asbestos, hornblende, amphibole, tremolite

adamantine: [Greek *adamas* = invincible, diamond, hard steel < *a* = not + *damao* = to tame; < IE *dem* = to constrain, to force, especially to break in horses.] adj. Incapable of being broken. Of minerals, having a high hardness, brilliant luster and a high index of refraction. In ancient Greek and Roman times the term was applied equally to steel, the hardest known metal, and to a legendary 'hardest known crystalline gem.' Theophrastus first applied the term to the Emery Stone of Naxos, a form of corundum. Hardness should not be confused with toughness, the hard and brittle diamond shatters more easily than some other less hard minerals. IE derivatives include dauntless and indomitable.
See also: diamond, corundum

adamellite: [Toponymous term after *Monte Adamello*, a mountain in Italy.] n. An igneous intrusive rock that contains 10-50% quartz and an alkali feldspar to total feldspar ratio of 35-65%. Adamellite is the equivalent of the former U.S. definition of quartz monzonite (the present IUGS

definition allows for 5-20% quartz). Cathrein coined this term in 1890 for the orthoclase-bearing tonalites found on Mount Adamello, but Brogger first used it in its modern sense in 1895.
See also: tonalite, monzonite

adapical: [Latin *ad* = toward, directed to + *apicis* < *apex* = tip, top, pointed end.] adj. Directed toward the apex, especially when referring to gastropods, cephalopods, or echinoderms. The genitive of apex is apicis, and this explains the adjectives apical, adapical, and abapical.
See also: abapical, adaxial, apex

adaxial: [Latin *ad* = to, toward + *axis* = axle.] adj. In structural contexts, toward the axis, as in a synform or antiform. In paleontology, toward the ventral (belly) portion of an organism.
See also: abaxial, abapical, adapical, apex

adhesion: [Latin *ad* = to, toward + *haerere* = to stick.] n. The action of one thing attaching itself to another by physical or chemical attraction (as opposed to cohesion). In soils, used with reference to a film of molecularly bonded water molecules attached to mineral particles in the zone of aeration. In biology, with reference to animals that adhere to a surface, either by chemical secretion, as in barnacles, or physical tenacity, like the holdfasts of algae.
See also: cohesion

adiabatic: [Greek *a* = not + *diabatos* = to pass < *dia* = through, across + *bainein* = to go.] adj. Involving neither loss nor gain of heat. Used with reference to changes in temperature of a fluid mass as a function of a change in pressure. Of, relating to, or being a reversible thermodynamic process that occurs without exchange of heat and without a change in entropy. Used largely in meteorology and oceanography to explain why temperatures of discreet volumes of fluid rise and

fall with changes in elevation but no input or output of heat. In an adiabatic process, compression causes a rise in temperature, whereas expansion causes a drop.
Cited usage: A certain curve, which may be called a curve of no transmission, or adiabatic curve [1859, W. J. M. Rankine, A Manual of the Steam Engine III. iii. 302].

adit: [Latin *aditus* = access, approach < ad = to, toward + *itus* = going < *ire* = to go.] n. A horizontal opening by which a mine is entered or drained, and an ore-body is accessed. The attitude of an adit is level, as opposed to the vertical orientation of a shaft.
Cited usage: At the exit of a typical qanat you see a tunnel adit similar to a mine entrance, which is exactly what it is. The 'mineral' mined is water [2000, Rod Baird, Ancient Routes].

adobe: [Spanish *adobar* < Latin *adobare* = to daub, to plaster; < (possibly) Arabic *altuba* = the brick < *al* = the + *tuba* = a brick; < Egyptian *dbt* = brick.] n. A sun-dried, unburnt brick made of clay, mud, and straw. In geology, used with reference to clay-rich soils or clayey material in desert playas.
Cited usage: The houses in Costa Rica are built of adobes or undried bricks two feet long and one broad, made of clay mixed with straw to give adhesion [1841, J. L. Stephens, Incidents of Travel in Cent. Amer., Chiapas and Yucatan 224]

adsorption: [Latin *ad* = to, toward + *sorbere* = to suck.] n. Accumulation and adherence of ions or molecules (gases, liquids, or solutes) to the surface of a solid or liquid.
Cited usage: One property of colloids is their capacity for taking dissolved substances out of solution and retaining them. This phenomenon is now generally called 'Adsorption' [1912, E. Hatschek, Introd. Physics & Chem. Colloids (1913) i. 5].

adularia: [Toponymous term after Italian *Adula*, a mountain group of southeast Switzerland.] n. A variety of transparent or translucent potassium feldspar formed at low temperatures in hydrothermal veins, especially in the Swiss Alps. Folklore suggests holding adularia beneath the tongue in one's mouth at the full moon when wanting to divine the future, refresh one's memory, or awaken tender passions in the heart; thus the common name moonstone.
Synonyms: moonstone
See also: orthoclase, sanidine

advection: [Latin *ad* = to, toward + *vehere* = to carry; < IE *wegh* = to go.] n. In meteorology, the transfer of a property of the atmosphere, such as heat, cold, or humidity, by the horizontal movement of an air mass. In oceanography, the movement of water, as in an ocean current. Also, any advectable quantity, such as those created by disequilibrium of heat, salt, or oxygen. In ore deposits, the unidirectional mass transport of solutes by flow of hydrothermal fluid through permeable rock (as opposed to diffusion). Indo-European derivatives include graywacke, weight, away, and wagon.
See also: graywacke, diffusion

aegirine: [Eponymous term after Norwegian *aegir* = Norse god of the sea.] n. Sodic pyroxene found in alkalic igneous rocks associated with sodic amphiboles, alkali feldspars, and nepheline. The mineral is common in alkali granites, quartz syenites, and nepheline syenites (all alkalic plutonic rocks) and is also found in sodic volcanic rocks such as peralkaline rhyolites. Berzelius named this form of pyroxene in 1835. The type locality is in Norway.
Alt Spelling: aegerine
See also: augite

aeolian: [Greek *Aiolos* = mythological god of the winds < *aiolos* = quick, changeable.] adj.

Pertaining to action of the wind, as in sand dunes, seracs, and the reg.
Alt Spelling: eolian
See also: eolian

agate: [Latin *achates* < Greek *akhates* = river in Sicily where agates were found.] n. A translucent cryptocrystalline variety of quartz, frequently mixed or alternating with opal, and characterized by alternating colored bands, clouds, or mosslike forms; a type of variegated chalcedony. The river *Achates* could just as easily have been named after the stone.

agglomerate: [Latin *ag-(ad-)* = to, toward + *glomerare* = to wind up, gather into a ball < *glomus* = ball.] n. A mass consisting of volcanic or eruptive fragments, which have united by the action of heat; as opposed to a water deposited conglomerate.
Cited usage: This great overlying deposit is a white tufaceous agglomerate [1830, Lyell, Principles of Geology II. ii. xxvii. 72].
See also: conglomerate

aggradation: [Latin *ag-* = to, at + *gradus* = to step, walk.] n. The process of raising a surface by the deposition of detritus. Opposite of degrade.
Cited usage: In its course near the sea a river aggrades during the mild phases and erodes during the cool phases [1946, F. E. Zeuner, Dating the Past v. 130].

aggregate: [Latin *aggregare* = to add to the flock <*ag- (ad-)* = to + *gregare* = to collect, to herd; < Greek *grex* = flock.] n. In general, collected particles or units constituting a single mass or body. In geology, distinct minerals combined into a single rock, such as granite. Also, particulate matter, such as variously sized particles or sediment combined together into a composit mass, an example being sand and gravel aggregate for making concrete.

agonic line: [Greek *a* = without + *gonia* = angle; < IE *genu* = knee + Latin *linea* = line < *linum* = flax.] n. An imaginary line through all points on the earth's surface at which the magnetic declination is zero; the locus at which magnetic north coincides with true north. The IE root gives us diagonal, amblygonite, orthogonal, angle, and corner.
Cited usage: In certain parts of the earth the magnet coincides with the geographical meridian. These points are connected by an irregularly curved imaginary line, called a line of no variation or agonic line [1863, E. Atkinson, trans. of Ganot's *Elements do physique* ("Elementary Treatise of Physics, Elementary and Applied") IV].
See also: orthogonal

aiguille: [French *aiguille* < Latin *acus* = needle; < IE *ak* = sharp.] n. A slender, sharply-pointed peak (of rock); especially the numerous peaks of the Alps, which are so named.

alabaster: [Toponymous term from Egyptian *Alabaster*, an ancient city in Egypt < *Bast* = Egyptian lion goddess of the sun.] n. A fine-grained massive variety of gypsum. Because the rock limestone (compound of calcite) is only a little harder than gypsum, and is also mined in Egypt, the translucent or variegated varieties of stalagmitic limestone (travertine) are often called Oriental alabaster, or incorrectly alabaster. The Wadi Gerawi and Helwan quarries in Upper Egypt served as sources of alabaster since the 4th millennium BCE. Egyptian alabaster has unique and prized coloration due to iron and copper rich solutions that percolated through the soft gypsum.
Cited usage: The Oriental alabaster, or alabaster of the ancients, is to be carefully distinguished from the mineral now commonly known as alabaster; the former is a carbonate, the latter a sulphate of lime. [1875 Andrew Ure, Dictionary of Arts, Manufactures, and Mines I. 41]

alb: [Latin *alba* = white vestment or tunic <*albus* = white; < Greek *alphos* = white leprosy; < IE *albho* = white; (possibly influenced by Celtic *alp* = high mountain).] n. A nearly flat shelf separating the vertical side of an alpine glacial trough from the mountain slope above.
See also: alp

albedo: [Latin *albus* = white.] n. The ratio of the amount of energy reflected by a surface to the amount of energy that falls upon it, said especially of a celestial body. The relation to white being the visible color of the highest possible amount of reflected radiation, as opposed to a black body.

albite: [Latin *albus* = white + Greek *-ites* < *-lite* = pertaining to rocks < *lithos* = rock, stone.] n. A colorless or milky-white, sodium-rich member of the plagioclase series of feldspars with composition ranging from Ab100 An0 to Ab. It is also an alkali feldspar. Etymologically related terms are albino and albumin.

Algonkian: [Eponymous term after the Algonquian language spoken by the tribes of indigenous 'Maliseet' people of North America.] adj. An epithet designating a geologic era of time between the Archaean and the Paleozoic. Synonymous with Proterozoic. *Algonquian* is a French term derived from the Maliseet word meaning dancer. The rocks that make up the Algonkian series are a feature of the region of Lake Superior, a territory of the Algonquian Indians. Invented by the U.S.G.S. to replace the Agnotozoic eon.
Synonyms: Proterozoic
See also: Proterozoic

alidade: [Arabic *al-idadah* = the revolving radius of a graduated circle < *adud* = the humerus or upper arm.] n. An indicator or sighting apparatus on a plane table that measures degrees of arc. The connection to the humerus because of the way in which it revolves in its socket.

alkali: [Arabic *al-* = the + *qaliy* = ashes < *qalay* = to fry, roast.] n. In chemistry, a strong base with high ph. In mineralogy, said of silicate, halide, or carbonate minerals that contain alkali metals (Na, K) but little calcium. Early chemists obtained their high ph reagents by burning marine plants (Salicornia, pickleweed) and separating the solute from the precipitate by the percolation of water through wood ashes.

Allegheny orogeny: [Algonquian *welhik* = most beautiful + *heny* = stream; the name *Welhik-heny* was Anglicized to *Allegheny* + Greek *oros* = mountain + *genesis* = origin, creation < *genesthai* = to be born, come into being; < IE *gen* = to give birth, beget.] n. The third and last part of the Appalachian Orogeny, it formed as a result of the Eastern Laurentia-Gondwana (Africa) collision from mid-Late Pennsylvanian into Early Permian Periods. Resultant mountain range was at least as high as present Himalayas. The Iroquois people, who inhabited western Pennsylvania, considered the Allegheny River to be the upper part of the Ohio River. *Ohio* in Iroquois means beautiful river, from *oh*, meaning river, and *io*, meaning good, fine, or beautiful. When the Delaware, an Algonquian people, moved to western Pennsylvania in the 18th century, displacing the Iroquois, they translated the Iroquoian *Ohio* into the Delaware language, yielding *welhik-heny*.
See also: taconite

allo-: [Greek *allos* = other, different; IE *al-* = beyond.] Combining prefix meaning other or different. IE derivatives include allegory, parallax, parallel, alien, and ultimate.

allochthon: [Greek *allos* = other, foreign + *cthon* = earth, soil.] n. In tectonics, a mass of rock that has been moved from its place of origin by tectonic process, as in a thrust sheet or nappe. Similar in meaning to allogenic, which refers to constituents rather than whole formations. As an adjective, consisting of or formed from transported detritus; not formed in situ, as opposed to autochthonous. Term coined by K. W. von Gümbel circa 1888.
Cited usage: Nappes characteristic of the true geosyncline which have often been moved far from their place of origin, and to which the term allochthonous is applied [1932, A. Steers, The Unstable Earth iii. 114].

allotriomorphic: [Greek *allostrios* = belonging to another <*allos* = other, different + *trios* = belonging + *morphos* = shape, form.] adj. Refers to a holocrystalline texture of an igneous or metamorphic rock with anhedral crystals not bound by their own crystal faces but with their forms controlled by adjacent mineral grains. Term coined by H. Rosenbusch in 1886.
Cited usage: Allotriomorphic, a term applied to crystals which have taken their shape from their surroundings and have not been allowed to develop the contours prescribed by their crystalline form [1958, A. D. Merriman, Dictionary of Metallurgy 5/1].
Synonyms: xenomorphic, anhedral
See also: xeno-, xenocryst

allotropic: [Greek *allotropia* = changeableness, of mind or manner < *allos* = other, different + *tropos* = turning.] adj. Polymorphism applied to an element; a property of an element that can exist in different forms at differing temperatures and pressures.
Cited usage: Sulphur becomes allotropic by the continued application of heat [1869, M. Somerville, Molecular Science i. i. 16].
Synonyms: polymorphism
See also: polymorphism

alluvium: [Latin *alluvius* = washed against; < *al-* = to + *luere* = to wash; < IE *leu* = to wash.] n. A deposit of silt, sand, gravel, and other transported matter deposited by streams and other bodies of flowing water. Terms derived from the IE root *leu* include: colluvium, dilute, eluvium, latrine, and launder.

Cited usage: The Mississippi, by the continual shifting of its course, sweeps away considerable tracts of alluvium [1830, Lyell, Principles of Geology 187].

See also: Holocene, fluvial, proluvium, eluviation, illuviation, diluvium

almandine: [Toponymous term from Turkish *Alabanda*, a city in ancient Asia Minor.] n. A deep violet-red garnet, $FeAl_2Si_3O_{12}$, found in metamorphic rocks and used as a gemstone. Said to be a corruption of Pliny's *alabandine*, a term applied to garnets cut and polished at Alabanda.

Synonyms: almandite

See also: ugrandite, pyralspite, pyrope, uvarovite, andradite, grossularite, garnet

alp: [Gaelic *alp* = a high mountain; (possibly) Latin *albus* = white;.] n. As a plural Alps is a proper name of the tallest mountain range in central and southern Europe. As singular, a single peak, or in Switzerland, also the green pasture land on a mountain side. The general term *alp* derives as a back-formation from the Alps Mountains. The Latin historian Servius first related the term to Celtic roots. The relation to Latin *albus* comes from the highest peaks being snow-covered, and therefore white.

Cited usage: Between the forests and the cap of perpetual snow is a zone of alp, pasture land with drought-resisting bushes. Also, any high, especially snow-capped, mountains [1922, W. G. Kendrew, The Climates of the Continents xi. 54].

altiplano: [Spanish *alto* = high + *plano* = flat; < Latin *altus* + *planus*.] n. Specifically, the high tableland between the Western and Eastern Cordilleras of the Andes, extending over 200,000 square km from Peru through Bolivia to Argentina. Generally, a high plateau or basin.

altitude: [Latin *altus* = high, deep; < IE *al* = to grow, nourish + Latin *tudo* = the quality of being.] n. The height of a point above a reference level, especially above sea level or above Earth's surface. In astronomy, the angular distance of a celestial object above the horizon.

Cited usage: The Himalaya chain has a mean altitude of about 18,000 feet [1880, Haughton, Phys. Geogr. ii. 43]. The apparent altitude of the heavenly bodies is always greater than their true altitude [1849, M. Somerville, The Connection of Physical Sciences xviii. 172].

Synonyms: elevation

See also: elevation, altocumulus

altocumulus: [Latin *altus* = high; < IE *al* = to grow + *cumulo-* = heap; < IE *keu* = to swell.] n. A cloud-formation made up of rounded masses similar to cumulus clouds but at a higher altitude. IE derivatives from *keu* include cave, cavern, excavate, and coelenterate.

alum: [Latin *alumen* = alum, bitter salt < *aluta* = tawed pig or goat skin (treated with aluminum salts to whiteness); Greek *aludoimos* = bitter.] n. A white to colorless, water soluble double sulfate, potassium sulfate aluminum sulfate hydrate, $K_2SO_4 \cdot Al_2(SO_4)_3 \cdot 24H_2O$. The name was used in ancient times as a general term for hydroxy aluminum sulfates.

aluminum: [Latin *alumen* = alum, bitter salt < *aluta* = tawed pig or goat skin (treated with aluminum salts to whiteness); Greek *aludoimos* = bitter.] n. A whitish to silver-colored ductile metallic element, aluminum is relatively light and corrosion resistant. It is the most abundant metal in the earth's crust, but is not found native, rather always in combination with other elements particularly oxygen and silicon. Term coined by Sir Humphry Davy in 1812, calling it *alumium* in adherence to the convention of ending element names in *-ium*. His colleagues convinced him to

change it to *aluminum*, but the International Union of Pure and Applied Chemistry (IUPAC) later changed it to *aluminium* so that it once again conformed with conventional chemical nomenclature. The U.S. reverted back to *aluminum* in 1925, creating yet another British-American linguistic anachronism. The bitterness of many fermented barley beverages has led to a particular type being called *ale*, alluding to the meaning of the term's root.
Cited usage: Aluminium, for so we shall take the liberty of writing the word, in preference to aluminum, which has a less classical sound [1812, Quarterly Review]. Until 1925, the word was 'aluminium' even in the U.S., but in that year the American Chemical Society decided to change it. We also got "sulfur" in that same year, which still looks silly, and was not universally adopted by the engineering world. Fortunately, the urge for simplified spelling did not result in Fosforus [1925, James B. Calvert].
Alt Spelling: aluminium

alunite: [Latin *alumen* = alum, bitter salt < *aluta* = tawed pig or goat skin (treated with aluminum salts to whiteness); Greek *aludoimos* = bitter.] n. A white to gray to tan-colored trigonal mineral, hydroxy potassium aluminum sulfate, $KAl_3(SO_4)_2(OH)_6$. It is isomorphous with natroalunite and forms by hydrothermal alteration of potassium feldspar-bearing igneous rocks, and has been mined as a source of alum. A similar mineral species is jarosite, the iron analogue of alunite. The name is derived from the contraction of the Latin *alumen*. First usage in English c. 1865.
Synonyms: alumstone, alum rock

amazonite: [Toponymous term after discovery locality along the Amazon River in Brazil.] n. An apple-green, bright-green, or blue-green variety of microcline, with the color imparted by trace amounts of lead. It is found in granitic pegmatite. First usage in English c. 1600.

amber: [Arabic *anbar* = ambergris, amber.] n. A hard translucent yellow, orange, or brownish-yellow fossil resin, used for making jewelry and other ornamental objects. The term in Arabic originally applied only to ambergris, the residue from the intestines of a whale, but was later extended to mean amber, the fossil resin, due to some confusion of the substances. In French, the two are distinguished as grey amber (ambergris) and yellow amber (fossil resin).

amblypod: [Greek *amblyos* = blunt + *poda* = foot.] n. A now-defunct order of large, mainly hoofed ungulates that were widely distributed in North America, Europe, and East Asia in the Early Tertiary Period. Amblypoda is now separated into the orders Pantodonta and Dinocerata. Term coined c. 1875 by E.D. Cope.
Cited usage: All ungulates, in passing from the taxeopodous to the diplarthrous stages, traversed the amblypodous [1887, E. D. Cope, Amer. Naturalist XXI. 987].

amethyst: [Greek *amethustos* = not intoxicating, remedy for intoxication < *a* = not + *methuein* = to be drunk < *methu* = wine; < IE *medhu* = honey, mead.] n. A variety of quartz violet in color, of different degrees of intensity, and colored by trace amounts of manganese. The mineral amethyst is named from the supposition that it was a preventive of intoxication.

ammonia: [Eponymous name from the gas originally obtained from sal ammoniac found in a region near the temple of Amen in Libya; < Greek *Ammon* < Egyptian *Ammun*.] n. A colorless gas with pungent smell and strong alkaline reaction. Chemically, a compound of three equivalents of hydrogen with one of nitrogen (NH_3). First isolated in the laboratory by Bergman in 1782. Originally called 'spirit of hartshorn' in allusion to an early source, the dry distillation of nitrogenous matter in the hoofs and horns of animals.
Synonyms: spirit of hartshorn, animal alkali

ammonite: [Latin *cornu Ammonis* = horn of Ammon; < Greek *Amen*.] n. A fossil genus of extinct cephalopods characterized by a large, chambered shell coiled in a spiral plane. The ammonoids, important as zone fossils in the Paleozoic, were widespread and diverse from the Devonian to the Cretaceous. In 1792 biologist Bruguiere first described their resemblance to the involuted Horn of Jupiter Ammon. Ammon, the god of industry and oracles of ancient Egypt, had the curled horns of a ram on his head. The similar curled shape of the cephalopod genus Ammonoidea led people to belive they were petrified coiled snakes, hence their common name of 'snake-stones.' There is no relation to another group of Ammonites, the Semitic descendants of Ben Ammi (son of Lot), who in ancient biblical times lived east of the river Jordan. The Chinese have a parallel imagery for ammonites, there called *jiao-shih*, meaning horn stones. The 'Horn of Ammon' also refers to the hippocampus, a portion of the human brain.

amoeba: [Greek *amoibe* = change < *ameibein* = to change.] n. Any of various one-celled aquatic or parasitic protozoans of the genus Amoeba or related genera classified in the domain of Eukaryota and the kingdom Protista. Known for its perpetually changing shape due to both its highly flexible outer membrane plus mobile inner protoplasm. It moves by means of pseudopods.
Cited usage: Amoebae are little masses of protoplasm, moving and taking food by means of pseudopodia [1878, Macalister, Invertebr. 22].
Alt Spelling: amoebae (pl.)

amorphous: [Greek *a* = without, not + *morph* = shape + *ous* = abounding in, full of, characterized by, of the nature of.] adj. Having no determinate shape. Not composed of crystals in physical structure; uncrystallized; massive.

amphibian: [Greek *amphibios* = living a double life, an animal living in both realms < *amphi-* = on both sides + *bios* = life.] n. A cold-blooded, smooth-skinned vertebrate of the class Amphibia, such as a frog or salamander, that characteristically hatches as an aquatic larva with gills. The larva then transforms into an adult having air-breathing lungs. Etymologically related terms include: amphineuron, amphoteric and amphitheater.

amphibole: [Greek *amphibolos* = ambiguous, doubtful < *amphiballein* = to throw around doubt < *amphi* = on both sides + *ballein* = to throw.] n. Any of a large group of structurally similar hydrated double-chain silicates, such as hornblende, containing various combinations of sodium, calcium, magnesium, iron, and aluminum. The entire group of amphiboles are closely related in crystal structure, but crystallize in two different mineral systems, orthorhombic and monoclinic. Named in 1801 by René-Just Haüy because of the mineral's variety of composition and visual ambiguity (the word ambiguity also derives from the same Greek roots).
Cited usage: Amphibole is a general term under which hornblende and actinolite may be united [1833, Lyell, Elements of Geology 592]. The varieties of amphibole are as numerous as those of pyroxene [1868, James D. Dana, A System of Mineralogy 233].
See also: actinolite, asbestos, hornblende, tremolite

amphidromic: [Greek *amphi* = going both ways + *dromos* = running.] adj. In oceanography, said of numerous points in the ocean that, due to the frictional component of Earth's tidal wave cycle, have little vertical fluctuation, but often significant currents. Radiating in all directions, tidal flux increases with distance from these points. Besides tides, there are numerous other systems having amphidromic points, such as seiches and surface

gravity waves. An amphidromic point has maximum horizontal flow and zero vertical displacement. Because the tide neither rises or falls at amphidromic points, it would appear from this term's meaning that the etymology should read "running in no direction," rather than both ways; however, it is the cancelling effect of one tidal wave cycle being low while the other is high that causes a net vertical change of zero ("running both ways").
Cited usage: The effect of uniting the two waves which will have the higher part of their crests on opposite shores is to produce a set of amphidromic points. These are points at which there is no tidal rise and fall of the water level [1938, Nature 11 June 1067/1].
Synonyms: nodal point

amphineuran: [Greek *amphi-* = on both sides + *neuron* = sinew, cord, tendon, nerve.] n. A bilaterally symmetrical marine mollusc belonging to the class Amphineura, having a flattened body covered by eight dorsal plates. The class comprises the chitons and related forms.

amphoteric: [Greek *amphi-* = on both sides + *tero* = comparative suffix + *ic* = adjectival ending.] adj. Having the characteristics of an acid and a base and capable of reacting chemically either as an acid or a base.

amplitude: [Latin *amplus* = large + *tudo* = quality of being.] n. In oceanography, half the wave height, or half the vertical distance from wave trough to adjacent crest. In structural geology, used in describing folds. In seismology, used to describe earthquake waves. And in ecology, amplitude describes the degree of adaptability of a species.
See also: magnitude

amygdule: [Latin *amygdala* = almond, tonsils; < Greek *amygdal* = almond.] n. In igneous petrology, a gas cavity or vesicle in a volcanic rock

that is filled with secondary hydrothermal minerals such as, quartz, calcite, chalcedony or zeolite. In sedimentology, an agate pebble. As an adjective, amygdaloidal is used as a mineral habit. Named for the common almond shape of the mineral-filled cavity.
Alt Spelling: amygdale

ana-: [Greek *ana-* = on, up, back, again or anew.] Combining prefix with multiple meanings relating to the original Greek. The term 'anadromous' conveys two seemingly different meanings of *ana-*; anadromous fish swim upstream to return back to their original hatchery.

anabatic wind: [Greek *anabatis* = ascent, upward movement < *ana* = up + *batis* = walk, going.] n. A wind caused by the local upward motion of warmed air due to convection. This term is derived from the narrative account *Anabasis* by the Greek historian Xenophon. It tells about the memorable retreat by the army of Cyrus up and over the high tablelands of Armenia to Trapezus on the Black Sea.
Cited usage: A local wind is called anabatic if it is caused by the convection of heated air, as, for example, the breeze that blows up valleys when the sun warms the ground [1919, Faraday, Glossary of Aeronautical Terms 31].
See also: katabatic

anabranch: [Greek *ana* = up, in place or time, back, again, anew + *branca* = paw of an animal.] n. A branch stream that turns out of a river and reenters it lower down. This is a portmanteau term composed of two words: *anastomosis*, meaning interconnected mouths and *branch*.
Cited usage: Thus, such branches of a river as after separation re-unite, I would term anastomosing-branches; or, if a word might be coined, ana-branches [1834, J. R. Jackson, Jrnl. Geogr. Soc. IV. 79].
Synonyms: billabong, anastomosing river
See also: billabong

anaerobic: [Greek *an-* = not + *aer* = air + *bios* = life.] adj. Said of an organism, especially a bacteria, that can live without free oxygen. Term coined by Louis Pasteur in 1863 in *Comptes Rendus LVI* ("Report LVI").
Cited usage: The decomposition of the organic remains by anaerobic bacteria results in the formation of sulphureted hydrogen [1959, Clegg, Freshwater Life 2nd ed. 68].

analogy: [Greek *ana* = up, back again + *logos* = word, speech, discourse, reason < *legein* = to speak, to say; IE *leg* = to collect.] n. A comparison based on similarity in some respects between things that are otherwise dissimilar. In biology, a correspondence in function or position between organs of dissimilar evolutionary origin or structure.

anamorphism: [Greek *ana-* = up, back, anew + *morphos* = shape, form + *ism* = suffix that creates a noun of action from a verb.] n. Intense metamorphism in the zone of ductile flow where simple minerals of low density are recrystallized into complex minerals of higher density by dehydration, carbonation, and silication.
Cited usage: The zone of anamorphism may be defined as the zone in which alterations of rocks result in the production of complex compounds from simple ones [1904, Van Hise, Treatise on Metamorphism iv. 169].

anaplasis: [Greek *ana* = up, back, again + *plastos* = molding, formation.] n. An evolutionary state showing increased diversification of organisms, considered to mark the initial stage of a developing evolutionary line.
See also: cataplasis

anastomosing: [Greek *ana* = on, up + *stoma* = stomach.] adj. Said of any network of branching and rejoining lines or surfaces, such as the channel pattern of a braided stream, branching and rejoining fault traces, or irregular solution tubes along a bedding plane.
Cited usage: The African name for a central lake is Tanganyika, signifying an anastomosis, or a meeting-place [1859, R. Burton, Jrnl. of Royal Geographical Society (RGS) XXIX. 234].

anatexis: [Greek *ana* = on, up + *texis* = to melt.] n. Melting of preexisting rock, usually by ultra-metamorphism.
See also: eutectic, palingenesis, syntexis

andalusite: [Toponymous term from Spanish *Andalusia*, a province in southern Spain.] n. A variably colored alumino-silicate mineral, $(Al_2Si_2O_5)$. Andalusite is trimorphic with kyanite and sillimanite. It generally occurs in rhombic crystals found in metamorphic rocks. The name was given by Jean Claude Delametherie in 1798, after Andalusia, a province in Spain where it was first found.
Cited usage: Andalusite occurs in gneiss in England [1843, W. Humble, Dict. Geol.].
Similar Terms: chiastolite
See also: kyanite, sillimanite, chiastolite

andesine: [Toponymous term after the Andes Mountains in South America, where it occurs in andesite lavas.] n. A white to grayish triclinic mineral of the plagioclase feldspar solid solution series, with composition ranging from $Ab_{70}An_{30}$ to $Ab_{50}An_{50}$. It occurs almost exclusively as grains in intermediate igneous rocks, such as andesite and diorite.

andesite: [Toponymous term from Incan *Andes*, the largest mountain range in South America.] n. A volcanic rock of intermediate silica content that commonly contains phenocrysts of intermediate plagioclase (e.g., andesine) and mafic minerals (e.g., hornblende), chiefly plagioclase feldspar. Also, a silicate of alumina, lime, and soda, found at Marmato in the Andes and elsewhere;

perhaps only a mixture of labradorite with soda-felspar. An etymologically related term is andesine.
Cited usage: The name andesite was first used by L. von Buch in 1829 [1879, Rutley, Stud. Rocks xii. 234].

andradite: [Eponymous term after Portuguese mineralogist from Brazil *J.B. d'Andrada* (1763-1838).] n. A variety of calcium-iron garnet.
Cited usage: Named Andradite by the author after the Portuguese mineralogist d'Andrada, who described and named the first of the included subvarieties, allochroite. The included kinds vary so widely in color and other respects that no one of the names in use will serve for the group [1868, James D. Dana, A System of Mineralogy (ed. 5) 268].
See also: ugrandite, pyralspite, pyrope, uvarovite, almandine, grossularite, garnet

anglesite: [Toponymous term after its discovery locality on the Isle of Anglesey, Wales.] n. A white, yellowish to dark gray orthorhombic mineral, lead sulfate ($PbSO_4$). It has a adamantine luster when found in macroscopic crystals, but dull luster when in fine-grained massive aggregates. Anglesite is a common supergene mineral formed by oxidation of galena and often forms concentric layers around a galena core. The Isle of Anglesey is the site of Parys mine, where anglesite was first found. First usage in English c. 1835.
Cited usage: Sardinian is distorted anglesite from Monteponi white and like anglesite in lustre [1837, James D. Dana, A System of Mineralogy 624].
Synonyms: lead vitriol

angstrom unit: [Eponymous term after the name of Anders Jonas Angstrom (1814-74), a Swedish physicist.] n. One angstrom unit equals 10 to the minus 10 power meters. They are frequently used in measuring the wavelength of electromotive force (emf) near the visible spectrum such as: actinic light, UV, IR, and X-rays.

Angstrom was defined in 1907 in terms of the wavelength of cadmium, which at STP is 6438.4696 angstroms, or 643.84696 nanometers, the unit that has largely replaced the angstrom (10 Angstroms = 1 nanometer).
Synonyms: International Angstrom (I.A.), absolute Angstrom

anhedral: [Greek *an* = not + *hedra* = seat, base + *al* = pertaining to.] adj. Applied to crystals that lack planar faces, generally as a result of crowding of adjacent grains during crystallization or recrystallization.
Cited usage: 'Anhedral', 'euhedral', and 'subhedral', in mineralogy, are probably to be preferred to 'allotriomorphic', 'idiomorphic', and 'hypidiomorphic' as avoiding any confusion between mineralogical terms and petrographical-textural terms [1961, J. Challinor, Dict. Geol. 7/2].
Synonyms: allotriomorphic, xenomorphic

anhydrous: [Greek *an* = not + *hydro* = water + -*ous* = abounding in, full of, plentiful.] adj. Said of any substance such as a mineral or body of magma that is essentially or completely lacking in water. Used especially in discriminating compounds and minerals that can be either hydrous (hydroxy) or anhydrous. Hydrous calcium sulphate is the mineral gypsum, whereas anhydrous calcium sulphate is anhydrite. Anhydrite is less soluble than gypsum.

anion: [Greek *ana-* = up, on + *ienai* = to go.] n. A negatively charged ion, such as the ion that migrates to an anode (positive electrode) in electrolysis.
Cited usage: If you take anode and cathode, I would propose for the two elements resulting from electrolysis the terms anion and cation, and for the two together you might use the term ions [1834, W. Whewell, Let. 5 May in I. Todhunter W. Whewell (1876) II. 182].
See also: anode, cathode, cation, ion

anisotropic: [Greek *an* = not + *iso* = equal + *tropos* = a turning + *-ic* = in the manner of.] adj. Having a physical property that varies with direction. Optically, all minerals except those in the isometric system are anisotropic (as opposed to isotropic).
Cited usage: Minerals which exhibit double refraction are anisotropic [1879, Rutley, Stud. Rocks ix. 77].
Synonyms: aeolotropic

ankerite: [Eponymous term after Austrian mineralogist M.J. Anker.] n. A white, gray, or red iron-rich mineral, $Ca(Fe,Mg,Mn)(CO3)2$, closely allied to dolomite. Forms two isomorphous series with magnesium- bearing dolomite and manganese-bearing kutnohorite. First usage in English c. 1840.

annabergite: [Toponymous term after its discovery locality at Annaberg, Germany.] n. An apple-green monoclinic mineral, cobalt arsenate hydrate, $Ni_3(AsO_4)_2 \cdot H_2O$. It forms an isomorphous series with cobalt-bearing erythrite and occurs as encrustations as an alteration product of nickel arsenides.
Synonyms: nickel bloom, nickel ochre

annelid worm: [French *anneler* = to ring; < Latin *anulus* = ring.] n. Any of various worms or wormlike animals of the phylum Annelida, characterized by an elongated, cylindrical, segmented body. Includes the earthworm and leech. Because annelids lack hard body parts (except for their chitinous jaws), their fossil remains usually consist of burrows and trails. In 1801 Lamarck named Annelida as his first subdivision of the Invertebrates.

anode: [Greek *ana* = up + *odos* = way.] n. The path by which an electric current leaves the positive pole (electron/anion attractor), enters the electrolyte, and moves to the negative pole (proton/cation attractor). The terms anion/cation and anode/cathode are confusing. Simply put, the ion that has a negative charge (giving off electrons) is the anion, and the ion that has a positive charge (taking in electrons) is the cation. The anode attracts anions (negatively charged electrons), so it is positively charged; conversely, the cathode attracts cations (positively charged protons), so it is negatively charged. To further confuse this matter, the terminals of a reversible electrochemical reaction (such as at the terminals of a rechargeable battery) change roles depending on whether the system is charging or discharging.
Cited usage: I have considered the two terms you [Faraday] want to substitute for eisode and exode, and recommend instead of them anode and cathode [1834, W. Whewell, Let. 25 Apr. in Todhunter W.W. (1876) II. 179].
See also: cathode, ion, anion

anomaly: [Greek *an* = not + *omallos* = even.] n. A departure from the expected or norm. In astronomy for example, it is the angular deviation of a planet from its perihelion as observed from the sun.
Cited usage: The motions of the sun and moon had other anomalies or irregularities [1837, Wm. Whewell, History of Inductive Sciences I. iii. ii. 175].

anorthite: [Greek *an* = not + *orthos* = straight, right, correct.] n. A white or grayish triclinic mineral of the plagioclase feldspar solid solution series with composition ranging from $Ab_{10}An_{90}$ to Ab_0An_{100}. It occurs in anorthosite plutons and mafic igneous rocks, such as gabbro and norite. Not to be confused with the monomineralic rock, anorthosite. First usage in English c. 1830.
Cited usage: Anorthite, so called for the oblique interfacial angles of its rhombohedral prisms [1833, Lyell, Elements of Geology (1865) 590].
See also: anorthosite

anorthoclase: [Greek *an* = not + *orthos* = straight, right, correct + *klaisis* = fracture.] n. A sodium-rich triclinic mineral of the alkali feldspar group, typically occurring in volcanic rocks, ranging in composition from a proportion of 40% orthoclase & 60% albite to 10% orthoclase & 90% albite, and displaying cryptoperthitic exsolution lamellae. Named in 1823 by Heinrich Rose, who also named the element niobium.
See also: orthoclase

anorthosite: [Greek *an* = not + *orthos* = straight, right, correct.] n. An essentially monom-ineralic, plutonic, igneous rock composed almost entirely of plagioclase feldspar (usually labradorite, but can be a more calcic feldspar). Anorthosite occurs as stratiform units in layered mafic intru-sions, and as large nonstratiform plutonic bodies. The rock is also common in the highland areas of the Moon. Not to be confused with the plagio-clase feldspar mineral, anorthite. First usage in English c. 1860.
See also: anorthite, labradorite

Antarctica: [Greek *anti* = opposite, against + *arctic-us* = of the (constellation) bear.] n. A continent lying chiefly within the Antarctic Circle and asymmetrically centered on the South Pole. 95% of Antarctica is covered by an icecap aver-aging 1.6 km (1 mi) in thickness. The region was first explored in the early 1800s, and although there are no permanent settlements, many coun-tries have made territorial claims. The Antarctic Treaty of 1959, signed by 12 nations, prohibits military operations on the continent and provides for the interchange of scientific data.

antecedent: [Latin *ante* = before + *cedere* = to go.] adj. Anything happening prior to another. Said of a stream established before onset of local uplift, which maintains its original course despite subsequent deformation. Term first applied in 1875 by J.W. Powell in reference

to the Colorado River cutting through a rising plateau.
See also: consequent

antediluvial: [Latin *ante* = before + *diluvium* = flood.] adj. Pertaining to things or events before Noah's flood, as described in the Old Testament of the Bible. Applied by Buckland in 1823.
Cited usage: I have felt myself fully justified in applying the epithet antediluvial to the state of things immediately preceding the Universal Deluge [1823, Wm. Buckland, Reliquiae Diluvianae 2].
See also: diluvium

anthozoa: [Greek *anthos* = flower + *zoa* = plural of *zoon* = animal, living being.] n. Any marine organism of the class Anthozoa, such as the corals and sea anemones, that have radial segments and grow singly or in colonies. Another name for the zoophytes called Actinozoa.

anthracite: [Greek *anthrakitis* = coal < *anthrax* = charcoal, carbuncle.] n. Coal of the highest metamorphic grade having a semimetallic luster, a high carbon content, and little sulfur or volatile matter. The bacterial disease *anthrax* causes ulcerative lesions in many species of mammals, and was equated by the ancients to coal because the red pustules of anthrax glow like burning coal embers.
Synonyms: hard coal

anticline: [Greek *anti* = opposite, against + *clinos* = to slope, bend.] n. A fold that is convex upward with strata sloping downward on both sides away from the fold axis. In regions of compression, the folds tend to occur in repeating patterns, with limbs being both synclinal and anticlinal, depending on the fold axis being referenced.
Cited usage: When strata dip like the roof of a house the strata are spoken of as forming an anticline or saddleback [1876, Page, Advd. Textbook Geol. iv. 83].

anticlinorium: [Greek *anti*= opposite, against + *klinein* = to slope, bend + Latin *orium* < *arium* = place associated to.] n. A composite anticlinal structure of regional extent composed of subordinate anticlines and synclines, the whole having the general contour of an arch. First usage in English c. 1875.

antimony: [Latin *antimnium* = antimony; < Greek *stimmi* = black antimony; perhaps < Arabic *al-* = the + *imid* = antimony.] n. Rarely found in nature as the native element, this metallic element is hard, extremely brittle, lustrous, and silver-white. Its elemental symbol Sb comes from the Latin root for stibnite (Sb_2S_3), the primary ore of antimony. A folk etymology for antimony exists in French, *antimoine* derives from *anti-*, meaning against, and *moine*, meaning religious monks. Its source being Chaucer's 'cursed monk' in his book The Canterbury Tales.
Cited usage: Sometimes there are veins of a red or golden colour intermixed, from which it is called male antimony; that without them being denominated female [1751, Chambers Cyclopedia].
See also: stibnite

antipodes: [Greek *anti-* = against, opposite + *pod-* = foot.] n. Any two places or regions that are on diametrically opposite sides of the Earth.

antithetic fault: [Greek *anti* = against + *thetikos* = to place.] n. A fault that is subsidiary to a major fault, and for strike-slip faults displacement is opposite to its associated major synthetic fault, whereas in normal faults the antithetic fault dips in the opposite direction to the major fault. Term introduced in 1967 by H. Cleoso Dennis.
See also: synthetic fault

apatite: [Greek *apat* = deceit + *-ites* < *-lite* = pertaining to rocks < *lithos* = rock, stone.] n. A common accessory mineral in igneous, metamorphic and hydrothermal rocks, Ca5(PO4)3(F,Cl,OH).

Apatite commonly forms six-sided prismatic crystals and occurs in many colors, from shades of green to blue, brown to yellow, and violet to red, as well as colorless. It is too soft for a gemstone, but is the chief inorganic component of most bones and teeth. The etymological connection to *deceit* results from apatite's many forms and colors, causing it to be often mistaken for other minerals such as: beryl, quartz, nepheline, and tourmaline. A. G. Werner named it in 1786 after the Greek *apatao*, meaning I am misleading.

apex: [Latin *apex* = peak, tip < *ap* = to fit to.] n. The tip of anything, the top or peak of a mountain, pyramid, spire, or shell, for example.
See also: abapical, adapical, adaxial

aphanitic: [Greek *a* = not + *phan-* = to appear, be seen.] adj. A volcanic rock texture with crystal constituents so fine that they cannot be distinguished by the naked eye. Also said of the groundmass showing such a texture. An etymologically related term is *phenomenon*.

aphelion: [Greek *ap-* = off, from + *helios* = sun.] n. That point in the orbit of a planet, comet, or artificial satellite at which the object is farthest from the sun. Term coined by Johannes Kepler in his Laws of Planetary Motion (1609). The first law defines *aphelion* and the opposite *perihelion* in terms of the elliptical orbit of satellites.
Cited usage: The apogaeum of the sun or the aphelium of the earth ought to be about the 28th degree of Cancer [1656, Hobbes' Elem. Philos. Concerning Body 443].
See also: apogee

aphorism: [Greek *aphorizein* = to mark off, define < *apo* = away from + *horos* = boundary, limit.] n. A concise statement of a principle such as, 'The present is the key to the past.' Originally from the 'Aphorisms of Hippocrates,' then applied to the terse and potent statements of the prin-

ciples of physical science, and now also to statements of principles in general. An *axiom* is a statement of self-evident truth; a *theorem* is a demonstrable proposition in science or mathematics; an *epigram* is akin to an aphorism, but not as important.
Synonyms: maxim, saying

aphyric: [Greek *a* = not + *phyre* = combining suffix form of *porphyry* = purple.] n. Said of a glassy or aphanitic volcanic rock devoid of phenocrysts. The Greek word for purple became a textural term in geology because the carved purple stone *porphyry* busts made in ancient Greece were hewn from igneous intrusive rocks with large interlocking phenocrysts.
See also: porphyry

aplite: [Greek *aplos* = single, simple.] n. Consisting primarily of quartz and alkali feldspars, aplite is a light-colored hypabyssal igneous intrusive rock characterized by fine-grained allotriomorphic-granular texture. Composition can be granitic to gabbroic, but the term usually is applied to the granitic end members. As in haploid or single-cells, the simple homogenous texture of aplite merits its name.
Cited usage: Aplite or haplite, also termed semi-granite or granitell, is a rock consisting of a crystalline-granular admixture of feldspar and quartz [1879, Rutley, Stud. Rocks xii. 211].
Synonyms: haplite

apo-: [Greek *apo-* = off, away from.] With sense of standing off or away from something; detached, separate, as in apogee.

apogee: [Greek *apo* = off, away from + *ge* = earth.] n. The point in the orbit of the Moon, an artificial satellite, or of any planet, at which it is at its greatest distance from the Earth. It is also the greatest distance of the sun from the Earth; at this point, the earth being furthest from the sun is in aphelion.

Cited usage: The apogaeum of the sun or the aphelium of the earth [1656, Hobbes' Elem. Philos. Concerning Body p.443].
See also: aphelion

apophysis: [Greek *apo-* = off, away from + *phuein* = to grow.] n. An intrusive tongue of an igneous or hydrothermal intrusion that extends outward from the main intrusive body.
Cited usage: The apophysis is a branch from a vein or fracture that has been filled by the injection of a larger intrusive body [1989, Geology & Soils, Geotechnical Consultants, County of San Diego and the U.S. Department of Interior].
Synonyms: tongue

Appalachian: [Eponymous term after the *Appalachee* tribe of Muskhogean Indians, who called their homeland in NE Georgia *Apalatchi*.] adj. Designating or forming part of an extensive system of mountain ranges in the eastern United States, stretching from the northern border with Canada southward to Alabama. Until the latter 19th century the Appalachian Mountains were called the Alleghenies.

applanation: [Latin *ap-* < *ad-* = to + *planus* = level.] n. All processes that tend to reduce the relief of an area toward that of a level highlands plain, including such processes as erosion of highlands and sediment accumulation in lowlands.

apron: [French *naperon* = diminutive of *nape* = tablecloth; < Latin *mappa* = napkin.] n. An extensive deposit of unconsolidated material accumulated at the base of a mountain, often at the terminus of a glacial moraine.

aqueous: [Latin *aqua* = water.] adj. Pertaining to water as an aqueous solution.

aquiclude: [Latin *aqua* = water + *claudere* = to close.] n. A relatively impermeable body of rock that may absorb water slowly but does not transmit it fast enough to supply a spring or well.
See also: aquifer, aquifuge

aquifer: [Latin *aqua* = water + *ferre* = to bear.] n. Generally of considerable volume, a body of rock that is either consolidated or unconsolidated porous material that conducts water. A water-bearing stratum. Etymologically related terms are aquitard and aquifuge.
Cited usage: There are only two methods of (water) storage: in surface reservoirs and in the aquifers from which we pump a part of our supplies [1966, Listener, 1 Sept. 309/2].
See also: aquiclude, aquifuge

aquifuge: [Latin *aqua* = water + *fugere* = to flee.] n. A body of rock having no interconnections or interstices and therefore can neither store nor transmit water.
See also: aquiclude

arachnid: [Greek *arakhn* = spider.] n. Any arthropod belonging to the class Arachnida, comprising spiders, scorpions, and mites; closely allied to insects and crustaceans, but distinguished by the possession of eight legs, the absence of wings and antennae, and by breathing by means of tracheal tubes or pulmonary sacs. Although there are aquatic spiders, as yet no arachnids have been found to inhabit the ocean.

aragonite: [Toponymous term after Spanish *Aragon*, a province of Spain.] n. A white to yellowish orthorhombic mineral ($CaCO_3$), dimorphous with calcite. Named in 1800 by René-Just Haüy (1744 -1822), French mineralogist at the Sorbonne.
Cited usage: Carbonate of calcium, in its two forms of calc spar (rhombohedral) and aragonite (rhombic or right prismatic) exhibits one of the most striking examples of dimorphism [1863, Watts Dict. Chem. I. 358].

Archean: [Greek *arkhaios* = ancient < *arkh* = beginning.] adj. Said of the middle eon of the three eons of Precambrian or Cryptozoic time. The Archean Eon lasted from 3.96-2.5Ga. There is controversy between the International Union of Geological Sciences (IUGS) and the International Commission on Stratigraphy whether or not to call the Archean an era or eon. If the Precambrian is considered an eon, then the Hadean, Archean, and Proterozoic are all eras.
Cited usage: In Precambrian nomenclature almost every geochronologic and chronostratigraphic subdivision should probably be regarded as informal, except for Archean and Proterozoic and their subdivisions into Early, Middle, and Late, which have been formalized as geochronometric units (Harrison and Peterman 1980, 1982). However, there is even some disagreement whether Archean and Proterozoic should be ranked as Eons/Eonothems or Eras/Erathems [1987, Donald E. Owen, Journal of Sedimentary Petrology, Vol.57, No.2, March pp.363-372].
Alt Spelling: Archaean
Synonyms: Archeozoic
See also: Hadean, Cryptozoic, Proterozoic, Precambrian

archeocyathid: [Greek *archeao* = ancient < *archi* = beginning + *cyanos* = dark-blue.] n. Archeocyathids are an extinct group of sponge-like animals that, due to their brief but diverse appearance in the Lower Cambrian, are a good chronostratigraphic index fossil. This group takes its name from the possibly symbiotic presence of photosynthesizing cyanobacteria found in fossil remains.

archi-: [Greek *archi* = chief, principal, first in authority.] Greek prefix indicating chief position, as in archrival or archbishop.

archipelago: [Italian *Arcipelago* < Medieval Latin *Egeopelagus* < Latin *Aegaeus* < Greek *Aigaios* = Aegean Sea < *archi-* = chief, principal + *pelagos* = sea.] n. In general usage, any large group of islands. Also, any sea or body of water, in which there are numerous islands. Although now used primarily in reference to land masses, the etymology (as well as a modern secondary meaning) refers to the body of water in which islands reside.
Cited usage: A Skärgaard archipelago, or 'garden of rocks,' as it is picturesquely termed in Norsk [1857, B. Taylor, North. Trav. xx. 206].

areal: [Latin *rea* = a vacant piece of level ground in a town < *aridus* = dry.] n. A part of any surface, such as part of the Earth's surface. The adjective *areal* (as in 'of areal extent') should not be confused with aerial.

aren-: [Latin *arena* = sand.] Prefix pertaining to sand or sandstone. The ancient *arena* was the sand-strewn place of combat in an amphitheater.
See also: arenite, lutite, rudite

arenite: [Latin *arena* = sand.] n. The general name for sedimentary rocks with sand-sized fragments, having clasts smaller than those of a rudite and larger than those of a lutite. The Greek-derived equivalents to the Latin rudite (conglomerate), arenite (sandstone), and lutite (claystone) are psephite, psammite, and pelite, respectively.
Synonyms: psammite
See also: lutite, rudite, psephite, psammite, pelite

arete: [French *areste* = ridge, sharp edge; < Latin *arista* = fish bone, spine.] n. A narrow, steep serrate mountain crest above the snow line of rugged mountains like the Swiss Alps, Rockies, and Sierra Nevada. They are sculpted by glaciers resulting from the continued backward growth of the walls of adjoining cirques. Some geologists assert that aretes were not cut by glacial ice, but instead mark the vertical limit of glacial advance. Term developed by mountain climbers in the French-speaking Swiss Alps. More generally, the related term *arris* is defined as the sharp edge formed by the angular contact of two plane or curved surfaces.
See also: arris, horn

argentite: [Latin *argentum* = silver; < Greek *argillos* = white clay < *argos* = white; < IE *arg* = to shine + Greek *-ites* < *-lite* = pertaining to rocks < *lithos* = rock, stone.] n. A dark gray cubic mineral (Ag_2S) that is dimorphous with acanthite.
Synonyms: Silver-glance, argyrose

argillaceous: [Greek *argillos* = white clay < *argos* = white; < IE *arg* = to shine, white, silver.] adj. Pertaining to or containing an appreciable amount of clay-sized particles. The prefix *argil-* is used to make clay-related terms such as argillite, a metamorphic rock, intermediate between shale and slate, that does not possess true slaty cleavage.
Cited usage: The argillaceous odour given out by minerals containing alumina [1841, Joshua Trimmer, Practical Geology & Mineralogy 88].

argon: [Greek *argos* = idle, inactive < *a* = not + *ergou* = work.] n. A colorless and odorless elemental gas (symbol Ar) occurring in trace amounts in the atmosphere. Discovered in 1894 by Rayleigh and Ramsey; named for its chemical inertness.
Cited usage: The gas deserves the name 'argon', for it is a most astonishingly indifferent body, inasmuch as it is unattacked by elements of very opposite character [1895, Lord Rayleigh & W. Ramsay, Phil. Trans. CLXXXVI. A. 187].

arid: [Latin *aridus* = dry < *arere* = parched with heat.] adj. Dry desert-like conditions. Said of a climate too dry to support plant life, less than 25 cm annual rainfall, or an evaporation rate much greater than the precipitation rate.

arkose: [French *arkose* probably < Greek *archaios* = ancient.] n. A sandstone characterized by a high percentage of feldspar fragments relative to quartz, derived from igneous intrusive rocks that have disintegrated rapidly. The type locality is found in the Vosges Mountains of Northeast France along the Rhine River near Strasbourg and the German border.
Cited usage: Amongst the carboniferous and triassic rocks of some countries a sandstone occurs to which the name Arkose is given [1879, Rutley, Stud. Rocks xiv. 280].

armalcolite: [Eponymous term after the Apollo 11 astronauts; *arm* < Neil Armstrong + *al* < Edwin Aldrin + *col* < Michael Collins.] n. A black orthorhombic mineral in the pseudo-brookite group, magnesium iron titanium oxide, $(Mg,Fe)Ti_2O_5$. The astronauts were first to collect samples of this mineral on the Moon. First usage in English c. 1970.

arroyo: [Spanish *arroyo* = intermittent stream; < Latin *arrugius* = gold mine, underground passage < *arrugia* = a galleried mine.] n. A term applied in the SW United States to a flat-bottomed gully in an arid environment that is usually dry but may be occupied by runoff water during intermittent periods of heavy rainfall. Gold found in auriferous gravels of active stream beds led to the excavation of paleo-stream channels, thus mining for ore deposits in underground passages.

arsenic: [Latin *arsenicum* < Greek *arhenikon* = yellow orpiment; alteration of Old Iranian *zarna* = golden; < IE *ghel* = to shine.] n. A poisonous metallic element (symbol As) having three allotropic forms, yellow, black, and gray, of which the brittle, crystalline gray is the most common. Aristotle's student Theophrastus of Eresos (370-286 BCE) named the element *arhenikon*. A folk etymology has it from Greek *arsen*, meaning male, strong, virile, supposedly in reference to the powerful properties of the substance. Whereas arsenic is the element, orpiment (As_2S_3) is the mineral from *auri pigmentum*, meaning golden pigment of the ancient Greeks and Romans. Realgar (As_2S_2) was known as red orpiment or *sandaraca* by the ancients after Mt. Sandaracurgium in Pompeiupolis.
Cited usage: The Greek word is The mineral (as opposed to the element) is properly orpiment, from Latin auri pigmentum, so called because it was used to make golden dyes.
See also: orpiment

arsenopyrite: [Greek *arsenikou* = yellow + *pyr* = fire.] n. A silver-white to gray orthorhombic (FeAsS) mineral arsenic ore.
Synonyms: mispickel, arsenical pyrites

artesian: [French *Artois* = a province of France (formerly *Arteis*).] adj. Pertaining to Artois, or resembling the wells made there in the 18th century, in which water from a well drilled through impermeable strata of a synclinal fold was capable of rising to the surface by internal hydrostatic pressure, thereby producing a constant surface supply of water. By extension, also applied to water obtainable by artesian-type drilling. Arteis, the Old French name of the province, derives from *Atrebates*, meaning a tribe who lived in northwestern Gallia.
Cited usage: Here, in every village, centuries before the principle of the artesian well was acknowledged in Europe, the Rouar'a have been in the habit of boring simple artesian wells [1860, Tristram, Gt. Sahara xvii. 287].

articulite: [Latin *articulus* = joint + Greek *-ites* < *-lite* = pertaining to rocks < *lithos* = rock, stone.] n. A micaceous sandstone or metamorphic quartzite with loosely interlocking interstitial grains of phyllosilicates (mica, talc, chlorite) that impart flexibility to the rock when split into thin slabs.
Synonyms: itacolumite, flexible sandstone

assimilation: [Latin *assimilare* = to liken < *ad* = to, toward + *similis* = like.] n. In geology, the incorporation and absorption of extraneous solid or fluid foreign matter into magma, such as in the contamination or hybridization of a magma by assimilation of wall rock into it.
See also: dissimilation

asteroid: [Greek *astro* = star.] n. Name given to the numerous minute planetary bodies revolving round the sun between the orbits of Mars and Jupiter.
Cited usage: From this, their asteroidical appearance, if I may use that expression...I shall take my name, and call them Asteroids [1802, Herschel, Phil. Trans. XCII. 228].
Synonyms: planetoids

asthenosphere: [Greek *a* = no, not + *sthenos* = strength + *sphairos* = sphere.] n. The concentric layer of the Earth directly beneath the lithosphere. It is part of the upper mantle, and is usually from 70-250 km deep. Relative to the Earth's crust and lithosphere, it is weak and plastic (semimolten). It is the zone of lithospheric isostatic rebound and strongly attenuates seismic waves. Term coined by J. Barrell in 1914 for a zone of high thermal conductivity that he thought existed between 50-1200 km deep. Presence of the asthenosphere as a tectonically plastic zone was suspected as early as 1926, but global occurrence of this plastic zone was confirmed through analysis of seismic waves from the Chilean earthquake of May 22, 1960.
Cited usage: Its comparative weakness is its distinctive feature. It may then be called the sphere of weakness, or the asthenosphere [1914, J. Barrell, Jrnl. Geol. XXII. 659]. The weakest part of the asthenosphere is of the order of one-hundredth of the maximum strength of the lithosphere 1915, J. Barrell, Jrnl. Geol. XXIII. 43].
Synonyms: zone of mobility

atoll: [Maldavian *atollon* = atoll < Malayalam *adal* = closing, unity.] n. A coral island consisting of a ring-shaped reef enclosing a lagoon in which no pre-existing land rises above sea level. Darwin's theory, now generally accepted, is that the coral reef builds upward from a submerging island, whose fringing reef developed adjacent to the circumference of the cinder cone. Adoption of the native name *atollon*, meaning atoll, applied to the Maldive Islands, which are typical examples of this structure.
Cited usage: In the centre of each atoll there is a lagoon from fifteen to twenty fathoms deep [1832, Lyell, Principles of Geology II. 285]. Such sunken islands are now marked by rings of coral or atolls standing over them [1859, Darwin, Origin of Species xii. (1873) 324].
Synonyms: lagoon island, ring reef
See also: lagoon

attrition: [Latin *ad-* = against + *terere* = to rub; < IE *ter* = to rub, turn.] n. A rubbing away or wearing down by friction, especially of loose rock fragments worn down by wind, water, or ice with resulting increase in roundness and decrease in size. The IE root refers to the agrarian practice of rubbing cereal grain to remove the husks. Not to be confused with abrasion.
See also: abrasion platform

augen: [German *augen* = eyes < *auge* = eye.] n. Large lenticular mineral grains in schists and gneisses that are eye-shaped masses of feldspar or quartz. In structural geology, a relatively competent porphyroclast or porphyroblast contained within a mylonitic shear fabric.
Cited usage: When large feldspars, of rounded or elliptical form, are visible in the gneiss, it is said to have augen structure. [1910, Encycl. Britannica XII. 149/2]. Augengneiss, a general term for gneissose rocks containing 'eyes', i.e., phacoidal or lenticular crystals, or aggregates, which simulate the porphyritic crystals of igneous rocks. [1940, C. M. Rice, Dict. of Geological Terms 27/2].
Synonyms: porphyritic gneiss

augite: [Greek *augites* = an inferior variety of turquoise < *auge* = luster + *-ites* < *-lite* = pertaining to rocks < *lithos* = rock, stone.] n. A dark-green to black monoclinic mineral of the clinopyroxene group (Ca,Na)(Mg,Fe,Al,Ti)(Si,Al)$_2$O$_6$, occurring as prismatic crystals. It is a common mineral of the clinopyroxene group that is an essential component of many igneous and metamorphic rocks, such as: basalt, gabbro and dolerite.
Cited usage: Augite has also considerable resemblance to the olivin. [1807, J. Murray, Chem. III. 574].
Synonyms: basaltine

aulacogen: [Greek *aulakos* = furrow, hollow + *genesis* = origin, creation < *genesthai* = to be born, come into being; < IE *gen* = to give birth, beget.] n. In plate tectonics, a failed arm of a rift. At the junctions of tectonic plates, three intersecting lithospheric plates typically are separated by arms. Of the rifting arms, the one along which the motion that spreads the plates apart ceases is termed the failed arm, or aulacogen. Spreading or rifting along the other arms of the triple junction can form new oceanic basins, whereas the aulacogen can become a sediment-filled graben, often resulting in a large river valley, such as the Mississippi River. The term was coined by Schatsky and Bogdanoff in 1960. In 1978, Sengor coined the follow-up term *impactogen*, being a continental collision that leads to high-angle rifting at collision fronts.
Cited usage: Such 'aborted pull-aparts' most commonly occur at places in continental lithosphere where three directions of sea-floor spreading are tending to occur at the same time. The common occurrence is for two of these directions to become predominant and for the third axis to become an aulacogen [1974, J.F. Dewey and K. Burke, Hot spots and continental break-up, Geology 2, 57-60].
See also: graben

aureole: [Latin *aureolus* = somewhat golden, diminutive of *aureus* < *aurum* = gold.] n. A zone where the surrounding country rock has metamorphosed in response to an intrusive igneous body. The etymological implication being that contact metamorphism causes the surrounding rock to burn and become brownish. In addition, the Latin term *aureola* relates to *corona*, meaning crown, which encircles the head.
Cited usage: By French writers the term aureole has been introduced to designate the concentric zone of metamorphosed rock which surrounds an intrusive mass of igneous rock [1884, A. J. Jukes-Browne, Student's Handbk. Physical Geol. II. xii. 442].

aurora borealis: [Latin *Aurora* = Roman goddess of dawn; < IE *aus* = to shine + Latin *borealis* = northern.] n. A multicolored, variegated, luminous atmospheric phenomenon, now considered to be electrical discharges in ionized air, occurring in the vicinity of the Earth's northern or southern magnetic pole. Thinking that sunlight reflecting in the upper atmosphere was the cause for the northern lights, Galileo first used the term aurora borealis in 1619 to suggest an early dawn in the northern sky. His term *aurora borealis* became common in usage by philosophers and scientists of the 17th century. It is now common to refer to the southern phenomena as *aurora australis*, and both northern and southern lights are referred to generally as *aurora polaris*.
Synonyms: northern lights, aurora polaris

authigenic: [Greek *authi-* = self, in situ + *genesis* = origin, creation < *genesthai* = to be born, come into being; < IE *gen* = to give birth, beget.] adj. Formed in place. Pertaining to a mineral that crystallizes in place at the site where it is now found, rather than having been transported and deposited. Examples include quartz, chlorite, glauconite, evaporites, and other pore-filling minerals or cements that grow in situ during

diagenesis. Said of a sediment that is formed directly out of solution. Term applied by E. Kalkowsky in 1880 to a mineral constituent, to denote that it came into existence with, or after, the rock containing it.
Cited usage: Such crystals, which are obviously more ancient than those forming the general mass of the rock, have been called allogenic, while those which belong to the time of formation of the rock, or to some subsequent change within the rock, are known as authigenic [1893, A. Geikie, Geology 3rd ed. 65].

autochthon: [Greek *autos* = self + *khthon* = earth < IE *dhghem* = earth.] n. A body of rock that remains at its site of origin rooted to the Earth's crust. The adjective *autochthonous* is similar in meaning to *authigenic*, but refers to a large body of rock rather than the constituents of the rock. Term coined by Naumann in 1858. IE derivatives include: humus, exhume, chameleon and chamomile.
See also: allochthonous

autolith: [Greek *auto* self, own + *lithos* = rock, stone.] n. An inclusion in an igneous rock to which it is genetically related. As opposed to xenolith.
Synonyms: cognate inclusion, endogenous inclusion
See also: xenolith

autotroph: [Greek *autos* = self + *trophos* = nourishment.] n. An organism capable of synthesizing its own food from inorganic substances, using either light, heat, or chemical energy. They are subdivided into photo-autotrophs and chem-autotrophs. Green plants, algae, and certain bacteria are autotrophs, which make up the fundamental trophic layer of the food web.

axinite: [Greek *axine* = axe.] n. A violet, brown, gray, yellow, or green triclinic mineral, hydroxy calcium iron manganese aluminum boron silicate, $(Ca,Fe,Mn)_3Al_2BSi_4O_{15}(OH)$. Axinite occurs in the contact zones around granitic intrusions. The name alludes to the sharp, wedgelike shape of the crystals. First usage in English c. 1800.

azurite: [Arabic *al* = the + *lazward* = lapis lazuli, dark-blue.] n. An azure blue monoclinic mineral of basic copper carbonate, $Cu_3(CO_3)_2(OH)_2$, which is a copper ore and semiprecious gemstone. Closely allied to malachite. The Arabic article *al-* is frequently kept as an essential part of many words, such as: algebra, alcove, and alchemy. In the term azurite however, the *a* sound is retained while the *l* is dropped.
Synonyms: chessylite
See also: malachite

bacterium: [Greek *bakterion* = diminutive of *baktron* = rod, stick, staff < IE *bak* = staff used for support.] n. In the kingdom Protista, any of the unicellular prokaryotic microorganisms of the class Schizomycetes; they vary widely in morphology, metabolic and nutritional needs, and motility. They may be free-living, saprophytic, or pathogenic in plants or animals. An IE derivative is baguette.

badlands: [Toponymous term after The Badlands of South Dakota.] n. Regions of eroded land on which most of the surface is covered with ridges, gullies, and deep channels, having sparse vegetation. Term coined by French explorers in North America around 1700 for the particularly inhospitable, heavily eroded and arid region of southwest South Dakota and northwest Nebraska. The Badlands National Monument in South Dakota was established in 1939 to protect the area's colorful rock formations and prehistoric fossils.

baguette: [French *baguette* = rod, long loaf of bread.] n. A step cut used for gemstones that are small, narrow, and rectangular in shape.

bahr: [Arabic = *sea*.] n. A small but deep, crater-shaped, spring-fed lake as found in some Saharan oases.

baiu: [Japanese *bai* = plum + *uu* = rain.] n. A cloudy and rainy period in early summer in Japan. The baiu begins in the southern Nansei Islands, and proceeds northward, affecting all areas except the northern island of Hokkaido. This effect lasts from early June to late July, starting and ending earlier in the south. The term means the same as the Chinese *meiyu*, and is written using the same kanji characters.
Cited usage: In Japan, this rainy season is commonly called the 'Bai-u', meaning the plum rains, as it comes when plums are getting ripe [1922, W. G. Kendrew, The Climates of Continents iii. xxii. 142].
Synonyms: tsuyu
See also: tsuyu

baiuzensen: [Japanese *bai* = plum + *uu* = rain + *zen* = before + *sen* = line.] n. Over the Pacific Ocean there develops a high-pressure mass of warm, moist air. When this high-pressure mass meets a mass of cold air, a front develops. Along this front line, there often develop areas of low-pressure warm air, giving rise to the baiuzensen. This line of rain clouds extends from southern China over the Japanese archipelago, causing prolonged periods of continuous rainfall. Because the Japanese archipelago extends from latitude 30N to nearly 50N, the baiuzensen, starting in southern China, moves progressively northward as a rainy season front.

bajada: [Spanish *bajada* = descent, slope.] n. Especially in the U.S. Southwest, that section of a piedmont slope formed by aggradation, composed of rock debris and detritus that deposit in outwash slopes with long straight profiles.
Cited usage: The whole slope from the range to the infilled playa lake is usually termed a piedmont. The upper part is often, but not always, a rock-cut surface, normally termed a pediment, while the lower part is an aggradation feature formed of detritus from the ranges and termed a bahada [1960, B.W. Sparks, Geomorphology xi. 256].

ballena: [Spanish *ballena* = whale.] n. An erosional remnant of a once more extensive alluvial fan having a distinctively lenticular form in plan view and a rounded surface in cross section.

banket: [Dutch *banket* = hard-baked almond-containing confection.] n. General term for a compact siliceous conglomerate of vein-quartz pebbles embedded in quartzite matrix. Because the pebbles in matrix closely resemble the nuts in Dutch almond cake, gold-bearing conglomerates of the Witwatersrand of South Africa were named *banket* by Dutch Boers (*boer*, meaning farmer) who migrated to the area.
Cited usage: The conglomerate is a peculiar formation of almond-shaped pebbles, pressed into a solid mass in a bed of rock of an igneous nature, and is called 'banket' on account of its resemblance to a favourite Dutch sweetmeat known in England as almond rock [1887, Chambers' Jrnl. Apr. 284].
Synonyms: pigeon-egg conglomerate

bar: [Latin *barra* = long, narrow and thin object.] n. In oceanography, an elongate offshore accumulation of unconsolidated material, usually sand, built up by the action of waves and/or current that is submerged at least at high tide and may be a hazard to navigation. In fluvial geomorphology, a ridge of sand or gravel deposited at that part of a stream where decreased velocity results in deposits of sediment, such as in channel, meander or point bars.

barchan: [Arabic *barkhan* = crescent-shaped dune; < Kazhak *Baro Khan* = the founder of the Barozai family.] n. A crescent-shaped dune of shifting sand forming an arcuate ridge, the horns of which point downwind. The type barchan dune occurs in the deserts of Turkestan in the Barkan district, where the dune has a distinct crescent shape with wings, or horns, pointing downwind, a wind-packed windward slope of about 15 degrees,

and a steep lee slope at the angle of repose inside the horns. Barkhan is an eponymous term derived from *Baro Khan* of the Barozai family who ruled the Barkhan District on behalf of the Governors of Sibi. Simple barchan dunes may appear as larger, compound barchan or megabarchan dunes, which may migrate with the wind. Barchans and megabarchans may coalesce into ridges that extend for hundreds of kilometers.
Cited usage: The barchan types [of dune] found only where the wind is predominantly from a single direction [1953, Sci. News XXVII. 14].
Alt. spelling: barchane, barkhan

barite: [Greek *barus* = heavy.] n. A yellow, white, or colorless orthorhombic mineral of barium sulfate (BaSO$_4$) with an unusually high specific gravity (4.5) for a nonmetallic mineral. It is the chief ore of barium. It forms an isomorphous series with strontium-bearing celestite. Named as early as 1640, and adopted by Dana in 1895, the term for both the element barium and mineral barite alludes to their unusually high density.
Synonyms: heavy spar

barium: [Greek *barus* = heavy.] n. A soft, silvery-white alkaline-earth element (Ba) not occurring in the native state on Earth. Named for its high density (3.5).

barranca: [Spanish (Iberian) *barranca* = steep gorge.] n. A deep, stream-cut gorge or ravine with steep sides, common in plateau regions of the U.S. Southwest. The term *barranca* also denotes a rift in a piedmont glacier or in an ice shelf. But *barranco*, the masculine form of *barranca*, is exclusive to the U.S. Southwest, and therefore only refers to a stream-cut gorge.
Alt. spelling: barranco (masc.)

barrow: [Middle English *bergh* < Old English *beorg* = hill, burial site; < IE *bhergh* = high, hill.] n. A large mound of earth or stones, especially when placed as a memorial, as at a burial site. The English *barrow* and German suffix *berg* (as in inselberg) have the same root. *Barrow*, meaning hill or mountain closely resembles the term *burrow* but has the opposite meaning. This similarity of antonyms is more prevalent in British English than it is in American English, and was the cause of some harrowing maritime incidents due to the coincidental terms *starboard* (right) and *larboard* (left). Larboard was first changed to *port* in America, followed by England

barysphere: [Greek *barus* = heavy + *sphaira* = ball.] n. The interior of the Earth beneath the lithosphere, including the mantle and core, but sometimes used to refer only to the mantle or core. *Cited usage*: The bulk of the earth consists of a nickel-iron mass, the barysphere, which is enclosed by a rocky crust, the lithosphere [1926, Encycl. Britannica New Suppl. II. 172]. *Synonyms*: centrosphere

basalt: [Latin *basanites* = very hard stone; < Greek *basantes* = touchstone; < Egyptian *basanos* = touchstone for gold < *bauhan* = slate.] n. A hard, dense, dark volcanic rock composed chiefly of plagioclase, pyroxene, and olivine, which is commonly extrusive, but may form near-surface intrusions as dikes, sills, or laccoliths. Basalt occurs with striking columnar structure at Devils Postpile in California, the Giants Causeway in Ireland, and Fingals Cave in the Hebrides. Named circa 1600 as *basaltes* from a misspelling of Latin *basanites*. The Greek and Egyptian 'touchstone' is not basalt but rather a type of dark slate that was used as a streak plate to test for gold.

basanite: [Greek *basantes* = touchstone, test; < Egyptian *basanos* = touchstone for gold < *bauhan* = slate.] n. In igneous petrology, a basaltic rock composed of calcic plagioclase, clinopyroxene, the feldspathoids nepheline and leucite, and olivine. In sedimentary petrology, a hard, black, flint-like

variety of jasper used in the past as a streak plate or touchstone. The sedimentary rock was named c. 1750, alluding to its usefulness as a streak plate for testing the purity of metals, such as gold. The igneous rock of the same name wasn't coined until 1813 by Alexandre Brongniart. *Cited usage*: Basanites, the touchstone used for trying gold [1753, Chambers, Cyclopedia Suppl.]. *Synonyms*: Lydian stone

base: [Greek *basis* = a stepping, pedestal < *ba* = walk, go.] n. Obsolete term relating to the supposed lack of silicic acid (H_4SiO_4) in a molten rock. Often loosely and incorrectly equated with mafic. In one silica-content classification scheme, rocks with less than 66% silica are termed basic. In general, the gradation from acid to basic corresponds to an increase in color index (percent of dark minerals), which has led to the development of a parallel classification system from felsic to mafic (i.e., light to dark). The original 4-part classification system for igneous rocks, going from low to high silica content, is: ultrabasic, basic, intermediate, and acidic. There is no direct relation to ph, only to silica percent and color. *Cited usage*: In salts the taste is determined more by the base than by the acid [1855, Bain, Senses & Int. II]. *Synonyms*: alkali *See also*: alkali

batholith: [Greek *bathos* = depth + *lithos* = rock, stone.] n. A large, irregularly shaped plutonic intrusion with a surface exposure greater than 100 km^2 whose floor can only be approximated gravimetrically and seismically. Term coined by Suess in 1892 in *Das Antlitz der Erde* (*The Face of the Earth*). *Cited usage*: The magma simply filled the space...forming a cake of rock or true batholite [1904, Sollas, translation of Suess's 'Face of the Earth' I. I. iv. 168]. *Synonyms*: abyssolith

bathy-: [Greek *bathys* = deep < *bathos* = depth.] Combining prefix meaning deep, used in terms such as: bathymetry, bathysphere, and bathyscaphe (name given by Auguste Piccard to his deep-sea diving vessel, now applied generally to such deep-diving crafts).

bathyal: [Greek *bathos* = deep.] adj. Pertaining to the ocean environment or depth zone between 200m (approximate shelf-slope break) and 2000m; the region of the ocean bottom between the neritic and abyssal zones. (There is difference of opinion regarding the base of the bathyal zone; if the focus is the water column the base is considered to be 2000m, if the focus is the benthos it is often considered to be 4000m). Etymologically related terms are: bathysphere, bathymetry, bathyorographical, bathyscope, bathypelagic, and bathylite.
Cited usage: The bathyal environment of the sea bottom is that portion between 100 and 1000 fathoms. [1926, W. H. Twenhofel, Treat. Sedimentation vii. 612].
See also: abysso-, hado-

bauxite: [Toponymous term after French *Les Baux de Provence*, a location near Arles, France.] n. A rock composed principally of a mixture of hydrogen aluminum oxide and aluminum hydroxides, including: gibbsite, boehmite, diaspore, along with iron hydroxides and clay minerals. The principle ore of aluminum. Used collectively for lateritic aluminous ores.
Cited usage: This substance, obtained from Baux, near Arles, consisted of small round grains buried in a perfectly crystallised pure limestone. I then recognised it as the mineral to which M. Berthier gave the name of Bauxite. It is a hydrate of alumina which M. Dufrénoy has ranked with gibsite, or rather diaspore [1861, H. Sainte-Claire Deville, Chem. News 9 Nov. 241].

bay: [Latin *baia* = bay; possibly influenced by Latin *badata* = an opening, agape.] n. A body of water partially enclosed by land but significantly affected by tidal flux through its mouth; affords access to the sea but may be occluded by a bar.

bayou: [American Choctaw *bayuk* = small stream + French *boyau* = gut, intestine.] n. The name given (chiefly in the lower Mississippi River basin and Gulf Coast region of the U.S.) to the marshy, sluggish, stagnant tributaries and distributaries connected to larger rivers. Native American term adopted and altered during French occupation of Louisiana.

beach: [(possibly) Old Norse *Beck* = stream; or (possibly) Swedish *backe* = bank of a river.] n. The shore of a body of water usually covered by sand or gravel and lacking a bare rocky surface. Extends landward from the low waterline to the point where there is permanent vegetation or a definite change in physiography, such as a cliff.
Cited usage: Rowling pebble stones, which those that dwell neere the sea do call Bayche [1597, Gerard, Herbal xxxvi. 16. 249].

bearing: [Middle English *ibere* = to carry, sustain, bring forth; contemporary and synonymous with German *gebaren*.] n. The compass direction from any fixed point to another fixed point, usually expressed as an angle of less than 90 degrees east or west of a north or south reference meridian.
Cited usage: By means of the direction-finding apparatus a bearing to the source may be obtained [1920, Discovery May 131/2]

belemnite: [Greek *belemnon* = dart, javelin.] n. An extinct member of the class Cephalopoda, having a smooth, cigar to dagger-shaped, internal shell. Now extinct, their range was from the Carboniferous to Eocene. Before being recognized as a fossil, belemnites were once identified as thunder-bolts, thunder-stones, or elf-bolts.

Cited usage: The belemnite, one of the cephalopodes not found in any tertiary formation [1833, Lyell, Principles of Geology III. 325]. Ed. Note: now known to have existed into the Eocene Period.

ben: [Gaelic *beann* = peak; < Celtic *bendo* = horn, conical point.] n. A Scottish term for high hill or mountain, it is now used only in proper names of the higher peaks in a range such as: Ben Dorain, Ben Lomond, and Ben Nevis (Scotland's highest peak at 1,343m or 4406 ft).
Cited usage: Lowlanders and Irishmen who never climbed a ben [1884, Manchester Examiner 13 Sept. 5/3].

benitoite: [Toponymous term after the Benito Gem Mine in San Benito County, California, where the type locality is located.] n. A blue to colorless hexagonal barium titanium silicate ($BaTiSi_3O_9$). It is strongly dichroic, resembles sapphire in appearance, and is the California state gem.
Cited usage: It is a new mineral species, it has been called benitoite, as it occurs near the head waters of the San Benito River in San Benito County [1907, G. D. Louderback, Bull. Dept. Geol. Univ. Calif. V. 149].

benthos: [Greek *benthos* = depth of the sea.] n. The bottom of a sea or lake; also said of the organisms dwelling at the bottom, or within the substrate making up the bottom. The deep-sea benthos is further subdivided into the archibenthic (upper) and abyssal benthic (lower) zones. The Greek prefix *archi*, meaning chief, principal, or first in authority is appropriate because the archibenthos is frequently in the photic zone and most nutritive detritus stays in the upper layer.
Similar terms: demersal
See also: demersal

bentonite: [Toponymous name after Fort Benton, Montana.] n. A highly absorbent aluminum silicate clay formed from volcanic ash containing montmorillonite. The Fort Benton strata is found in the Cretaceous Benton Formation of the Rock Creek district in eastern Wyoming.
Cited usage: In a recent article the writer described briefly a new variety of clay found in Wyoming....the clay will hereafter be known as Bentonite [1898, W. C. Knight, Engineering & Mining Jrnl. 22 Oct. 491/1].
Synonyms: amargosite, soap clay
See also: montmorillonite

bergschrund: [German *berg* = mountain + *schrund* = cleft, crevasse.] n. The relatively wide, deep crevasse that separates the moving ice and snow of the main glacier or snowfield from that of the relatively immobile ice apron adhering to the confining headwall of a cirque.
Cited usage: This slope was intersected by a so-called Bergschrund, the lower portion of the slope being torn away from its upper portion to form a crevasse [1860, Tyndall, Glaciers of Alps i. xiv. 98].

berm: [Dutch *berm* = strip of gravel < Middle Dutch *baerm* < IE *bherem* = to form an edge.] n. In oceanography, a raised, landward-sloping to nearly horizontal surface on the back shore of a beach, formed of material first built up by smaller summer waves, then cut away by high-energy winter storm waves. Not to be confused with berm crest, which is the seaward edge of the berm separating the foreshore from the backshore, and usually marking the highest point on the beach.

beryl: [Greek *beryllos* = green gem; (possibly) < Dravidian *Velur (now Belur)* = city in South India; < Tamil *veiruor* = to whiten, become pale; (perhaps linked to Arabic *ballur* = crystal).] n. A green, bluish-green, yellow, or pink, hexagonal beryllium aluminum silicate ($Be_3Al_2Si_6O_{18}$). The green gem variety of beryl colored by chromium impurities is

emerald. Other gem varieties are: aquamarine, golden beryl, and morganite. Beryl is the main source of the element beryllium. The Greek term *beryllos* was originally applied to all green gemstones, but later used only for beryl. Of the two variant spelling forms, *beryl* is commonly used for the mineral name and *beryll* usually refers to varieties of the gem.
Alt. spelling: beryll
Similar terms: emerald

beryllium: [Greek *beryllos* = green gem; (possibly) < Dravidian *Velur (now Belur)* = city in South India; < Tamil *veiruor* = to whiten, become pale; (perhaps linked to Arabic *ballur* = crystal).] n. A lightweight, corrosion-resistant, hard, silver-white metallic element, having a high melting temperature (symbol Be).
See also: beryl

bhangar: [Toponymous term after *Bhanger*, a large geo-climatic region of India.] n. A high terrace, scarp, or hill formerly an alluvial river plain, but now beyond reach of floodwaters. Bhangar, along with Khadar and Nardak, is one of three agro-climatic regions in the Yamuna River valley, which courses from the Himalayas to its confluence with the Ganges River. Term used primarily in India.
See also: khadar

bif: [Acronym for *banded iron formation*.] acr. Banded iron formation is iron-rich chert, usually found in expansive tracts formed circa 2Ga. The banded coloration is usually on a centimeter scale. Color variation results from different amounts and oxidation states of iron-containing minerals such as: hematite, magnetite, grunerite, limonite, siderite, and sometimes pyrite. Organisms in the sea gave off oxygen as a waste product of photosynthesis and bonded with the abundant dissolved iron to yield minerals like magnetite and hematite. Banded iron formations are an exception to the Law of

Uniformitarianism in that they are not forming during present time. Although they can be found in Archean strata, most deposits of bif were formed in the subsequent Proterozoic Era around 2 Ga.

billabong: [Wiradhuri (Aboriginal language of southeast Australia) *bilaba* = stream bed filled only after rain + *bong* = dead.] n. A blind channel of a main stream leading to a seasonally filled stream bed or oxbow lake, or a stagnant pool in a river valley. Billabong is the Wiradhuri name for the modern Australian Bell River, due in part to its braided pattern or anabranch zone with many abandoned channels and oxbows which are seasonally-filled. Early English settlers in Australia popularized this term by naming ponds or artificial water holes created for livestock *billabongs*.
Cited usage: A small watercourse, then dry, and named Billibang, skirts the eastern side of the hill [1836, T. L. Mitchell, Three Expeditions into the Interior of Eastern Australia].
Similar terms: anabranch
See also: anabranch

bioherm: [Greek *bios* = life + *herm* = sunken rock.] n. A reeflike mass of carbonate rock, often in the shape of a mound, dome, or lens, and formed from calcareous remains of sedentary marine organisms, such as corals, molluscs, and algae. Commonly surrounded by rock of a different lithology. First coined by E.R. Cummings and R.R. Shrock in 1928.
Cited usage: Coral reef encourages the misconception that reefs are largely made of coral, whereas many of them were formed by other organisms.... The authors have for some time used the term 'bioherm' [1928, E. R. Cumings & R. R. Shrock, Bull. Amer. Geol. Soc. XXXIX. 599].
See also: hummock

biome: [Greek *bios* = life + *ome* = group, mass, structured order.] n. An extensive community of plants and animals whose makeup is determined

principally by physical factors such as soil and climate (e.g., grassland, desert). The suffix -ome occurs chiefly in botanical terms and usually signifies a group of individual species, or cells forming a normal part of the whole, as in biome or rhizome. This is in contrast with an abnormality implied by the suffix -oma, as in carcinoma.
Cited usage: In paleo-ecology the concept of the biome, or biotic community, seems to have peculiar value, as it directs special attention to the causal relations and reactions of the three elements: habitat, plant, and animal [1918, Bull. Geol. Soc. Amer. XXIX. 372].
See also: influent, chaparral

biostrome: [Greek *bios* = life + *stroma* = bed, rug.] n. A distinctly bedded, blanketlike mass of rock formed chiefly from remains of sedentary organisms, which does not swell into a mound or lens. Term coined by E. R. Cumings in 1932.
Cited usage: For purely bedded structures, such as shell beds, crinoid beds, coral beds, etcetera, consisting of and built...by sedentary organisms...I propose the name biostrome.... Biostrome means literally an organic layer [1932, E. R. Cumings, Bull. Geol. Soc. Amer. XLIII. 334].

biota: [Greek *bios* = life.] n. A collective term for all the living organisms of a region.

biotite: [Eponymous term after French physicist Jean Baptiste Biot (1774–1862).] n. A brown to black form of mica, $K_2(Mg,Fe,Al)_6(Si,Al)_8O_{20}(OH)_4$. Biotite is an important rock-forming mineral, is widely distributed, and is commonly found in igneous and metamorphic rocks.

birefringence: [Latin *bi* = two + *refringere* = to refract < *re* = back, again + *frangere* = to break.] n. The ability of an anisotropic crystal to split a light ray into two components that travel at different velocities and are polarized at right angles to each other. All crystals except those in the isometric

system show birefringence and double refraction.
Cited usage: The wide separation of the two refracted rays by calcite is a consequence of the large difference in the values of its indices of refraction, in other words, as technically expressed, it is due to the strength of its double refraction, or its birefringence [1898, E. S. Dana, A Textbook of Mineralogy (ed. 2) ii. 172].
Synonyms: double refraction

bitumen: [Latin *bitumen* = asphalt; < Celtic *betulla* = birch tree.] n. A generic term for a group of naturally occurring, inflammable mineral substances containing volatile hydrocarbons, which are essentially free from oxygenated compounds. Term adopted by the Roman Pliny from the Celtic language in reference to the birch tree, which was a supposed source of bitumen. The adjective bituminous, as used to describe a variety of minable coal, came into vogue in England around 1620. The Southern Dead Sea region has been known since antiquity to produce large masses of high-quality bitumen ("natural asphalt") that float like icebergs on the water. They are often dislodged from their subsurface strata by earthquakes. So valuable was Dead Sea bitumen that, after large quakes, inhabitants of the area would jump onto rafts, row frantically out to the floating islands with armed guards at the ready, and mount the asphalt "bulls" wielding axes. They cut up the bitumen, rolled it in sand to preserve the volatile compounds, stuffed them in leather bags, then loaded them onto camels for sale at the market in Alexandria. This phenomenon was chronicled by Hieronymus of Cardia in 312 BCE. In addition to using bitumen for embalming and as a strong glueing agent, it was also used as a base for cosmetics and perfume. Cleopatra had a cosmetics works set up on the Dead Sea shores after Marc Antony gave her the annexed kingdom as a gift.
Cited usage: Niépce produced his best results by coating his materials with a solution of bitumen of

Judea, an asphalt compound dating back to the time of the Egyptians [2004, Photographic Exhibition Website, University of Texas, Austin].
Similar terms: naphtha, petroleum, asphalt
See also: maltha, ozocerite

bituminous: [Latin *bitumen* = mineral pitch from the Near East.] adj. Pertaining to coal that is dark brown to black in color, burns with a smoky flame, and contains more than 14% volatile matter. Bituminous coal is the most abundant type of coal and ranks between sub-bituminous coal and anthracite.
See also: anthracite

bivalve: [Latin *bi-* = two + *valvae* = folding-doors.] n. A mollusc, such as an oyster or a clam, that has a shell consisting of two, not necessarily symmetrical, hinged valves.

blasto-, -blast: [Greek *blasto-* = sprout, bud.] As a prefix used in metamorphic petrology to signify a relict texture inherited from the protolith, such as: blastogranular, blastoporphyritic, and blastopelitic. As a suffix, it implies a textural feature produced entirely by metamorphism (e.g. porphyroblast, crystalloblast).

bleb: [English *bleb* < *blob* = the action of producing a bubble with the lips.] n. In igneous petrology, a small rounded inclusion with no preferred orientation, such as olivine enclosed in pyroxene. Blob first appeared as a verb in the 15th century; the noun followed a century later. In relation to blob, bleb is the more diminutive term.
Similar terms: blob

blende: [German *blendendeserz* = deceiving ore < *blenden* = to deceive < *blentan* = to blind; < IE *bhel* = to blind, shine.] n. In mineralogy, any of several sulfide minerals having a bright, resinous, nonmetallic luster, such as zinc blende and sphalerite, (Zn,Fe)S. It occurs in usually yellow-brown to black crystals or cleavage masses. Miners considered zinc blende to be a worthless intrusion into valuable galena-silver ores, but they often were initially deceived by the bright luster.
Cited usage: Blende called by some mock-lead [1753, Chambers, Cyclopedia Suppl.].
Synonyms: sphalerite, pseudogalena, mock lead
See also: sphalerite, galena

blizzard: [German *blitzartig* = like lightning < *blitz* = lightning.] n. A severe snowstorm with heavy winds, driving snow, and reduced visibility. In meteorology, a violent snowstorm with winds blowing at a minimum speed of 35 mph (56 kph) and visibility of less than one-quarter mile (400 meters) for three hours. The term *blizz* was used by early American settlers in Virginia to refer to wind-driven rain or snow that reduced visibility. In the Dakotas, early German settlers may have derived the English word blizzard from the German *blitzartig* (lightning-like) to characterize the sudden, severe winter storms they experienced. The word became popularized in American newspapers during the severe winter of 1880-81.
Similar terms: buran

bluff: [(probably) Dutch *blaf* = flat, broad; akin to German *blaff* = smooth.] n. Any cliff with a steep, broad and precipitous face, used especially for a high riverbank or a steep headland along a coastline (promontory). First used in North America, and still mostly said of American landscapes.

boca: [Spanish *boca* < Italian *bocca* < Latin *boca* = mouth.] n. The point where a stream channel emerges from a canyon or gorge, and flows out onto a plain. Also used as a variant spelling of the volcanology term 'bocca.'
Alt. spelling: bocca
See also: bocca

bocanne: [French *bocanne* = smoke.] n. A naturally burning, hydrocarbon-rich shale bank. Term originated in French-speaking Canada.

bocca: [Italian *bocca* < Latin *boca* = mouth.] n. An aperture from which magma and gas escapes from a volcano. Also used as a variant spelling of the hydrological term *boca*.
See also: boca

bog: [Irish Gaelic *bogach* = bog, soft; < IE *bheug* = to bend (pliable objects).] n. An area of water-logged, spongy, acidic ground without through-flowing water, composed chiefly of sphagnum moss along with sedges and plants in the heather family. The accumulated vegetation may develop into peat and eventually into coal. It is unable to support much weight in one spot. Etymologically related terms include bowline and boat.
Similar terms: fen, swamp, marsh
See also: marsh, muskeg

bogue: [Choctaw *bouk* = stream, creek.] n. Term used in the area of the lower Mississippi River for a stream mouth or outlet for a bayou, or for the stream itself.

bolide: [Latin *bolis* < Greek *bolid* = missile, flash of lightning < *ballein* = to throw; < IE *gwel* = throw, reach, pierce.] n. An exploding meteor or meteorite. IE derivatives include belemnite and ballet.
Cited usage: Not a space equal in extent to three diameters of the moon, which was not filled every instant with bolides and falling stars [1852, Ross, Humboldt's Trav. I. x. 352].

bolson: [Spanish *bolson* = large purse < *bolsa* = purse.] n. In the desert regions of the U.S. Southwest and Mexico, a flat, arid, alluvium-floored basin from which ephemeral streams flow. Bolson displays the Spanish superlative suffix -*on*, implying something relatively large. This suffix is used in contrast to the diminutive -*ita*, as in *bolsita*, meaning a small purse.
Cited usage: The bolson plains may be considered as sections of an upraised peneplain surface in its earliest infancy, at a stage in which they are as yet untouched by stream-action [1904, Amer. Geol. Sept. 164].
Synonyms: playa basin

bonanza: [Spanish *bonanza* = fair weather, prosperity; < Latin *bonacia* = calm sea < *bonus* = good + *malacia* = calm sea; < Greek *malakos* < IE *mel* = soft.] n. In mining, a rich vein or pocket of ore; therefore, in general and by extension, a source of great wealth or prosperity. A mine is considered "in bonanza" when it is making a profit. IE derivatives include mollusk, smelt, and amblygonite.

bora: [Greek *boreios* = coming from the north < *Boreas* = god of the North Wind.] n. A strong, cold, northeasterly katabatic winter wind that blows across the Upper Adriatic Sea. Etymologically related terms include Russian *burya* and the *Illyrian bura*.
Cited usage: A violent wind began to blow. 'The Bora! the Bora!' resounded on all sides, in tones of terror and dismay [1864, Viscountess Strangford (aka Emily Anne Smythe), E. Shores Adriatic 263].

borax: [Arabic *braq* = to glisten; also Persian *burah* = white.] n. Transparent to white mono-clinic, hydrated sodium borate, $Na_2B_4O_7 \cdot 10H_2O$. On exposure to air, borax dehydrates to chalking tincalconite, $Na_2B_4O_7 \cdot 6H_2O$. Borax is the chief ore of the element boron. It commonly occurs as crystals in alkaline lakes fed by volcanic hot springs.

bore: [Middle English *bare* < Old Norse *bara* = wave; < IE *bher* = to carry, bear children.] n. In oceanography, a high tide-generated wave caused by flood surge up a narrowing estuary or by

colliding tidal currents. IE derivatives include: fertile, fortune, and billow
Synonyms: eagre, higre
See also: eagre

boreal: [Greek *Boreas* = god of the north wind.] adj. Of or pertaining to the north; in particular, relating to the forest areas of the northern temperate zone, dominated by coniferous trees such as spruce, fir, and pine. The Greeks personified the North Wind (Boreas) as an old, gray-haired man, strong-bodied and harsh-willed. Various cultures also personify the north wind as strong. The Egyptians called it *bai*, meaning ram, and the Algonquin people *Kibibonokko*, meaning the fierce one.

bornhardt: [Eponymous term in honor of the German explorer, Wilhelm Bornhardt.] n. A residual peak protruding through the surrounding alluvium in a desert area, especially a peak composed of granitic gneiss. Bornhardt first described this feature in Tanzania (formerly Tanganyka).
Synonyms: inselberg
See also: inselberg

bornite: [German *bornit*, eponymous term after Austrian mineralogist Ignaz von Born (1742–1791).] n. A brittle, brownish-bronze, submetallic copper iron sulfide (Cu_5FeS_4). Bornite tarnishes to iridescent purple or blue when exposed to air, explaining the synonym erubescite.
Cited usage: The commonest ores of copper are sulphides of copper and iron: chalcopyrite and bornite or erubescite [1946, J. R. Partington. Gen. & Inorganic Chem. xiii. 325].
Synonyms: erubescite, horseflesh, peacock or purple copper

boron: [Portmanteau word *bor-* < borax + *-on* < carbon.] n. A nonmetallic brownish element (symbol B). Boron resembles carbon and silicon in some properties. First discovered in 1807 by Sir

Humphry Davy and J. L. Gay-Lussac.
Cited usage: I first procured boron in October, 1807, by the electrical decomposition of boracic acid [1812, Sir H. Davy, Chem. Philos. 315].

bort: [Dutch *boort* < (probably) Old French *bourde, bourt* = bastard.] n. An aggregate of granular to finely crystalline diamond, often in spherical forms with a radial fibrous structure. Bort is primarily only suitable for crushing into abrasive powder for industrial tools, such as saws and drill bits. The fine-grained diamond fragments produced in cutting that are too small for cosmetic usage as gems; also diamonds that are too coarse for cutting. The etymology alludes to various references made by miners, who have used terms such as bastard coal and bastard quartz when referring to geologic materials that are of marginal economic value.
Alt. spelling: boart, bortz, boort, boartz, borts, bowr
Synonyms: ballas, industrial diamonds
Similar terms: carbonado

boscage: [Latin *boscum* = wood + *-age* = that which belongs to, functionally related to.] n. A natural growth of trees or shrubs; a thicket, grove, or wooded place.
Cited usage: Als he went in that boskage, He fond a letil ermytage [1400, Ywaine & Gaw]. The sombre boskage of the wood [1830, Tennyson, Dream Fair Women 243].

boss: [Old French *boce* = hump, swelling; < Italian *boccia* = ball.] n. In geomorphology, a smooth and rounded hillock or knoll of resistant bedrock, usually bare of vegetation. In igneous petrology, an intrusion less than $100km^2$ in surface exposure, and circular in plan.
Cited usage: In the midst of a tract of mica-schist a boss of granite rises [1863, A. Ramsay, Phys. Geog. 31].
Synonyms: stock

botryoidal: [Greek *botryos* = bunch of grapes.] adj. referring to the habit of mineral aggregates that have a surface of hemispherical shapes; a globular surface having the form of a bunch of grapes. The surface marks the termination of domains of radiating crystals, which are so thin as to be barely visible in cross-sections of the botryoides. Mineral aggregates commonly showing this habit are: agate, hematite (variety kidney ore), and cave coral.

boudinage: [French *boudin* = a blood-sausage, a black pudding.] n. Common in severely de-formed rocks, an originally competent and continuous rock layer that has been stretched, thinned, and broken into sausage-shaped seg-ments called *boudins*. Unlike French sausage and pudding, which must be identified as 'boudin blanc' if it is light-colored, boudinage rock structures are not distinguished by color.

boulangerite: [Eponymous term after French mining engineer Charles L. Boulanger, who first described it.] n. A bluish-gray or lead-gray monoclinic mineral, lead antimony sulfide (Pb_5Sb_4Sn). It commonly occurs in plumose aggregates in hydrothermal veins.

boulder: [Middle English *bulder* = noisy, short for *bulder stan* = noisy stone; < Swedish *bullra* = to roar + *sten* = stone.] n. A loose mass of rock larger than a cobble with a diameter greater than 256 mm (about the size of a soccer ball). *Alt. spelling*: bowlder

Bouma sequence: [Eponymous term after Dutch sedimentologist Arnold H. Bouma.] n. A characteristic succession of five intervals of sediment that comprises a complete turbidite. One or more of the intervals may be missing. These intervals from the bottom up are: a) graded bed, b) lower parallel laminates, c) current ripple laminates, d) upper parallel laminates, e) pelitic.

bourne: [Scottish *bourne* = small stream or river; < Old English *burna* = spring, fountain, stream.] n. A small stream, a brook; often applied (in this spelling) to the winter bournes or winter torrents of the chalk downs of Southern England. Applied to northern streams it is usually spelled burn. Originally pronounced like adjourn, now also like mourn; hence the two spellings. The Old English meaning of spring implied a welling up or upward commotion, in a wider sense applicable to both water emanating from a spring as well as fire, possibly the reason for the new spelling *burn*. *Alt. spelling*: bourn, burn *Synonyms*: chalk stream

brachio-: [Latin *bracchium* = arm; < Greek *brakhion* = upper arm.] Combining prefix meaning arm, used in terms such as: brachiopod, brachiate, brachial, and brachiosaurus. Not to be confused with *brachy-*, meaning short. *See also*: brachy-, brachiopod, brachium

brachiopod: [Greek *brakhion* = upper arm + *pod* = foot.] n. Any solitary marine invertebrate in the phylum Brachiopoda characterized by a mouth segment with two armlike projections. There are two main divisions of brachiopods, Articulata and Inarticulata. The most abundant brachiopods are the terebratulids (class Terebratulida). The common name 'lamp shells' alludes to their resemblance to ancient oil lamps. *Synonyms*: lamp shells

brachium: [Latin *bracchium* = arm: < Greek *brakhion* = upper arm.] n. Any armlike body part of an invertebrate such as: the ray of a starfish, the tentacles of a cephalopod, or an armlike coiled muscular projection from the mouth segment of a brachiopod.

brachy-: [Greek *brachys* = short.] Combining prefix meaning short, used in terms such as: *brachycephalic*, meaning short or broad head,

brachyanticline, meaning a short, broad convex upward fold, or *Brachyura*, meaning short tail, a family of sponge crabs. Not to be confused with *brachio-*, meaning arm.
See also: brachio-

brady-: [Greek *bradys* = slow.] Combining prefix meaning slow or delayed, used in terms like *bradyseism.*

bradyseism: [Greek *brady* = slow + *seismos* = earthquake.] n. A slow rise and fall of the Earth's crust. Such movement was originally recognized by Lyell in the Phlegraean volcanic district west of Naples, Italy, where over the past two thousand years there have been vertical movements ranging from 6 m below sea level to 6 m above sea level in the Phlegraean bradyseism.
Cited usage: The Campi Flegrei caldera is well known for a phenomenon named "bradyseism" which is the alternating uplift and sinking of the ground within the caldera [2004, Italy's Volcanoes, the Cradle of Volcanology].

brae: [Old Norse *bra* = eyebrow; akin to German *braue*.] n. A steep slope, a hill-side, or the steep bank bounding a river valley (as opposed to a hill crest). In southern England a hill includes both crest and slope, but in the north a *hill* is always the mount or summit, with slopes called *braes* on all sides of it.
Cited usage: By yon bonnie banks and by yon bonnie braes Where the sun shines bright on Loch Lomond Where me and my true love were ever wont to gae On the bonnie, bonnie banks of Loch Lomond [Classic Scottish Folk Song]

branchi-, -branch: [Combining prefix and suffix < Greek *bragkia* = gills < *bragkion* = fin.] As a prefix, such as in the crustacean subclass Branchiopoda, which includes brine shrimp and water fleas with many pairs of leaf-like gills extending out from the body. As a suffix, meaning gill or fin, as in pterobranchia (winged gills), pleurobranchia (lung gills), and ctenobranchia (comblike gills).

breccia: [Italian *breccia* = broken stones;< German *brachen* = break; <IE *bhreg* = to break.] n. A coarse-grained clastic rock of angular fragments embedded in a fine-grained matrix or a mineral cement, which is often of the same chemical composition as the fragments (e.g., calcite cement enclosing a marble or limestone breccia). Breccia may originate as a result of: tectonic processes (fault breccia), talus accumulation (landslide breccia), cavern collapse (collapse breccia or solution breccia), or volcanic processes (volcanic breccia or breccia pipe). Originally used as an adjective in the specific rock name Breccia Marble. IE derivatives include: breach, fraction, and suffrage (because in ancient times a broken piece of tile was used to cast one's vote).
Cited usage: The name of Breccia is derived from the well-known Breccia Marble, which has the appearance of being composed of fragments joined together by carbonate of lime [1836, Penny Cyclopedia V. 374].
See also: conglomerate

brine: [Old English *bryne* = a burning, salt liquor < *brinnan* = to burn; < Dutch *brijn* = water of the sea.] n. As used in geology, denotes warm to hot highly saline water containing chiefly Na, K, Ca, Cl. Major occurrences are found in hydrothermal fluids, oil-field waters, connate pore water in marine sediments, and stratified hot fluids in restricted marine basins, such as the Red Sea. The Latin root for salted brine is *salsilago*, meaning salted lake. The Spanish equivalent *muria* gives us the root for muriatic acid.
Cited usage: At Lymington in Hampshire, the reservoirs of concentrated brine are always peopled by a sort of shrimp commonly known as the brine shrimp [1860, Gosse,The Romance of Natural History 74].
Synonyms: salsugo, muria, salsilago

Brocken: [Toponymous term after the highest peak (1,143m) in the Harz Mountains of Central Germany.] n. A granite peak, Brocken is the legendary site of the witches' revels on Walpurgis Nacht, the eve before May Day (May 1). The original Celtic eve of Bealtaine, meaning 'bright fire,' was a pagan Druid ritual that was Christianized to the eponymous Walpurgis Nacht after St. Walpurga, the Abbess of Heidenheim.
See also: Spectre of the Brocken

bromine: [Greek *bromos* = stink + *-ine* = of the nature of.] n. A chemical element (symbol Br) occurring in the form of a reddish-brown, corrosive liquid, which volatilizes to a smelly vapor that is irritating to the mucous membranes. Along with fluorine, chlorine, iodine, and astatine, Bromine is one of the five halogen elements; discovered by Balard in 1826.
Cited usage: The name first applied (by Balard) is muride; but it has since been changed to brome from the Greek signifying a strong or rank odour [1827, Edward Turner, Elements of Chemistry add. 695].

brook: [Old English *broc* = brook; < German *bruch* = marsh, moor, bog, or fen.] n. A small stream, considered by some to be more diminutive than creek. Even though the etymology of brook has been related to German *breken*, meaning to break, it is unlikely that there is a linguistic connection. The folk etymology likely comes from the allusion to a watery spring that "breaks or bursts forth" from the ground. Debate over the differences between a brook, stream, creek, burn, gill, race, rindle, rivulet, and wadi are due to local variations and literary influence.
Similar terms: rivulet, creek, wadi, stream, rindle

brucite: [Eponymous term after Archibald Bruce (1777-1818).] n. A white trigonal mineral, magnesium hydroxide, $Mg(OH)_2$. It commonly occurs in pearly folia in fibrous aggregates in serpentinite.

Bruce, a professor at Yale University, discovered the mineral, and established the first geological journal in the U.S., *The American Journal of Science*. First usage in English c. 1865.

bryochore: [Greek *bruon* = moss < *bruein* = to swell, teem + *khrein* = to spread about < *khros* = place, room.] n. That part of the Earth's surface defined as tundra, especially as a climatological distinction.

bryozoan: [Greek *bryo* = moss + *zoa* = plural of *zoon* = animal, living being.] n. Any invertebrate of the phylum Bryozoa, characterized by colonial growth and formation of mosslike or branching colonies permanently attached to stones or seaweed. Bryophytes are plants, which include the colonial mosses.
Cited usage: Bryozoans, which produce delicate corals, sometimes branching and moss-like [1872, James D. Dana, Corals and Coral Islands i. 19].
Synonyms: Polyzoa

buccal: [Latin *bucca* = mouth, cheek + *al* = pertaining to.] adj. Pertaining to the mouth parts of echinoderms, decapod crustaceans, and some molluscs. Buccal refers specifically to an animal's feeding cavity, as opposed to the more general term for body cavity, *orifice* from Latin *or*, meaning mouth, and *facere*, meaning to make.

buhrstone: [French *bourre* = rough hair + Old English *stan* = stone.] n. A rough, hard, cohesive, silicified rock suitable for use as a millstone. Other terms that are cognate with buhrstone, and also allude to its etymology include: *burr*, meaning any rough flower head of a plant; the Danish *borre*, meaning burdock; and the Swedish *borre*, meaning sea urchin.
Alt. spelling: burrstone
Synonyms: whetstone, burr

bullion: [Middle English *bullion* = ingot of precious metal; < Dutch *buliouen* < Old French *billion* = small coin < *bille* = stick, bar.] n. Gold or silver ingots in a form before coinage.

bund: [Hindi *band* < IE *bhendh* = to bind.] n. In India, a constructed dike, quay, or levee running parallel to, and constraining the course of, a waterway. Also, especially in the Far East, a road running atop a levee.
Cited usage: To remove the dykes, or bunds, by which the ancient kings of Persia or Assyria had obstructed the navigation [1839, Thirlwall, Greece VII. 83].

buran: [Russian *buran* = blizzard.] n. A tempestuous windstorm of the Eurasian steppes, accompanied in summer by dust and in winter by snow. The buran of Russia, the purga of northern Siberia, and the boulbie of southern France all are similar in intensity to the American blizzard. There may be a connection to the Greek *boreas*, meaning north wind and Middle Eastern *bora* = strong wind in the Stara Planina mountains.
Synonyms: myatel, blizzard

bushveld: [Dutch *bosc* < Latin *boscum* = wood + Dutch *veld* = field.] n. An elevated open grassland in an otherwise low-lying, wooded region of southern Africa, especially in the Transvaal.
Cited usage: For big game, the low country and Bushveld is that part of the Transvaal which the hunter must seek [1879, Chambers's Jrnl. 1 March].

butte: [French *butte* = a hillock or rising ground; < Old French *butt* = the mound behind a practice target.] n. A conspicuous, usually isolated hill or small mountain with steep slopes and generally a flat top; the summit is usually smaller than that of a mesa. In volcanic areas, an isolated hill with steep sides and a craggy, round, pointed, and irregular summit, as in Black Butte near Mt. Shasta, California or Menan Buttes in Idaho.

Cited usage: The word butte is applied to the detached hills and ridges which rise abruptly, and reach too high to be called hills or ridges, and not high enough to be called mountains. [Frémont, 1845, The Rocky Mountains, p.145].
Synonyms: knob

bysmalith: [Greek *bysma* = plug + *lithos* = rock, stone.] n. A roughly vertical, cylindrical igneous intrusion that is bounded by steep faults; has been interpreted as a type of laccolith.
Cited usage: By this mode of intrusion, the vertical dimension of the intruded mass becomes still greater as compared with the lateral dimensions, so that its shape is more that of a plug or core. Such an intruded plug of igneous rock may be termed a bysmalith [1898, J. P. Iddings, Jrnl. Geol. VI. 706].
Similar terms: laccolith
See also: laccolith

bytownite: [Toponymous term from *Bytown* = the old name of Ottawa, Canada.] n. A dark gray mineral of the plagioclase feldspar group with composition ranging from $Ab_{30}An_{70}$ to $Ab_{10}An_{90}$ Bytownite is a rare form of feldspar. Recently discovered, it is best found near Ottawa, Canada, or in Crystal Bay, Minnesota.
Alt. spelling: Bytownit, Bytownita

caatinga: [Tupi *caa* = forest + *tinga* = white.] n. In Brazil, a forest consisting of thorny shrubs and stunted trees. A semiarid area in NE Brazil, which results in a low, sparsely forested region. The bare spots between plants are bleached laterite soils, which appear bare and white in the distance, hence the name *caatinga*, meaning white forest.

caballing: [Latin *cabbala* < Hebrew *qabbalah* = received lore, tradition < *qababal* = to receive, accept, admit.] v. In oceanography, the mixing of two water bodies having nearly the same density but different temperatures and salinities. After mixing, the resultant water mass sinks because its density becomes greater than either of the two original components. The cabbala, which has over two dozen modern spellings was the name given in postbiblical Hebrew to the oral tradition handed down from Moses to the Rabbis of the Mishnah and the Talmud.
Alt. spelling: cabbeling, cabelling

cabochon: [French *caboche* = cabbage, head; < Latin *caput* = head.] n. A polished, unfaceted gem with dome or convex upper surface and flat base, commonly made from jade, turquoise, or jasper. Prior to introduction of the cabbage to the rest of Europe by the Celts, *caboche* referred only to the head of an animal, not to the head of a vegetable.

cacimbo: [Angolan *cacimbo* = dew, dry season.] n. Characterized by dense, warm, early morning fog, the dry season in Angola and much of southern Africa is relatively cool and is called cacimbo. The cool fogs create a morning dew, which explains both the wet and dry meanings of cacimbo. Of Angola's two seasons, the dry *cacimbo* season occurs from May to September; the rainy season is warmer, and normally lasts from September to April.

cadastral: [French *cadastre* = register; < Greek *katastikhon* = register < *kata* = down, downwards + *stikhos* = row, line, line of verse; < IE *steigh* = stride, step.] adj. Pertaining to the delineation or recording of property boundaries as in a cadastral map. Originally used strictly as a basis for calculating taxation, the term *cadastral* is practically obsolete; current usage is *land survey* or *property survey.*. Etymologically related terms include catazonal and catatonic.

cadmium: [Latin *cadmia* = zinc carbonate (calamine); < Greek *kadmeia ge* = cadmean earth < *Kadmeia* = fortress of Thebes < *Kadmos* = Phoenician prince.] n. A soft, bluish-white metallic element (symbol Cd). Cadmium occurs primarily in zinc, copper, and lead ores. Legend states that Kadmos was the first person to find a zinc rock and notice that it gave a golden tinge to copper during smelting. He is also attributed with the knowledge of extracting gold from the deposits of Pangaios.

cafemic: [Composite mnemonic term from English *Ca* = calcium + *Fe* = ferrum (iron) + *M(a)* = magnesium + *ic* = of, pertaining to.] adj.. Pertaining to rocks rich in calcium, iron, and magnesium. The term mafic, coined to describe dark igneous rocks rich in magnesium and iron, inspired creation of this less common term.
See also: mafic

cairn: [Gaelic *carn* = heap of stones.] n. A usually conical mound of rocks, made intentionally for the purpose of marking a route, or for identifying a point, boundary, benchmark, claim, or memorial.
Alt. spelling: carn, carne

cala: [Spanish *cala* = cove, inlet, creek; < French *cala* = shelter.] n. A term for a small, semicircular, shallow bay; also, a creek running lateral to a main drainage. The two distinct meanings of this term, small bay and small stream, may be connected by the fact that, geomorphologically, streams that flow to the sea create natural coves as the canyon sidewalls become progressively eroded at the stream mouth. The related term *caleta*, meaning ultimate and smallest headwater, is a double diminutive because cala is already diminutive for bay.

calaverite: [Toponymous term after *Calaveras*, a gold mining district in California; < Spanish *calaveras* = skulls.] n. A pale, yellow to white, monoclinic mineral, $AuTe_2$. It is an important ore of gold. The name *Calaveras* was first applied to the river in north-central California because a number of skeletons were found there around 1837. The Spanish word *calavera*, meaning skull and crossbones, the international pirate symbol. An etymologically related term is *Calvary*, the Hebrew *Golgotha*, meaning the 'hill of skulls' outside ancient Jerusalem where Jesus Christ was crucified.

calc-: [Latin *calcem, calx* = lime, limestone, pebble; < Greek *khalix* = pebble.] Combining prefix indicating content of calcium, calcium carbonate, or lime, as in calcareous, calcrete, calc-alkaline. Etymologically related terms include chalk and concrete
See also: calcarenite, calcilutite, calcirudite, calcite, calcium, calcrete

calcarenite: [Latin *calcem, calx* = lime, limestone, pebble; < Greek *khalix* = pebble + *arena* = sand + *-ites* < *-lite* = pertaining to rocks < *lithos* = rock, stone.] n. A type of limestone composed of greater than 50% detrital (recycled) sand-sized calcite particles. Introduced in 1904 by Amadeus William Grabau, along with a number of other sedimentologic and stratigraphic terms such as calcirudite and calcilutite. In defiant response to Grabau's proclivity toward deriving new terms from classical roots, Joseph Barrell suggested that "perfectly clear and quite as sharp English terms" could be used instead. For example, Barrell suggested *shallow-sea deposits* over Grabau's *thalassigenic deposits*. In 1913, Grabau dedicated his exhaustive 1,185-page treatise *Principles of Stratigraphy* to Johannes Walther (Walther's Law).
See also: rudite, lutite, psephite, psammite, pelite, calcilutite, calcirudite

calcilutite: [Latin *calcem, calx* = lime, limestone, pebble; < Greek *khalix* = pebble + *lutum* = mud.] n. A limestone composed of greater than 50% detrital (recycled) silt or clay-sized calcium carbonate particles. Term coined in 1904 by American geologist A. W. Grabau.
Synonyms: calcipelite, calcareous mud
Similar terms: micrite
See also: rudite, psephite, psammite, pelite, calcarenite, calcirudite

calcine: [Middle English *calcinen* < Old French *calciner* < Latin *calcinare* = in alchemy, to change to powder by heat.] v. To heat a substance to its temperature of dissociation, for example, to change a solid phase into a gas, as in limestone to $CaO + CO_2$ (gas), or gypsum to $CaSO_4 + H_2O$ (gas).
See also: lime

calcirudite: [Latin *calcem, calx* = lime, limestone, pebble; < Greek *khalix* = pebble + *rudus* = rubble, broken stone.] n. Any consolidated breccia or conglomerate consisting of greater than

50% detrital (recycled) calcium carbonate particles larger than sand size. Term coined in 1904 by A.W. Grabau.
Similar terms: calcibreccia
See also: psammite, lutite, psephite, pelite, calcarenite, calcilutite

calcite: [Latin *calcem, calx* = lime, limestone, pebble; < Greek *khalix* = pebble.] n. A common white, colorless, or pale rock-forming rhombohedral mineral, $CaCO_3$. It is trimorphous with aragonite and vaterite and is the principal constituent of the rocks limestone, marble, chalk, tufa, and travertine.
Synonyms: calcspar

calcium: [Latin *calcem, calx* = lime, limestone, pebble; < Greek *khalix* = pebble + *ium* = suffix used to form the names of many metallic elements; < Greek *ion* = a condition of being.] n. A highly reactive, silver-colored, metallic element (symbol Ca). Calcium is of moderate hardness, and constitutes about 3% of the Earth's crust. It is a significant component of most animals and plants.
Cited usage: I shall venture to denominate the metals from the alkaline earths barium, strontium, calcium, and magnium [1808, Sir H. Davy, Phil. Trans. XCVIII. 346].

calcrete: [Latin *calcem, calx* = lime, limestone, pebble; < Greek *khalix* = pebble + (con)crete < *concrescere* = united or connected by growth.] n. Any size of detrital clasts bound together by calcareous material to form concretionary rock, such as: breccia, conglomerate, and calcareous sandstones. Any dissolved element or mineral in an aqueous solution that is capable of cementing unconsolidated detritus can be used as a prefix with '-crete' in order to make a new term for a specialized type of concretion, such as ferricrete and silcrete.
Cited usage: In the Gulf of Manaar, calcareous masses ('calcretes') of great extent are formed in

situ on the sea-bottom by the cementing of sand and other loose material by calcareous incrusting Polyzoa [1903, Nature, 22 Oct. 614/1].
See also: silcrete

calculus: [Latin *calculare* to reckon < *calculus* = a pebble (used in doing arithmetic), diminutive of *calx* = limestone.] n. In mathematics, a system of analysis devised simultaneously by Sir. Isaac Newton and Gottfried Wilhelm Leibnitz. In medicine, any abnormal stony mass formed in the body, as in a gallstone, kidney stone, or dental plaque. Etymologically related terms are compute, calculate, cipher, and reckon.

caldera: [Spanish *caldera* = cauldron, kettle; < Latin *caldarium* = hot bath; < *calor* = heat; < IE *kel* = warm.] n. A large, basin-shaped, more or less circular volcanic depression, with a diameter many times greater than that of the included vents (generally exceeding 1 mile / 1.6 km), and most often formed by summit collapse. The Spanish *caldera*, Portuguese *caldeira*, and French *chaudiere* all mean cauldron, kettle, or boiler. Because the Oxford English Dictionary definition of caldera is "a deep cauldron-like cavity on the summit of an extinct volcano," it's easy to see why there is confusion between the two terms crater and caldera. The terms should not be used interchangeably, because a caldera is an order of magnitude larger than a crater. IE derivatives are calorie and scald. The volcano of Campi Flegrei west of Naples, Italy, which had a dramatic eruption in 1538, may have been the first collapsed volcano to be called a caldera.
See also: crater

Caledonian: [Toponymous term after *Caledonia*, the Latin term for Scotland.] adj. Used in reference to the orogenic belt extending north from Ireland and Scotland through Scandinavia. These mountains formed in the early Paleozoic Caledonian Orogeny about 430Ma, when England

and Scotland collided. In British 19th-century literature, *Caledonian* was the term referring to the Scottish Highlands. The first use of the word in a tectonic sense was in 1885 when Suess named "the pre-Devonian mountains" extending from Norway, through Scotland to Ireland and Wales as the *Caledonians*.
Cited usage: Haug (in 1900) was, as far as we can ascertain, the first geologist to use the words 'Caledonian' and 'Orogeny' in the same sentence [2000, W.S. McKerrow et al., Jrnl. Geol. Soc. of London v. 157 pp. 1149].

caliche: [American Spanish *caliche* = pebble in a brick, flake of lime < *cal* = lime; < Latin *calc* = lime + Greek *khalix* = pebble.] n. In pedology, characteristic of stony soils in arid regions and composed of an accumulation primarily of calcite as friable to well-indurated layers over one meter thick. In the study of industrial mineral deposits, caliche is recognized in the Atacama Desert in Chile as a complex mixture of nitrates, chlorides, iodates, sulfates, and borates of sodium and potassium, acting as a cement for gravel deposits.
Cited usage: In the district of Tarapaca, northern Chili, the dry pampa is covered with beds of this salt (caliche) several feet in thickness [1892, E. S. Dana, J. D. Dana's A System of Mineralogy 6th ed. 871].

calina: [Uncertain, possibly Spanish or aboriginal from the Canary Islands.] n. A summer haze prevalent in Spain and Ecuador, caused by air filled with dust that is swept up from dry ground by strong winds. Like the larger and stronger sirocco, the calina blows out of a high-pressure zone over the North African Sahara and is normally drawn northwards ahead of a passing cold front or depression. This type of atmospheric turbidity (haze) is a common meteorological phenomenon that has acquired different names around the world, for example: harmattan (W. Africa), gobar (E. Africa), haboob (Arabia), kosa (Japan), whangsa (Korea), and huangsha (China).
Cited usage: In July and August the plains of New Castile are sunburnt wastes; the atmosphere is filled with a fine dust, producing a haze known as calina [1887, Encycl. Britannica XXII 296/2].
Alt. spelling: calima
See also: sirocco

callus: [Latin *callus* = hardened skin.] n. A thickened part of the shell of a gastropod or brachiopod shell. An etymologically related term is callous.

calm: [Latin *cauma* = resting place in the heat of the day; < Greek *kauma* = burning heat < *kaiein* = to burn.] n., adj. A condition of no wind or a wind with a speed of less than 1 mile (1.6 km) per hour; zero on the Beaufort wind scale of 0-12.

calorie: [Latin *calor* = heat; < IE *kel* = warm.] n. The quantity of heat required to raise the temperature of 1 gram of water by 1°C at one atmosphere of pressure. This amount of heat depends somewhat on the initial temperature of the water, which results in various different units having slightly different energy values, but sharing the name *calorie*. Confusion arises over calorie (small calorie) and Calorie (large calorie). The Calorie (large calorie) is actually one kilo-calorie, and is used for measuring the energy produced by food when oxidized in the body. Etymologically related terms are caloric, calorimeter.
Similar terms: 15°C calorie, 4°C calorie, mean 1°C to 100°C calorie, thermochemical calorie

calotte: [French diminutive of *caul* = close-fitting headcap; < Greek *kalyptra* = hood < *kalytein* = to cover.] n. Ring of weakened rock surrounding the entrance, shaft, or adit of a tunnel, primarily caused by removal of rock, and the resulting response to pressure relief.

calx: [Latin *calx, calcem* = lime, limestone; < Greek *khalix* = pebble.] n. The crumbly residue left after a mineral or metal has been so thoroughly heated as to drive off all its volatile constituents ('calcining'), as lime is calcined in a kiln. The *calx* of a mineral was considered by early alchemists to be the essential substance or 'alcohol' of that mineral after all its volatile parts had been dispelled. The *calx* of a metal was supposed to be the result of the expulsion of *phlogiston* (in reality it was usually the metallic oxide, but in some cases the metal itself in a state of sublimation).
Cited usage: Calcination is solution of bodies into Calx or Alcool [1617, John Woodall, The Surgeon's Mate].

calyx: [Latin *calyx* = cup, goblet; < Greek *kalyx* = shell, husk, pod < *kalyptein* = to cover; < IE *kelk* = cup.] n. In paleontology and biology, the small cup-shaped structure in which a coral polyp sits. In botany, the whorl of sepals, forming the envelope in which a flower bud is enclosed. An etymologically related term is chalice.
Cited usage: Calyx .the cup enclosing or containing the flower [1671, Malpighi, Anatomia Plantarum].

cam: [German *kamm* = crest of a hill; < Old English *comb* = tooth.] n. The crest or serrated ridge of a hill or mountain. The term is cognate with: Scottish *kame*, Swedish *kam*, Welsh *comb*, and German *kamm*.
Synonyms: kame, kam , comb
See also: kame

camber: [Middle English *caumber* = curved; < French *cambrer* = to bent, curve; < Latin *camur* = crooked, arched; < Greek *kamara* = vault.] n. A surface that is slightly curved convex upward, as a ridge or road.

Cambrian: [Toponymous term after Cambria, Welsh name for Wales (Latinized form of Cymry); < Old Celtic *combroges* = compatriots.] adj. The first period of the Paleozoic Era, 542Ma-488.3Ma. Cambria, a synonym for the modern Wales, and Cumbria, presently a county in northwest England, were originally considered synonymous, but became differentiated as the name Cambria was equated strictly to Wales by Geoffrey of Monmouth in the 12th century. The terms are both Latinized derivatives of the Welsh *Cymry*, The geologic term Cambrian was first applied by Rev. Adam Sedgewick in 1836 to that system of Paleozoic rocks lying below the Silurian in both Wales (Cambria) and Cumbria.

camera: [Latin *camera* < Greek *kamara* = chamber.] n. In optics, a device for taking photographs, consisting of a closed box containing a sensitized plate or film on which an image is formed when light enters through a hole or lens. In biology, a space enclosed between adjacent septa in a cephalopod shell. Etymologically related terms are: camera obscura, camera lucida, or as an adjective, bicameral legislature.
Synonyms: chamber

camouflage: [French *camoufler* = to disguise < *camouflet* = puff of smoke < *ca* = collective prefix + *moufler* = to cover up.] n. Any device, action, protective coloration, or morphologic mimicry employed to conceal, mislead, or deceive.

campanulate: [New Latin *Campanula* = botanical genus; < Latin *campana* = bell.] adj. In botany, referring to any of various plants of the genus Campanula, including Canterbury bells. In geology, occasionally used to mean bell-shaped. In general, as the noun campanile, a bell tower, especially one that stands apart from any other building.
Cited usage: Calicles tubular or campanulate [1842, James D. Dana, Zoophytes 686].

campus: [Latin *campus* = plain, field, level surface.] n. The grounds of a school, college, or large corporation. Etymologically related terms include camp and campaign.

canal: [Italian *canali* < Latin *canalis* = channel, pipe, tube.] n. In general usage, an artificial waterway for irrigation or transportation. In astrogeology, a term for dark, linear markings on the surface of Mars, interpreted in 1877 by the Italian astronomer Giovanni Schiaparelli as channels (Italian *canali*), in reference to the natural flow of water. *Canali* was mistranslated into English by the American astronomer William Henry Pickering as *canals*, which implied features of artificial origin dug by intelligent beings on Mars. H.G. Wells' *War of the Worlds* was inspired by Pickering's claim. This novel, about an invasion of Earth by Martians, was a runaway best-seller that fired people's imagination of the day, all of it stemming from a mistranslation.

canga: [Portuguese *canga* = a large, wooden yoke, invented in China, and used to fasten around the neck as punishment for petty crimes; < Annamese *conga* < *gong* = stock, yoke.] n. A term used in Brazil for an iron-rich, unstratified, surficial sedimentary rock. Canga is composed of fragments derived from oxide facies in BIF (itabirite), and cemented by limonite. As used in Sierra Leone, a ferruginous, frequently pisolitic, limonite-cemented laterite soil. This term came into common use because of the many laborers who carried heavy loads of iron-ore on their shoulders in yoke-like containers while working for bosses of European descent at the numerous strip mines in Brazil. The term canga is also a local name in Brazil for vegetation growing on iron-rich laterite soils. Brazil, after China, is the world's second leading producer of iron. An etymologically related term is congue.
Cited usage: This conglomerate, consisting largely of rounded quartz grains and pebbles,

occurs in extensive beds or in isolated blocks known as "tapanhoancanga" or "canga", which may enclose crystals of diamond [2004, R. Swiecki, Diamond in Brazil, Alluvial Exploration and Mining Website].
See also: BIF, itabirite

cannel coal: [English, a corruption of *candle coal*.] n. A compact, tough, bituminous coal characterized by a dull to waxy luster and a composition dominated by sapropel (decomposed pollen and algal sludge). It has a high volatile content and burns with a bright flame and much smoke. With increasing content of clay and silt, cannel coal grades into cannel shale, also called oil shale.
Cited usage: Mr. Bradeshau hath a place caullid Hawe a myle from Wigan. He hath founde moche Canel like Se Coole in his Grounde very profitable to hym [1538, Leland, Itin. VII. 47]. Famous for yielding the Canal (or Candle) coal. It is so termed, as I guess, because the manufacturers in that country use no candle, but work by the light of their coal fire [1734, North, Lives I. 294]. Cannel coal does not soil the fingers [1878, Green, Coal I. 30].
Alt. spelling: canel-coal, kennel-coal
Synonyms: candle coal, parrot-coal (Scotland)
Similar terms: boghead coal, sapropelic coal, tasmanite

cant: [Latin *canthus* = outer rim of a wheel, < Greek *kanthus* = corner of the eye; < IE *kantho* = corner, bend.] n. Any inclination or divergence from a given line. Numerous northern European languages have adopted this word with some variation in meaning. In Welsh, *cant*, meaning the circumference of a circle, conforming closely to the original meaning in Latin. Both the German, *kante* and the Italian *canto*, meaning edge, corner, or border, conform to the Greek root. Not to be confused with the homograph *cant* and its etymologically related terms cantor, canto, or

incantation, all derived from Latin *cantus*, meaning song.
Similar terms: dreikanter, decant, canton

canyon: [Spanish *cañon* < *caña, caño* = tube, cane; < Latin *canna* < Greek *kanna* < Semitic *qanaw* = reed.] n. A long narrow valley between high cliffs, often with a stream flowing through it. Canyon is the anglicized form that best represents the Spanish cañon. Etymologically related terms include: cane, cannon, and cannister, as well as the Italian food terms cannelloni and cannoli.
Cited usage: Two cañons ran up into the bosom of the ridge (by which word cañon the Spaniards express a deep, narrow hollow among the mountains) [1834, A. Pike, Sketches 20].
Similar terms: gorge
See also: gorge

cape: [Latin *caput* < IE *kaput* = head.] n. An extensive protrusion of land jutting into the sea. A cape can be in the form of a continental terminus (Cape Horn and Cape of Good Hope), a peninsula (Cape Cod), a projecting part of a coastline (Cape Hatteras), a promontory, or a headland. A cape is larger than a point. Etymologically related terms include: captain, capital and chief.
Cited usage: Cape Fly-away, a cloud-bank on the horizon, mistaken for land, which disappears as the ship advances [1867, Smyth, Sailor's Word-book].
Similar terms: point

capillary: [Latin *capillus* = hair.] adj. In mineralogy, a habit of minerals that form exceedingly slender hairlike or threadlike crystals. Examples being millerite and humboldtine. In hydrology, referring to tubes or interstitial openings with such a small diameter that water or other fluids can be drawn up against the force of gravity as a result of surface tension effects. In volcanology, referring to hairlike pyroclastic material, such as Pele's hair. In general, as a noun, a tube with a very

small internal diameter.
Cited usage: The slower rise of capillary water in a dry soil [1895, F.H. King, Soil v.176].

carapace: [Spanish *carapacho* = carapace; < Latin *capa* = cape, hood.] n. The chitinous to bony protective covering over all or part of the back of certain animals, such as: turtles, crustaceans, arachnids, trilobites, and eurypterids.
Cited usage: A continuous covering for the body, like the carapace of the Arthropoda [1878, Carl Gegenbauer, Vergleichende Anatomy (Comparative Anatomy) 38].
Similar terms: dorsal exoskeleton
See also: isopod, exoskeleton

carat: [Arabic *qirat* = pod, husk, weight of four grains, < Greek *keration* = carob seed.] n. A unit of weight for precious stones and pearls, equal to 200 milligrams (approx. 4 grains). The Greek equivalent of one carat was the *keration*, a weight equal to that of a single carob bean. When beans weren't available, one carat was equivalent to 1/3 the weight of a single *obol*, the smallest unit of ancient Greek currency. The Romans related the carat, or *siliqua* to 1/24 the weight of their gold coin, the *solidus* (itself weighing 1/6 of an ounce). This is close to the modern metric weight of 200 milligrams per carat. In 309 C.E. Emperor Constantine reduced the weight of the *solidus*, meaning "solid bit" to 72 per Roman pound (4.48 grains each). Its name changed to the *aureus*, but its weight remained the standard until the eleventh century. In gemology, a *"point"* equals 1/100th of a carat, or 2 mg. Not to be confused with *karat*, a measure for the purity of gold.
Cited usage: These pearles are prised according to the caracts (sic) which they weigh, every caract is 4 graines [1598, Hakluyt, Voy. II. i. 225].
Similar terms: karat
See also: karat

carbon: [French *carbone* = carbon < *charbon* < Latin *carbo* = coal, charcoal < IE *ker* = heat, fire, to burn.] n. A nonmetallic element (symbol C) occurring in nature in two allotropic forms, diamond and graphite; in many common carbonate minerals combined with other elements; and as amorphous carbon in coal, bitumen, asphalt, and petroleum. As an adjective meaning carbon-rich, the word carbonaceous is employed. Of the 88 elements occurring in natural substances on Earth, four of them - hydrogen, oxygen, carbon, and nitrogen - comprise more than 95% by weight of Earth's living matter. Excluding the inert gases helium and neon, these four elements are the most abundant elements in the universe. They are not, however, the four most abundant elements in the entire Earth (iron, oxygen, silicon and magnesium), or in the Earth's crust (oxygen, silicon, aluminum, and iron). This comparison reveals that the composition of Earth's living matter reflects the average composition of the universe more than of Earth itself, a fact that has given rise to theories that life first arrived on Earth from an extraterrestrial source rather than having originated on the planet itself.
Cited usage: Carbon in its amorphous state, is charcoal; when crystallised in prisms, it becomes black and opaque graphite; and when crystallised in octohedrons, it is etherealized into the limpid and transparent diamond [1862, R. H. Patterson, Essential Hist. & Art 8].

Carboniferous: [from Latin *carbo* = coal; < IE *ker* = heat, fire + *-iferous* = suffix meaning bearing < *ferre* = to bear + *os* = abounding in, full of, or characterized by.] adj. In European usage, referring to a single period, divided into Early and Late, which extends from 359.2Ma-299Ma. Thick coal beds were deposited on most continents during the Carboniferous Period. In North American usage the Carboniferous Period is replaced by the geochronologic units Mississippian Period and Pennsylvanian Period. In chronostratigraphic

nomenclature, the Carboniferous System lies above Devonian strata and below Permian strata. Etymologically related terms include: carbon, carbuncle, cremate, and ceramic.
Cited usage: The Scar Limestone, a member of the carboniferous series [1830, Charles Lyell, Principles of Geology (1850) II. iii. xlv. 529].
Synonyms: Mississippian-Pennsylvanian
See also: Mississippian, Pennsylvanian, Paleozoic

Carborundum: [Portmanteau word from *carb(on)* + *(c)orundum*.] n. Trade name for a hard, synthetic, crystalline compound silicon carbide (SiC), used as an abrasive and refractory lining in furnaces. The synthetic compound was discovered accidentally by Edward G. Acheson in 1891 while he was attempting to make artificial diamonds. According to the official gazetteer of U.S. patents, the trademark was registered by the Carborundum Company of Monongahela City, Pennsylvania, in 1892. Synthetic carborundum is identical to the hexagonal mineral moissanite (SiC), discovered in the Canyon Diablo meteorite by Ferdinand Moissan in 1893.
Cited usage: While examining the hardness of 'carborundum', a carbide of silicon, made by Mr. Acheson of Pittsburg, it was found that it readily scratched red, blue, white, pink, and yellow corundum in the form of fine gems [1893, Amer. Jrnl. Sci. XLVI. 472].

cardinal: [English *cardines* = poles; < Latin *cardinalis* = principal, chief < *cardo* = hinge.] adj. Principal, chief. In geography *cardinal points* refers to the four principal points of the compass: north, south, east, and west. In zoology, cardinal axis refers to the hinge axis of a bivalve mollusc. The use of the adjective *cardinal* for describing the four principal directions led to an association with the number four, as in the four cardinal winds from Greek Mythology: Boreas (north), Notos (south), Euros (east), and Zephyros (west), the

four cardinal virtues of ancient Greek philosophy: justice, prudence, fortitude, and temperance, and the four cardinal humors of the body formerly considered responsible for one's health and disposition: blood, phlegm, choler (yellow bile), and melancholy (black bile). The Scottish equivalent word is *airt* < Celtic *aird*, meaning point of the compass.
Cited usage: Four of them are called the Cardinal Points S, the South; W, the West; N, the North; E, the East sometimes called the four Winds of Heaven [1755, Benjamin Martin, The General Magazine of Arts & Sciences iii. ii. 179].
Synonyms: airt

cardioid: [Greek *kardia* = heart + *eidos* = shape.] adj. Heart-shaped. An etymologically related term is cardiac

cargneule: [French *corgneule* < Latin (possibly) *corneus* = horny < *corn* = horn.] n. A porous or pitted limestone or dolomite breccia, in which carbonate clasts are enclosed in a soft, friable matrix of evaporitic material. French geologists acquired this term from the *patois* (dialect) spoken in the area *d'Aigle* (The Eagle) in the Alps near the border between Switzerland and France, where these "spongy dolomitic limestones" are common. These rocks, most commonly found in the European Alps, have a complex origin likely derived from diagenesis of calcareous fragments in a fine-grained matrix, but are as yet not fully explained.
Alt. spelling: cornieule, corgneule, carnieule
Synonyms: cellular dolomite
Similar terms: rahwacke
See also: rahwacke

caries: [Latin *caries* = decay; < Greek *ker* = death.] adj. In ore microscopy, the term *caries texture* refers to a scalloped contact between two minerals in which the apparently younger mineral has replaced the older mineral in a series of convex incursions which resemble filled dental cavities.
Synonyms: cusp and caries texture

carina: [Latin *carina* = keel; < IE *kar* = hard.] n. A keel-shaped ridge or flange, especially on part of the skeleton, such as: that on the breastbone of a bird, the dorsal shingle plate of the shell of certain crustaceans, the vertebral column of an embryo, the septum of a rugose coral, or the tests of some forams.

Carlsbad twin: [Toponymous name after Karlsbad, in Bohemia, Czech Republic, known for the granite there.] n. Large, chiefly orthoclase feldspar crystals in porphyritic texture, having a regular composition surface. They are common as simple, contact, or penetration twins along the c-crystallographic axis. The *Carlsbad twin law* describes two intergrown crystals growing in opposite directions. Karlsbad is an important natural hot springs and spa in the Krkonose (Giant) Mountains, the largest mountain chain in the Czech Republic. The source water for these springs comes from sedimentary roof rock called *Sprudelschale* (spurting, spritzing shale) underlain by intrusive granite at this type locality for Carlsbad-twinned orthoclase crystals.
Alt. spelling: Karlsbad twin

carnelian: [Latin *cornus* = bunch-berry, dogwood, cornel cherry < *carnen, carneus* = flesh, meat.] n. A red, orange-red to flesh-colored, translucent variety of chalcedony (cryptocrystalline quartz). The root is not related to the much used Latin homograph *cornus*, meaning horn, as evidenced by the Italian *corniola*, which refers to both the gemstone carnelian and to the fruit of the dogwood.
Cited usage: The common carnelion has its name from its flesh colour...which is, in some of these stones, paler, when it is called the female carnelion; in others deeper, called the male [1695, Woodward, Nat. Hist. Earth].
Alt. spelling: cornelian
See also: sard

carnotite: [Eponymous name after Marie Adolphe Carnot (1839-1920), French inspector-general of mines.] n. A canary-yellow, monoclinic, hydrated vanadate of potassium and uranium, $K(UO_2)_2(VO_4)_2 \cdot 3H_2O$. It occurs as powdery encrustations on joints and bedding planes in sandstones, and is an important ore mineral of vanadium, uranium, and radium. Radium is a product in the radioactive decay series of uranium-238; it is therefore found in uranium ores, usually in a ratio of one part radium to three million parts uranium. Although carnotite sands from Colorado are a famous source of radium, the largest supplies are presently in the Kinshasa district of the Democratic Republic of Congo (formerly Zaire) and in pitchblende from western Canada.
Cited usage: A new mineral containing uranium and vanadium, to which the authors give the name carnotite, is found in yellow, friable masses, mixed with very variable quantities of silica, together with malachite and chessylite, in pockets at the surface of a grit in Montrose Cty., Colorado [1899, Journal of the Chemical Society LXXVI. ii. 434].

carse: [Scottish *carse* = plural of *carr* = low wetlands; < Old Norse *koer* = pool, pond.] n. A flat area or stretch of low alluvial land along the banks of a river or adjacent to an estuary. Although the etymology is somewhat vague, it is certain that the present meaning of the term originated for the lowlands along the banks of rivers in Scotland e.g., the Carse of Gowrie, the Carse of Stirling and the Carse of Falkirk.
Alt. spelling: kars
Similar terms: bog, fen, carsen reed, korsen

Cartesian: [An eponymous term from the Latinized form of Rene Descartes' name (*Cartesius*).] adj. Referring to the two and three axis coordinate systems developed by French philosopher and mathematician Rene Descartes (1596-1650). In the two-dimensional graphic system, any point can be located by measuring its distance from two axes at right angles to each other on a single plane. In the three-dimensional system, any point can be located by measuring its distance from the intersection point of three planes at right angles to each other.
See also: cartography

cartography: [Latin *carta* = paper; < Greek *cartos* = leaf of paper + Latin *graphicus* < Greek *graphikos* < *graphe* = writing < *graphein* = to write; < IE *gerbh* = to scratch.] n. The art, technique, and production of charts or maps. Etymologically related terms include: carton, cartoon, cartouche, cartology, and cartridge.

cascade: [Italian *cascare* = to fall; < Latin *cadere* = to fall; < IE *kad* = to fall.] n. A small waterfall or a series of small waterfalls descending over steeply sloping rocks. IE derivatives include: cadaver, decay and deciduous.

cascajo: [Spanish *cascajo* = shard, rubbish, gravel < *cascar* = to break; < Latin *quatere* = to shake.] n. Reef-derived sediment composed of coral fragments and other detritus occurring in older deposits.
See also: cascalho

cascalho: [Portuguese *cascalho* = gravel.] n. In Brazil, a alluvial deposit of pebbles, gravel, and ferruginous sand that contains diamonds or gold.
Cited usage: The gold lies in a stratum of rounded pebbles and gravel, called cascalhão [1812, Mawe, Trav. Brazil v. 77].
Alt. spelling: cascalhao
See also: cascajo

casehardening: [Latin *capsa* = box, sheath + Middle English *heard* < Old English *heard* = hard.] v. In metallurgy, the process by which metal, especially iron or soft steel, is hardened on its

surface by coating the outer layer of the metal with carbon during heat treatment. This process creates a sheath or case of hard carbon steel on the outside. In geology, the process by which the surface of a porous rock, especially tuff or sandstone, is coated with desert varnish or cementing agents, possibly by evaporation of mineral-bearing solutions, or by the action of microorganisms.
Alt. spelling: case hardening

cassiterite: [Greek *kassiteros* = tin + *-ites* < *-lite* = pertaining to rocks < *lithos* = rock, stone.] n. A yellow to reddish-brown to black tetragonal tin oxide, SnO_2. It is the principal ore of tin, and occurs as prismatic crystals or as botryoidal aggregates with a concentric structure known as *wood tin* or *stream tin*. The principle source of tin for the ancient Greeks were the *Kassiterides*, meaning 'tin islands'. The Greek poet Homer referred to tin as "the metal from the land of the Kassi."
Cited usage: Stannic (di)oxide, occurs native as tin-stone or cassiterite [1873, H. Watts & G. Fownes, Elem. Chem. 445].
See also: tin

cast: [Middle English *casten* = to throw.] n. In paleontology and biology, the replica of the external details of a fossil shell or skeleton, or other organic structure. A cast is formed by mineral material that has filled the cavity formed by dissolution or decomposition of the original hardparts of the organism. In sedimentology, a structure, such as a flute cast or a sole cast, formed by the infilling of an original mark or depression made on top of a soft bed, and preserved as a solid form on the underside of an overlying, more durable bed. The etymological connection to the verb *cast* arose because material is 'thrown' into the space originally occupied by the organism, not unlike casts made in ceramics and metal sculpture.
See also: mold

cata-: [Latin *cata-* < Greek *kata-* = down, against, inferior, reduced; < IE *kat-* = down.] Combining prefix meaning down, used in such terms as: catadromous, catabatic, cataract, and catarrh; meaning against, as in catapult; inferior, as in catacomb; or meaning reduced, as in catastrophe and cataclysm. Sometimes used to strengthen or enhance the meaning of a term, an intensive used for emphasis, such as in the term catalyst.
See also: kata-, ana

cataclasite: [Greek *kataklastos* = broken down < Greek *kata* = down + *klastos* = broken < *klan* = to break.] n. A fault-metamorphosed rock that is cohesive, but that lacks or has poorly developed schistosity. Or, a non-cohesive fault rock characterized by generally angular porphyroclasts and lithic fragments in a fine-grained matrix. Until viewed microscopically, it is difficult to distinguish foliated fault rocks that are formed by brittle deformation (cataclasite) from those formed by plastic deformation or grain-boundary sliding (mylonite), therefore the terms are often used synonymously in the field.
Synonyms: mylonite
See also: mylonite

cataclastic: [Greek *kataklastos* = broken down < Greek *kata* = down + *klastos* = broken < *klan* = to break.] adj. In petrology, pertaining to the texture and structure produced in a rock, such as a tectonic breccia or crush breccia. In a cataclastic rock, angular fragments have been produced by severe mechanical deformation, and the resultant bending, breaking, rotating, crushing, and granulation of the original rock occurs without chemical recrystallization. Also, pertaining to a type of local metamorphism confined to the vicinity of faults, and involving purely mechanical crushing and granulation of the preexisting rock (cataclasis). Term coined by the Norwegian geologist Kjerulf in 1885. Etymologically related terms are: catalepsy

(*lambanein*, meaning to seize), catalogue (*falicum*, meaning scaffolding), and catapult (*pallein*, meaning to hurl).
Cited usage: The structures are of the kind for which Prof. Kjerulf has proposed the term cataclastic. I venture to suggest...that we should distinguish between the three types of clastic rocks at present recognized by using the terms epiclastic, cataclastic and pyroclastic [1887, J. J. H. Teall, in Geol. Mag. IV. 493].
Synonyms: crush breccia
See also: mylonite

cataclinal: [Greek *kata* = down + *klinein* = to bend.] adj. Said of a stream valley that dips in the same way as the underlying strata. Now rare, introduced by John Wesley Powell in 1875.
Cited usage: I have classified these valleys in the following manner cataclinal, valleys that run in the direction of the dip [1875, J. W. Powell, Exploration of the Colorado River of the West II. xi.160]. Terms originally proposed by Powell, such as cataclinal have found no wide usage, as they are unnecessarily obscure [1960, B. W. Sparks, Geomorphol. vi. 104].
Synonyms: conclinal

cataclysm: [Greek *kataklysmos* = deluge < *kata*= down + *klyzein* = to wash away.] n. Any violent event that causes a sudden and extensive change in the Earth's surface, such as a violent flood or earthquake. An etymologically related term is catastrophism.
Cited usage: For the proofs of these general cataclysms we have searched in vain [1833, Lyell, Principles of Geology III. 101].
Synonyms: catastrophe

catacomb: [Latin *cata* < Greek *kata* = down + Latin *tumbas* = tomb.] n. Any of a series of vaults or galleries in an underground burial place.

catadromous: [Greek *kata* = down, against + *dromous* = moving, running.] adj. Said of certain freshwater fish, such as eels, which spend most of their life in fresh water and enter salt water to spawn; as opposed to anadromous fish, such as salmon, that migrate upstream into fresh water as adults in order to spawn.

catagenesis: [Latin *cata* < Greek *kata* = reverse + Latin *genesis* < Greek *genesis* = origin, creation < *genesthai* = to be born, come into being; < IE *gen* = to give birth, beget.] n. Evolutionary change toward decreased vigor and simpler forms, and away from complex forms which are highly specialized for a particular set of environmental conditions.
Cited usage: The process of creation by the retrograde metamorphosis of energy, or, what is the same thing, by the specialization of energy, may be called catagenesis [1884, E. D. Cope, Amer. Assoc. for Advancement of Science XXXIII. 468].
Synonyms: Retrogressive evolution

catalyst: [Greek *kata* = down + *lyein* = a loosening.] n. A substance that, when present, increases the rate of a chemical reaction, but itself undergoes no permanent change by the reaction. The term for the catalytic process, catalysis, was coined by the chemist Jons Jakob Berzelius in 1836. Catalysts are generally present in small amounts relative to the reagents. If a reaction rate is retarded by the added sub-stance, that substance is called a negative catalyst.
Cited usage: By means of this action they produce decomposition in bodies, and form new compounds into the composition of which they do not enter. This new power, hitherto unknown, is common both in organic and inorganic nature. I shall call it catalytic power. I shall also call Catalysis the decomposition of bodies by this force [1836, Berzelius, Edinburgh New Philosophical Jrnl. XXI. 223].
Synonyms: contact action

cataract: [Greek *katarakts* = downrush, waterfall < *kata* = down + *arassein* = to strike.] n. A waterfall, especially one of great volume in which the vertical drop has been concentrated in a single descent over a precipice.
Cited usage: From the steppes of Scythia to the cataracts of the Nile [1839, Thirlwall, Greece II.185].
See also: cascade

catastrophe: [Greek *katastrephein* = to overturn < *kata* = down + *strephein* = to turn.] n. A sudden violent disturbance of nature variously explained by either exceptional natural causes or supernatural causes. An example of the former would be the mass extinction of an entire fauna by asteroid impact. An example of the latter would be the Noachian flood. Etymologically related terms include catastrophism and catastrophist.

catazonal: [Latin *cata* < Greek *kata* = down + Latin *zona* < Greek *zone* = girdle < *zonnai* = to gird.] n. The deepest of the three commonly delineated metamorphic zones (epi-, meso-, and cata-zonal), where temperature exceeds 800°C and pressure is 3-10 kilobars (kbs). The zone is characterized by the granulite facies, composed of less hydrous minerals, such as garnet, hypersthene, and cordierite. There is some difference of opinion among meta-morphic petrologists over the exact temperature and pressure boundary conditions of the three metamorphic zones.
Cited usage: Entiat and Seven Fingered Jack plutons. Cater (1982, pp.90-91) considered the migmatite a contact zone (see also Cater and Wright, 1967) related to igneous intrusion of the pluton at catazonal depths where the distinction between metamorphism and melted rocks is not well defined (see also Buddington, 1959, p.714)[1980, R.W. Tabor, et. al., Geologic Map of the Chelan Quad, WA, U.S.G.S.].
See also: epizonal, mesozonal

catena: [Latin *catena* = chain; < IE *kat* = twine, to twist.] n. In general, a linked or connected series of anything. In soils, referring to a sequence of soils with differing characteristics, due to differing relief and drainage but with the same age, parent material, and climatic conditions. Etymo- logically related terms include: catenary, catenate, and cateniform.

cathode: [Greek *kata* = down + *hodos* = way.] n. In an electrolytic cell or vacuum tube (cathode-ray tube), the negatively charged electrode from which electrons (negatively charged) flow. When the cathode is immersed in an electrolyte with both positively and negatively charged ions, the positive ions (cations) are attracted to the cathode while the negative ions (anions) flow to the anode. Cathode and anode (therefore cation and anion) are frequently confused. The following definition of cathode from the American Heritage Dictionary is one of many sources helping to perpetuate the confusion: a) A negatively charged electrode, as of an electro-lytic cell or storage battery, that is the source of electrons in an electrical device. b) The positively charged terminal of a primary cell or a storage battery that is supplying current.
Cited usage: The cathode is that surface at which the current leaves the decomposing body, and is its positive extremity [1834, Faraday, Experimental Researches in Electricity v.1 ser.7p.197 para.663].
See also: anode, ion, cation, anion

cation: [Greek *kata* = down + *ion* < *ienai* = to go.] n. An ion or group of ions having a positive charge that move towards the cathode (negative electrode) in an electrolytic cell. The term was coined by Michael Faraday in 1834. Some IE derivatives include: entrance, exit, circuit, companion, and January (from the Roman god Janus, guardian of portals and of beginnings and endings).

Cited usage: I require a term to express those bodies which can pass to the electrodes, or, as they are usually called, the poles...I propose to distinguish such bodies by calling those anions which go to the anode of the decomposing body; and those passing to the cathode, cations [1834, Michael Faraday, Experimental Researches in Electricity v.1 ser.7 p.198 para. 665].
See also: ion, cathode, anode, anion

catlinite: [Eponymous term after George Catlin (1796-1872), American painter of Native Americans.] n. A red, silicious, lithified clay used as a pipestone by the Dakota People of the Upper Missouri Valley.

catoctin: [Toponymous term after the Catoctin Formation, on Mt. Catoctin in the Blue Ridge Mtns. of Maryland and Virginia; < Native American *catoctin* = the land of many deer.] n. A hill or ridge of resistant material that rises above a pene-plain, and preserves at its summit a remnant of a former peneplain.
See also: monadnock, peneplain

Catskill Mountains: [Dutch *kat* = cat + *kill* = river, stream; < IE *gleubh* = tear apart, cleave.] n. A mountain range in southeastern New York, just west of the Hudson River. The name was derived during the time of the Dutch domination in the 17th and 18th centuries. So called due to the many catamounts found there, such as: cougars, pumas, panthers, and mountain lions. The term "kill" is frequently used for place names in the Catskill region; for example Freshkill and Fishkill. An IE derivative is hieroglyphics (from Greek *gluphein*, meaning to carve, thus the term hieroglyphics).

cauldron: [Old French *caudiere* = cooking pot; < Latin *caldarium* = hot bath < *calidus* = heat; < IE *kel* = warm.] n. A term for all volcanic subsidence structures regardless of shape or size, depth of erosion, or connection with the surface. The term can thus be applied to volcanic depressions, such as collapse calderas and cauldron subsidences. As an adjective, used as cauldron subsidence, referring to the process where roughly cylindrical blocks sink into a magma chamber, the blocks being bounded by steep ring fractures, sometimes reaching the surface. The ring fractures are normally occupied by ring dikes. Term coined in 1968 by Robert Smith and Robert Bailey.

causse: [French *causse* < Latin *calx* = lime.] n. A karst limestone plateau of small size, similar to limestone plateaus of the Grands Causses region in southeastern France (e.g., the causses duLarzac, Severac and Noir). Term derived from the unique dialect spoken in that region of France.
Cited usage: These plateaux are called Causses in the vernacular dialect [1827, G. P. Scrope, Memoir on the Geol. of Central France i. 11]. By the Dordogne and Lot the surface is divided into a number of limestone plateaus known by the name of causses [1883, Encycl. Britannica XV. 8/2].

cave: [Latin *cava* = plural of *cavus* = hollow; < IE *keu* = vault, hole, to swell.] n. A natural under-ground opening, usually with a connection to the surface that is large enough for a person to enter. The most common caves are formed in limestone or marble by the dissolving action of acidic ground water. The term has also been applied to subsurface openings beneath rockfalls. As a compound adjective, it is the first word in terms like: cave breccia, cave blister, cave bubble, cave coral, cave cotton, cave flower, cave onyx, cave pearl, and cave raft. The related term cavern usually implies a larger size than a cave; it may also indicate a series of connected caves. Some IE derivatives include:excavate, cavity, cumulus, and church.

cay: [Spanish *cayo* = shoal, barrier reef; < Taino *cay* = island.] n. A low, coastal island or emergent reef composed of local detritus, such as sand or coral fragments, that are built up just above high-tide level. Widely used in the West Indies to describe islets and reefs fringing the coast. The Taino, indigenous tribes of the Caribbean, were probably the first to coin the term. In British English, the word key was originally pronounced "kay;" the geologic term, cay, therefore, came to be written k-e-y, due to their identical pronunciation. (This was due to the process called reverse assimilation, the revision of spelling and pronunciation between already existing words in a language, and newly adopted terms.) The meaning of cay may have transferred to the Celtic and Welsh *cae*, meaning hedge, and the Breton *kae*, meaning embankment. An etymologically related term is quay.
Cited usage: Called by the Spaniards Cayos, whence by corruption comes the English word Keys [1707, Sloane Jamaica I. Introd. 86]. Caies, a ridge of rocks, or sand-banks; called in the West Indies, keys [1769, Falconer, Marine Dictionary].
Alt. spelling: key

ceja: [Spanish *ceja* = eyebrow, summit.] n. The jutting edge along the top of a plateau or mesa. Also used for the escarpment or cliff at the edge of a plateau or mesa, and for the steeper of the two slopes of a cuesta. Used largely in the U.S. Southwest.
See also: cuesta

celadonite: [Eponymous term after the sea-green garment worn by the shepherd Celadon in the 17th-century pastoral novel *L'Astrée* by Honoré d'Urfé < French *celadon* = delicate green.] n. A soft, green or gray-green earthy monoclinic mineral of the mica group, hydroxy potassium magnesium iron aluminum silicate, $K(Mg, Fe)(Fe,Al)Si_4O_{10}(OH)_2$. It usually occurs as vesicle fillings in amygdaloidal basalt and has a structure and composition similar to glauconite. Celadon, hero of d'Urfe's romance *L'Astree* (1610), was the lover of the heroine Astree. Due to the success of the stage play, his dress became all the rage in Europe. At the same time, the green, jadelike, Chinese form of pottery called qingci made its debut in Paris, winning acclaim. People compared its colour to Celadon's apparel and began to refer to the porcelain as 'celadon.' Term coined in mineralogy by E.F. Glocken in 1847.
See also: amygdule

celestite: [Latin *caelestis* = celestial < *caelum* = sky.] n. A white to pale blue orthorhombic strontium sulfate, $SrSO_4$. It commonly occurs in evaporite deposits with gypsum and halite, and is the chief ore of strontium. Name altered by Dana from celestine, and originally chosen for the common sky-blue color of the crystals.
Cited usage: Wittstein finds that the blue color of the celestite of Jena is due to a trace of phosphate of iron. Celestite is usually associated with limestone [1854, James D. Dana; A System of Mineralogy 620].

cell: [Latin *cella* = a small room or hut, especially one of many in the same edifice.] n. In biology, a minute unit of protoplasm enclosed in a semiper-meable membrane usually with a nucleus (often used with a specifying adjective, as in brain-cell). With the possible exception of viruses, all known life forms are made of one or more cells. Originally used for a store-closet, slave's room, prison cell, ora monk's cell, the term took on biologic meaning in the 17th century. As the smallest subdivision of living things with the possibility of independent existence, the cell in the life sciences is loosely analogous to the atom in the physical sciences. There is a

possible etymological connection to *cera*, meaning the wax of a honeycomb.
Cited usage: The Microscope shews that these Pores are all, in a manner, Spherical, in most Plants; and this Part an infinite Mass of little Cells [1672,Nehemiah Grew, The Anatomy of Plants 64].

Celsius: [Eponymous term after the Swedish astronomer, Anders Celsius (1701-44).] n. The name of a temperature scale invented by Celsius in 1742, which registers the freezing point of water as 0° and the boiling point as 100° under normal atmospheric pressure.
Synonyms: centigrade

cement: [Latin *caementum* = rough-cut stone < *caedere* = to cut; < IE *ka-id* = to strike.] n. As an industrial material, cement is a gray powder made from calcined limestone, that is heated until dissociated into CaO, CO_2, and clay. It is then ground into a fine powder that, when mixed with water, forms a viscous fluid that will harden to a solid mass. Cement is combined with crushed rock or gravel to make concrete. Portland cement, which comprises nearly all of today's cement production, is so called because its color resembles Portland stone, a yellowish-white oolitic limestone from the Isle of Portland on the southern coast of England, and used extensively as a building stone. Portland cement was originally prepared by calcining a mixture of chalk with the clayey mud of the Thames.

cenote: [American Spanish < Mayan *conot* = reservoir < Yucatec Mayan *tzonot* = well.] n. In the karst terrain of Yucatan, Mexico, a cylindrical, smooth-walled sinkhole open to the surface and containing water, the level of which is controlled by the regional water table.
Cited usage: A cenote. It was a large cavern or grotto, with a roof of broken, overhanging rock and at the bottom water pure as crystal resting upon abed of white limestone rock

[1841, J. L. Stephens, Incidents of Travel in Cent. Amer., Chiapas and Yucatan II. xxiii. 408]. The forms most typical of karst are the cenotes [1963, H. G. Termier, "Erosion & Sedimentation" translation by D. & E. Humphries xiv. 308].

Cenozoic: [Greek *kainos* = new, recent + *zoikos* = pertaining to animals < *zoa* = plural of *zoon* = animal,l iving being; < IE ken = fresh, young, new + *gyo* = to live.] adj. Designating the present era of geologic time (65.5Ma - Present), which is divided into the Tertiary Period (65.5-1.81Ma) and Quaternary Period (1.81Ma-0); alternatively, into the Paleogene (65.5-23.0Ma), Neogene (23.0-1.81Ma), and Quaternary Periods. It is characterized by the formation of modern continents, glaciation, and the diversification of mammals, birds, and plants.
Cited usage: Some geologists have introduced the term Cainozoic, for Tertiary [1865, Lyell, Elements of Geology 92].
Alt. spelling: Cænozoic, Cainozoic (Europe), Kainozoic
See also: Tertiary, Quaternary, Paleogene, Neogene

centigrade: [Latin *centum* = one hundred + *gradus* = step, rank, degree; < IE *ghredh* = to walk, go.] n. Divided into a hundred degrees; usually applied to divisions on Celsius's thermometer, in which the space between the freezing and boiling points of water is divided into 100 degrees. This term was largely discontinued after adoption of the term Celsius for the same scale, at an international conference on weights and measures held in1 948.
Synonyms: Celsius
See also: Celsius

centrosphere: [Latin *centrum* < Greek = center + *sphairos* = ball.] n. The central core of the earth.
Synonyms: barysphere
See also: barysphere

cephalopod: [Latin *cephalo* = head + *poda* = foot.] n. Any of the various marine molluscs of the class Cephalopoda, such as the present-day octopus, squid, cuttlefish, and nautilus, having a definite head with large eyes, and a mouth with a beak surrounded by part of the foot that is modified into armlike or lobelike tentacles having suckers or hooklets. The shell may be external, as in the nautiloids and the extinct ammonoids, and shaped like a hollow cone, which may be straight, curved, or coiled and divided into chambers, separated by septa of varying complexity that join the outer shell wall in lines called sutures. The shell may also be internal, as in the extinct pelemnites, and all present-day cephalopods, except the nautilus. With the exception of the primitive nautilus, cephalopod species have a gland that ejects a defensive ink that obscures their presence and confuses their predators.

ceratite: [Greek *kera* = horn; < IE *ker* = top of the head.] n. Any ammonoid cephalopod belong-ing to the order Ceratida, characterized by a shell having sutures with serrate (sawlike) lobes, and in some groups an ornamental shell with hornlike projections. As an adjective, ceratitic describes this type of suture (as opposed to ammonitic sutures, characterized by other orders of ammonoid cephalopods). The ceratite lineage flourished with great evolution-ary radiation during the Triassic Period.

cerussite: [German *zerussit* < Latin *cerussa* = ceruse, white lead carbonate; < Greek *keroessa* = waxy < *keros* = wax + *-ites* < *-lite* = pertaining to rocks < *lithos* = rock, stone.] n. A colorless to white to yellowish or grayish orthorhombic lead carbonate, $PbCO_3$. Cerussite is a common alteration product of galena and a valuable lead ore mineral. Named in 1845, probably by Dana.

Cited usage: Cerusite isomorphous with aragonite [1850, James D. Dana, A System of Mineralogy 498].
Alt. spelling: cerusite

chalcanthite: [Greek *kalkos* = copper + *anthos* = flower.] n. A blue triclinic mineral, hydrated copper sulfate $CuSO_4 - 5H_2O$. Chalcanthite commonly forms secondary encrustations (flowers of copper) on primary copper sulfides.
Synonyms: blue vitriol, copper vitriol, bluestone

chalcedony: [Latin *chalcedonius* < Greek *khalkedon* = a mystical stone mentioned in the bible < *Khalkedon* (Chalcedon) = ancient Greek city in Asia Minor.] n. A variously colored cryptocrystalline variety of quartz, SiO_2. Chalce-dony usually has a waxy luster. It can be translu-cent to transparent, and frequently displays distinctive microscopic crystals. These crystals are arranged in slender fibers, that are perpendicular to the parallel bands or botryoidal surfaces found in agates. Chalcedony often precipitates as a major component of low temperature hydrothermal veins or cavity fillings. These precipitates are found as the varieties: carnelian, agate, onyx, sardonyx, jasper, prase, bloodstone, and sard. It is also the major component of the sedimentary rocks chert and flint. Chert is often used synonymously with chalcedony, but chert is generally a less pure form of SiO_2, as well as having a larger average crystal size. A biblical reference to chalcedony occurs in the New Testament in Revelation 21:19. Chalcedony was one of the foundation stones used to build the walls of New Jerusalem (along with jasper, sapphire, emerald, and topaz). The source of this toponymous root is the City of Chalcedonin

northwestern Asia Minor, across the Bosphorus Strait from Byzantium, near the site of the modern city of Istanbul.
See also: chert, flint, agate

chalcocite: [Greek *khalkos* = copper.] n. A black to dark lead gray orthorhombic mineral, copper sulfide, Cu_2S. It often occurs in a massive habit and is an important copper ore. Term coined by J. D. Dana as an alteration of the older name chalcosine.

chalcophile: [Greek *khalcos* = copper + *philos* = loving.] n. Said of an element concentrated in the sulfide phase of meteorites, rather than the silicate phase (lithophile) or metallicphase (siderophile). They are probably concentrated in the mantle, rather than in the crust or the core, according to V. M. Goldschmidt's threefold scheme of element partition in the early Earth. Also said of elements, such as: Cu,Pb, Zn, Cd, Ag, S, Se, Te, and As, that tend to concentrate primarily in sulfide and related mineral deposits within the Earth's crust. Goldschmidt's fundamental laws of geochemistry, which include his famous tripartite division of Earth's elements, appear in his series of monographs titled *Geochemische Verteilungsgesetze der Elemente* (*Geochemical Laws of Distribution of the Elements*). These were published between 1923 and 1937.
Synonyms: sulphophile
See also: lithophile, siderophile

chalcopyrite: [Greek *khalkos* = copper + *pyrites* = fire-stone, flint, a mineral that strikes fire < *pyr* = fire + *-ites* < *-lite* = pertaining to rocks < *lithos* = rock, stone.] n. A brass-yellow colored tetragonal mineral, copper iron sulfide, $CuFeS_2$. It is the most important ore of copper, and is one of several minerals considered as fool's gold.

Cited usage: Chalcopyrite resembles iron pyrites, but is of a deeper yellow color, much softer, being scratched with a knife [1862, James D. Dana, Manual of Geology].
Synonyms: yellow pyrites, copper pyrites, yellow copper ore
See also: pyrite

chalk: [Old English *cealc* < Latin *calx* = lime.] n. A soft, earthy, porous, fine-grained, white to light-gray to buff-colored variety of marine limestone, consisting almost completely of calcite tests of pelagic planktonic micro-organisms, such as foraminifera and calcareous algae (coccoliths). Chalk beds often contain nodules of flint that have formed during diagenesis by migration of silica, originally in sponge spicules. The most widespread chalk deposits are of Cretaceous age, and are exposed in cliffs along the English channel. In a prepared form, with the addition of a binding agent, chalk is used to write on blackboards. In the 17th to 18th centuries, it was prescribed orally to young women suffering from chlorosis.
See also: cretaceous

chalybeate: [Latin *chalybs* < Greek *khalyps* = steel < *Khalybes* = Chalybes, people of Asia Minor famous for their steel.] adj. Referring to water flavored with iron compounds, or to a well or spring yielding such water.
Cited usage: The chalybeate waters form the best tonics [1816, James Smith, The Panorama of Science & Art II. 385].

channel: [Middle English *chanel, canel* = channel; < Latin *canalis* = groove, channel, pipe < *canna* = reed.] n. In oceanography, a comparatively long, narrow stretch of water between two land masses in close proximity, connecting two larger bodies of water, usually seas. In fluvial processes, a channel is the deeper part of a body of water, such as an estuary, bay, or strait through which a moving current flows. In streams, the channel wall and floor form the periphery of the troughlike depression for flowing water. There is no definite size distinction between a channel and a strait. In confined waters, a channel is usually larger than a strait, whereas, in waters opening to the sea, there is no general trend (e.g., The Strait of Juan de Fuca or The Strait of Gibralter vs. The Golden Gate Shipping Channel, as opposed to the English Channel or Santa Barbara Channel vs. the Dardanelles).

chaos: [Greek *khaos* = formless primordial space < *Khaos* = god considered to be the most ancient deity by the Greeks.] n. A term for a gigantic fault breccia, usually associated with the hanging wall of major thrust faults, and consisting of a mass of disordered, irregularly sized and shaped blocks with little fine-grained matrix. The term chaos was coined in geology by Levi Noble in 1941, when he first described the Amargosa Chaos, a thrust-formed mega-breccia in Death Valley, California, with blocks ranging in size from 1m - 800m square. *Cited usage:* Where eldest Night And Chaos, Ancestors of Nature, hold Eternal Anarchie [1667, John Milton, Paradise Lost. ii. 895]. *Similar terms:* megabreccia, melange

chapada: [Toponymous term after the Chapadas Veadeiros and others in Brazil < *chapa* = sheet, any broad expanse or surface.] n. A steep mesa, elevated plateau, or wooded ridge, especially in the savanna areas of Brazil. The type locality, Chapada do Veadeiros, is near the city Alto Paraiso de Goiás in Brazil at elevations of 700 - 900 meters. *Synonyms:* mesa *See also:* mesa

chaparral: [Spanish *chaparro* = evergreen oak; < Basque *txapar* = diminutive of *saphar* =

thicket.] n. An ecological biome characteristic of the Mediterranean climates of parts of Greece, Italy, California, Texas, Mexico, and small regions in South Africa and Australia. This Mediterranean type climate is typified by hot dry summers and cool moist winters, and dominated by a dense growth of mostly small-leaved evergreen shrubs. The word came in touse about 1846 in the United States during the Mexican-American War. *Cited usage:* This word, chaparral, has been introduced into the language since our acquisition of Texas and New Mexico, where these bushes abound [1860, Bartlett's Dictionary]. *See also:* biome

charnokite: [Eponymous term after Job Charnok, agent of East India Company during the 17th century.] n. A granofels consisting mostly of pathetic alkaline feldspar and quartz plus minor Fe-rich orthopyroxene and plagioclase. The feldspars are commonly dark green, brown, red, and even black. Garnet and clinopyroxenes may also occur in charnokites. The name charnokite was coined in 1900 by T. H. Holland, from the tombstone of Job Charnok, which was made of this rock. This tombstone is located at St. John's Church, Calcutta, India. Charnok chose Calcutta as a settlement for British trade. Some geologists believe that charnokites originate by recrystal-lization in the solid state from a granitoids protolith. Alternatively, other geologists believe that charnokites are magmatic rocks crystallizing directly from hot dry magmas. *Alt. spelling:* charnockite

chasm: [Latin *chasma* < Greek *khasma* = yawning hollow < *khainein* = to yawn, gape; < IE *gheu* = to gape.] n. A deep, steep-sided opening in the Earth's surface, such as an abyss, deep fissure, or narrow gorge. An etymologically related term is chaos.

chatoyancy: [French *chatoyer* = to change luster or shimmer like cats' eyes < *chat* < Latin (Vulgar) *cattus* = cat.] n. An optical property of certain minerals, which, when turned, display a silky sheen of reflected light, concentrated in a narrow band. This is a result of inclusions of parallel fibers or acicular crystals. Chatoyancy is visible in stones such as: chrysoberyl, cat's-eye, and tiger's eye, especially when cut into cabochons.
Cited usage: The Moonstone a variety of pearly adularia presenting chatoyant rays [1859, Tennent, Ceylon 38].
See also: asterism

chattermark: [Echoic word imitating the natural sound < Middle English *chaterea* + *mearc* = sign, mark.] n. In glacial geology, oneo f a series of small crescentic cracks in a polished bedrock surface, made by vibratory chipping of brittle bedrock by rocks embedded in the base of a glacier. The horns of the crescents point in the direction of ice movement.

chelation: [Latin *chele* = claw; < Greek *khele* = crab's claw.] n. In chemistry, the reaction between a metallic ion and a complexing agent, where the central metal ion links with at least two bonds to the complexing agent, thus forming a ring structure. Chelation is important in mobilization of metal ions during chemical weathering. In biology and paleontology, chela is the term for the prehensile and pincerlike claws of arthropods, especially crabs, lobsters, and scorpions.

chemoautotrophic: [French a*lquemie* < Latin *alchymia* < Arabic *al* = the + *kimiya* = chemistry; < Greek *khemeia* = chemistry < *Khemia* = Egypt + *auto* = self + *trophikos* = nursing, tending.] n. Referring to an organism that obtains nourishment from chemical reactions of inorganic substances. In ancient Egypt, the word Khem was used in reference to the fertility of the flood plains of the Nile River. This was an ideal region for agrarian science to prosper. In addition, the ancient Egyptians' goal of immortality necessitated their development of at least rudimentary chemical knowledge. After Alexander the Great conquered Egypt in 332 BCE, Greek philosophers combined their system, based on the four elements of nature (earth, air, fire, water), with the Egyptian 'sacred science of the floodplain' to make the word *khemeia*.
Synonyms: chemotrophic
Similar terms: methanotrophic

chenier: [French *chenier* = oak ridge < *chene* = oak.] n. A long, narrow, wooded beach ridge and berm located seaward of marshes and mudflats, about one to six meters in height, and formed along a prograding shoreline, such as along the SW Louisiana shore of the Gulf of Mexico. Despite their low profile, cheniers can be hundreds of kilometers long. The fertile, porous nature of the sediment overlying mudflats creates a fertile zone for vegetation. The lowland maritime deciduous forests known as cheniers represent some of the highest ground found in coastal Louisiana, testimony to the low relief of the Mississippi floodplain. The countless cheniers found on the fringes of the bayou account for Louisiana being the type location for this feature, and the fact that the French populated this area accounts for the term's root.

chernozem: [Russian *chernyi* = black + *zemlya* = earth, sand; < IE *dhghem* = earth.] n. A great soil group of the 1938 Soil Conservation Service (SCS) classification system, characterized by a dark, humus-rich surface horizon, below which is a lighter-colored horizon having accumulated calcium carbonate. Chernozems are typical of the steppe-grasslands in central and southern Russia, and in central Canada. Most chernozems are now classified as mollisols. IE derivatives include: humus, homo, allochthonous, and autochthonous.

Cited usage: Tchernozem, a local name for the black earth of the south of Russia, which covers the whole of the Aralo-Caspian plain [1859, Page Handbk. Geol. Terms].

Alt. spelling: chernosem, tchernozem

chert: [Uncertain origin, perhaps Old English *cert* = rough ground or Irish *ceart* =stone.] n. A hard, dense, dull to semi-vitreous, micro-crystalline to cryptocrystalline sedimentary rock, consisting dominantly of chalcedony, but possibly containing small amounts of opal, iron-oxide, calcite, clays, and remains of various micro-organisms. Chert occurs extensively as bedded deposits of ribbon chert that are chiefly biologic. It is the final result of diagenesis of radiolarite and diatomite. It also occurs as chert nodules (flint) in limestones and dolostones, where it originated by the mobilization and concretionary precipitation of silica during diagenesis of opaline components, such as sponge spicules. Though it bears the closest resemblance in sound and spelling to chart, the term chert is probably not derived from the rough, brushy charts of rustic England (e.g., Brastred Chart); the evidence being that chart describes a type of ground cover obscuring the country rock. Chert is often used synonymous-ly with chalcedony, but chert is generally a less pure form of SiO_2, as well as having a larger average crystal size.

Synonyms: hornstone, silex
See also: chalcedony, flint

chiastolite: [Greek *khiastos* = arranged cross-wise, < *khazein* = to mark with the letter *chi* (X), < *khiasma* = crossing of two lines (sticks, etc.) as across piece + *lithos* = rock,

stone.] n. An opaque variety of the ortho-rhombic mineral andalusite, aluminum oxide, Al_2Si2O_5. Chiastolite usually has carbonaceous impurities that are regularly arranged during metamorphic recrystallization, so that a section perpendicular to the C-axis of the crystal displays a black cross, which is visible under a microscope. Named in 1800 by Karsten, after a variety of andalusite, a transverse section of which often exhibits the figure of a cross when viewed in a petrographic microscope.

Cited usage: Karsten, on account of the resemblance of its surface to the letter X, has denominated it Chiastolith [1804, R. Jameson, Min. I. 547].

See also: andalusite

chine: [Old English *cine* = fissure, crack < *ki* = to split, burst; < IE *skei* = to cut, split.] n. A fissure in the surface of the earth; specifically, a deep ravine or gorge cut into a seaside cliff by stream erosion. Term used especially on the Isle of Wight and on the Hampshire coast of England, for a deep and narrow ravine or cleft cut in soft rock strata by a stream descending steeply to the coast. IE derivatives include science, and schism.

Cited usage: One of these chines, near Boscomb, has been deepened twenty feet within a few years [1830, Lyell, Principles of Geology I.281].

chinook: [Eponymous term after Chinook =snow eater, the name for the Native American people living in the Pacific Northwest along the Columbia River.] n. Originally, a warm, moist wind blowing from the southwest up the Columbia River Valley. Later the term was broadened to mean a warm, dry wind that descends from mountains, specifically the eastern slopes of the Rocky Mountains, causing a rapid rise in temper-ature. The Chinook wind of the Rocky Mountains, the

Santa Ana of Southern California, and the foehn of the Alps are warm, dry chinooks that descend a mountain slope and blow across the flat lands below.
Synonyms: foehn
See also: foehn

chitin: [French *chitine* = chiton; < Greek *khiton* = frock, tunic, coat of mail armor.] n. A resistant organic compound, similar in structure to cellulose, but with some hydroxyl (OH) groups replaced by nitrogenous acetamide groups, CH_3CONH_2. Chitin is secreted by arthropods informing the exoskeletons of crustaceans and insects. It also occurs in the spores of fungi and in the shells of certain brachiopods. Not to be confused with chiton, an invertebrate mollusc named for its calcareous "tuniclike" carapace, commonly called "coat of mail shell."
Cited usage: Chitin ($C_9H_{15}NO_6$) is the principal constituent of the horny cover of beetles and crustaceans [1874, C. Schorlemmer, A Manual of the Chemistry of the Carbon Compounds 467].
See also: chiton

chiton: [Greek *khiton* = tunic; < Akkadian *kitu* = flax, linen.] n. Any of various marine molluscs of the class Polyplacophora that live on rocks and have shells consisting of eight overlapping calcareous plates. Also, a tunic worn by men and women in ancient Greece. Not to be confused with chitin, a term having the same root, but being a noncalcareous nitrocellulose-based compound in arthropod exoskeletons.
See also: chitin

chlorine: [Greek *khloros* = yellowish or light green.] n. A greenish-yellow gaseous element (symbol Cl). Chlorine is a member of the halogen series, along with fluorine, bromine, iodine, and astatine.

Cited usage: It has been judged most proper to call it Chlorine, or Chloric gas [1810, Davy, Trans. Royal Soc. 15 Nov. (1811) 32].

chlorophyll: [Greek *khloros* = green + *phyllon* = leaf.] n. A complex, waxy, micro-crystalline pigment, $C_{55}H_{72}MgN_4O_5$. Chlorophyll is a catalyst in photosynthesis, and a primary chromophore in green plants. Generally, any of a group of green pigments that are found in the chloroplasts of plants and in other photosynthetic organisms such as cyanobacteria. Chlorophyll exists naturally in many types: chlorophyll-*a* occurs in all organisms exhibiting aerobic photosynthesis (green plants, algae, and cyano-bacteria), chlorophyll-*b* in higher plants, chlorophylls-*c1* and -*c2* in diatoms and brown algae, and chlorophyll-*d* in red algae.
Cited usage: When I want to know why a leaf is green, they tell me it is coloured by 'chlorophyll', which at first sounds very instructive; but if they would only say plainly that a leaf is coloured green by a thing which is called 'green leaf', we should see more precisely how far we had got [1869, John Ruskin, The Queen of the Air sec.57]. The chlorophyll granule consists of two parts; a colourless solid portion derived from the protoplasm...and a green colouring matter, the chlorophyll, which is diffused through and colours the granule [1883, McNab, Botany (Lond. Sc.Class-bks.) I. 17].

chondrichthyes: [Greek *chondros* = cartilage.] n. A class of vertebrate fish, especial-ly the sharks and rays, having skeletons composed of cartilage instead of bone.
See also: chondrite

chondrite: [Greek *chondros* = granule.] n. A stony meteorite characterized by small (about 1mm diameter), spheroidal chondrules of olivine and orthopyroxene. In paleontology,

a common trace fossil that consists of tunnel structures that radiate outward from a central vertical tube. Also, the term chondrite has been used for feeding burrows made by marine worms. The Greek root chondros also means cartilage, which accounts for the related term chondrichthyes having an unrelated meaning.
Cited usage: Most of the stony meteorites are made up of chondri or chondrules of silicate minerals and are therefore classed as chondrites. [1944, C. Palache, Dana's A System of Mineralogy (ed. 7) I. 121].
See also: chondrichthyes

chonolith: [Greek (possibly) *khoane* = funnel + *lithos* = rock, stone.] n. An igneous intrusion so irregular that it can't be classified as a laccolith, lopolith, sill, or dike.
Cited usage: A cactolith is a quasi-horizontal chonolith composed of anastamosing ductoliths whose distal ends curl like a harpolith, thin like a sphenolith, or bulge discordantly like an akmolith or ethmolith. [1953, C.B. Hunt, Geology and geography of the Henry Mountains region, Utah]. (Ed.note: This definition by C. B. Hunt was a satirical jab at the unbridled proliferation of geologic terms.)

chordate: [Latin *chorda* = cord.] n. Any of numerous animals belonging to the phylum Chordata, having at some stage of development a dorsal nerve cord, a notochord, and gill slits. Includes all vertebrates and certain invertebrate marine animals, such as salps and tunicates.

choropleth: [Greek *choros* = place, area + *plethore* = fullness, quantity, repletion < *plethein* = to become full.] n. An area on a thematic map that is usually shaded lighter or darker; this is done to distinguish differences in frequency or density of a particular characteristic relevant to that map's theme. Term coined by J. K. Wright in 1938. He based the term on the idea of

"quantity in area" and derived it from the older word isopleth, which is a line of equal quantity.
See also: isopleths

chromite: [Greek *chromos* = color.] n. A black to brownish-black, isometric mineral of the spinel group, iron chromium oxide, $FeCr_2O_4$. It occurs commonly as disseminated isometric crystals in mafic and ultra mafic igneous rocks. Also, as monomineralic segregations in the ultramafic units in layered mafic intrusions, such as the Bushveld complex in South Africa. It is also found with other heavy minerals in placer deposits. The name was chosen because of the high chromium content of the mineral. Term was coined by Vauquelin in 1797 after he isolated the mineral chromic oxide from Siberian red lead.
See also: chromium

chromium: [Greek *chromos* = color.] n. A grayish-white, hard, metallic element (symbol Cr). The mid-18th-century analysis of the mineral crocoite, $PbCrO_4$, showed that it contained a lot of lead (not surprising because it was then known as Siberian red lead). Crocoite also contained an additional mineral identified in 1797 by Louis-Nicholas Vauquelin as chromium oxide. The following year, he isolated the element chromium from the oxide, and named it for its remarkable panorama of colors.
See also: chromite

chron-: [Greek *chronos* = time.] Combining form meaning time, used as a prefix, suffix, or base. It is common in such geologic terms as: chronocline, chronomere, chronozone, isochron, geochronologic, chronostratigraphic, chronolitho-logic, and chron.

chronomere: [Greek *kronos* = time + *meros* = part, share.] n. A unit of geologic time regardless of duration.

chronotaxy: [Greek *kronos* = time + *taxia* = arrangement.] n. Different fossil assemblages or stratigraphic sequences composed of units that are of equivalent age. Term coined by L. G. Henbest in 1952 in his article on the significance of evolutionary explosions for diastrophic division of Earth history.
Synonyms: chronotaxis

chrysocolla: [Greek *chrysos* = gold + *kolla* = glue.] n. A blue to blue-green orthorhombic mineral, hydrated hydroxy copper aluminum silicate, $(Cu,Al)_2H_2Si_2O_5(OH)_4 - nH_2O$. It usually occurs as cryptocrystalline encrustations, and accumulates as seams in the oxidized zone of copper deposits. The name has been used since antiquity in reference to the mineral being used asa flux in soldering gold ornaments (malachite of borax was used for this purpose as well).

chrysotile: [Greek *krysos* = gold + *tilos* =fiber.] n. A white, gray, or greenish, fibrous monoclinic mineral of the serpentine group, trimorphic with lizardite and parachrysotile, hydroxy magnesium silicate, $Mg_3Si_2O_5(OH)_4$. It occurs in hydrothermal veins in bodies of serpentinite, and is one of the minerals classed as asbestos. Chrysotile is used industrially for its fibrous nature. Chrysotile is termed white asbestos, in contrast to crocidolite (blueasbestos), and amesite (brown asbestos). In reflected light, thin veinlets of chrysotile have a golden sheen, thus inspiring the name.
Cited usage: Chrysotile is fine asbestiform [1850, James D. Dana, A System of Mineralogy255].
Synonyms: white asbestos

chthonic: [Greek *khthonic* = referring to the earth and the realms beneath it < *khthonios* = dwelling within the Earth < *khthon* < IE *dhghem*= earth.] adj. In general, dark, primitive, mysterious. In sedimentology, refers to deep-sea sediment and clastic debris formed from pre-existing rock. In Greek mythology,

chthonic gods and spirits lived in the underworld and ruled the dead. IE derivatives include: humus, homo-, autochthonous, and chernozem.
Cited usage: The chthonic divinity was essentially a god of the regions under the earth [1882, C.F. Keary, Outl. Primit. Belief v. 215].

chuco: [Eponymous term < Spanish; < Aymara *chuquicamata* = sulfate ore.] n. Sodium sulfate and other sulfur salts comprising the upper part of a caliche deposit. It is one of the primarymineable sulfur deposits in Chile. Chuqui is the name for an indigenous group of people living in Chile who speak Aymara. The sulfur salt ores (chuco) were accessory to the copper ores mined by the Chuqui people for use in making tools and weapons.
Synonyms: chuqui

cienega: [Spanish *cienega* = marsh.] n. A low marshy area with a local source of water, such as a spring or creek. They are usually found in arid regions, such as the spring-fed Quatro Cienega in Northern Mexico. Previously, a term commonly found in place names in America when located in an area of meadows; now rare, and used more for creeks and spring-fed flats in arid regions with surrounding marshes.
Alt. spelling: cienaga

cima: [Italian *cima* = mountain peak.] n. A dome, summit, or peak of a mountain.

cinnabar: [Latin *cinnabaris* < Greek *kinnabari* = cinnabar; < Arabic (possibly) *zinjafr* = dragon's blood.] n. A brilliant red, trigonal mineral, dimorphous with black metacinnabar, mercury sulfide (HgS). It varies from a massive, earthy habit to adamantine acicular crystals. It occurs primarily in epithermal vein deposits associated with Tertiary volcanics or silicified, carbonated serpentinite. It is the principal ore of mercury.
Cited usage: Mercuric Sulphide exists both amorphous and crystallized; in the former state

it is black; in the latter, it has a fine red colour and constitutes the well-known pigment called cinnabar or vermillion [1863-72, Watts, Dict. Chem.III. 912].

circadian: [Latin *circa* = about + *diem* = day.] adj. Referring to the rhythms associated with the 24-hour cycle of the Earth around its axis of rotation.

circle: [French *cercle* < Latin circus < Greek *kirkos, krikos* = circle; < IE *sker* = to turn, bend.] n. A two-dimensional plane curve described by a circumference line upon which all points are equidistant from a given fixed point, the center. The term is often applied to the circumference line alone, without the included space. This shape-related term was popular in Roman antiquity, because the circus was a large building, generally oblong or oval, and surrounded with rising tiers of seats for the exhibition of public spectacles. The Circus Maximus, often referred to as "The Circus," was the largest and most celebrated in Rome.
Cited usage: The circumference or periphery itself is called the circle, though improperly, as that name denotes the space contained within the circumference [1796, Hutton, Mathematical Dict.]. If you can draw a perfect circle on the blackboard in front of an audience, you are well on your way to becoming an artist [1992, Mario Savio, oral communication].

cirque: [Latin *circus* < Greek *kirkos, krikos* = ring, circle; < IE *sker* = to turn, bend.] n. In glacial geomorphology, a steep-walled, horseshoe-shaped basin or recess at the head of a glacial valley, formed by the erosive action of a mountain or alpine glacier. It may contain a cirque lake or tarn. This term's many synonyms from different languages indicate the global distribution of similar glacial features.

Synonyms: corrie, cwm, coire, kar, botn, oule, van, zanoga
See also: cwm, kar

cirriped: [Latin *cirrus* = curl, ringlet + *ped* = foot.] n. Any of the saltwater crustaceans of the subclass Cirripedia, including barnacles and related organisms that are sessile (attached to objects), or become parasitic in the adult stage.
Cited usage: So loaded with cirripeds, and with numerous ova, that all the upper part of its shell is invisible [1832, Lyell, Principles of Geology].

cirrus: [Latin cirrus = curl, fringe, curly lock of hair.] n. In meteorology, a high-altitude cloud formation composed of detached filaments, narrow bands, or patches of thin, generally white, diverging wisps. In biology, a tendril or slender flexible appendage, such as the fused cilia of certain protozoans. Cirrus clouds often resemble a curl or lock of hair or wool; particular varieties are known as cat's tails or mare's tails. Most of the names given to clouds (cirrus, cumulus, stratus, nimbus, and their combinations) were coined in 1803 by the English meteorologist Luke Howard.
Alt. spelling: cirri (pl.)
See also: cumulus, nimbus, stratus

cistern: [Latin *cisterna* = reservoir for water < *cista* < Greek *kiste* = a box, basket.] n. In geomorphology, a natural reservoir containing water, in the form of a hollow.
Cited usage: Throw him into the olde sisterne, that is in wildernes (sic) [1382, Wyclif, Gen. xxxvii].

citrine: [Latin *citrus* = citron tree.] n. A transparent, pale-yellow variety of crystalline quartz resembling topaz in color. Synthetic citrine can be produced by heat-treating amethyst or smoky quartz. The etymology relates to the pale-yellow color of the fruit of the citron tree.

Synonyms: topaz quartz, false topaz, yellow quartz

cladistics: [Greek *klados* = branch + *ikos* = of, pertaining to.] n. The systematic classification of groups of organisms (clades) based on their phylogenetic relationships and evolutionary history. Groups are classified on the basis of shared characteristics thought to derive from a common ancestor. This leads to the study of the branching of evolutionary lines of descent, and the comparative relations between those branches. Etymologically related terms include cladogramand clade.

Cited usage: Cladistics is a systematic method of classification adopted by some taxonomists. It implicitly assumes evolution by descent with modification, but it avoids statements about thes peed or the mechanism of evolution [1982, Times11 Sept. 7/7].

clarain: [Latin *clarus* = clear + *-ane* = in organic chemistry, the formative element of the names of the saturated hydrocarbons.] n. A lithotype of coal with a semibright silky luster and a sheetlike irregular fracture. It is distinguished from vitrain because it contains intercalations of a duller lithotype, durain.

Cited usage: These four distinguishable ingredients, all of which are to be found in mostordinary bituminous coals, I name provisionally as follows: (i) Fusain. (ii) Durain. (iii) Clarain. (iv) Vitrain [1919, M. C. Stopes, Phil Trans Royal Society Bull. XC. 472].

See also: durain, fusain, vitrain

class: [French *classe* < Latin *classis* = class of citizens; < IE *kel* = to shout.] n. In biology, a taxonomic category ranking below phylum and above order in the hierarchy of classification of

organisms developed by Carl Linnaeus (1707-1778). In mineral chemistry, the more than 3,800 recognized minerals are divided into classes devised by James D. Dana in 1837, based on the characteristic anion or anionic group, such as the sulfide class, oxide class, or silicate class. In Roman history, a class was each of the six divi-sions or orders of the Roman people in the constitu-tion ascribed to Servius Tullius. In crystallography, minerals and synthetic crystalline substances are organized into six crystal systems and 32 possible crystal classes based on outward symmetry.

Cited usage: He divided the Romans into six great Armies or Bands which he called Classes; The valua-tion of those in the first Classe was not under two hundred pounds [1656, Blount, Glossogr. s.v. Classical].

clast: [Greek *klastos* = broken < *klan* = to break; < IE *kel* = to strike.] n. An individual rock frag-ment or grain usually resulting from the mechani-cal weathering of a larger rock mass. As an adjective, clastic pertains to a rock or sediment composed principally of broken fragments that are derived from preexisting rocks, and that have been transported from their place of origin. The term also applies to the texture exhibited by such rocks. Etymologically related terms include: clastic, epiclastic, and clastogen (an agent that causes breaks in chromosomes).

clathrate: [Latin *clathratus* = furnished with a lattice < *clathri* = lattice; Greek *klethra* = door < *kleiein* = to close.] n. In chemistry, a non-stoichiometric, composite molecular structure formed by the inclusion of molecules of one kind located in the cavities of the crystal lattice of another kind. Clathrate is commonly used to describe gas molecules (CO_2, CH_4, H_2S) occupying cavities in a lattice of water

molecules. In geology, clathrates may provide a major source of fossil fuel as they have been discovered in large amounts in recent sediments in the Gulf of Mexico, where gas molecules of methane (CH_4) are enclosed in a lattice framework of water molecules. The word clathrate is also often used to describe compounds paired up in host-guest complexes. Term coined by H. M. Powell in 1948.
Cited usage: It is suggested that the general character of this type of combination should be indicated by the description clathrate compound [1948, H. M. Powell, Jrnl. Chem. Soc.].

clay: [Old English *clei, claeg* = to stick, cleave; < Greek *gloios* = sticky matter < *glia* = glue.] n. A fine-grained, firm, earthy material that is plastic and sticky when wet, and hardens permanently when heated. Clay consists primarily of hydroxy aluminum silicates, and is widely used in making bricks, tiles, pottery, and other ceramics. In geology, the term is used for both a family of the silicate class of minerals (including such common clay minerals as kaolinite and montmorillonite), and as a size term applied to any material smaller than very fine silt, with a diameter less than 1/256mm or 4 microns (0.004mm). Clay size is the approximate upper limit for size of particles that can display colloidal properties.

cleavage: [Middle English *cleven* < Old English *cleopan* < German *klieben* < IE *gleubh* = to cut, slice, cleave.] n. In mineralogy, the property of a crystal to break along planar surfaces, varying from perfectly flat and mirrorlike (perfect cleavage) to stepped and somewhat irregular (imperfect) cleavage. The directions of cleavage are controlled by crystallographic planes within the crystal. In structural geology, the property of a rock, such as slate, to split along surfaces controlled by metamorphic alignment of platy or prismatic

minerals. Etymologically related terms include clover and cleft.

cleft: [Middle English *clift* < Old English *clyft* < Teutonic *cleof* = to cleave; < IE *gleubh* = tear apart, cleave.] n. An abrupt chasm, cut, or open-ing, such as a wave-cut gully in a cliff, an open fissure in a rock mass, or a notch in the rim of a volcanic crater. As an adjective, cleft deposit refers to a pocket of projecting crystals, often of clear quartz, which crystallized from a hydrothermal fluid in an open fissure, usually in alpine regions. In the 16th to 18th centuries cleft, a variant form of clift appears to have been almost completely con-founded with cliff; all three terms being used inter-changeably as synonyms.
See also: cliff

cliff: [Old English *clif, cleof* = high steep rock wall; < IE (possibly) *gleibh* = to adhere, be attached.] n. A perpendicular or steep rock face usually produced by erosion. Beginning in the 16th century, cliff was confused with clift (the original spelling of cleft).
See also: cleft

climate: [Middle English < French *climat* < Latin *clima* = region, slope of the Earth; < Greek *klima* = region, zone, slope < *klinein* = to slope, lean, incline; < IE *klei* = to lean.] n. The meteorological conditions, including temperature, precipitation, and wind, that characteristically prevail in a particular region over a given period of time. Climate can be determined through recordings taken over many years, or from paleo-meteorological evidence. The term 'normal' in climatology refers to the average conditions over the past 30 years. The term climate describes a more permanent condition,in contrast to the more temporary conditions called weather. The implication that climate occurs at the 'surface of the Earth' comes from the Greek root meaning 'slope or inclination of the earth and sky from the equator to the poles.' Early geographers defined zones of the Earth according to the angle of the sun on aslope. The zone or region of the Earth occupying a particular elevation on this slope was called a clime. The meteorological conditions at the various climes were distinct, and therefore had unique climates. The meaning of the term changed from the geographic "region" to the meteorological "weather of that region" around 1600. The term climatology was first recorded in 1843. Some IE derivatives include: decline, incline, climax, and clitoris. *Cited usage:* Climate may be defined as the complex effect of external conditions of heat and moisture upon the life of plants and animals [1880, Haughton, Phys. Geog. iii. 74]. *See also:* weather

clinker: [Echoic term from German *klinken* =to sound, ring.] n. In volcanology, a rough, jagged pyroclastic fragment or flow-broken lava block that resembles, in both shape and, when moving, sound, the incombustible residue or slag from a coking oven or blast furnace used in the smelting process. *Cited usage:* A kind of Clinker, extracted from the ashes of sea coal [1769, Phil. Trans. LXI.70]. The hardened crust breaks up like ice on a pond, but makes black and rough cakes and blocks 100 to 10,000 cubic feet in size, which lie piled together over acres or square miles. Such masses are called clinkers [1862, James D. Dana, Manual of Geology 694].

clino-, -cline: [Greek *klin* = sloping, inclining < *klinian* = to slope; < IE *klei* = to bend, to lean.] Combining form meaning slope or sloping. As a prefix, used in terms such as climate and clinocloro. As a suffix, as in monocline and syncline.

clinometer: [Greek *kleinein* = to slope + *metron* = to measure.] n. Any of various instruments for measuring angle of slope or inclination, especially the drip of a planar structure, such as a fault surface or geologic stratum. For geologic applications, the clinometer is usually combined with a compass, as in the Brunton compass.

clinopyroxene: [Greek *kleinein* = to slope; <IE *klei* = to lean + Greek *pyro* = fire + *xenos* = stranger.] n. Group name for pyroxenes that crystallize in the monoclinic system, which usually contain calcium (e.g., augite, hedenbergite, acmite, diopside, jadeite, omphacite, and spodumene) as contrasted to the orthopyroxenes, that crystallize in the orthorhombic system, e.g., enstatite, bronzite, hypersthene. *See also:* pyroxene, orthopyroxene

clinozoisite: [Greek *kleinein* = to slope; < IE *klei* = to lean + eponymous term after Baron Sigismund Zois von Edelstein of Slovenia, discoverer of the zoisite mineral group.] n. A light-colored, grayish-white, pink, or green mono-clinic mineral of the epidote family, hydroxy calcium aluminum silicate, $Ca_2Al_3Si_3O_{12}(OH)$. Clinozoisite isdimorphic with zoisite. Originally named saualpite by Zois in 1805 when he discovered it in the Sau-Alp Mountains, it was later renamed in recogni-tion of him by Abraham Gottlob Werner in 1811. *See also:* zoisite

clint: [Middle English *clint* < Old Swedish *klinter* < Icelandic *klettr* = rock; < Old Norse (possibly) *klettr* = cliff, crag.] n. In karst topography, a relatively flat slab of limestone that is separated from adjacent slabs by solution fissures along joints (called grikes). Also, sometimes used to denote a hard rock projecting on the banks, or in a stream bed.

65

Contemporary usage has "clint and grike" topography used as a compound term, much like basin and range, where clint is the higher-elevation interfluve and grike is the lower, water-worn fissure of a karst region. The limestone clints were originally named in the karst regions of the N.W. Pennines in Britain.
Alt. spelling: clynt, klynte
See also: grike

cloaca: [Latin cluere = to cleanse.] n. In biology, the passage into which both the intestinal and genitourinary tracts empty in amphibians, reptiles, birds, and many fishes.

clone: [Greek *klon* = twig.] n. In biology and paleontology, all genetically identical descendants derived asexually from a single parent individual.

cloud: [Middle English *cloud* = hill, cloud < Old English *clud* = rock, hill.] n. A visible aggregate of minute water droplets or ice particles above the Earth's surface that have condensed around particles of dust or pollen (condensation nuclei). There are ten main classification categories of clouds, with 27 subtypes. These types are classified according to height of cloud base, shape, color, and associated weather. Many of the cloud group names can be shortened into prefixes to more precisely define a cloud type, for example: cirr- and alto- before the terms cumulus or status further delineate these clouds by height; strato- before cumulus, and cumulo- before nimbus help define them by shape.

clove: [Old English *cleofan* = to split; < Dutch *kloof* = cleft; < IE *gleubh* = tear apart, cleave.] n. A term used in the Catskill Mountains of New York State for a deep ravine or gorge. Related IE terms include: cloven, petroglyph, and hieroglyphics.
See also: cleft, Catskill Mountains

cluse: [Old English *clus* = a narrow passage; < Latin *clusa* = a closed place.] n. A narrow, deep gorge or water gap that is cut through an other-wise continuous ridge.
Alt. spelling: clow, clouse

cnidarian: [Greek *cnida* = nettle.] adj. Any invertebrate animal belonging to the phylum Cnidaria, characterized by a radially symmetrical body, a saclike internal cavity, and stinging cells called cnidae containing threadlike toxin bearing organelles called nematocysts. This phylum includes: jellyfish, ctenophores, corals, sea anemones, and hydras. When it was discovered that all coelenterate and siphonophore species possessed cnidae, the new phylum Cnidariare placed the now obsolete phylum Coelenterata.
Synonyms: coelenterate (obs.)

co-: [Latin *cum* = with.] Combining prefix meaning with. Co- is a back-formation of com- or con-. The prefix may mean together, as in company and common; or it may mean jointly, equal, reciprocally, or mutually, as in coaxial, coeval, and coseismic. Like the prefixes com- and con-, co-combines with verbs, adjectives, adverbs, and nouns.

coal: [Old English *col* = charcoal; < IE *geulo* = live coal.] n. A combustible rock containing more than 50 weight-percent of carbonaceous material.
Coal is formed by compaction and induration of plant remains under anaerobic conditions. It is classified according to: a) its degree of metamor-phism (rank), b) amount of impurities (grade), c) kinds of plant material (type).
Synonyms: black diamond

coast: [Middle English *coste* = coast; < Old French *coste* = coast, shore, rib; < Latin *costa* = rib, side; <Indo-European *kost* = bone.] n.

The strip of land adjacent to the sea. First used c. 1125 in France, meaning shoreline. The intercostal muscles run adjacent to and along the length of the ribs of all mammals.

cobalt: [German *kobold* = goblin of the mines; < Greek *kobalos* = rogue.] n. A hard, lustrous, steel-gray, ductile element (symbol Co). Term used by miners in the 1600s in reference to silver-colored, cobalt-bearing minerals that were considered worthless. Miners believed that mischievous goblins inhabited mines and substituted such worthless material for silver.
See also: knockers

cobble: [Middle English *cobel* = diminutive of *cob* = round object.] n. In sedimentology, a rock fragment larger than a pebble but smaller than a boulder, with a diameter between 64 and 256 millimeters (comparable to a size between a golf ball and a soccer ball), especially one that has been naturally rounded. In geomorphology, regional usage in the U.S. Northeast defines cobble as a rounded hill of moderate size. In mining, the term cobbing is the process of separating worthless from desirable minerals, usually with a small tool such as a hammer.

coccolith: [Latin *coccus* = spheroidal seed; < Greek *kokkos* = grain, kernal, seed + *lithos* = rock, stone.] n. A general term for the microscopic, calcareous plate composed of minute calcite or aragonite crystals, which interlock to form the spheroidal test of the marine nanoplanktonic algae called coccolithophores. Coccoliths are a major component of chalk and deep-sea oozes of temper-ate and tropical oceans. The term was coined by Professor Thomas Henry Huxley for small disk-shaped fossils discovered during deep-sea dredging, as well as in terrestrial-based chalk deposits. In his day, Huxley was known as "Darwin's Bulldog" for his defense of Charles Darwin's theory of evolution by natural selection. In a celebrated exchange, Huxley went head to head with Samuel Wilberforce, Bishop of Oxford; with Huxley declaring he would rather be descended from an ape than a bishop. A species of coccolith, *Emiliania huxleyi* (commonly known as Ehux, was named to honor Huxley. The cocco- in this term relates to the shape of the calcareous plate, not to its composition.
Cited usage: The chalk, like the soundings, contains these mysterious coccoliths and coccospheres [1868, Huxley, Lay Serm. 206].

cockade: [French *cocarde* < *coq* = cock, rooster; < Old Norse (echoic origin) *kokr*.] adj. As used in ore deposits, the texture common in open-space hydrothermal vein filling. In cockade fillings, the long axes of minerals are directed radially outward, and are deposited in comblike crusts around breccia fragments. The term alludes to the resemblance of the texture to a cock's comb.

coelacanth: [Greek *koilia* = body cavity < *koilos* = hollow + *akantha* = spiney thorn.] n. Any of mostly extinct, primitive crossopterygian, lobe-finned fishes known only as fossils until a single living species, *Latimeria chalumnae*, was caught and identified in a fisherman's catch off the island republic of Mada-gascar (formerly Malagasy Republic) in 1938.
Cited usage: The Coelacanths have changed very little since their first known appearance in Upper Devonian formations [1940, Nature, 13 July53/1]. The first living coelacanth (*Latimeria chalumnae*) was trawled in 40 fathoms [1957,Encycl. Britannica V. 934/1].

coelenterate: [Greek *koilia* = body cavity; < *koilos* = hollow + *enteros* = intestine.] n. Former term (now replaced by cnidarian and

ctenophore) for invertebrate animals belonging to the phylum Coelenterata (now replaced by the phyla Cnidaria and Ctenophora).
See also: cnidarian, ctenophore

coesite: [Eponymous term named after Loring Coes Jr. (1915-).] n. A dense, colorless, mono-clinic, high-pressure polymorph of silica, SiO_2. Coesite is stable at room temperature only above 20 kilobars pressure. Found naturally in impact craters, in xenoliths in kimberlite, and as inclusions in garnet and pyroxene in eclogite facies rocks. Based on high-pressure phase equilibrium studies, quartz should change to coesite at depths of 60-100km below the Earth's surface. While working for Norton Salt Company, U.S. chemist Loring Coes, Jr. first synthesized coesite in 1953. Norton Salt Co., along with General Electric, kept the findings confidential while trying to use the process to synthesize diamonds. Coesite was first discovered in nature in 1960 at the Barringer Meteorite Crater in Arizona. Prior to this discovery, the crater was thought to be volcanic in origin.
Cited usage: I have sought and obtained permis-sion from Coes and Keat, both of whom are now at the Research Laboratories of the Norton Company in Worcester, Mass., to propose names for the new phases–namely, coesite for the phase produced at 35,000 atm., and keatite for the phase produced hydrothermally [1954, R. B. Sosman, Science CXIX. 738].

cognate: [Latin *co-* = together + *gnatus* = born of < *nasci* = to be born < *gen-* = to produce; < Indo-European *gen* = to be born.] adj. In struc-tural geology, referring to fractures or joints that originated at the same time and by the same de-formational process. In igneous petrology, referring to an inclusion or autolith that originated at essentially the same time and by the same process as the

host rock. In linguistics, certain word sin genetically related languages descended from the same ancestral root, such as English *petrology* and French *pétrologie.*

cohesion: [Latin *cohaerre* < *co-* = together + *haerre* = to cling.] n. The attractive force by which molecules are held together. The strength of any material to resist failure by shearing is a function of both intermolecular cohesive forces and frictional forces.
Similar terms: coherence

coke: [Middle English *colk* = charcoal.] n. A solid, combustible substance, used since the 17th century, produced by heating coal in the absence of oxygen in order to drive off volatile compo-nents. It resembles charcoal obtained from wood by a similar process.

col: [French *col* = neck; < Latin *collum* = neck; < IE *kwel* = to revolve.] n. In geomorph-ology, a saddle-like pass or depression in a mountain ridge, generally between two adjacent peaks and usually resulting from headward erosion and head-to-head juxtaposition of two cirques. In meteorology, a region of low pressure between two anticyclones. The term, originated by people speaking the Romanic dialects of the Alps, is now used by climbers and geologists in reference to other regions as well. IE derivatives include the term cyclone.
Cited usage: Between every two anticyclones we find a furrow, neck or 'col' of low pressure [1887, R. Abercromby, Weather 26].
See also: halse

colemanite: [Eponymous term named in 1884 after William T. Coleman (1824-1893).] n. A white to colorless, monoclinic mineral, hydrated calciumborate, $Ca_2B_6O_{11} \cdot 5H_2O$. Colemanite is a major source of borax, commonly occurring as nodules in playa lake sediments. Coleman, longtime friend of Death Valley Scotty, produced 2 million pounds per year of "cottonball" borax (ulexite) from his Harmony and Amargosa borax mines in Death Valley, California. In 1882, the Lee Brothers discovered a new source of borax in Death Valley along Furnace Creek Wash, calling it colemanite.

colloid: [Greek *kolla* = glue + *eidos* = form.] n. In mineralogy, one of the basic forms in which minerals occur. Generally, a peculiar state of aggregation in which substances exist. Used in contrast to crystalloid. Applied generally in chemical terms by Graham in 1861.
Cited usage: Minerals occur in four conditions, according to the circumstances under which they have been produced 1. Crystalline 2. Vitreous 3. Amorphous 4. Colloid, as a jelly-like though stony substance, deposited from aqueous solution. The most abundant mineral in nature which takes the colloid form is silica [1885, Geikie, Geol. ii. ii. ii. 62].

colluvium: [Latin *colluere* = to wash thoroughly < *com-* < *cum* = together + *luere* = to wash; < IE *leu* = to wash.] n. A loose deposit of rock debris accumulated through the action of gravity and mass transport, formed at the base of a cliff or slope and coalescing together. IE derivatives include eluvium, effluent, lye, and latrine.

colonnade: [French < Italian *colonnato* < *colonna* < Latin *columna* = column, pillar.] n. In volcanology, the lower zone of columnar jointing in a lava flow or intrusive sill, with thicker and better-formed columns than the upper zone, or entablature. In architecture, a series of columns set at regular intervals, usually supporting a roof.

Cited usage: The Two Sisters is made of basalt with large vertical columns. Slow cooling causes fewer cracks and produces the colonnade; more rapid cooling causes more cracks and produces the entablature [2004, Steve Reidel, Northwest Geology: Two Sisters Website].

comb: [Middle English < Old English *camb* = toothed object; < German *kamm, kamb* < IE *gombhos* = tooth.] adj. As used in ore deposits, the texture of a open-space, hydrothermal, vein filling, in which subparallel crystals, generally of quartz, have grown perpendicular to the vein.

combe: [Old English *cumb* = small valley.] n. In England and France, a deep, narrow valley or ravine leading to the sea. Also, used in Scotland and some parts of England for a glacial valley or cirque. Used as an adjective, such as in combe-rock, the debrisfilling the combe. The variant form *cum-* is a commonly used prefix in such place-names as: Cumbria, Cumwhitton, Cumdivock, and Cumloden.
Cited usage: Various deposits indicate that this southern fringe of England had its own glacial conditions. Among these is the 'Coombe-rock' of Sussex–a mass of unstratified rubbish [1903, A. Geikie, Textbook. Geol. 4th ed.].
Synonyms: comb, coom, coombe, coomb, cwm

combustible: [Latin *combustus* < *comburere* = to burn.] adj. Said of a substance that catches fire and burns easily, as in combustible shale. First usage in English c. 1520.

comminution: [Latin *comminuere* = to reduce into smaller parts, lessen < *com-* < *cum* = together + *minuere* = to make small < *minimus* = least; < IE *mei* = small.] n. Reduction or breaking up of rock fragments into increasingly smaller, more minute

fragments until pulverization is complete. Etymologically related terms include Miocene, minestrone, and minuet.
Synonyms: pulverization, trituration

compaction: [Latin *compactus* < *compingere* = to fasten together < *com-* < *cum* = together- + *pangere* = to fasten, fix; < IE *pak* = to fasten.] n. Reduction in bulk volume or thickness of a body of fine-grained sediment in response to: a) the increasing weight of overlying material being deposited, or b) pressure resulting from tectonic movements within the Earth's crust. It is characterized by a decrease in pore space resulting from tighter packing of sedimentary particles. Along with cementation, it is one of the two major processes involved in lithification of sediment. IE derivatives include: palisade, pegmatite, pacific, and peace (a binding together by treaty).

compass: [Middle English *compas* = circle < *compassen* = to go around; < Old French *compasser* = to measure; < Latin *compasare* = to pace off < *com-* < *cum* = together + *passus* = step.] n. In surveying, a device for determining horizontal reference directions relative to a fixed point on the Earth's surface, such as the rotational North Pole, by means of a magnetic needle swinging freely on a pivot and pointing to the magnetic North Pole. In drafting, a simple instrument for describing arcs or circles, usually consisting of two pointed legs joined at the top by a pivot (thus, by extension, one's hips and legs).

competent: [Latin *competere* = to strive together < *com-* < *cum* = together with + *petere* = to seek.] adj. In structural geology, said of rock layers that act as brittle solids, and can therefore support a tectonic force; in contrast to incompetent layers that deform in a ductile manner and cannot support a tectonic force.

concave: [Middle English < Latin *concavus* = hollow < *con-* < *cum* = intensive prefix used for emphasis + *cavus* = hollow.] n. Hollow and curved like the inside of a hollow sphere, or the meniscus formed by water in a cylinder. In sedimentology, concave cross-bedding is very common in foreset beds that are downward-arching; they are used as a criterion for distinguishing top from bottom in sedimentary layers.
See also: convex

conch: [Latin *concha* < Greek *konkh* = mussel, shell.] n. Any of the large spiral-shelled marine gastropods, such as members of the genera Strombus or Cassis. The term is also used more generally for the shell of any marine invertebrate, such as a mollusc or brachiopod. Etymologically related terms include conchoidal and conchology.
See also: conchoidal

conchiolin: [Latin *concha* = shell + *-ol* = diminutive.] n. The organic constituent of the shells of molluscs, a fibrous protein that is closely allied to keratin.
Cited usage: The nacre consists of alternating lamellae of conchiolin, and of conchiolin-containing calcareous deposits [1888, Rolleston & Jackson, Animal Life 126].
See also: pearl

conchoidal: [Latin *choncha* < Greek *konkh* = shell, mussel + *eidos* = form , shape.] adj. A type of fracture of either a rock (e.g., obsidian) or a mineral (e.g., quartz). Conchoidal fracture that produces a series of smooth convexities and concavities, similar to those on the surface of a clam shell, on broken glass, or in scooped out Jell-O.
See also: conch

concordant: [Latin *concordia* = of the same mind, agreeing < *con-* < *cum* = together + *cordis* = heart.] adj. In stratigraphy, said of rock stratadis playing parallelism of bedding or structure.

Cited usage: Concordant plutons include sills, laccoliths, lopoliths, and phacoliths. Discordant plutons include dikes, necks, chonoliths, batholiths, stocks, and bosses [1961, F. H. Lahee, Field Geol. (ed. 6) vi. 141-2].
See also: discordant, pluton

concretion: [Latin *concrescere* = to grow together < *con-* < *cum* = together + *cresco* = to grow.] n. A hard, compact, mass of material that varies shape from spheroidal to oblate to disc-shaped to irregular. Concretions have a composition distinctly different from the host rock. Concretions form by precipitation of mineral matter, usually around a nucleus made of bone, shell, vegetation, or sand. They vary in size from a few millimeters to several meters in diameter. A concretion grows from the center outward around a seed nucleation site. Not to be confused with a secretion, in which layers are deposited inward from the walls of a cavity or fracture surface forming a geode, nodule, amygdule, or vein.
See also: secretion

conduction: [Latin *conducere* = to lead together < *con-* < *cum* = together (intensive used for emphasis) + *ducere* = to lead; < IE *deuk* = to lead.] n. In physics, the transmission or transfer of energy by the passing of energy from particle to particle in a conducting medium without perceptible motion of the medium itself, such as observed in electricity and heat.
See also: convection

confluence: [Latin *confluere* = to flow together < *con-* < *cum* = together + *fluere* = to flow.] n. In geomorphology, the junction of two or more streams or glaciers.

conformable: [Latin *con-* < *cum* = together + *formare* = to form, shape < *forma* = form, shape.] adj. In stratigraphy, referring to a sequence of strata formed by uninterrupted deposition, one strata above the other, neither showing evidence of time lapses, nor evidence that the lower beds were folded, tilted, or eroded before the upper beds were deposited. In general English usage, to be in harmony or agreement.

conglomerate: [Latin *con-* < *cum* = together + *glomerare* = to roll, wind, or form into a ball < *glomus* = ball.] n. A coarse-grained, clastic sedi-mentary rock composed of rounded fragments larger than sand size (> 2mm diameter), and enclosed in a fine-grained matrix of sand, silt, or clay. Conglomerate is often classified by its composition of fragments, type of cement, and/or environment of formation.

congruent: [Latin *congruere* = to come together, agree, correspond.] adj. In structural geology, said of a minor parasitic fold with a fold axis and axial plane parallel to that of the related main fold. In igneous petrology, said of melting in which a crystal-line solid melts to a liquid of the same composition as the solid. In general use, a term meaning agreeing, consistent. First usage in English c. 1400.

conjugate: [Latin *conjugare* = to join together < *con-* < *cum* = together + *jugare* = to join < *jugum* = yoke.] adj. In structural geology, said of a system of joints that originated by the same deformational process, usually compression. Also, referring to two sets of minor folds with axial planes that are inclined toward one another; or, to two faults that originated at the same time and during the same deformational episode.

connate: [Latin *connatus* = born at the same time, twins < *con-* < *cum* = together + *nasci* = to be born.] n. In sedimentology, designating water trapped in the interstices of a sedimentary rock during deposition.
Cited usage: Geologists also recognize two further underground groups connate waters' which were trapped in some sedimentary rocks during their deposition and 'juvenile waters' which are of plutonic or magmatic origin [1955,J. Brown & A.K. Dey, India's Mineral Wealth 3rd ed. 647].
Synonyms: fossil water

conodont: [Greek *konos* = conical + *odont* = tooth.] n. Any of a large number of phosphatic fossil elements that are toothlike in form, but not in function, produced by marine animals of un-certain affinity. A member of an extinct group of small primitive fishlike chordates, preserved primarily in the form of their cone-like teeth. Conodonts were first discovered in 1856 by Russian naturalist Christian Pander (1794–1865). His view, largely disregarded until the 1990s, was that they were a primitive form of fish. Most paleontologists now agree that they are in the phylum Chordata, whereas a cladistic analysis by Donoghue in 1998 suggests that conodonts be further classified in the sub-phylum Vertebrata.
Cited usage: Much difficulty was felt by scientific men in accepting Pander's view that the Conodonts were the teeth of fishes [1872, Nicholson, Palaeont. II. 122].

consanguine: [Latin *con-* < *cum* = together + *sanguin* = blood.] adj. In petrology, said of rocks that have the same origin, especially of igneous rocks having the same parent magma.
Synonyms: comagmatic

consequent: [Latin *consequentia* < *consequi* = to follow after < *con-* < *cum* = similar, together, alike + *sequi* < IE *sek* = to follow.]

adj. In geomorphology, referring to a stream, valley, or drainage basin having a course or direction controlled by the original form or slope of the pre-existing land surface. In particular, a stream course that develops in response to local uplift. Also said of other topographic features that originated as a result of preexisting conditions, such as a consequent ridge, which is coincident with a preexisting anticline, or a consequent waterfall that is located at the site of a preexisting monoclinal fold. In general English usage, referring to anything that follows as a result of a prior action or event. Term coined in geology by John Wesley Powell in 1875.
See also: antecedent, obsequent

contact: [Latin *contactus* < *contingere* = to touch < *con-* < *cum* = similar, together, alike + *tangere* = to touch.] n. The planar to irregular surface between two types or ages of rock. Examples of contacts are: bedding planes, unconformities, faults, and intrusive contacts.

contemporaneous: [Latin *con-* < *cum* = similar, together, alike + *tempus* = time + *aneous* = adjectival suffix.] adj. Formed, existing, or occurring at the same time. In structural geology, an example of a contemporaneous fault is a growth fault that forms at the same time and continues to grow with deposition. Contemporaneous deformation takes place in sediments during or after their deposition, and includes many types of soft sediment deformation, such as slumps.
Cited usage: Volcanic rocks contemporaneous with the sedimentary strata of three of the above periods [1833, Lyell, Principles of Geology III. p.xiii].

continent: [Latin *terra contins* = continuous land < *continere* = to hold together.] n. One of the principal land masses of the Earth, usually regarded as Africa, Antarctica, Asia, Australia,

Europe, North America, and South America. Continents are areas underlain by continental crust of average density 2.7g/cm^3, including both dry land and the submerged continental shelf and slope. In contrast, ocean basins are underlain by oceanic crust with average density of 3.0g/cm^3.

contour: [Italian *contornare* = to outline, to go around; < Latin *con-* < *cum* = intensive used for emphasis + *tornare* = to turn < *tornus* = lathe; < Greek *tornos* = tool used to make a circle; < IE *ter* = turn, rub.] n. In map-making, the line connecting points of equal value, generally elevation. Contours are commonly used to represent topographic or structural surfaces. They can also be used to depict other values, such as grade of ore in an ore body, thickness of sediments in a geologic formation, or severity of shaking on an earthquake intensity map.

convection: [Latin *convehere* = to carry together < *con-* < *cum* = together + *vehere* = to carry, bring; < IE *wegh* = to go, transport.] n. In physics, the transfer of energy, particularly heat, by mass movement of molecular material, usually liquid or gas, generally in response to differences in density. In tectonics, the mass movement of mantle material as a result of temperature variation and resultant density differences. In oceanography, the density driven movement of water masses in the ocean, resulting from differences in temperature and salinity. In meteorology, the transfer of heat within the atmosphere by vertical movement of air parcels, owing to temperature-induced density differences. *See also:* conduction

convergence: [Latin *convergere* = to incline together < *con-* < *cum* = similar, together, alike + *vergere* = to bend, turn, incline; < IE *wer* = to turn, bend.] n. In evolution, the acquisition of similar characteristics by organisms of differing taxonomy, as a result of similarity in habit or environment. In oceanography, the meeting of ocean currents having differing densities, temperatures, or salinities. In meteorology, contraction of an airmass toward a central region. In petrology, the production of petrographically similar rocks from different original rocks. In stratigraphy, reduction in thick-ness of beds caused by changes in rate of deposi-tion or by unconformities. In plate tectonics, used as an adjective to describe plates that are moving toward each other. IE derivatives include: vertebra, versatile, vermicular, transverse, universe, wrinkle, rhombus, and vermicelli.

convergent: [Latin *con-* < *cum* = together + *vergere* = to incline, turn, bend.] adj. Referring to bodies, processes, or conditions that approach one another. In tectonics, said of a plate boundary where the two plates are moving toward one another. In paleontology and biology, said of evolution where similar forms develop in unrelated lineages. *See also:* divergence

convex: [Latin *convexus* = vaulted, arched < *con-* < *cum* = similar, together, alike + *vehere* = to bring.] adj. Curving outward like the outer surface of a sphere.

convolute: [Latin *convolutus* < *convolvere* = to roll together < *con-* < *cum* = similar, together, alike + *volvere* = to roll, turn.] adj. Coiled or wound together. In paleontology, said of a coiled gastropod shell or a coiled foram test in which the outer whorl entirely conceals the innermost whorls. In stratigraphy, said of laminations in silt and fine-sand beds that are intricately crumpled, twisted, and folded, or rolled into jelly-roll-like structures, whereas the beds above and below are undisturbed. The term 'convolute laminations' was coined by P. H. Kuenen in 1953 for structures produced by deformation contemporaneous with deposition and related to liquefaction of the deformed bed.

coordinate: [Latin *co* = with < *con-* < *cum* = similar, together, alike + *ordinare* = to set in order, to arrange < *ordo* = order.] n. A set of values used to locate the position of a point, line, curve, or plane in space, with respect to a specific reference system. For example, latitude and longitude are coordinates that can be used to define the location of any point on the Earth's surface.

copepod: [Greek *kope* = oar; < IE *kap* = to grasp + Greek *podos* = foot.] n. Any crustacean belonging to the class Copepoda, characterized by minute size, four or five pairs of feet chiefly used for swimming, and the absence of a carapace, or compound eyes. Along with the more familiar gastropods (*stomach foot)*, pelecypods (*axe foot*) and decapods (*10 feet)*, the oar-footed copepods contribute to the wide variety of terms in which we use the suffix *-pod.*

copice: [Middle English *copis* < Old French *copeis* < *couper* = to cut.] adj. In geomorphology, referring to mounds in arid environments that are protected from aeolian deflation by the anchoring effect of shrubs or bunch grass, as in Devils Cornfield in Death Valley. In general English usage, employed as a noun synonymous with copse, for a thicket of small trees or shrubs.

copper: [Middle English *coper* < Medieval (Late) Latin *cuprium* = copper; < Latin *cyprium aes* = Cyprian metal; < Greek *kupros* = Cyprus Island.] n. A ductile, malleable, reddish-brown to salmon-colored metallic element (symbol Cu). Copper occurs in the native state as a cubic mineral, often in large

dendritic to massive aggregates, especially in association with Precambrian basalt flows. It was the first metal to be widely used in tool-making, later being alloyed with tin to make bronze, and zinc to make brass. The original Latin *Cyprium aes*, meaning Cyprian metal stems from the fact that the island of Cyprus was the main source of copper in Greco-Roman times. The Teutonic tribes adopted this specific term for copper and its alloys rather late; originally they used the Latin cognate *aes*, meaning metal as a general term for all metals. Alchemists assigned copper to the planet Venus, because the island of Cyprus was regarded as the domain of Venus-Aphrodite. Botticelli depicted this Cyprian goddess as being born from the sea on the shores of Cyprus in his painting *The Birth of Venus*.

coprolite: [Greek *kopros* = excrement, dung; < IE *kek* = dung + Greek *lite* < *lithos* = rock, stone.] n. Any fossilized excrement of vertebrates, such as dinosaurs and mammals, being larger than a fecal pellet, and of potentially enormous size. Coprolites are variously colored (usually brown to black), elongate to ovoid in shape, and largely composed of calcium phosphate. In 1823, William Buckland first compared some enigmatic white fossil lumps with fresh hyena feces and deduced a fecal origin. He developed great expertise in identifying fossil feces, which he termed coprolites in 1829. Coprolites up to 40cm in diameter are commonly found, probably from sauropods. Coprologists, experts in the discipline of coprology, have proposed the synonym scatolite.
Cited usage: On the Discovery of Coprolites, or Fossil Feces, in the Lias at Lyme Regis, and in other formations [1829, Buckland, Trans. Geol. Soc. (1835) III. 223].
Synonyms: scatolite

coquina: [Spanish *coquina* = shellfish, cockle; < Latin *cocca* < *concha* = shell.] n. A porous, detrital limestone composed principally of loosely aggregated shells and shell fragments, weakly to moderately cemented. Coquina lithifies readily, so is often of recent age, especially in beach rock. There is a false cognate between the Spanish *coquina*, meaning shellfish and the Latin *coquina*, meaning kitchen < *coquere*, meaning to cook.

coral: [Latin *corallium* < Greek *korallion* = coral.] n. A general name for any sessile-benthic, marine organisms of the class Anthozoa, and the phylum Cnidaria. Coral is characterized by a calcium carbonate exoskeleton secreted by polyps, existing as solitary individuals, or as colonies. Coral is common in warm modern seas below 30° latitude, as well as in the fossil record since the Ordovician Period. It is often deposited in extensive masses forming reefs, barrier islands, and atolls. Modern coral polyps are heterotrophic, and have a mutually beneficial symbiotic biochemical relationship with the autotrophic algae Zooxanthellae.

cordillera: [Spanish *cordillera* = mountain chain < *cordilla* < Latin *chorda* = rope.] n. A comprehensive term for a broad belt of roughly parallel mountain ranges and chains, together with their intervening plateaus, basins, and valleys. The term is used especially for the main mountain belts of a continent. Originally applied by the Spanish (c.1700) to the parallel chains of the Andes in South America (Las Cordilleras de Los Andes), subsequently extended to the continuation of the same system through Central America,Mexico, the U.S. Southwest, Western Canada, and Alaska.

core: [Middle English < Old French < Latin *cor* = heart.] n. In geology, the central zone of the Earth's interior, separated from the overlying mantle by the Gutenberg discontinuity at a depth of 2,900km. It is primarily composed of iron and nickel, and is divided into a molten outer core and solid inner core. First usage in English c. 1300.

Coriolis effect: [Eponymous term named for Gaspard G. Coriolis (1792-1843), French civil engineer and mathematician.] n. Of bodies in motion, an apparent force but a real deflection of a body in motion with respect to the Earth. The Coriolis effect is noticed in bodies moving through water or air due to limited frictional contact with solid earth. It is the result of making observations in a noninertial reference frame of objects moving in relation to the rotating Earth. It appears as a deflection to the right in the Northern Hemisphere (clockwise), and to the left in the Southern Hemi-sphere (counterclockwise). The Coriolis effect is often thought to be a physical force, but the rotating Earth imparts no acceleration to the moving mass, therefore no force is applied to it.
Cited usage: The Coriolis acceleration is twice the vector product of the angular velocity of the Earth and the velocity of the body [1950, Metr. Glossary].

cornice: [Italian *cornice* < Latin *cornix* = crow; < (possibly) Greek *kornis* = curved line.] n. An overhanging ledge of snow or ice on the edge of a steep cliff face. So named for its resemblance to a crow's overhanging beak.
Synonyms: corniche

corona: [Latin *corona* < Greek *karone* = crown.] n. In petrology, a zone of minerals usually arranged radially around another mineral. The term has been applied to reaction rims and kelyphytic rims. In astronomy, the outermost part of the Sun's atmosphere, which can be seen during a total solar eclipse. The corona has an extremely high temperature.

corrie: [Gaelic *coire* = cauldron, kettle.] n. In Scotland, the term is a synonym for cirque. Also in Scotland, the term means a whirlpool, having gotten this meaning from the poetically-named Corrivreckan Cauldron, located in the Hills of Craignish.
Alt. spelling: currie, correi, corri, corry
See also: cirque

corrosion: [Latin *corrosio* < *corodere* = to gnaw < *co* = intensive used for emphasis + *rodere* = to gnaw; < IE *red* = to scrape, scratch, gnaw.] n. In geomorphology, the combined processes of chemical weathering by solution, hydrolysis, or oxidation, with subsequent removal of material by some transporting agent. In petrology, corrosion is the removal of a portion of all early formed crystals. It occurs by reaction with residual magma, resulting in resorption, dissolution, or fusion modification. Etymologically related terms include: erode, corrode, rodent, rostrum, radula, and raster.

corundum: [Tamil *kurundam* = corundum; < Hindi *kurund* < Sanskrit *kuruvinda* = ruby.] n. A variously colored, trigonal mineral, aluminum oxide Al_2O_3. Corundum is tough and hard, having a Mohs hardness rating of 9 on a scale of 1 to 10. It occurs as the gem varieties ruby and sapphire. Known from antiquity, the original gem-quality specimens came from India. Specimens were sent from India to England in 1798 by C. Greville-Williams, who coined the name corunduma.
Cited usage: The sapphire the ruby and the Oriental topaz are all mere coloured varieties of the mineral substance known as corundum [1870, H. Macmillan Bible Teach. xiv. 273].
Synonyms: adamantine spar, diamond spar

coseismic: [Latin *co* = similar, together, alike + Greek *sei-ein* = to shake.] adj. Relating to a line that connects points on a map that indicate places simultaneously affected by an earthquake.
Cited usage: These points will lie in circles called 'isoseismic' or 'coseismic' circles [1886, J. Milne, Earthquakes 10].
Synonyms: isoseismic

cosmogony: [Greek *kosmos* = universe, world + *gonio* = a begetting.] n. A scientific theory or cultural mythology regarding the origin of the entire universe, or a limited portion of the universe, such as the Earth and its solar system. As humanity's understanding of the universe broadens, the meaning of *kosmos* follows suit.
Cited usage: It was a most ancient tradition amongst the Pagans that the cosmogonia or generation of the world took its first beginning from a chaos [1678, Cudworth, Intell. Syst. 248].
Similar terms: cosmology
See also: cosmology

cosmology: [Greek *kosmos* = universe, world + *logos* = word, speech, discourse, reason < *legein* = to speak, to say; IE *leg* = to collect.] n. In astronomy, the study of the general structure and evolution of the universe. In philosophy, the study of the universe, particularly dealing with such characteristics as space, time, and causality. An etymologically related term is cosmogeny. First usage in English c. 1655.
Cited usage: By Cosmology is implied a philo-sophical or physiological Discourse of the World, or Universe in general [1735, B. Martin, Philos. Gram. 101].
See also: cosmogony

cosmopolitan: [Greek *kosmos* = universe, world + *polis* = city.] adj. In paleontology and biology, referring to organisms that are widely distributed throughout the world in various geographic and ecologic regions. First usage in English c. 1840.

cosmos: [Greek *kosmos* = the universe or world, order, harmony.] n. The universe, regarded as an orderly harmonious system. In botany, the genus Cosmos is comprised mostly of Mexican herbs having radiate heads and pinnate leaves. As the adjective cosmic, the terms is used in geology to refer to matter or processes with extraterrestrial sources, such as cosmic dust, cosmic radiation, cosmic erosion, and cosmic spherules. The related term cosmetic literally means to put in good order. First usage in English c. 1175.

costean: [Cornish *cothas* (possibly) = dropped, or < *cos* < *coid* = wood + *stean* = tin.] v., n. As a prospecting technique in search of hydrothermal veins, to dig shallow pits through surficial deposits into solid bedrock below, and then to drive a passage from one pit to another across the direction of the veins in such manner as to cross all the veins between the two pits. The possible etymology from Cornish *cothas*, meaning dropped comes from a quote by W. Pryce in 1778; the other possible etymology cos, meaning wood is attributed to the Cornish miner Jago who described the costean ore of tin as having a structure like wood.
Cited usage: This way of seeking the Tinners call costeening, from Cothas Stean; that is fallen or dropt tin [1778, W. Pryce, Min. Cornub].

coteau: [Canadian French *coteau* = hill, slope.] n. A term used in parts of the U.S. for a variety of landscape features, including: a high plateau, a dividing ridge between two valleys, a morainal hill, an elevated, pitted plain

(Missouri), and a low dry ridge in a swamp (Louisiana).
Cited usage: Topographically the coteaus maybe considered as extensive mesas standing some 500 to 700 feet above the Dakota Valley which separates them [1918, Visher, S. Dakota 36].

cotectic: [Greek *co-* < *con-* < *cum* = similar, together, alike + *tekein* = to melt.] adj. In petrology, referring to a line or surface in a P-T-X (pressure, temperature, composition) diagram, defining condi-tions under which two or more solid phases will crystallize simultan-eously from a single liquid.

coulee: [Canadian French *coulée* = coulee; < French *couler* = to flow; < Latin *colum* = strainer, sieve.] n. In volcanology, a flow of viscous lava, such as rhyolite or dacite, that has a blocky, steep-fronted form. The term is used both when the flow is molten and after solidifi-cation. In geomorphology, used in Western Canada and the United States for a deep ravine or gulch, usually dry in summer, and especially with reference to a large abandoned meltwater channel of an ice sheet. The term appears to have arisen among French trappers in the U.S. Northwest and led to the naming of the Grand C- related terms include colander and portcullis. First usage in English c. 1805.
Cited usage: Large stratiform and horizontal coulées of volcanic rock [1839, Murchison, Silur. Syst. i. xxxii. 428]. Every ravine short of an inhabit-able valley is called a 'cooley' [1881, N.Y. Times, 18 Dec. in Notes & Queries 6th Ser. V. 65/1].
Alt. spelling: coulie, cooley, coulee

couloir: [French *couloir* = colander, passage, lobby, steep incline for timber hauling < *couler* = to flow; < Latin *colare* = to filter *colum* = sieve.] n. A passage in a cave. In the Alps, a deep, narrow valley, especially a gorge on the side of a mountain. First usage in English c. 1850.

cove: [Old English *cofen* = cave, closet; < Old Norse *kofi* = hut.] n. In geomorphology, a small, sheltered, recess or bay in a coastline. Also, an opening cirquelike opening at the head of a valley resembling a cirque, such as is found in the blue Ridge foothills in Virginia.

covellite: [Eponymous term after Italian mineralogist Nicolas Covelli (1790-1829).] n. An indigo-blue, hexagonal mineral, copper sulfide (CuS). It is common in both hydrothermal veins and secondary deposits formed below the water table during weathering of primary copper sulfides. Covelli discovered the mineral on Mt. Vesuvius, in southwestern Italy near Naples.

crag: [Welsh craig < Celtic creag = rock.] n. In geomorphology, a steep, precipitous, rugged outcropping projecting from a rock mass. First usage in English c. 1300.

crater: [Greek *krater* = bowl, mixing vessel.] n. In volcanology, a basinlike depression at the summit or flank of a volcano. Craters may be formed by collapse, explosive eruption, or accumulation of pyroclastic material forming a rim. The distinction between a volcanic crater and a caldera is mainly one of size; a volcanic crater being smaller, generally less than 1.6km (1 mile) in diameter, whereas volcanic calderas can be tens of kilometers in diameter. Craters can also be formed by impact of meteorites, asteroids, and comets. Impact craters have no size limit, and may range in diameter from a few centimeters to hundreds of kilometers. First usage in English c. 1610.
Cited usage: The solution to the old question of whether the lunar craters have a volcanic, meteoritic or some other origin may be near at hand [1964 Yearbook of Astronomy 1965 113].
See also: caldera

craton: [German *kraton* = craton; < Greek *kratos* = strength, power; < IE *kar* = hard.] n. A major structural unit of the Earth's crust, being that part of the Precambrian continental crust that has become tectonically stable, and has remained relatively undeformed during the Phanerozoic Eon. It is characterized by a complete absence of volcanoes, and only rare occurrences of earth-quakes. The craton includes both the shield and stable platform, such as the Canadian Shield. Term coined by Hans Stille in 1940, shortly after he coined the terms miogeosyncline and eugeosyncline. IE derivatives include cancer, and the Greek *karkinos*, meaning crab.
Cited usage: For these platforms incorporating remnants of past orogenic belts Stille invented the term 'kraton', of obscure etymology but considerable convenience [1970, The Cambridge Ancient History 3rd ed. I. I. i. 5].
Alt. spelling: kraton
Synonyms: shield and platform
See also: hedreocraton

crenulation: [Latin *crenula* = diminutive of *crena* = notch.] n. In structural geology, extremely small scale structural folding (wavelength up to a few millimeters) superimposed on larger folds. In paleontology, minute and rounded teeth or crenations, such as those forming on the margins of whorls on ammonoid cephalopods. First usage in English c. 1845.

crest: [Latin *crista* = crest; < IE *sker* = to turn, bend.] n. A peak or summit, especially the ridge of a hill or mountain, or the topmost line of wave. First usage in English c. 1300.

Cretaceous: [Latin *creta* = chalk; *cretaceus* = chalklike, chalk-bearing.] adj. The final period of the Mesozoic Era, from 145.5 to 65.4Ma. Named for the English chalk beds of this age, the White Cliffs of Dover being a

classic example. The Cretaceous Period is noted for the extinction of dinosaurs, toothed birds, and ammonites, in addition to the development of early mammals and flowering plants.
Cited usage: The chalk and its associated sands have been termed the 'cretaceous system' [1854, F.C. Bakewell, Geol. 56].
Similar terms: Chalk Period (Brit.)
See also: Mesozoic, Jurassic, Tertiary, Paleocene

crevasse: [French < Old French *crevace* = crack, fissure; < Latin *crepa* = cracked < *crepare* = to crack, rattle, creak; < IE *ker* = echoic word imitating a hoarse cry.] n. A fissure or chasm in the ice of a glacier, usually of great depth, and sometimes of great width. Also, in the southern U.S., a breach in the bank of a river or canal; especially a breach in the levee or artificial bank of the lower Mississippi River. The French form crevasse, as used in the Alps, has found universal appeal because it signifies a feature of greater magnitude than the English crevice.
Cited usage: A breach in the levée, or a crevasse, as it is termed, is the greatest calamity which can befall the landholder [1819, The Edinburgh Review XXXII. 240].
Synonyms: crevice

crinoid: [Greek *krinoides* = lily-like.] n. Any echinoderm of the class Crinoidea, characterized by radial fivefold symmetry, calcareous plates enclosing a globular body, and branched append-ages, generally all supported by a single stem or column. Of approximately 2,800 known species, over 2,000 of them are extinct forms. Although their distribution is worldwide, living species of crinoids are found primarily in the East Indian Ocean. Most crinoids are sessile (attached), this is especially true for the extinct species. Sessile forms are commonly called sea lilies, whereas floating forms are called feather stars.

cristobalite: [German *cristobalit* = toponymous name after Cerro San Cristóbal in Hidalgo, Mexico.] n. A colorless to white, high-temperature polymorph of silicon dioxide, (SiO_2). Cristobalite is found in the cavities of silicic volcanic rocks, such as obsidian and rhyolite. One of the five polymorphs of silica (the others being quartz, tridymite, coesite, and stishovite). Cristobalite forms at high temperatures as isometric beta-cristobalite, and changes at lower temperatures to tetragonal alpha-cristobalite, which occurs both massive (as in opal), and as small metastable octahedral crystals. Term coined by G. von Rath in 1887 after Cerro San Cristóbal, its discovery locality.
Cited usage: (Cristobalite) Named by von Rathin 1887. It was first established as a distinct mineral by Mallard in 1890, but was not generally recognized until 1913, when Fenner proved that its indices of refraction are different from those of tridymite [1928, Austin F. Rogers, American Mineralogist v.13 73-92].
Alt. spelling: christobalite, crystobalite (both erroneous)

crocidolite: [Greek *krokis* < *krokidos* = nap or hairy surface on woolen cloth + *lithos* = rock, stone.] n. A lavender-blue to indigo-blue, fibrous, or asbestiform variety of the amphibole riebeckite, $Na_2(Fe_2, Mg)_3Si8O_{22}(OH)_2$. Inhalation of the silky fibers is known to cause asbestosis and mesothelioma.
Synonyms: blue asbestos

cromlech: [Welsh *cromlech* < *crom* < *crwm* =c rooked, bent + *llech* = flat stone.] n. A structure of prehistoric age, consisting of a large, flat, unhewn stone resting horizontally on top of three or more stones set upright; found in various parts of the British Isles, especially in Ireland, Wales, and the Devonshire and Cornwall areas of England. Also applied to similar structures in other parts of the world.
Synonyms: dolmen

crustacean: [Latin *crustaceus* = hard-shelled, <c *rusta* = shell; < IE *kreus* = to freeze, form a crust.] n. Any arthropod belonging to the subphylum Crustacea, characterized by: a) two pairs of antennae, b) a hard, segmented, and close-fitting, usually chitinous exoskeleton, c) a carapace that is periodically molted, and d) a habitat usually in a marine environment. Members include: lobsters,crabs, shrimps, barnacles, and a terrestrial isopod of the genus Oniscus, the common sow or pill bug. Term coined by Lamarck in 1801 as a name of the class of animals called by Cuvier Les Insectes Crustaces. Crustaceans are second only to insects in abundance. Etymologically related terms include crystal and crouton.

cryo-: [Greek *kryos* = frost, icy-cold.] Combining prefix meaning cold or freezing.

cryolite: [Greek *kryos* = frost + *lithos* = rock, stone.] n. A white or colorless monoclinic mineral, sodium aluminum fluoride, Na_3AlF_6. It is found chiefly in large masses in the pegmatites at Ivigtut, Greenland. It is an important ore of aluminum. The similarity of the mineral's appearance to ice inspired the name given in 1799. Etymologically related terms include: cryology, cryomorphology, and cryopedology.
See also: cryo-

crypto-: [Greek *kryptos* = hidden, concealed, secret.] Combining prefix meaning hidden or concealed, as in the terms cryptozoon, cryptozoa, and Cryptozoic.

Cryptozoic: [Greek *kryptos* = hidden, concealed, secret + *zoikos* = pertaining to animals < *zoa* = plural of zoon = animal, living being.] adj. A little used synonym for Precambrian time. The name was chosen because of the relatively little evidence of life in the fossil record during this time. First usage in English c. 1890.
Cited usage: The first major division of earth history determinable from rocks exposed at the surface is the Cryptozoic Eon. [1958, R. C. Moore, Introd. Hist. Geol. (ed. 2) iv. 59].
Synonyms: Precambrian
Similar terms: Eparchaic, Algonkian
See also: Archean, Hadean, Proterozoic, Precam-brian

crystal: [Latin *crystallum* = crystal ice; < Greek *krystallos* = clear ice, rock crystal; < IE *kreus* = freeze, form a crust.] n. A homogeneous solid made up of an element or compound. A crystal has an ordered atomic arrangement formed by a repeating, three-dimensional pattern that is outwardly expressed by planar faces, formed either during crystal growth (crystal faces), or by cleavage (cleavage faces). The connection to clear ice comes from the fact that the ancient Greeks thought that clear quartz crystals formed from deeply-frozen Alpine water.

cteno-: [Greek *ktenos* = comb.] Combining prefix meaning comb or comblike. Used in terms such as: ctenophore, ctenoid, ctenodont (comblike teeth), and ctenosaura (spiny-tailed lizards).

ctenophore: [Greek *ktenos* = comb + Greek *phoros* = bearing.] n. Any of various marine animals of the phylum Ctenophora, having transparent, gelatinous bodies, and bearing eight 'comb rows' of fused cilia that are used primarily for locomotion. These rows are arranged laterally along the sides of the animal. Also called comb jellies (also, sea gooseberries, sea walnuts, or Venus's girdles), ctenophores are vaguely similar in appearance to jellyfish. They are voracious predators of

microplankton. There are about 100 modern species, but due to their soft and fragile bodies, the fossil record is poor. They possibly date back to the Middle Cambrian.
Synonyms: comb jelly
See also: cnidarian, coelenterate

cuesta: [Spanish *cuesta* = slope of a hill; < Latin *costa* = rib, flank, side; < IE *kost* = bone.] n. In geomorphology, an asymmetric ridge or hill formed by the outcropping of an inclined sedimentary bed that is relatively resistant to weathering and erosion. As a result, one face of the cuesta (the dipslope of the bed) has a long gentle slope, whereas the opposite face of the cuesta is steep like a cliff. Etymologically related terms include cutlet and coast.
See also: ceja

cummingtonite: [Toponymous term after its discovery locality Cummington, Massachusetts.] n. A brownish monoclinic mineral of the amphibole group, hydroxy iron magnesium aluminum silicate, $(FeMg)_7Si_8O_{22}(OH)_2$. It is commonly found in medium grade regionally metamorphosed rock srich in iron. It can also be found as a late stage mineral in some mafic plutonic rocks, such as gabbro and norite.

cumulate: [Latin *cumulare* = to heap up, accumulate.] n. The general term for an often monomineralic igneous rock, formed by the accumulation of crystals that settle out from a magma by the action of gravity. Cumulates, such as dunite, pyroxenite, and chromitite form by the process of magmatic segregation at the base of layered mafic intrusions. First usage in English c. 1530.

cumulus: [Latin *cumulus* = heap; < IE *ken* = to swell, vault.] n. Along with cirrus and stratus, cumulus is one of the three basic cloud forms. Consisting of rounded masses heaped upon each other like cotton balls and resting on a nearly horizontal base, cumuli frequent temperate summer skies appearing in front of a blue-sky backdrop as dense, white, and fluffy flat-bottomed clouds with multiple rounded tops and well defined outlines. They are usually formed by the ascent of thermally unstable air masses. Most of the names given to clouds (cirrus, cumulus, stratus, nimbus, and their combinations) were coined in 1803 by the English meteorologist Luke Howard. He immediately gained international fame, becoming a cult figure among writers and painters. Goethe, Constable, and Coleridge revered him; their approval helped legitimize the science of meteorology. IE derivatives include: accumulate, cave, and church (< *kuro*, meaning swollen, strong).
Cited usage: Cumulus, convex or conical heaps, increasing upward from a horizontal base [1803, Luke Howard, On the Modification of Clouds].
Alt. spelling: cumuli (pl.)
See also: stratus, nimbus, cirrus, alto-

cuprite: [Latin *cuprum* = copper.] n. A vermillion-red isometric mineral, copper oxide, Cu_2O. It occurs as a secondary mineral in the oxidized zone of hydrothermal copper deposits.

cusp: [Latin *cuspis* = point.] n. A point at which two arcs meet with a common tangent. Also, the moving point describing a curve when its motion is exactly reversed. In astronomy, either point of a crescent moon. Also, figuratively, a transitional point or time, as the cusp day between two astrological signs. Typically, features such as scalloped beach fronts, barchan sand dunes, and river meanders exhibit pointed cuspate form.

cutan: [French *cutane* < Latin *cutis* = skin.] n. A fine claylike particulate skin or coating deposited on the surfaces of soil materials. A modification of the texture, structure, or fabric

at natural surface sin soil materials, due to concentration of particular soil constituents, or in situ modification of the soil body. Cutans may be composed of any of the component substances of the soil material.
Cited usage: Illuviation cutans are coatings of oriented clay on the surfaces of peds, mineral grains, and lining pores [2002, Glossary of Soil Science Terms Website].
Synonyms: clay films, clayskin, argillan, tonhäutchens

cwm: [Welsh *cwm* = cirque.] n. A steep-sided, crescent-shaped bowl located above a glacial valley at the head of the glacier, frequently with a lake in its steep-sided bowl. Occasionally used in Wales to also describe a deep, narrow non-glaciated valley.
Cited usage: The snow-patches are cwm-ice masses occupying deep scallops in an elevated position of the old erosion-surface [1933, Geography Jrnl. LXXXII. 202].
Similar Terms: coomb, coom, comb
See also: cirque

cyclosilicate: [Greek *cyclos* = round, circle + Latin *silex* or *silic* = flint, hard stone.] n. The structural group of silicates in which the SiO_4 tetrahedra are arranged in chains. Cyclosilicates are linked like in osilicates, but are also connected at the ends to form rings. The 1:3 silicon to oxygen ratio is generally the same as in the in osilicates. The rings have a variety of sizes and shapes. Three tetrahedra link to form triangular rings, as in benitoite; four tetrahedrons form a rough square, as in axinite; six tetrahedra form hexagonal shapes, as in beryl, cordierite, and tourmaline.

cyclothem: [Greek *kuklos* = circle + *thema* = something laid down, put, set.] n. A series of sedimentary beds laid down in a single sediment-tary cycle. A cyclothem is similar to a formation,

but is usually not as extensive nor as rigor-ously defined.
Cited usage: The word cyclothem is proposed to designate a series of beds deposited during a single sedimentary cycle of the type that prevailed during the Pennsylvanian period [1932, Wanless & Weller, Geol. Soc. Amer. Bulletin, XLIII. 1003].
Synonyms: formation
See also: formation

dacite: [Toponymous name after *Dacia*, the Roman province now part of Romania.] n. The igneous volcanic equivalent of the plutonic rock granodiorite. Compared with other volcanic rocks, dacite contains more silica than andesite and less silica than rhyolite. Term coined by Stache in 1863. *Cited usage:* Stache has given the name of Dacit(sic) to a quartzose trachyte [1878, Lawrence Cotta, Classification of Rocks 185].

dagala: [Italian *dagala* = steptoe.] n. An isolated hill of older rock surrounded by a younger lava flow. Originally used in the area around Mt. Etna, Sicily, where a distinctive isolated outcrop has well developed vegetation, characterized by stands of beech and birch trees. *Synonyms:* steptoe, kipuka

dale: [Old English *dael* < Teutonic *dalom, daloz* = deep, low place < *dalab(a)* = down, below; < IE *dhel* = a hollow.] n. A broad river valley. Although still used in a specific geographic sense, its use is chiefly poetical, as in the phrase, 'hill and dale.' The native English word appears to have been reinforced from the Norse, for it is in the northern British Isles that the word is commonly used in geographical names such as: Clydesdale, Annandale, Borrowdale, and Dovedale. *Cited usage:* By dale and eek by doune [1386, Chaucer, Sir Thopas 85]. Till over down and over dale All night the shining vapour sail [1847, Alfred Lord Tennyson, In Memoriam A.H.H. Obituary]. Around Whitby all the valleys are 'dales' [1876, Whitby Glossary].

dalles: [French *dalles* = plural of *dalle* = tube, gutter, trough; < Old Norse *doele*.] n. The rapids of a river where the channel is narrowed between rock walls. French fur traders of the Hudson's Bay Company originally coined this term for rapids they encountered in America, where rivers like the Columbia and Wisconsin are confined in long narrow troughlike channels. *Cited usage:* The Columbia River is there compressed into 'dalles,' or long, narrow, and broken troughs [1884, Harper's Mag. Feb. 364/1]. The Dalles of the Columbia. the Dalles of the Wisconsin [1890, M. Townsend, U.S. 137]. *Alt. spelling:* dells

datum: [Latin *datum* = given, that which is given < *dare* = to give.] n. Something known or assumed as fact; an assumption or premise from which inferences are drawn. As used in geology and oceanography, any fixed position (such as a point, line, or surface) in relation to which measurements of height, depth, etc., are referred. For example, *mean sea level* is a common datum used in topographic mapping; the top of a particular sedimentary bed may be chosen as a datum on which structure contours are drawn. Used as an adjective in such terms as: datum horizon and datum plane.

debris: [French *de* = of, from + *brise* = fragment < *briser* = to break.] n. Any surficial accumulation of loose material detached from bedrock by weathering and erosion. It may include rock fragments, soil material, and organic matter. As an adjective, used in terms such as: debris apron, debris avalanche, debris cone, debris fall, and debris flow. The term debris has a broader connotation than detritus. *See also:* detritus

decapod: [Greek *deka* = ten + *pod* = foot; < IE *ped* = foot.] n. Any shrimp-, crab-, or lobster-like Crustacean belonging to the order Decapoda. They are characterized by eight pairs of thoracic appendages; the first three pairs modified for feeding (maxillipeds), and the remaining five pairs as legs (the first pair of legs often enlarged as chelipeds).

Deccan: [Sanskrit *dakshina* = the south.] n. The broad plateau of south-central India between the Eastern and Western Ghats. It is underlain by over 200,000km³ of flood basalts, which erupted approximately 66 million years ago. The Deccan Plateau extends into four states of southern India: Andhra Pradesh, Karnataka, Kerala, and Tamil Nadu.

decke: [German *decke* = cover < *decken* = to cover; < IE *steg* = to cover.] n. A sheetlike allochthonous rock unit that has moved on a largely horizontal surface as a result of low angle thrust faulting or recumbent folding. IE derivatives include: stegosaurus, tegument, deck, and tuille.
Synonyms: nappe

declination: [Latin *de* = from + *clinare* = to bend.] n. The horizontal angle at any particular location between true north and magnetic north. In marine navigation, declination is synonymous with variation or magnetic deviation.

décollement: [French *décoller* = to unstick, disengage; < French < Latin *de* = from + French *colle* < Latin *colla* < Greek *kolla* = glue, paste.] n. A detachment structure usually associated with folding and thrust faulting where rock units above and below the detachment undergo distinctly different styles of deformation.
Cited usage: Owing to the push exerted by the Alps in formation, the sedimentary rocks of the Jura have been detached on the Middle Muschelkalk, of which the salt beds played the role of a lubricant, and folded. This kind of folding is a 'décollement' [1927, L. W. Collet, Structure of Alps ii. viii. 129].
Synonyms: detachment

decrepitation: [Latin *de* = from + *crepitare* = to crackle < *crepere* = to burst.] n. The audible crackling and disintegration of a mineral when exposed to sudden heat.
Cited usage: Decrepitation is generally occasioned by the expansion of the outer portions before the interior has had time to heat [1827, Michael Faraday, Chemical Manipulation v. 169].

decussate: [Latin *decussitatus* = the roman numeral X (ten), to cross in the form of a capital X < *decussis* < Greek *dekas* = the number ten + *assis* = monetary unit, often divided into ten parts.] adj. In general, having the form of an X. In geology, referring to a texture in which adjoining platy or columnar crystals lie in criss-cross directions. In botany, refers to an arrangement of leaves, which, in a three-dimensional reference viewed from above, forms a 90° cross pattern (right angles). This pattern occurs from leaves arranged on the stem in successive pairs, the directions of which cross each other at right angles, so that the alternate pairs are parallel and in line.
Cited usage: Foliation, transversely coalescent or intersecting one another (decussately aggregated) [1846, James D. Dana, Zoophytes (1848) 329].

deflation: [Latin *de* = from + *flare* = to blow.] n. A process of wind erosion in which fine-grained particles (fine sand, silt, and clay) are removed by the turbulent eddy action of the wind. As an adjective, used to describe features formed by deflation, such as: deflation armor (desert pavement), deflation basin (blowout), and deflation lake (dune lake).

deliquescent: [Latin *de* = from + *liquescere* = to melt < *liquere* = to be liquid.] adj. Pertaining to minerals that become liquid by absorption of moisture from the air.
Cited usage: A salt is deliquescent, when it has a greater attraction for water than the air, as it will in that case take water from the air [1812, James Smith, The Panorama of Science & Art II. 482].

dells: [French *dalles* = plural of *dalle* = tube, gutter, trough; < Old Norse *doele*.] n. The rapids of a river where the channel is narrowed between rock walls, e.g., between steep precipices of a gorge or through a narrow valley. The term is a corruption of *dalles*. The Wisconsin River Dells are the most famous dells in the U.S., and were probably named by Pierre Marquette and Louis Joliet in 1673 during their expedition to the Mississippi River via the Wisconsin.
Alt. spelling: dalles
See also: dalles

delta: [Greek *delta* = forth letter of Greek alphabet in the shape of a triangle.] n., adj. An alluvial deposit at the mouth of a river, commonly forming a triangular-shaped, low plain crossed by many distributaries. First named by the Greek historian Herodotus in the fifth century BCE, after the resemblance between the area of land at the mouth of the Nile River and the shape of the Greek capital letter delta. As an adjective, delta is used in terms like *delta front* or *delta plain*.
Cited usage: Islands have become connected with the main land by the growth of deltas and new deposits [1830, Lyell, Principles of Geology I. 13].

deluge: [Latin *diluvium* = flood < *diluere* = to wash away < *dis-* = off, from + *luere* < *lavere* = to wash; < IE *leu* = to wash.] n. A great flood. The deluge as described in the Old Testament refers to the great flood in the time of Noah. Derivatives of IE *leu* include: dilute, effluent, lutefisk, and latrine.

Cited usage: We two [Deucalion and Pyrrha, after the deluge] form a multitude [1A.D., Ovid, Metamorphoses (translated by Dryden, 1717)].
Synonyms: Noachian flood
Similar terms: eluvial, illuvial

demersal: [Latin *demergere* = to submerge < *de-* = down, from + *mergere* = to immerse.] adj. Pertaining to aquatic life forms that are bottom dwelling; for certain fish, near bottom dwelling.
Cited usage: In the Pacific Ocean, however, demersal eggs are quite common among species whose Atlantic counterparts have pelagic eggs [1968, F. R. H. Jones, Fish Migration ii. 13].
Synonyms: benthic
See also: benthic

demoiselle: [French *damsel* = girl; < Old French *dameisele, damoiselle* = young lady.] n. A pillar, usually made of volcanic breccia or glacial till, capped by a large boulder. Term originated in the French Alps in reference to this type of conical pillar that supports a large boulder, and resembles a girl in a long dress.
Alt. spelling: damoiselle
Similar terms: hoodoo
See also: hoodoo

dendrite: [Greek *dendron* = tree.] n. Either a surficial deposit or an inclusion of manganese oxide that has crystallized in a branching pattern. Dendrites are usually composed of the mineral romanechite.
Cited usage: Crystallizations called dendrites usually consisting of the mixed oxides of iron and manganese, forming extremely delicate brownish sprigs, resembling the smaller kinds of sea-weeds [1863, Sir Charles Lyell, The Geological Evidence of the Antiquity of Man vii. (ed. 3) 116].
See also: dendro-, dendrochronology

dendro-: [Greek *dendron* = tree.] Combining prefix meaning tree or treelike. As the adjective

dendritic, the term is used in geomorphology in such terms as dendritic drainage pattern and dendritic tufa. In mineralogy, as in dendritic habit. As the adjective dendroid, the term is used in paleontology, as in dendroid colony (graptolites or archeocyathids). In nouns, the prefix is used in dendrogram (branching diagram of evolutionary connection of fossil organisms) and dendrology (the study of trees).
See also: dendrite, dendrochronology

dendrochronology: [Greek *denron* = tree + *chronos* = time + *logos* = word, speech, discourse, reason < *legein* = to speak, to say; IE *leg* = to collect.] n. The study of annual growth rings of trees as a method for dating past events, such as climatic changes of the recent past.
See also: dendro-, dendrite

density: [Latin *densus* = compact; < IE *dens* = thick.] n. In physics, the mass per unit volume of a substance under specified conditions of pressure and temperature. More generally the number or quantity per unit, as in density of a population in a city or density of grains in a rock.

dentate: [Latin *dentis, dens* = tooth; < IE *dent* = to bite.] adj. Toothed, or having small cone-shaped or toothlike projections. IE derivatives include: dentition, mastodon, odonthalia, and dandelion.

desert: [Latin *desertus* = abandoned, deserted < *deserere* = to desert.] n. A terrestrial region that receives less than 25cm (10in) annual precipitation and is so devoid of vegetation as to be incapable of supporting large populations. As an adjective, the term is used for: desert varnish, desert pavement, desert armor, desert rose, and desert soil.
Cited usage: The loose salts are blown away, but oxides of iron, accompanied by traces of manganese and other similar oxides, form a red, brown, or black film which is firmly retained. The surfaces of long-exposed rocks and pebbles thus acquire a characteristic coat of 'desert varnish' [1944, A. Holmes, Princ. Physical Geol. xiii. 270].

desiccate: [Latin *de-* = from, down + *siccere* = to dry up < *siccus* = dry.] v. To dry completely; to rid, deprive, or exhaust of moisture.
Synonyms: dehydrate

detritus: [Latin *detritus* = past participle of *deterere* = to lessen, wear away.] n. Collective term for loose rock and mineral material that has been eroded by mechanical means such as: abrasion, frost wedging, or thermal expansion.
Cited usage: I have nowhere said that all the soil of this earth is made from the decomposition or detritus of these stony substances [1795, Hutton, Theory of Earth (1797) I. 115]. The fine earthy material deposited by streams or their sediment, is called silt or detritus [1862, James D. Dana, Manual of Geology 643].
See also: debris

deuteric: [Greek *deuteros* = second, after + *ikos* = akin to.] adj. Pertaining to the reactions that take place in igneous rock between primary magmatic minerals and the aqueous solutions that separate from the same magma at a late stage in its cooling history. Etymologically related terms include: deuterogenic, deuteroglacial (the glacial period following the last great, or proteroglacial, glaciation), and Deuteronomy (Moses' law set down for a second time).
Cited usage: It would be advisable to discriminate between metasomatic changes which belong to a later period of metamorphism, i.e. are secondary in the strictest sense of the word, and those which have taken place in direct continuation of the consolidation of the magma of the rock itself. I propose to call the latter deuteric, as distinct from secondary changes [1916, J. J. Sederholm, Bulletin de la Commission Géologique de Finlande IX. xlviii. 142].

deuterium: [Greek *deuteros* = second, after.] n. One of the three isotopes of hydrogen (symbol = H^2. Deuterium differs from the commonest isotope, protium, which has a mass of one atomic mass unit (amu), in having a neutron as well as a proton in the nucleus, and resulting in a mass of 2amu. In naturally occurring hydrogen deuterium comprises about one part in 6,000. Term coined by Harold C. Urey in 1933.
Cited usage: We wish to propose that the names for the H1 and H2 isotopes be protium and deuterium [1933, Harold C. Urey, Jrnl. Chem. Physics].
Synonyms: heavy hydrogen
See also: tritium, hydrogen

devitrification: [Latin *dis-* = away < *de* = down from, from, off + *vitrum* = glass.] n. The conversion of glass to crystalline material, such as that which occurs in obsidian to form snowflake texture and lithophysae.

Devonian: [Toponymous term after Devonshire < Old English *Devon*, a county in southwest England.] adj. Pertaining to a period of the Paleozoic Era covering the span of time from 416-359.2Ma. Also the corresponding system of rocks. The Devonian Period was coined in 1838 by geologists Sir Roderick Impey Murchison and Adam Sedgwick in reference to Devonshire, England, where they first investigated rocks formed during the period.

dextral: [Latin *dextra* = right, right-hand.] adj. In general, relating to the right-hand side. In paleontology and biology, pertaining to the right-hand direction of coiling in gastropods as viewed from the apex (apical view). In structural geology, pertaining to a strike-slip fault in which the motion of the block on the opposing side of the fault from an observer is towards the right (right-lateral). First usage in English c.1645.

Cited usage: Conklin concluded from his observations that this difference between dextral and sinistral snails is rooted in the structure of the undivided egg-cell [1938, M. J. Sirks, The Botanical Review, Genetisch Instituut, Groningen, Holland]. *See also*: sinistral

dia-: [Greek *dia* = through, across, apart.] Combining prefix used in several contexts. The prefix means thoroughly or completely, as in diagnosis, diorama, and diamictite; apart, as in diastrophism and diagenesis; across, as in diapir and diachronous; going apart, as in dialysis; and opposed, as in diamagnetism.

diabase: [Greek *dia* = through, across, apart — in this term erroneously used in place of *di* = two + Latin *basis* < Greek *basis* = step, pedestal, foundation.] n. In the U.S., a hypabyssal intrusive, igneous rock with essentially the same chemical composition as basalt, and consisting principally of augite and labradorite. Diabase is characterized by an ophitic texture. In England, the term is applied to altered basalt or greenstone. In Germany, it is used for any pre-Tertiary basalt. The name was originally coined by A. Brongniart in 1807 for the rock now recognized as diorite. The term diabase was erroneously formed because (according to Littré) it was meant to signify "rock with two bases." The name was abandoned by its author, Brongniart, in preference to René-Just Haüy's name diorite, but it was reintroduced by Hausmann in 1842 with the sense of 'crossing over' with regard to the internal alteration that diabase undergoes.
Cited usage: Diabase, a massive hornblende rock,...It is like diorite in composition, except that the feldspar is less abundant, and is either labradorite or oligoclase [1862, James D. Dana, Manual of Geology].

diachronous: [Greek *dia* = across, through, apart + *kronos* = time.] adj. Said of a rock unit that

is of different ages in different geographic locations, or cuts across time planes or biozones. An example of a diachronous unit is a marine sandstone formed during shoreline advance (progradation); it becomes younger in the direction in which the sea transgresses.
Cited usage: It is now proposed to introduce the term diachronous to describe a bed having such relations to the zonal succession [1926, W. B. Wright, Report of the British Association for the Advancement of Science 1926 355].
Synonyms: time-transgressive

diagenesis: [Greek *dia* = across, apart, through + *genesis* = origin, creation < *genesthai* = to be born, come into being; < IE *gen* = to give birth, beget.] n. The processes of chemical, physical, and biological changes undergone by a sediment after its initial deposition, and continuing through lithification, but excluding metamorphism and surface weathering. Diagenesis, therefore, includes such processes as: compaction, cementation, replacement, leaching, hydration, bacterial action, and concretion. There is no universally accepted definition of the limits of diagenesis. Some U.S. and Russian workers restrict the term to changes occurring while the sediment is still unconsolidated, and consider it complete at the point of lithification. Other geologists, however, include changes occurring after lithification. Term coined by Gümbel in 1868.

diagnostic: [Greek *dia* = through + *gnosis* = to know.] adj. Serving to characterize or identify, as in to know thoroughly. In petrology, used as a diagnostic mineral, such as quartz or olivine, whose presence in an igneous rock indicates oversaturation or undersaturation with respect to quartz. In pedology, used as a diagnostic horizon for classification of soils using Soil Taxonomy. First usage in English c. 1680.

diallage: [Greek *diallage* = interchange < *dia* = through, across + *allassein* = to change < *allos* = other.] n. A dark to grass-green, brown, bronze, or gray-colored clinopyroxene, usually a variety of augite or diopside occurring in lamellar or foliated masses, often with a metallic or brassy luster. Diallage is characterized by a conspicuous parting, and is typically found in gabbro. Named by the French mineralogist René-Just Haüy in 1801 after the philosophical term *diallage* being a figure of speech concerning arguments, first considered from various points of view, then brought to bear upon one point. He related this to a parting, dissimilar to the cleavages exhibited by the mineral variety.

diamagnetic: [Greek *dia* = apart (in the sense of opposed) + *magnete* = Toponymous term after natural lodestone found in the land of the Magnetes in Thessaly.] adj. Pertaining to substances with a small negative magnetic susceptibility that are slightly repelled by magnets, such as quartz, feldspar, graphite, and water; used in contrast to substances with a positive magnetic susceptibility that are magnetic or paramagnetic, such as iron and neodymium. Not to be confused with magnetic polarity. First usage in English c. 1860.

diamictite: [Greek *dia* = through (in the sense of thoroughly) + *miktos* = of mixed parentage or descent.] n. A general term for any nonsorted to poorly sorted, noncalcareous, clastic, terrigenous sedimentary rock containing a wide range of particle sizes. Glacial tillite and pebbly mudstone are good examples of diamictite. This term, along with diamicton (referring to the nonlithified version of diamictite), was proposed by glacial geologist R. F. Flint in 1960.
Cited usage: The Cayo ejecta are found in the Teakettle Diamictite, a 10 to 30-m thick heterogeneous deposit of pebbles, cobbles, and boulders supported in a matrix of red clay and silt [2002, Kevin Pope, Lunar and Planetary Science].

diamond: [Latin *diamant* = alteration of *adamas* = hard metal, diamond; < Greek *a* = not + *daman* = to tame, conquer.] n. A colorless to variously tinted to opaque black isometric mineral composed entirely of carbon (C). It is the hardest substance known (H = 10), and is polymorphous with graphite, chaoite, and lonsdaleite (the latter two polymorphs were found in the Ries and Canyon Diablo Craters, respectively). Diamond is formed commonly as octahedrons with rounded edges and faces at extreme temperatures and pressures in the Earth's mantle. It is commonly found in ultramafic breccia pipes and in placer deposits. In the year 16 CE, the Roman Manlius was the first to apply the term *adamas* to the hardest crystalline gem known. This was confirmed by Pliny the Elder in approximately 100 CE, who described six different varieties of diamond: the Indian, the Macedonian, the Arabian, the Cyprian, etc. In 400 BCE, the Greeks made mention of an 'unsubduable' hard stone they also called *adamas,* but this was probably not diamond, rather the emerystone of Naxos (a mixture of corundum and iron oxide, usually magnetite). The name was later transferred to the polymorph of carbon used today. Pure diamond is usually colorless, with colors imparted by impurities. When flawless and free from inclusions, it is among the most highly valued gemstones. Its high refractive index and dispersion results in remarkable brilliance and play of prismatic colors when facetted. The term diamond has also been attached to various occurrences of the rock crystal variety of quartz, such as: 'Herkimer diamond,' 'Arkansas diamond,' 'Lake George diamond,' and 'Clear Lake diamond.' First usage in English c. 1300.

diaphaneity: [Greek *diaphans* = transparent < *dia* = through, across + *phan* = to show, visible.] n. In mineralogy, the manner in which light is transmitted through a mineral, ranging from opaque to translucent to transparent. In general usage, as the adjective diaphanous, the meaning is restricted to a state allowing passage of light from translucent to transparent. Etymologically related terms include: phaneritic, aphanitic, and epiphany. First usage in English c. 1665.
See also: phanerozoic

diaphthoresis: [Greek *diaptherio* = to destroy utterly < *dia* = through (in the sense of thoroughly) + *ptherio* = to destroy.] n. Retrograde metamorphism. Term coined by Austrian mineralogist Friedrich Johann Karl Becke in 1909.
Cited usage: Retrogressive metamorphism, or diaphthoresis, is the mineralogical adjustment of relatively high-grade metamorphic rocks to temperatures lower than those of their initial metamorphism [1948, Geol. Soc. Amer. Bulletin XXX. 299].

diapir: [Greek *diapeirein* = to push through < *dia* = through, across + *peirein* = to pierce; < IE *per* = to lead, pass over.] n. In sedimentary rocks, a dome or anticlinal fold in which plastic core material, such as salt or gypsum, has pierced through the more rigid overlying rock. Igneous intrusions may also show diapiric structure.
Cited usage: A peculiar characteristic is the tendency of the Miocene beds, and especially the salt formation, to pierce through the overlying rocks at certain points on anticlinal axes, thus giving rise to a structure so common that Mrazec (1906) has coined the term *diapir fold* to describe it [1918, Economic Geol. XIII. 467].

diaschistic: [Greek *diaschisis* = a division < *dia* = through, across, apart + *skhistos* = split, divisible < *skhizein* = to split.] adj. Said of rocks in a minor intrusion, such as a dike, that consist of a differentiate whose composition is not the same as that of the parent magma. In particular, commonly referring to dikes genetically related to one parent magma, but with widely separated compositions, such as granite pegmatite and biotite lamprophyre. The sense of *diaschistic* is in opposition to that of *aschistic*, which refers to rocks of a minor intrusion that have the same composition as the parent magma.

diaspore: [Greek *diasphora* = dispersion < *diaspheirein* = to disperse, to spread about, to scatter < *dia* = through, apart + *spheirein* = to sow, scatter < *spora* = seed; < IE *sper* = to strew.] n. A white, gray, or yellowish orthorhombic mineral, hydroxy aluminum oxide, AlO(OH). It usually occurs in lamellar or scaly masses, is dimorphous with boehmite, and is a principal constituent of bauxite, as well as an important aluminum ore mineral. The name of the mineral alludes to its decrepitation when strongly heated. An etymologically related term is diaspora (to scatter), which relates to the *Semitic Diaspora*, the dispersion of Hellenistic Jews from Israel to live amid the Gentiles following their escape from captivity in the sixth century BCE. IE derivatives include: spew, spread, sprawl, spritz, spore, spray, and sperm. First usage in English c. 1800.

diastem: [Greek *diastema* = space between, interval.] n. In stratigraphy, a relatively short interruption in sedimentation involving only a brief interval of time with little or no erosion before resumption of deposition. A depositional hiatus of lesser magnitude than a paraconformity. Such gaps are common at bedding planes and represent missing intervals of time that are too short to measure quantitatively, and can only be deduced by paleontologic evidence. The term derives from ancient Greek music where a diastem is an interval, especially one forming a single degree of the scale. Also used in modern medicine to describe a vacant space or gap between the teeth. Term coined by the American geologist Joseph Barrell in 1917.
Cited usage: Found in every sequence everywhere...(ed. diastems) result from the fact that sedimentary rocks only account for 0.1 to 1% of geologic time [1995, Boggs, S., Principles of Sedimentology and Stratigraphy: 2nd ed. Prentice Hall].

diastrophism: [Greek *diastroph* = distortion < *diastrephein* = to distort < *dia* = apart, across, through + *strephein* = to twist; < IE *strop* = to wind, turn.] n. The processes of movement and deformation of the Earth's crust by which the major tectonic features of the Earth's surface are formed, including: continents, ocean basins, mountain ranges, trenches, and ridges. IE derivatives include catastrophism. First usage in English c. 1880.
Cited usage: Regions sink and regions rise and the upheaval and subsidence may be called diastrophism, and we have diastrophic processes [1895, J. W. Powell, Physiogr. Processes, Nat. Geogr. Monogr. I. I. 23].
See also: epeirogeny

diatom: [Greek *diatomos* = cut in half < *dia* = through, across + *temnein* = to cut; < IE *tem-* = to cut.] n. Any of various microscopic aquatic organisms of the class Bacillariophyceae in the phylum Heterokontophyta. Diatoms are found both as free-floating single-celled organisms, and as colonial organisms linked together in linear chains. Diatoms, which secrete walls of opaline silica called frustules, are the most common of the eukaryotic algae, and grow in both marine and freshwater environments. They are divided into two main groups: the pennate diatoms, which are bilaterally symmetrical, and the centrate diatoms, which are radially symmetric. The name may allude either to the fact that most diatom frustules are composed of two valves that overlap one another like the two halves of a petri dish, or to the fact that colonial diatoms chains are easily separated. IE derivatives include atom and anatomy.
Cited usage: The name Diatoma has reference to the readiness with which the strings or chains in which most of the forms are aggregated may be separated [1865, Gosse Land & Sea (1874) 158].

diatreme: [Greek *dia* = across, apart, through + *trema* = hole.] n. A carrot-shaped, breccia-filled volcanic pipe, sometimes originating at mantle

depths and formed by explosive gaseous escape. The famous diamond-bearing kimberlite pipes are examples of diatremes. Diatremes are often composed of the rock known as kimberlite. This rather unusual rock consists of olivine, serpentine, mica, ilmenite, carbonates, and other minerals. Kimberlite is a toponymous term named after its discovery at Kimberley, South Africa.
See also: kimberlite

dichroism: [Greek *di* = two + *khroma* = color.] n. The property possessed by some crystals of exhibiting two different colors when viewed along different axes, so that the mineral appears to change color when viewed at different angles. It is a type of pleochroism, which is more easily seen under polarized light than with the unaided eye. Dichroism is a two-color change. The greatest change, however, is limited to three colors and is called trichroic. Pleochroic, which means "many colors," is often the term used to cover both. First usage in English c. 1815.
Cited usage: This dichroism, as it may be called so far as I know, has never been observed in any other minerals than iolite and mica [1819, Brewster, Phil. Trans. 17].

diffraction: [Latin *diffringere* = break apart < *dis* = apart + *frangere* = to break; < IE *bhreg* = to break.] n. Change in the direction and intensity of a group of waves after passing by an obstacle, such as a reef, shoal, or breakwater; or after passing through an aperture, whose size is approximately the same as the wavelength of the waves. IE derivatives include fraction and fragment. First usage in English c.1670.
See also: diffusion

diffusion: [Latin *diffusio* = to spread, to pour out < *dis* = out, apart + *fundere* = to pour.] n. In geochemistry, a process of mass transfer in which dissolved ionic or molecular particles move through a stationary aqueous solution occupying

pores in a rock. In physics, the spontaneous intermingling of particles of two or more substances as a result of random thermal motion. With reference to radiation, the scattering of incident waves by reflection from a rough surface. In evolution, a dynamical system in which a population of structured individuals, monitored as a set of genotypes, diffuses through the space of all possible genotypes. First usage in English c. 1350.
See also: diffraction

digital: [Latin *digitus* = finger, toe.] adj. Referring to a measurement system in which information is stored and manipulated as a series of discrete units, as opposed to an analog system. Modern digital logic is based on the binary system, having two possible values: 0 and 1. This system is analogous to an electric switch being either on or off, thus making it the basis for nearly all computing machinery. First usage in English c. 1425.

dike: [Old English *dic* = trench, ditch < Old Norse *diki* = ditch.] n. In geology, a tabular igneous intrusion that is discordant with respect to its host rocks so that it cuts across the bedding or foliation of the country rock (local bedrock). Intrusions of similar structure composed of sediment, such as sandstone or conglomerate, are referred to as clastic dikes or pebble dikes. In engineering, an artificial wall, embankment, or mound usually made of earth or rock, and built along a stream course, lakeshore, or seacoast to prevent flooding. Also known as a levee. Conversely, though, the term also means an artificial watercourse, trench, or fosse excavated especially for use as a drainage ditch. The contrasting meaning of the term can be traced to the Dutch and Scandinavian cognate *dijk*, meaning mound, dam. This cognate shares the same root as the Old English *dic*, which is possibly from the Greek *teichos*, meaning wall. First usage in English before 900.
Alt. spelling: dyke

diktytaxitic: [Greek *dikty* = net + *taxis* = arrangement < *tassein* = to arrange, put in order.] adj. Referring to a texture of volcanic rocks, particularly olivine basalts, in which angular, jagged, and irregular voids are bounded by crystals, some of which protrude into the cavities. It has been suggested that the cavities rather than being gas vesicles, may result from egress of residual melt by filter pressing. Term coined by biologist Richard F. Fuller in 1930 in his PhD dissertation on the Columbia River Basalts (CRBs).

diluvium: [Latin *diluvium* = flood, inundation, deluge < *diluere* = to wash away, dissolve by water.] n. An archaic term applied during the early 1800's to widespread surficial deposits of unsorted boulders and other sediment that could not be explained by the normal action of rivers and seas, but were thought to be produced by extraordinary global floods, such as the Noachian Flood described in the Bible. Such deposits are now recognized as glacial drift. In continental Europe the term continues to be used for glacial deposits of Pleistocene age, as contrasted to younger Holocene alluvium. In reference to geologic time, the *Diluvium Epoch* (1.81Ma-10,000ybp), as used in continental Europe, is synonymous with the Pleistocene Epoch. First usage in English c.1815.
Cited usage: Diluvium as used by Elie de Beaumont and the modern foreign geologists, means precisely what I term drift [1839, Murchison, Silur. Syst. i. xxxvii. 509]. The term 'diluvium' was for a time the popular name of the Boulder Formation, because it was referred by many to the deluge of Noah, while others retained the name as expressive of their opinion that a series of diluvial waves raised by hurricanes or by earthquakes had swept over the continents, carrying with them vast masses of mud and heavy stones [1874, Lyell, Students' Geol. xi. (ed. 3) 145].
Alt. spelling: diluvian, deluvium
Synonyms: drift
See also: Pleistocene, alluvium, eluviation, illuviation, proluvium

dimorphism: [Greek *di* = two + *morphos* = shape, form.] n. In crystallography, the type of polymorphism in which two crystalline species known as dimorphs occur with the same chemical composition but different crystal structures, such as the dimorphs of iron sulfite, pyrite, and marcasite, both having the same composition, FeS_2. In biology, and paleontology, the characteristic of having two distinct forms in the same species of organisms, such as male and female, or polyp and medusa. First usage in English c. 1830.
See also: polymorphism

dinoflagellate: [Greek *dinos* = whirling < *dinein* = to whirl + Latin *flagellum* = diminutive of *flagrum* = whip.] n. Any of various single-celled chiefly marine protozoans of the order Dinoflagellata, characteristically having two flagella, and forming one of the chief constituents of plankton. They include brilliant bioluminescent forms and those that cause red tides. The etymological connection to a 'whirling whip' refers to the fact that, in addition to a posterior flagellum, dinoflagellates have one long transverse flagellum that encircles the entire body. First usage in English c. 1900.
See also: plankton

dinosaur: [Greek *deinos* = fearful, terrible, monstrous + *sauros* = lizard.] n. Any member of the superorder Dinosauria in the class Reptilia, an extinct group of largely terrestrial animals that represented a dominant taxon during the Mesozoic Era (251-65.5Ma). Dinosaurs could be either bipedal or quadrupedal; carnivorous, omnivorous, or herbivorous; and small to giant-sized. Dinosaurs are divided into two orders: Ornithischia having a bird-like pelvis (Stegosaurus, Triceratops), and Saurischia having a lizard-like pelvis (Tyrannosaurus, Brontosaurus). *Dinosauria* was originally coined by Richard Owen in 1842 as a suborder of Saurians. Although Owen gained fame for coining the term, Gideon Mantell contributed a

lot toward understanding the ancient reptiles. Though they started out on friendly terms, Mantell and Owen became bitter enemies over the years, squabbling over pterosaurs, Mesozoic birds, moas, and, finally, dinosaurs in general. In a bold effort to discredit Mantell, Owen tried to prevent the Royal Society from awarding him the Royal Medal. Mantell's discoveries of Iguanodon and Hylaeosaurus had linked his name with the Age of Reptiles for two decades. But, by coining the term Dinosauria, Owen effectively made the "terribly great reptiles" his own. Mantell could not cope with playing second fiddle to Owen in society, especially since he felt deserving of the honor. Ultimately, Mantell lost his personal fortune, family, and health, finally succumbing to an overdose of opiates in 1852.
Cited usage: The combination of such characters, some, as it were, from groups now distinct from each other, and all manifested by creatures far surpassing in size the largest of existing reptiles, will, it is presumed, be deemed sufficient ground for establishing a distinct tribe or suborder of Saurian Reptiles, for which I would propose the name of Dinosauria [1842, Richard Owen, Report on British Fossil Reptiles Part II, Report of the British Association for the Advancement of Science].

diopside: [Greek *di* = two + *opsis* = appearance, aspect.] n. A white to green, monoclinic, often transparent mineral of the clinopyroxene group, calcium magnesium silicate $CaMgSi_2O_6$. Diopside occurs most commonly in metamorphic rocks, especially as a result of contact metamorphism of limestone, as well as in some mafic igneous rocks. Diopside is the magnesium-rich end member of an isomorphous solid solution series with iron-rich hedenbergite. Term coined by René-Just Haüy in 1800, alluding to two possible appearances of its prismatic form, or as he stated the "l'antihémiédrie" of its crystals (hemihedrism is the tendency to crystallize with only half the

planes of full symmetry). Some authors state that the term comes from the Greek *diopsis*, meaning a view through < *dia*, meaning through, and *opsis*. This explanation derives from the fact that the clinopyroxene diopside is the most common of the transparent and translucent pyroxenes.
See also: hemihedral

dioptase: [Greek *dia* = through + *optasia* = view.] n. An emerald-green trigonal mineral, hydroxy copper silicate, $CuSiO_2(OH)_2$. It occurs as a secondary mineral near the water table as a result of weathering of primary copper-bearing ore deposits. The name was chosen because the cleavage plane can be seen on looking through the transparent crystal. First usage in English c. 1800.

diorite: [French *diorite* < Greek *diorizein* = to distinguish < *dia* = apart, between + *horizein* = to divide, limit.] n. An igneous plutonic rock, intermediate in composition between silicic granite and mafic gabbro, being the approximate equivalent of volcanic andesite. It is composed principally of plagioclase feldspar and hornblende with lesser pyroxene. Term coined by René-Just Haüy in 1820, it alludes to the fact that the characteristic mineral hornblende can usually be identified megascopically. The type locality of diorite comes from the Haute-Savoie in France.
Cited usage: The diabase, diorite, and amphibolite of French authors, seems to include both greenstone and hornblende rock [1826, William Phillips, Outline Mineral. & Geol. 151].
See also: mafic, silicic

dip: [Old English *duppen* = dip, plunge, pitch; akin to Danish *dyppe* = baptism; < IE *dheub* = deep.] n. In structural geology, the vertical angle that a planar surface, such as a bedding plane or fault plane, makes with the horizontal. It is

measured perpendicular to the strike of the structure. IE derivatives include: deep, depth and dive. First usage in English before 1000.
Synonyms: pitch

diploid: [Greek *diplous* = double, two-fold + *oidos* = having the likeness of.] n. A crystal form in the isometric system with 24 similar quadrilateral faces in paired arrangement. In biology and paleontology, a cell having double the basic haploid number of chromosomes. First usage in English c. 1905.
Synonyms: diplohedron

dipole: [Greek *dis* = two, twice, double + Latin *polus* < Greek *polos* = axis, pole, sky.] n. In physics, a pair of separated equal and opposite electric charges or magnetic poles. In chemistry, a molecule or atomic particle having such electric charges or poles. In geology, with reference to the Earth itself, having such magnetic poles. First usage in English c. 1910.

dis-: [Latin *dis* < *bis* = apart, away, between < *duo* = two, in twain; < Greek *dis* = two, twice, double.] Combining prefix meaning apart or separated. Not to be confused with the prefix *di-*, meaning two.

disconformity: [Latin *dis* = apart, away, between + *conformare* = to shape after < *con-* < *cum* or *cum* = with, together + *formare* = to shape < *forma* = shape.] n. A type of unconformity in which the bedding planes of strata above and below the break are essentially parallel. A disconformity indicates a considerable interval of erosion or nondeposition, usually marked by an uneven erosion surface, but without deformation prior to deposition of subsequent overlying layers. First usage in English c. 1600.
Synonyms: paraconformity, discontinuity
Similar terms: paraconformity, discontinuity
See also: discordant, concordant, paraconformity, nonconformity, unconformity

discontinuity: [Latin *dis* = apart, away, between + *continare* = to hold together < *com-* < *cum* = together + *tenare* = to hold; < IE *ten* = to stretch.] n. In stratigraphy, any interruption in sedimentation of any duration, as a result of nondeposition with accompanying erosion. In geophysics, a surface at which seismic velocities change abruptly. In structural geology, any surface separating two unrelated groups of rocks. First usage in English c. 1565.
See also: disconformity, discordance

discordant: [Latin *dis* = apart, away, between + *cordis* = heart.] adj. In igneous petrology, referring to an intrusion with contacts that crosscut bedding planes or foliation in the country rock. In stratigraphy, referring to strata that lack conformity or parallelism. In geomorphology, referring to related topographic features with differing elevation, such as a discordant stream entering the main valley via a waterfall. First usage in English c. 1275.
Cited usage: Concordant plutons include sills, laccoliths, lopoliths, and phacoliths. Discordant plutons include dikes, necks, chonoliths, batholiths, stocks, and bosses [1961, F. H. Lahee, Field Geol. (ed. 6) vi. 141-2].
See also: disconformity, discontinuity, pluton, concordant, unconformity

dispersion: [Latin *dis* = apart + *spergere* = to scatter < Greek *spherein* to sow, scatter.] n. The property of a transparent crystal to separate white light through diffraction or refraction and diverge it into spectral colors as a result of differences in refractive index for different wavelengths of light. First usage in English c. 1375.

disphenoid: [Greek *dis* = double, twice + *sphen* = wedge + *oidos* = having the likeness of.] n. A closed tetrahedral crystal form consisting of two sphenoids in which the two faces of the upper

sphenoid are rotated at 90° to the two faces of the lower sphenoid. A sphenoid differs from a simple tetrahedron in that the tetrahedron has four equilateral triangular faces, while the sphenoid has four isosceles triangular faces. First usage in English c. 1890.
Synonyms: bisphenoid

dissection: [Latin *dissecare* = to divide by cutting < *dis* = apart, away between + *secare* = to cut; < IE *sek* = to cut.] n. The process of erosion by which a relatively even topographic surface is gradually sculptured and subdivided into flat upland areas by formation of networks of gullies, ravines and canyons. This process is especially applicable to erosion of uplifted plains or peneplains and creation of rejuvenated topography. IE derivatives include scythe, sickle, and Saxon (warrior with knives). First usage in English c. 1630.

disseminated: [Latin *dis* = apart, away + *semen* = seed < *seminare* = to sow.] adj. Referring in general usage to anything scattered widely, as though sowing seed. In ore deposits, referring to a deposit in which the desired minerals occur as scattered particles in the rock, but in sufficient quantities to reach ore grade. First usage in English c. 1600.

dissociation: [Latin *dissociare* = to divide, sever < *dis* = apart, away + *associatus* = joined to, united with.] n. The process by which a compound breaks up reversibly to form two or more substances, such as $CaCO_3$ separating into CaO and CO_2. First usage in English c. 1610.

dissolution: [Latin *dis* = apart, away + *solvere* = to loosen, to break up, to dissolve.] n. In general usage, the separation into parts or constituent elements, as in the reduction of a body into elements or atoms, or, figuratively, the undoing or breaking of a partnership. In geology and chemistry, synonymous with solution, the process where a solid, liquid, or gas becomes homogeneously

dispersed within a different solid, liquid, or gaseous phase, without chemical change or reaction. First usage in English c. 1375.

distal: [Latin *dis* = apart, away + *stare* = to stand.] adj. In general usage, referring to a location situated away from the point of origin or attachment. In stratigraphy, said of a sedimentary deposit formed farthest from the source area and generally consisting of relatively fine clastic material, such as a distal turbidite. In ore deposits, said of a mineral deposit formed at considerable distance, often tens of kilometers, from a volcanic source to which it is related. In paleontology and biology, referring to a location situated away from the point of attachment, the point of origin, or the center of growth.

diurnal: [Latin *diurnalus* < *diurnus* = daily < *dies* = day; < IE *dyeu* = to shine, god of the sky + Latin *alis* = pertaining to.] adj. Pertaining to a day, daily, occurring within a 24-hour period, or in opposition to nocturnal, belonging to the daytime. In oceanography, referring to a tide with only one high water and one low water during a 24-hour period, as in the Gulf of Mexico. Originally, the term was used chiefly with regard to the motion of celestial bodies before also being used in reference to terrestrial events and animal behavior. First usage in English c. 1425.

divergent: [Latin *dis* = apart, away + *vergere* = to bend, to incline; < IE *wer* = to turn, bend.] adj. Referring to movement in different directions from a common center so that the intervening distance continually increases. Used to describe plate boundaries at spreading centers, such as the mid-ocean ridges (e.g., Mid-Atlantic Ridge) and continental rift zones (e.g., East African Rift Zone). In biology and paleontology, said of species characteristics that become progressively more distinct from each other over time. The opposite of convergent. First usage in English c. 1695.
See also: convergent

divide: [Latin *dividere* = to separate, divide < *dis* = apart, away + *videre* = to separate.] n. In geomorphology, the line of separation marking the boundary between two adjacent drainage basins that divide surface waters that flow in one direction from those that flow in the opposite direction. It can be a ridge, summit, or narrow belt of high ground. In hydrogeology, a ridge in the water table or other potentiometric surface from which ground water flows in opposite directions. In North America, the Great Divide refers to that part of the Rocky Mountains that divides the continent into eastward Atlantic-bound drainages and westward Pacific-bound drainages. The noun, with reference to separation of watersheds, was first used in 1807. First usage in English c. 1350.
Cited usage: Struck and passed the divide between the Grand River and the Verdegris River [1807, Pike, Sources of the Mississippi II].

doab: [Persian and Urdu *doab* = two waters.] n. The tongue or tract of alluviated land between two rivers that join in confluence. Specifically, in India, the *Doab* is the tongue of land between the Ganges and Jumna Rivers (also, generally used to describe similar tracts of land in the Punjab Valley). *Cited usage*: Khutu proper begins with a Doab. Note. This useful word, which means the land about the bifurcation of two streams, has no English equivalent [and] might be naturalized with advantage [1859, Sir Richard F. Burton, Jrnl. Geog. Soc. XXIX. 72].

dodecahedron: [Greek *dodeka* = twelve < *duo* = two + *deka* = ten + *hedra* = seat, base, face.] n. A closed crystal from having twelve equivalent faces that are either rhombic (rhombic dodecahedron) or pentagonal (pyritohedron). First usage in English c. 1565.

dolarenite: [A portmanteau word derived from *dol(omite)* and *arenite*.] n. A dolomite rock composed mainly of detrital dolomite particles of sand-size. In this portmanteau word, 'dol' refers to the rock's composition, while 'arenite' refers to its particle size.

doldrums: [English *doldrum* = dullard, sluggish fellow; < Middle English *dold* = stupid, dull.] n. In general usage, a state of inactivity or stagnation. In oceanography, a belt of calms and light winds near the equator between the northern and southern trade winds where they meet and neutralize each other, and where ships are likely to become becalmed. This term's evolution from a human condition to a meteorological state or geographical location is apparently due to a misunderstanding of the phrase *in the doldrums*, a condition or state of being that sounds like a place or locality. First usage in English c. 1800.
Cited usage: The 'equatorial doldrums' is another of these calm places. Besides being a region of calms and baffling winds, it is a region noted for its rains [1855, Maury, Physical Geography of the Sea].

dolerite: [French *dolerite* < Greek *doleros* = deceitful, deceptive < *dolos* = trick + *-ites* < *-lite* = pertaining to rocks < *lithos* = rock, stone.] n. In European usage, the preferred term for what in the U.S. is called diabase. Dolerite is a hypabyssal intrusive igneous rock allied to basalt, composed primarily of labradorite and augite. Named by the French mineralogist René-Just Haüy, alluding to the difficulty in identifying the rock megascopically as a result of its fine-grained texture. First usage in English c. 1835.
Cited usage: The basalts vary considerably in structure: the coarsely crystalline varieties, and those in which the different mineral constituents are sufficiently well developed to be distinguished by the naked eye, are termed dolerites [1879, Rutley, Study of Rocks xiii. 253].
Synonyms: diabase
See also: diabase

dolina: [Slovene *dolina* = valley; < Russian *dolina* = valley, plain.] n. A synonym for a sinkhole. A circular depression in a karst region that is usually conical or funnel-shaped and has subterranean drainage. The sides may be gently sloping to vertical or overhanging, and the size ranges from a few meters to several hundreds of meters in diameter.
Alt. spelling: doline
Synonyms: ponor, sinkhole
See also: uvala, karst, polje

dolomite: [Eponymous term after French geologist Déodat Guy de Dolomieu.] n. A white to colorless to grayish, pinkish, or yellowish trigonal mineral, calcium magnesium carbonate, $CaMg(CO_3)_2$. It forms a partial isomorphous series with ankerite, $CaFe(CO_3)_2$ and kutnohorite, $CaMn(CO_3)_2$. It forms saddlelike crystals and occurs in extensive deposits of dolostone and dolomitic marble, as well as occurring as hydrothermal veins in serpentinite and other magnesium-rich rocks. The Dolomite Mountains in the Italian Alps were named after this rock found there and described in 1791 by Dolomieu (1750–1801). Captured and imprisoned in Messina, Sicily, during Napoleon's expedition to in 1798, Dolomieu wrote his main treatise on minerals on the margins of a bible. Term coined in 1794.
Cited usage: The kind of marble which had been called dolomite, from Monsieur Dolomieu, who first remarked its peculiarity in dissolving slowly [1799, John Tennant, Phil. Trans. LXXXIX. 309].

dolostone: [Portmanteau word from *dolo(mite)* + *stone*.] n. A carbonate sedimentary rock consisting primarily of the mineral dolomite. In order to avoid confusion with the name for the common mineral dolomite, the term dolostone was proposed by American sedimentologist R. R. Shrock in 1948.

dome: [Latin *domus* = house, home.] n. In geomorphology, a general term for any smoothly rounded landform or rock mass resembling the dome of a building, such as the granitic exfoliation domes of Yosemite National Park of California. In structural geology, an anticlinal structure or uplift, ranging from circular to elliptical in plan view in which rocks dip gently away in all directions, such as the salt domes of Louisiana, the Black Hills Uplift of South Dakota, and the Nashville Dome of Tennessee. In volcanology, a rounded to irregular lava dome typically formed by extrusion of viscous high-silica flows. First usage in English c. 1510.

donga: [Afrikaans < Zulu *donga* = river channel.] n. A small, narrow, steep-sided ravine or gully formed by water erosion, and which is usually dry, except during the rainy season.
Cited usage: A donga was safely crossed. A donga would be called...in Scotland, a gully [1879, Daily News, 20 June].
Synonyms: wadi, nullah, arroyo
See also: gully, ravine, wadi, arroyo

doodlebug: [German *dudel* = fool, simpleton + English *bug*.] n. A popular term for various kinds of geophysical prospecting equipment. Due to its slang cuteness, the word has had many applications. In 1866, doodlebug was applied to a type of insect, the tiger-beetle. During World War II, it was applied to German pilotless planes or flying bombs by the English in R.A.F. slang. In modern common parlance, an aimless scrawl made by a person while daydreaming.
Cited usage: The Healy River Coal Company rigged an old truck with railroad wheels to shuttle local residents across the railroad bridge. This conveyance, called the Doodlebug, also took ladies of the evening to the Goat Ranch near Suntrana on payday weekends. "Goat Mary" Thompson had opened a beer parlor there in 1935 [1993, Wiliam Brown, Denali, Historic Resource Study].

dorsal: [Latin *dorsum* = back.] adj. In paleontology and biology, pertaining to or situated at the back or upper surface of an animal, or to the back or outer surface of a plant; as opposed to ventral. First usage in English c. 1540.
See also: ventral

down: [Old English *dun* = hill; < Dutch *duin* = sandhill; < Celtic *dunum* = hilly place name.] n. A name for any upland area of southern England that is generally grassy, treeless, and underlain by chalk. Usually used in the plural, as in the North Downs or Meadowlark Downs, but also used in the singular, as in Watership Down (a location between the counties of Hampshire and Berkshire along the Enbourne River in England, as well as a story by Richard Adams). An etymologically related term is dune. First usage in English before 1000 CE.
Cited usage: A downe is a lytyl swellynge or arerynge of erthe passynge the playne grounde..and not retchyng to hyghnesse of an hylle [1398, Trevisa, Barth. De P.R. XIV. xlv.].

dowsing: [Uncertain origin, probably from a dialect of southwest England.] v. The practice of searching for ground water or mineral deposits by use of a divining or dowsing rod. First usage in English c. 1690.
Cited usage: The divining or dowzing rod is certainly not older than the German miners, who were brought over by Queen Elizabeth to teach the Cornish to work their mines [1865, R. Hunt, Popular Romances of West England: or the Drolls, Traditions, and Superstitions of Old Cornwall 2nd ed]. The 'dowsing' or 'divining' rod is a forked stick of some fruit-bearing wood, generally hazel, held by the extremity of each prong of the fork in a peculiar way .][1869, Eng. Mech. 12/31].
Synonyms: divining, water witching, rhabdomancy

draa: [Toponymous term after the Draa Valley in southern Morocco, cut by the Draa River that runs through part of the Sahara Desert in North Africa.] n. The largest class of aeolian sand bed forms, with lengths on the order of kilometers and heights from tens to hundreds of meters. Each sand ridge lies from 0.5km to 5km from its nearest neighbor and moves approximately 2-5cm per year. It is the largest landform of the erg, or sand desert. A star-shaped dune, known as a rhourd, develops at the site where two draa chains cross. The Sahara derives its name from the Arabic *sahra*, meaning desert.

dreikanter: [German *drei* = three + *kante* = edge.] n. An angular, wind-abraded pebble or boulder having three curved faces intersecting in three sharp edges. First usage in English c. 1900.
Cited usage: Such wind-faceted pebbles, which often resemble Brazil nuts, are known as dreikanter or ventifacts [1944, Arthur Holmes, Principles of Physical Geology 1st ed.].
Similar terms: pyramid pebble
See also: ventifact, einkanter

drewite: [Eponymous term after British scientist George Harold Drew (1881-1913).] n. A calcareous mud or ooze composed of tiny aragonite needles, a few microns in length, and believed to have been precipitated from seawater by nitrate- and sulfate-reducing bacteria. Drew studied these marine bacteria in shallow Bahamian lagoons.

drift: [Middle English *drove* = herd, act of driving; < IE *dhreibh* = to drive, push; snow.] n. In glacial geology, a general term for any sediment either deposited directly by glacial ice, or by melt water emanating from a glacier. The term therefore includes both glacial till (unstratified) and glacial outwash (stratified). In mining, a horizontal or nearly horizontal opening driven along an ore-bearing vein. In oceanography, the movement of an object in a current. In mechanics, the variation

in calibration of an instrument over time. In the glacial sense, the term was introduced by British geologist R. I. Murchison in 1839 as a synonym for diluvium because he thought that it had drifted into piles on the sea floor as a result of catastrophic global floods.
See also: moraine, till

drought: [Old English *drge* = dry.] n. An extended period of dry weather, especially one that is injurious to crops. First usage in English before 1000.

drumlin: [Scottish *drumlin* = diminutive of *drum* = ridge; < Gaelic *druim* = back, ridge.] n. A low, smoothly rounded, elongate hill or ridge composed of compacted glacial till, and formed under advancing glacial ice, or shaped from an older moraine by readvancing ice. The long axis is parallel to the direction of glacial flow, and the blunt, steeper nose of the drumlin slopes in the direction from which the ice approached, whereas the gentler-sloping tapered end point in the opposite direction. First usage in English c. 1830.
Cited usage: Despite concentrated research and all the progress in understanding subglacial environments achieved in the last 20 years, there is a profound disagreement regarding the fundamental question of just how and why drumlins formed [2001, Jan A. Piotrowski & Wojciech Wysota, Drumlins: The unsolved problem, 6th International Drumlin Symposium Handbook, Torun NCU Press].

drusy: [German *druos* = bump, tumor.] adj. Referring to the habit of a mineral aggregate in which a rock surface or cavity is covered with tiny crystals, often having a composition similar to that of the rock substrate.
Cited usage: A surface on which very minute crystals abound is called drusy [1794, Kirwan, Elements of Mineralogy 2nd ed. I.31].
Alt. spelling: druzy
See also: geode

dune: [Old English *dun* = ridge, môund; < Dutch *duin* = sandhill.] n. A mound, ridge, or hill of windblown sand or other granular material, such as volcanic ash. It always has a characteristic shape, with the steep slip-face in the downwind direction, and the gentle-sloping face in the upwind direction. First usage in English c. 1785.
Cited usage: When the singing sand from the beach is compared with dune sand or blowout sand under the microscope no difference is perceptible [1919, W. D. Richardson, The Singing Sands of Lake Michigan, Science Mag. Nov. 28, 493-495].

dunite: [Toponymous term after Mount Dun in northern South Island, New Zealand.] n. An igneous plutonic rock of ultramafic composition in the peridotite family, with greater than 90% olivine, and almost always accompanied by accessory chromite. It forms most commonly by crystal settling at the base of layered gabbro bodies, and thus is often found serpentinized at the base of ophiolite sequences. Term coined by Austrian geologist Baron Ferdinand von Hochstetter in the earliest geological atlas of New Zealand in 1864.
Cited usage: Dunite (so named from Dun Mountain in New Zealand, which consists in great part of this rock and serpentine) is a crystalline-granular aggregate of olivine and chromic-iron [1879, Rutley, Study of Rocks xiii. 265].

durain: [Latin *durus* = hard.] n. A type of coal characterized by a dull luster, gray to brownish black color and granular fracture. Durain occurs in layers many centimeters thick, found within bituminous seams.
Cited usage: Vitrain and durain belonged to the coking coals and fusain to the non-coking coal class [1930, Engineering 18 July 92/1]. These four distinguishable ingredients, all of which are to be found in most ordinary bituminous coals, I name provisionally as follows: (i) Fusain. (ii) Durain. (iii)

Clarain. (iv) Vitrain [1919, M. C. Stopes, Phil Trans Royal Society Bull. XC. 472].
See also: clarain, fusain, vitrain

duricrust: [Latin *durus* = hard + *crusta* = coating, crust.] n. A general term for a hard layer in the upper horizon of a soil, typically found in a semiarid climate. It is formed by accumulation of soluble minerals in the upper layer that are deposited by evaporation of mineral-bearing waters moving upward by capillary action. Term coined in 1928 by W. G. Woolnough.
Cited usage: From the fact that...the chemically formed covering...always appears as a relatively hard 'armour plate', protecting softer, decomposed rock residue beneath it, I propose to group together all the representatives of the hard crust...under the name of the duricrust... (Note) This name was suggested to me by Professor Todd, Professor of Latin in the University of Sydney [1928, W.G. Woolnough, Jrnl. and Proceedings of the Royal Soc. New South Wales LXI 27].

duripan: [Latin *durus* = hard + German *phanna* = pan; < Latin *patina* < Greek *patane* = shallow pan, platter; < IE *pete* = to spread.] n. A diagnostic subsurface soil horizon characterized by cementation by silica, especially opal or chalcedony, and occurring mainly in areas of volcanism that have arid or Mediterranean climates. IE derivatives include: compass, spawn, patina, and petasos (broad-brimmed hat).

dy: [Swedish *dy* = silt, mud, ooze.] n. A dark, jellylike, freshwater mud consisting of unhumified organic matter transported and precipitated in colloidal form in a nutrient deficient lacustrine environment. Term coined in 1862 by H. von Post. Sweden is notable for its muds and clays such that a few Swedish terms have entered the geologic lexicon: *slam*, meaning dirt, mud, ooze, silt, sludge; *gyttja*, meaning dirt, mire, mud; and *dy*. The elemental symbol Dy stands for the trivalent rare earth element dysprosium.
Cited usage: Gyttja is greenish brownish and the color gets lighter when the material dries. Dy is brownish blackish and does not become lighter when drying. The two are often found mixed together [2000, Bengt H. Fellenius, Internet Geoforum].
See also: slime, mud, clay, silt

dynamic: [Greek *dynamis* = force, power < *dunasthai* = to be able + *iko* = of the nature of, pertaining to.] adj. In physics, referring to force related to motion. In geomorphology, pertaining to the quantitative analysis of processes related to forces that produce characteristic varieties of failure, as in erosion, mass wasting and transport. In metamorphism, referring to the total of tectonic processes and orogenic movement producing, crushing, and shearing at low temperatures, and recrystallization at higher temperatures. Term coined in 1692 by physicist and mathematician Carl Leibnitz, in reference to a force producing motion (as opposed to static).

dyne: [Greek *dynamis* = power, force.] n. In physics, the standard unit of force defined as that force that, acting on one gram for one second, generates a velocity of one centimeter per second. The unit is written, gm-cm/sec^2). The term was originally coined in 1842 in France as the force required to raise one kilogram one meter in one second; this is a different sense than what it is today.
See also: dynamic

dystrophic: [Greek *dys-* = bad, ill, impaired + *trophe* = food.] adj. In limnology, referring to a lake characterized by a deficiency in nutrient matter and notably high oxygen consumption in the hypolimnion, often associated with peat bogs.

Dystrophic bodies of water are usually brownish
or yellowish, having much unhumified organic
matter in suspension. Etymologically related terms
include: dyslexia, dysphoria, and dysentery.

Earth: [Middle English *erthe* < Old English *oerthe* = earth; < IE *er* = earth.] n. The third planet from the Sun, located between Venus and Mars. It is the largest and most dense of the four inner or terrestrial planets, and has a sidereal period of revolution about the sun of 365.26 days at a mean distance of approximately 149 million kilometers (92.96 million miles), an axial rotation period of 23 hours 56.07 minutes, an average radius of 6,378 kilometers (3,963 miles), and an average density of 5.5g/cm^3. First usage in English c. 950.

earthquake: [Middle English *erthequake* < *erthe* = Earth + *cwacian* = to shake, temble.] n. A sudden motion or trembling of the Earth caused by the sudden release of slowly accumulated stress. First usage in English c. 1325.

ebb: [Old English *ebbe* = ebb; < IE *apo-* = off, away.] adj. In oceanography, pertaining to the receding or outgoing tide, as the water returns toward the ocean, as opposed to flood tide. First usage in English before 1000 CE.
Cited usage: Whiche the Mariners call nepe tydes, lowe ebbs or lowe fluddes [1561, Richard Eden, Arte of Navigation II. xviii. 50].

echinoderm: [Latin *echinus* = sea urchin; < Greek *ekhinos* = sea urchin, hedgehog < *ekhis* = adder, viper.] n. Any, usually sessile and benthic, marine invertebrate belonging to the phylum Echinodermata. They are characterized by radial symmetry, an endoskeleton composed of plates or ossicles made of calcite, and often display projecting spines (hence the phylum name). Modern members of the phylum include: sea stars, sea urchins, sea cucumbers, and sand dollars. First usage in English c. 1830.

echo: [Middle English *ecco* < Latin *echo* = echo; < Greek *eche* = sound < *echo* (eponymous term) = mountain nymph.] n. A repetition of sound produced by the reflection of sonic waves off of some obstructing surface, such as a mountain or the sea floor. Echo was a nymph who distracted Hera with her chattering, so that Hera could not catch her husband Zeus in the act of infidelity. In punishment, Hera cursed Echo so that she could only repeat the words of other people. Echo fell in love with Narcissus, who did not return her love as he loved only his own reflection, so Echo pined away until only her repeating voice remained. First usage in English c. 1325.

eclipse: [Latin *eclipsis* < Greek *ekleipsis* = eclipse < *ekleipen* = to fail to appear < *ek* = out + *leipein* = to leave.] n. The partial or complete obscuring, relative to a designated observer, of one celestial body by another. First usage in English c. 1275.

ecliptic: [Latin *linea ecliptica* = ecliptic line < *eclipticus* = of an eclipse; < Greek *ekleipein* = to fail to appear < *ek* = out + *leiphein* = to leave.] n. The great circle formed by the intersection of the plane of the Earth's orbit with the celestial sphere. From a terrestrial viewpoint, it is the apparent annual path of the sun. So called because solar eclipses can happen only when the Moon is on or very near this plane. First usage in English c. 1375.
Cited usage: It (is) impossible to conceive how a Sphere can be inclin'd to a Plane, passing thro' its Center as the Ecliptick does thro' the Center of the Earth [1698, John Keill, Examination of Dr. Burnet's Theory of the Earth].

eclogite: [Greek *eklogen* = selection < *eklegein* = to select < *ek* = out of + *legein* = to gather, to speak.] n. A granular, nonfoliated, metamorphic rock consisting principally of garnet (almandine-pyrope) and sodic pyroxene (omphacite), which is generally believed to form by recrystallization of a mafic protolith at high pressure. Rutile and kyanite are common accessory minerals. Term coined by French mineralogist René-Just Haüy in 1822. A possible etymological reference owes to the fact that the minerals in the rock are usually not found together, further alluding to the cognate term eclogue, a dialogue or pastoral poem between usually separated shepherds.
Cited usage: This rock, to which Haüy gave the name of eklogite, is usually very firm and coherent [1866, Lawrence, Cotta's Rocks Classification].
Alt. spelling: eclogyte, eklogite

ecology: [Greek *oikos* = house, dwelling + *logos* = word, speech, discourse, reason < *legein* = to speak, to say; IE *leg* = to collect.] n. The study of the relationships between organisms and their environments, including the study of communities, natural cycles, and relationships of organisms to each other. German biologist Ernst Von Haeckel (1834-1919) coined the term in 1866. Haeckel, who struggled to reconcile the differences between Darwinism and Lamarckian viewpoints, also coined the phrase "ontogeny recapitulates phylogeny" as a "biogenetic law" that was part of his *Recapitulation Theory*. During his career, Haeckel coined the terms phylum and phylogeny. Despite of his contributions to ecology, Haeckel's reputation became somewhat tarnished when he admitted to falsifying embryonic drawings. Also, as a scientific philosopher, he described human evolution in racial terms, which, along with writings by Spencer, later became a part of the pseudo-scientific basis for Nazism. Haeckel himself was a proponent for applying the biological concepts he called "selective breeding" and "racial hygiene" into human society

Cited usage: By ecology we mean the body of knowledge concerning the economy of nature - the investigation of the total relations of the animal both to its inorganic and its organic environment; including, above all, its amical and inimical relations with those animals and plants with which it comes directly or indirectly into contact - in a word, ecology is the study of all those complex interrelations referred to by Darwin as the conditions of the struggle for existence [1866, E. Haeckel, General Morphology of Organisms].
Synonyms: bionomics

ecoulement: [French *ecoulement* = flowing; < Latin *ex* = out of + *colare* = to creep.] n. Downward movement of earth and rock masses along slopes due to the force of gravity overcoming frictional forces. In hydrology, circulation and resultant flow of terrestrial water due to the effect of gravity. A pyroclastic ash flow in French is called an *ecoulement de cendres*.
Synonyms: gravitational sliding

ecto-: [Greek *ektos* = outside < *ek* = out.] Combining prefix meaning outer, outside of, or external, as in: ectoplasm, ectotherm, ectozoic, and ectoderm.

ectothermic: [Greek *ektos* = outside < *ek* = out + *therme* = heat < *thermos* = hot.] adj. Said of an organism that regulates its body temperature largely by exchanging heat with its surroundings. Ectotherms include all animals except birds and mammals.
Synonyms: poikilothermic, cold-blooded
See also: endothermic, poikilothermic, homoiothermic

edaphic: [Greek *edaphos* = ground, floor.] adj. Pertaining to those ecological factors produced or influenced by local physical conditions of the environment, such as the soil, bedrock, climate,

hydrology, and etc. Initially, the term pertained only to soil characteristics, but later was expanded to include other physical factors. First usage in English c. 1900.
Cited usage: The sand dune can be aptly described as an edaphic desert–a desert that is due, not to climatic but to soil conditions, since the coarse texture and very low organic content render sand a very poor retainer of water [1938, Nature 7 May 815].
Synonyms: edaphos

eddy: [Old English *ed* = turning + *ea* = water.] n. A current, such as one of water or air, which moves contrary to the direction of the main current whose force created it. Eddies usually develop circular motion in response to turbulence caused by aberrations in the flow regime. First usage in English c. 1450.

Ediacaran: [Toponymous term after the *Ediacaran Hills* in Southern Australia < Aboriginal Australian *ediacara* = spring of water.] adj. Said of the last period of the Precambrian Eon, from 600-542Ma. The Ediacaran Period (also called Vendian) is characterized by trace fossils of various multicellular, soft-bodied organisms, apparently lacking hard parts. These fossils are known collectively as the Ediacaran Fauna. Because there is evidence of macroscopic life during the Ediacaran Period (600-542Ma), some geologists say it should be included as the first period of the Paleozoic Era. This unique faunal assemblage was discovered and named by Australian geologist Sprigg in 1946. Confusion over the proper pronunciation of this term (ee-dee-ack'-a-rin) arises largely because of the mistaken variant spelling *Ediacarian*.
Alt. spelling: Ediacarian
Synonyms: Vendian

effervesce: [Latin *ef* < *ex* = out + *fervescere* = to begin to boil < *fervere* = to be hot, to boil.] v.

The release of small bubbles of gas from a liquid as a result of a chemical reaction. First usage in English c. 1700.
Cited usage: The carbonic acid froths up or 'effervesces' in small bubbles through the drop of liquid (on limestone) [1874, Lyell, Elementary Geology].

efflorescence: [Latin *efflorescere* = to blossom, produce flowers < *ef* < *ex* = out + *flora* = flower.] n. A whitish surface deposit of powdery to fluffy hairlike crystal, produced as a surface encrustation on a soil or rock, usually by evaporation of water brought to the surface by capillary action. The minerals are usually water-soluble salts, such as: epsomite, melanterite, natron, halite, or gypsum. First usage in English c. 1625.
Cited usage: (Nitrate of potassium) occurs as an efflorescence on the soil [1886, Roscoe, Elementary Chemistry].

effluent: [Latin *effluere* = to flow out < *ex* = out + *fluere* = to flow; < IE *bhleu* = to swell, well up, overflow.] n. In environmental geology, usually a liquid discharged as waste, such as water discharged from a storm sewer, contaminated water from a factory, or overflow form agricultural land after irrigation. Etymologically related terms include: fluid, fluvial, confluence, solifluction and influenza. First usage in English c. 1725.
Cited usage: The effluent is very clear and is not harmful to human or fish life [1930, Jrnl. Iron & Steel Inst. No.140].
See also: influent

effusion: [Latin *ef* < *ex* = out + *fundere* = to spread, to pour.] n. The extrusion of relatively fluid lava onto the Earth's surface. First usage in English c. 1875.
Cited usage: Eruptions were not explosive and probably consisted of gentle effusion of pillow lava [1985, J. G. Moore, Structure and eruptive mechanism at Surtsey Volcano, Iceland: Geol. Mag., v. 122].
Similar terms: extrusion

einkanter: [German *eins* = one + *kante* = edge.] n. A wind-abraded rock fragment or ventifact with only one face or sharp edge indicating a prevailing wind that comes from one predominant direction. *Cited usage*: The size of the original rock is an important control on the form of ventifacts. Classic faceted types are described as einkanter, zweikanter, and dreikanter for one-, two-, and three-ridged forms in fine-grained rocks. Surface features can include pits, flutes, grooves, or helical forms [2002, Journal of Geophysical Research Vol. 107 #E1]. *See also*: dreikanter, ventifact

ejecta: [Latin *ejectus* = thrown out < *e* < *ex* = out + *jacere* = to throw.] n. The material ejected from a volcano. *Cited usage*: The summit of the volcano usually has a well-defined crater from which ejecta escape to continue the building of the cone [1957, G. E. Hutchinson, Treatment of Limnology I. i. 25]. *Synonyms*: pyroclastics, ejectamenta *See also*: pyroclastic

elastic: [Latin *elasticos* = expanding spontaneously; < Greek *elastos* = beaten, ductile < *elaunein* = to beat out, to forge.] adj. In structural geology, pertaining to a solid body that is capable of recovering to its original shape or volume after the imposed force causing deformation is released. In such a body, strain (i.e., deformation) is totally recoverable. In strict modern use, as applied to solids, defined by James Bernouilli in 1694.

electrolyte: [Greek *elektron* = amber + *lytos* = loosed < *lyeo* = to loosen.] n. A substance which dissociates into ions when dissolved in water or other suitable medium, or when the substance is melted, thus making the solution a conductor of electricity. First usage in English c. 1830.

electron: [Latin *electrum* = amber; < Greek *elektron* = amber.] n. A subatomic particle that is a constituent of all matter, has a negative charge, can exist independently of the nucleus of an atom, and is a carrier of electric current. The term originates from the Greek *elektron*, meaning amber, a fossilized plant resin, which when rubbed with fur develops a negative static charge as a result of friction (triboelectricity). Term coined by Irish physicist G. J. Stoney in 1891.

element: [Latin *elementum* = rudiment, first principle, one of the four basic form of matter.] n. In the physical and life sciences, a pure substance composed entirely of atoms of one type, all having the same number of protons in the nucleus. Although of uncertain origin, the Latin *elementum* is perhaps ultimately from "lmn", the first three letters of the second half of the Canaanite alphabet. As awareness of the microscopic world developed, the term was redefined from the simple idea that an element is one of the four basic forms of matter: earth, air, water, and fire, to the more fundamental realization that it represents a substance that cannot be broken down any further by chemical means (1661, Robert Boyle in *The Sceptical Chymist*). The modern chemical sense of the term arose c. 1815.

elevation: [Latin *elevare* = raise up< *ex* = outside, away from + *levare* = to raise.] n. In cartography, the vertical distance above or below a datum, usually mean sea level, to a point or object on the Earth's surface. Whereas elevation is preferred to indicate heights or depths on the Earth's surface, altitude is usually used to indicate heights in space above the Earth's surface. *See also*: altitude

elutriation: [Latin *elutriare* = washed out < *elutirum* = vat, bath; < Greek *elutron* = cover, sheath, tank < *luere* = to wash out.] n. A method of mechanical analysis of sediment is which a slowly rising current of fluid (e.g., water or air) carries lighter particles upward and allows heavier ones to sink. Also, removal of material from a mixture in water by washing and decanting, leaving the heavier particles behind. First usage in English c. 1730.

eluviation: [Latin *eluere* = to wash out < *e* = out + *lavere* = to wash.] n. In pedology, the process of removal of soil material in suspension or in solution by downward groundwater percolation, particularly in the A or E horizons. First usage in English c. 1860.
Cited usage: We may refer to the translocation of material either mechanically or in solution, as eluviation, and two main types of eluviation may be distinguished: 1) mechanical eluviation, in which the finer fractions of the mineral portions of the soil are washed down to lower levels, and 2) chemical eluviation, in which decomposition occurs and certain products thus liberated are translocated in true or colloidal solution [1932 G. W. Robinson Soils iii. 53].
Similar terms: alluvial, diluvium, illuviation, proluvium
See also: alluvial, diluvium, illuviation, proluvium

embouchure: [French *emboucher* = to put or go into the mouth < *em* = before < *en-* = in + *bouche* = mouth; < Latin *bucca* = cheek.] n. The mouth of a river, or that part where it enters the sea. Also, the opening of a river valley into a plain. First usage in English c. 1755.
Cited usage: The city Foah so late as the beginning of the fifteenth century, was on this embouchure [1830 Lyell, Principles of Geology I. 238].

emerald: [Latin *esmeralda* = emerald;< Greek *smaragdos* = green gem; < Sanskrit *maragda* < Semitic *baraq* = shine.] n. A brilliant deep-green, transparent variety of the mineral beryl, valued as a precious gemstone. Chromium or possibly vanadium impurities cause the green coloration. The term has also been applied to various gem-

stones with a green color, such as 'oriental emerald' (sapphire), 'copper emerald' (dioptase), 'Brazilian emerald' (tourmaline), and 'emerald jade' (jadeite). First usage in English c. 1275.
Synonyms: smaragd

emery: [Latin *smericulum* < Greek *smiris* = polishing powder.] n. A fine-grained, granular, impure, gray to black variety of corundum containing varying amounts of iron oxides, usually magnetite or hematite. It may be formed by metamorphism by highly aluminous sediments, and is used in crushed form as a natural abrasive for grinding and polishing. First usage in English c. 1480.

emulsion: [Latin *emulgere* = to milk out < *ex-* = out + *mulgere* = to milk.] n. In chemistry, any colloidal suspension of minute droplets of one liquid in another. In ore microscopy, used as an adjective referring to a texture in which minute blebs or rounded inclusions of one mineral occur enclosed within another mineral. First usage in English c. 1610.

en-: [French *en-* or *em-* < Latin *in* = within, in, into, upon, or against, with the added sense of bringing something into a certain condition or state.] Combining prefix generally meaning within, in, or into, as in entice, envelop, and entrain; sometimes meaning upon or against, as in encrust. Many words can be formed with either *en-* or *in-*. The pronunciation of *en-* becomes *em-* before the letters b, p, and occasionally m.

en echelon: [French *en* = in + *echelle* = ladder.] adj. Said of geologic features, such as a series of fault traces that are in overlapping, steplike, or herringbone arrangement. In the case of fault traces, each is relatively short, but together they form a linear zone in which the strike of the individual fault traces is oblique to the zone as a whole. The French used the term extensively in

cavalry formations, in which successive divisions are parallel to one another, but laterally displaced so that each division is clear to advance in front.
Cited usage: The...rear-guard...was perceived drawn up *en echellon* [1812, Examiner 24 Aug.].
Alt. spelling: en echellon
See also: echelon

enantiomorphism: [Greek *enantios* = opposite + *morphos* = shape, form.] n. In crystallography, the characteristic of a mineral to develop into two possible crystal forms with identical chemical composition. They are mirror images of one another and are called enantiomorphs. Examples are right- and left-handed crystals of quartz. First usage in English c.1895.

enargite: [Greek *enargos* = brilliant, distinct, clear < *en* = in + *argos* = bright.] n. A grayish-black to iron-black orthorhombic mineral with bright metallic luster, copper arsenic sulfide (Cu_3AsS_4). The mineral is dimorphic with luzonite, and forms an isomorphous series with antimony-bearing famatinite. Enargite occurs in hydrothermal veins and is an important ore of copper. The name alludes to the mineral's distinct to excellent cleavage. First usage in English c. 1850.

encroachment: [French *encrochen* = to seize illegally < Old French *encrochier* to seize < *en-* = in + *croc* = hook.] n. In stratigraphy, the horizontal component of coastal onlap. The term is familiar to all who watch American football, being a defensive line penalty. First usage in English c. 1450.
Cited usage: The intervening strip of land, narrower now than then owing to the encroachment of the waves [1878, R. B. Smith, Carthage 433].

endemic: [Greek *endemos* = native < *en-* = in + *demos* = people.] adj. In ecology and paleontology, said of a group of organisms that are native to, or otherwise restricted to a particular region. In its more general and earlier cultural usage, referring to characteristics of a specific people, being synonymous with indigenous. Etymologically related terms include: democracy, epidemic, and pandemic. First usage in English c. 1660.
Cited usage: The term 'endemic' is relative in that it means, as now generally used in biology, a taxon or an ecological group limited in range to the geographical area under consideration [1951, Jrnl. Ecol. XXXIX. 215].

endo-: [Greek *endon* = within.] Combining prefix meaning inside or within. Common in biology and paleontology, as in endodermis and endosome; and in geology, as in endogenous, endolithic, endomorph, endorheic, endoskarn, and endodynamomorphic.

endogenous: [Greek *endon* = within + *genesis* = origin, creation < *genesthai* = to be born, come into being; < IE *gen* = to give birth, beget.] adj. In volcanology, referring to a lava dome that has grown chiefly by expansion from within, and is characterized by a series of concentric flow layers. In contrast to an exogenous dome. First usage in English c. 1830.
See also: exogenous

endoskarn: [Greek *endon* = within + Swedish *skarn* = rubbish.] n. During contact metamorphism, a skarn formed by replacement of the intrusive igneous rock itself by action of hydrothermal solutions. In contrast to and exoskarn.
See also: skarn, exoskarn, tactite, metasomatism

endoskeleton: [Greek *endon* = within + *skeletos* = dried up, mummy.] n. The internal skeleton of an animal, serving as a structural framework and as an attachment for tissues, especially the muscles. As opposed to an exoskeleton. First usage in English c. 1835.
See also: exoskeleton

endothermic: [Greek *endon* = within + *therme* = heat < *thermos* = hot.] adj. In biology and paleontology, said of an organism that generates and maintains internal body heat independent of environmental conditions (within certain extremes). Mammals and birds are the only present-day endotherms, but controversy continues as to whether dinosaurs were endothermic or ectothermic. In chemistry, pertaining to a reaction that occurs with absorption of heat, as contrasted to an exothermic reaction.
Cited usage: Most birds and mammals are endotherms, producing at rest an order of magnitude more heat than a reptile of the same weight at the same temperature [1972, Nature 14 July 81/1].
Synonyms: homoiotherm
See also: ectothermic, homoiothermic, poikilothermic

energy: [Latin *energia* < Greek *energeia* = energy < *energos* = active < *en* = at + *ergon* = work.] n. The capacity of system to cause change, or the capacity of a system to do work in changing from one state to another. First usage in English c. 1580.

enhydros: [Greek *en* = within + *hydrar* = water.] n. A hollow geode of chalcedony containing trapped water in the central cavity. Enhydros most commonly form in large vesicles in basalt flows by precipitation of silica from low-temperature hydrothermal solutions, some of the solution becoming trapped within the geode.

enigmatite: [Latin *aenigma* < Greek *ainigma* = riddle < *ainos* = fable.] n. A brown to black triclinic mineral, sodium iron titanium silicate ($Na_2Fe_5TiSi_6O_{20}$). It is a primary constituent in sodium-rich, silica-poor, alkalic pegmatites, syenites, and volcanic rocks. The name alludes to the formerly uncertain chemical composition of the mineral.
Alt. spelling: aenigmatite
Synonyms: cossyrite

enstatite: [Greek *enstats* = adversary, opponent < *en-* = in, at, near + *stats* = one that stands; < IE *sta* = to stand.] n. A grayish-green, yellowish, grayish to brown orthorhombic mineral in the orthopyroxene group, magnesium silicate ($MgSiO_3$). It is dimorphic with clinoenstatite, and forms an isomorphous series with ferrosilite. It is common in intermediate and mafic igneous rocks, and in stony meteorites. It was first described by G. A. Kenngott in 1855, who named it because the mineral has a refractory nature and is almost infusible before the blowpipe. IE derivatives include: arrest, static, prostitute, and insist.
Cited usage: Enstatite in prisms, resembling pyroxene or scapolite [1857, C. U. Shepard, Mineralogy 425].

entablature: [Latin *en* = within + *tabula* = plank, tablet.] n. In a lava flow, displaying columnar jointing. The upper zone is characterized by thinner and less regular columns than the lower zone or colonnade. The term originated in the architecture of a classical temple, referring to that part of the structure between the columns below and the eves above. First usage in English c. 1610.

enterolithic: [Greek *enteros* = intestines + *lithos* = rock, stone.] adj. Said of a ribbonlike sedimentary structure having intestinelike folds that resemble tectonic structures, but are actually formed by in situ chemical changes during diagenesis. They are developed particularly in evaporite sequences due to volume changes in the rock brought about by chemical transformations of the salts. Such structures are commonly formed by the swelling of anhydrite during its hydration to gypsum.

enthalpy: [Greek *enthalpein* = to heat in < *en-* = within + *thalpein* = to heat.] n. In thermodynamics, a quantity of energy equivalent to the sum of the internal energy of a system plus the product of the volume and the pressure of the system. Thus,

H = U + PV, where U is the internal energy, P is the pressure of the system, and V is the volume. The total enthalpy of a system can only be measured when pressure is constant and cannot be measured directly; the enthalpy change of a system is measured instead, and is usually measured relative to an arbitrarily defined zero. First usage in English c. 1925.
Cited usage: In a change occurring at constant pressure all the heat added to the system goes to increase the enthalpy [1967, Condon & Odishaw, Handbook of Physics 2nd ed.].
Synonyms: heat content, total heat

entisol: [Portmanteau word from English *(rec)ent* + *i* + Latin *sol* = soil.] n. One of the 12 soil orders used in the Soil Taxonomy of the U. S. Dept. of Agriculture. It is characterized by being of recent origin, and by an absence of distinct subsurface pedogenic horizons. First usage in English c. 1970.

entropy: [German *entropie* = entropy; < Greek *en-* = in + *tropy* = transformation, turning < *tropein* = to turn; < IE *trep* = to turn.] n. In physics, at the macroscopic scale, a thermodynamic quantity that is dependent on the following variables: temperature, pressure, and composition. It ultimately reflects the degree of atomic disorder of the system, representing the energy that is unavailable for work during a particular process. In a general sense, entropy is the steady and inevitable deterioration of a system into disorder and chaos, and, by extension, society. German physicist Rudolf Clausius first proposed the term *entropie* in 1865, but J. Willard Gibbs, who articulated the laws of thermodynamics, defined the term in its modern sense. Etymologically related terms include: tropism, tropic, and atrophy.

environment: [Middle English *environen* = environment; < Old French *environ* = around < *en* = in + *viron* = circle.] n. In paleontology and biology, all those external factors which may affect an organism or community. In sedimentology, a restricted region where sediment accumulates. It is characterized by certain physical, chemical, and biological conditions, along with particular geomorphic features. First usage in English c. 1600.

Eocene: [Greek *eos* = dawn + *kainos* = new, recent.] adj. Pertaining to the second epoch of the Tertiary Period from 55.8-33.9Ma, the beginning of which is characterized by the appearance of many new mammal species, and the end by a great extinction, possibly triggered by an asteroid impact that formed Chesapeake Bay on the U.S. East Coast. Term coined by Charles Lyell in 1833. Lyell devised a scheme for stratigraphically subdividing the entire Tertiary Period. His divisions were based on the percentage of fossil molluscs found in ancient strata that are represented by creatures still alive today. The oldest Tertiary rocks contain the fewest species living today, whereas the youngest would have proportionally more. In his division of the Tertiary Lyell created the Eocene (dawn of the recent), Oligocene (little recent), Miocene (middle ages of the recent), Pliocene (nearly recent), and Newer Pliocene (which he later termed the Pleistocene - Most Recent). Paleocene was later added, first as the lower part of the Eocene, and finally as a distinct epoch.
Cited usage: His Geology also is rather Eocene as I told him [1856, Charles Darwin, Letter 17 June].

eolian: [Eponymous term after Greek *Aeolus* < *Aiolos* = god of the four winds (direct translation = earth destroyer).] adj. Pertaining to the wind, said especially of deposits of wind-transported material, such as dune sand or loess; of sedimentary structures formed by the wind, such as ripples; and of geomorphic features created by wind erosion, such as blowouts and yardangs. The god Aeolus was the divine keeper and

custodian of the four winds. He lived on one of the rocky Lipara islands close to Sicily, where he kept the winds imprisoned in caves, letting them out as gentle or stiff breezes, gales, or massive storms, based on requests by other gods. First usage in English c. 1600.
Cited usage: The subaerial deposits now accumulating in the arid portion of the United States may be divided into four classes: 1, Eolian Sands; 2, Talus Slopes; 3, Alluvial Cones; and 4, Calcareous Clays [1889, I. C. Russell, Geology Mag. July 289].
Alt. spelling: aeolian

eolith: [Greek *eos* = dawn + *lithos* = rock, stone.] n. The name given to certain fragments of chalcedony found in Late Tertiary deposits in England, France, and elsewhere, which some have claimed to be the most primitive type of human-made stone implements, but whose origin is much disputed, and now generally considered to be the product of natural nonhuman processes. First usage in English c. 1890.
Cited usage: Flints of diverse shapes which have been termed 'eoliths', or stone implements of a dawning age [1907, T. R. Holmes, Ancient Britain 26].

eon: [Greek *aion* = age, space of time; < IE *aiw* = eternity, vital force, long life.] n. On the geologic time scale, the geochronologic unit of highest rank, above era, and corresponding to the span of time represented by the equivalent chronostratigraphic unit of highest rank, the eonothem. All geologic time on Earth has been divided into four eons. Precambrian time comprises three: the Hadean, Archean, and Proterozoic Eons, whereasthe most recent Phanerozoic Eon includes the Paleozoic, Mesozoic, and Cenozoic Eras. In astronomy, an eon represents one billion years, a meaning used more in England to avoid confusion over the fact that an English "billion" refers to the quantity "trillion" in the U.S.

Cited usage: The classification has developed traditionally on a hierarchical basis with eons (e.g. Phanerozoic), eras (e.g. Mesozoic), periods (e.g. Jurassic) [1982, W. B. Harland et al., Geologic Time Scale ii. 7/2].
Alt. spelling: aeon
See also: era, period, epoch, stage, age, cryptozoic

eonothem: [Greek *aion* = age + *thema* = deposit, proposition < *tithenai* = to put, set down.] n. In the rock record, the chronostratigraphic unit of highest rank, above erathem. It corresponds to the rocks that formed during the time span of the equivalent geochronologic unit of highest rank, the eon. The Phanerozoic Eonothem includes the Paleozoic, Mesozoic, and Cenozoic Erathems.

Eotvos correction: [Eponymous term after Hungarian physicist Roland Eötvös (1848-1919).] n. In gravity measurement, a correction for the centripetal acceleration caused by horizontal (east-west) motion on the surface of the rotating earth.

Eozoic: [Greek *eos* = dawn + *zoikos* = pertaining to animals < *zoa* = plural of *zoon* = animal, living being.] adj. An archaic term for all or part of the Precambrian.

epeiric: [Greek *epeiros* = continent, mainland.] adj. In geology, pertaining to seas located anywhere on the continent, including both the continental interior and the continental shelf. Originally, the term referred only to inland seas, but the meaning broadened to include continental shelves with advance of the plate tectonic theory. First usage in English c. 1920.
Cited usage: Epeiric seas, shallow marine waters connected with the oceans but wholly within the continental platform [1925, Joly, Surface-History of Earth].
Synonyms: epicontinental
See also: epicontinental, continent

epeirogeny: [Greek *epeiros* = continent, mainland + *genesis* = origin, creation < *genesthai* = to be born, come into being; < IE *gen* = to give birth, beget.] n. One aspect of tectonic or diastrophic processes that involves primarily vertical movements of large parts of the continents forming major uplifts, plateaus, and basins. In contrast to more localized orogenies, which produce mountain chains from significant horizontal movements and compression of crustal blocks. Term coined by American geologist G. K. Gilbert in 1890.
See also: diastrophism

ephemeral: [Greek *ephemeros* = short-lived < *epi* = over, on + *hemeros* = day.] adj. In geology, pertaining particularly to lakes that are short-lived, and to streams that flow briefly in direct response to local precipitation. In biology, said of certain plants and animals, especially insects that live for only one day. In particular, referring to insects of the genus Ephemeridae, including Day-flies and May-flies. In general usage, figuratively, one who or something that has a transitory existence. First usage in English c. 1575.
Cited usage: The insects...poor ephemeral things [1875, Helps, Essays and Exercises on Benevolence].

ephemeris: [Greek *ephemeris* = diary, calendar < *ephemeros* = short-lived, lasting one day < *epi* = over + *hemera* = day.] n. An astronomical almanac giving spacial coordinates for celestial bodies at regular intervals, often daily throughout one year. First usage in English c. 1550.
Cited usage: Since 1956 the ephemeris 'second', defined as the fraction 1/31 556,925.974 of the tropical year for 1900 January 0 at 12h ET, has been adopted as the fundamental invariable unit of time by the International Committee of Weights and Measure [1966, Kaye & Laby, Tables of Physical & Chemical Constants].

epi-: [Greek *epi-* = over, upon, above, on.] Combining prefix used with various nuances of meaning, such as: a) on or upon, as in epiphyte; b) over or above, as in epicenter; c) around, as in epicarp; d) close to or near, as in epicalyx; e) beside, as in epiphenomenon; and f) after, as in epilogue. The same meaning is conveyed by the prefix *ep-*, as in eponym and ephemeral.

epicenter: [Greek *epikentros* = on the center < *epi* = over, above + *kentron* = needle, point, centerpoint in drawing a circle < *kenteo* = to prick, goad, stab.] n. The point on the earth's surface directly above the focus, or hypocenter, of an earthquake. Because the term is properly a geologic one relating to an event of potentially catastrophic dimension, the English Usage Panel approves of figurative extensions of its use primarily in dangerous, destructive, or negative contexts. For example, 82% of the panel accepts the sentence "If Rushdie were not at the terrifying epicenter of this furor, it is the sort of event he might write about." But, the panel is less fond of accepting *epicenter* when it is used to refer to the focal point of neutral or positive events. Only 62% of the panel approves of the sentence, "The indisputable epicenter of Cotina's social life is the Hotel de la Poste, located squarely in the village center." First usage in English c. 1885.

epicontinental: [Greek *epi* = on, upon + Latin *contins* = continuous < *continere* = to hold together, to contain < *con* = together + *tenere* = to hold.] adj. Pertaining to seas located anywhere on the continent, including both the continental interior and the continental shelf. First usage in English c. 1900.
Synonyms: epeiric
See also: continent, epeiric

epidote: [Greek *epidotos* = increased < *epi-* = over + *dotos* = given < *didonai* = to give.] n. A pistachio-green, yellowish-green to blackish-green

monoclinic mineral, hydroxy calcium iron aluminum silicate, $Ca_2(Al, Fe)_3Si_3O_{12}(OH)$. Epidote usually has a bladed to acicular habit, forms an isomorphous series with iron-free clinozoisite, and is grouped crystallographically with zoisite and piemontite. It is commonly associated with albite and chlorite in low-grade metamorphic rocks, and in low-temperature hydrothermal veins and pipes. Term coined by French mineralogist René-Just Haüy c. 1805, probably an allusion to the base of the rhombo-hedral prism displaying one side with increased length over the other. Another possible reason for this name stems from a quote by Haüy, "*qui a reçu un accroissement*", denoting a greater length in the long axis of the crystal as compared with that of certain allied minerals with which it had been previously confused (to give in addition).
Synonyms: pistachite

epifauna: [Greek *epi* = above, over + Latin (eponymous) *Fauna* = rural, woodland god-dess.] n. Fauna (animals) living on the surface surface of the seafloor, rather than in the sediment below the seafloor. The Roman deity Fauna was the sister of Faunus, who was later identified with the Greek god Pan.

epigenetic: [Greek *epigens* = growing after < *epi-* = after, over, above + *genesis* = origin, creation < *genesthai* = to be born, come into being; < IE *gen* = to give birth, beget.] adj. In ore deposits, referring to a mineral deposit that formed later than the enclosing host rocks, such as hydrothermal veins. In contrast to syngenetic deposits. The related term epigene is used in reference to a process and its resultant features that form at or near the Earth's surface. First usage in English c.1805. Used in contrast to hypogene. Etymologically related terms include: genesis, generate, and indigenous.
See also: hypogene, impregnated, syngenetic

epilimnion: [Greek *epi-* = above, upon + *limnion* = small pond < *limnos* = lake, pool.] n. The uppermost layer of water in a lake, below which are located the mesolimnion and hypolimnion. The epilimnion is generally warmer and more oxygen-rich than the layers below, and is characterized by more uniform mixing due to wind and wave action. First usage in English c. 1905.
Cited usage: I employ two new words in this paper. These terms are epilimnion, for the upper warm layer of water and hypolimnion, for the lower colder water [1910, E. A. Birge, Trans. Wisc. Acad. Sci].

epipedon: [Greek *epi* = above, upon + *pedon* = soil.] n. In Soil Taxonomy of the U.S. Dept. of Agriculture, a diagnostic horizon that forms at the surface, and is used in taxonomic classification. It is not synonymous with the A horizon, and may extend deep into the B horizon.

epithermal: [Greek *epi* = above, upon + *therme* = heat.] n. In ore deposits, said of hydrothermal mineral deposits formed within about 1km of the surface at temperatures generally in the range of 50°-200°C, and typically associated with volcanic rocks. The term was applied to the classification of ore deposits by American geologist Waldimar Lindgren.

epizonal: [Greek *epi-* = above, upon + *zone* = belt, girdle < *zonnai* = to gird.] n. In igneous petrology, according to the classification of American geologist Arthur F. Buddington, formulated in 1959, the uppermost depth zone where stocks and batholiths often intrude into related volcanics, and may display miarolytic or aplitic textures. In metamorphic petrology, accord-ing to the classification of German petrologist Ulrich Grubenmann in 1904, the epizone is the uppermost depth zone of metamorphism. Modern usage of the term emphasizes temperature and pressure conditions of low metamorphic grade.
See also: catazonal, mesozonal, epi-

epoch: [Latin *epocha* = measure of time; < Greek *epokh* = a point in time.] n. On the geologic time scale, the geochronologic unit shorter than a period but longer than an age. For example, the Oligocene Epoch of the tertiary period includes the Chattian and the rupelian ages. The epoch is the span of time corresponding to the equivalent chronostratigraphic unit, the series. In astronomy, the point of time at which any celestial phenomenon takes place, such as an eclipse, supernova, or orbiting body passing a specific meridian. The International Geological Congress of 1881 implemented the use of the terms era, period, epoch, and age to denote successively smaller divisions of time.

epsomite: [Toponymous term after its discovery locality at Epsom, England.] n. A colorless to white, water-soluble, bitter-tasting orthorhombic mineral, magnesium sulfate hydrate, $MgSO_4 \cdot 7H_2O$. It is a natural form of commercial Epsom salts and occurs commonly as encrustations and efflorescences on outcrops of sedimentary rocks and in caverns. It was noted that during the drought of 1618, thirsty cattle would not drink water from the spring at Epsom in Surrey, England. It was discovered that these waters were rich in Epsom salts, or magnesium sulphate ($MgSO_4$). First usage in English c. 1810.

equant: [Latin *aequalis* = equal < *aequus* = even, plain.] adj. Said of a crystal having the same ore nearly the same dimension in all directions.

equator: [Latin *aequator* = equalizer *aequalis* = equal < *aequus* = even, plain.] n. The imaginary great circle around the Earth's surface, equidistant from the poles and perpendicular to the Earth's axis of rotation. It divides the Earth into the northern and southern hemispheres. First usage in English c. 1375.
Cited usage: When the Sunne cometh upon the Equator, then the daies and nights are of one length through the whole worlde [1594, John Davis,The Seaman's Secrets II].

equi-: [Latin *aequalis* = equal < *aequus* = even, plain.] Combining prefix meaning equal. *Equi-* was originally attached only to words of Latin origin, such as equiangular. It now occasionally forms Latin-Greek hybrids, such as equibalance. The majority of words so formed are parasynthetic adjectives (compound words that were previously hyphenated).

equilibrium: [Latin *aequalis* = equal < *aequus* = even, plain + *libra* = balance.] n. The condition of rest or balance as a result of equal action between opposing forces. In chemistry and experimental petrology, the state of a chemical system in which the phases do not undergo any change with passage of time, provided the initial phases are produced when the same conditions or pressure, temperature, and composition are reached by the reverse reaction. In geomorphology, a balance between form and process in which the form of some landscape feature remains essentially unchanged with continued operation of a geologic process (e.g., isostatic balance between uplift and erosion maintaining the elevation of a plateau, or stream erosion producing a convex upward, longitudinal, graded stream profile). First usage in English c. 1605.

equinox: [Latin *aequinoctium* = time of equal day and night < *aequalis* = equal < *aequus* = even, plain + *nox* = night; < IE *nekw* = night.] n. Either of the two times during a year when the sun crosses the Earth's equator and the length of day and night are approximately equal, e.g., the spring, or vernal, equinox and the fall, or autumnal, equinox. IE derivatives include night and nocturnal.
Cited usage: His vice is to his virtue, a just Equinox, The one as long as th' other [1604, Shakespeare, Othello II. iii. 129].
See also: ecliptic

era: [Latin *aera* = counters < *aes* = brass or money; < IE *ayes* = metal.] n. On the geologic time scale, the geochronologic unit of second highest rank, shorter than an eon but longer than a period. The Mesozoic Era includes the Triassic, Jurassic, and Cretaceous periods. The era is the span of time corresponding to the equivalent chronostratigraphic unit, the erathem. In ancient Rome, *aera* were metal counters used as money (*aes*).

erathem: [Latin *aera* = counters + *thema* = deposit.] n. In the rock record, the chronostratigraphic unit of second highest rank, below eonothem but above system. The erathem is the assemblage of rocks corresponding to the equivalent geochronologic unit, the era. The Mesozoic Erathem includes the Triassic, Jurassic, and Cretaceous systems.

erg: [Toponymous term after the Grand Erg Oriental, a region of the Sahara Desert in Algeria and Tunisia; < Arabic *irj*, of Berber origin.] n. In geomorphology, an extensive tract of sandy desert or a sand sea. This feature is common in the Nefud Desert of Northern Saudi Arabia. First usage in English c. 1870.
Cited usage: Two fairly distinct types of sand desert can be distinguished: the nebkas or small areas of sand, and the erg (pl. areg) or large dune system. The erg consists of drinn steppe with a drought-resisting grass as the dominant species [1943, Algeria Naval Intelligence Div. I. ii. 155].
Alt. spelling: ergh, areg (pl.)
Synonyms: koum (Asia)

erg: [Greek *ergon* = work.] n. In physics, the standard unit of energy or work in the C.G.S. (cm/gm/sec) system. It is equal to the work done by a force of one dyne acting over a distance of one centimeter (written: gm-cm^2/sec^2). First usage in English c. 1870.

erosion: [Latin *erosus* < *erodere* = to gnaw off < *e* = out + *rodere* = to gnaw.] n. The processes of removal or wearing away of materials of the Earth's surface by natural agents, such as water, ice, and wind. The term is not synonymous with weathering and transportation. Etymologically related terms include rodent and corrode.
Cited usage: Erosion inequalities, once commenced, tend to increase [1879, Le Conte, Elem. Geol. 251].

erratic: [Latin *erraticus* < *errare* = to wander, to err; < IE *ers* = to be in motion.] n. A rock fragment that has been transported by glacial ice (or floating ice) for considerable distance from its source outcrop and usually deposited onto bedrock of different lithology. Size may range from pebbles to huge boulders. Term coined in geology in 1779 by Swiss geologist, physicist, and meteorologist Horace Benedict de Saussure, for boulders of granite scattered over limestone bedrock in the Swiss Alps (even though he believed that the boulders were water transported). First use in English c. 1350.
Cited usage: To the unskilled eye Russia presents only monotonous undulations, chiefly covered by mud, sand, and erratic blocks [1849, Murchison, Siluria i].

eruption: [Latin *erumpere* = to burst out of < *e* = out + *rumpere* = to burst forth.] n. The ejection onto the Earth's surface of lava, pyroclastic material, and gas from a central vent or a fissure. First usage in English c. 1400.

erythrite: [Greek *erythros* = red, reddish.] n. A red or pink monoclinic mineral, cobalt arsenate hydrate, $Co_3(AsO_4)_2 \cdot 8H_2O$. Erythrite forms an isomorphous series with nickel-bearing annabergite and magnesium-bearing hörnesite. It most commonly forms as an earthy coating on cobalt ores as a result of weathering in the oxidized zone of hydrothermal veins.

Etymologically related terms include erythrophobia and erythrocyte. First usage in English c. 1840.
Synonyms: cobalt bloom, red cobalt, cobalt ochre

escarpment: [French *escarper* = to cut steeply < Italian *scarpa* = a slope + Latin *mentum* = suffix forming nouns from verbs and adjectives.] n. A long, generally continuous cliff or steep slope, facing in one direction and formed by differential erosion or by faulting. It breaks the general continuity of the land surface by separating level to gently sloping land surfaces. Often used synonymously with the term scarp, but escarpment more often implies a cliff formed by differential erosion, and marks the outcrop of a resistant bed occurring with a series of softer strata. First usage in English c. 1800.
Alt. spelling: scarp
Synonyms: scarp, cuesta (Great Britain)

esker: [Irish *eiscir* = ridge.] n. A long, narrow, sinuous, and steep-sided ridge composed of irregularly stratified sand and gravel deposited by a stream flowing in an ice tunnel within or under a glacial ice sheet, and becoming exposed after glacial retreat. They range in length from less than 100m to more than 500km. First usage in English c. 1850.
Cited usage: Ridges, known in Scotland as kames, in Ireland as eskers, and in Scandinavia as osar [1882, Geikie, Text of Geology].
Alt. spelling: eschar, eskar, eiscir
Synonyms: os, serpent kame

esplanade: [Latin *explanare* = to level < *ex* = out + *planus* = level, plain.] n. A term used in the U.S. Southwest for a broad bench or terrace bordering a canyon, especially in a plateau region, and often providing an overlook or vista. An etymologically related term is explain. First usage in English c.1680.

estuary: [Latin *aestuarium* = channel, creek, inlet < *aestus* = tide, surge, boiling, heat + *-arium* = suffix meaning a place for.] n. The seaward end of a river valley where freshwater comes into contact with seawater, and tidal effects are apparent. Estuaries can be classified by origin, as in: a) drowned river mouths, b) fjords, c) bar-built, and d) tectonically formed; or by pattern of seawater dilution, as in: a) stratified, b) partially mixed, c) well mixed, and d) wedge shaped. First usage in English c.1535.
Cited usage: Estuaries, a term which we confine to inlets entered both by rivers and tides of the sea [1830, Lyell, Principles of Geology I. 265].

eugeocline: [Greek *eus* = good, well + *ge* = earth + *klinos* = to bend, lean.] n. An accumulation of thick deposits of immature deepwater sediment and volcanic materials. Eugeoclines show evidence of crustal instability in turbidity flows and submarine landslides. They are deposited in trench or continental rise environments.

eugeosyncline: [Greek *eus* = good, well + *ge* = Earth + *syn* = similar, together, alike + *klinein* = to bend.] n. A narrow, downwarped basin in which volcanic materials are associated with clastic sediments. The volcanic part of an orthogeosyncline, away from the craton. First usage in English c. 1940.
Cited usage: The western, volcano-sedimentary 'eugeosyncline' of the North American Cordillera is a collage of terrains of varied origins that apparently became attached to the western continental margin of the continent at different times [1980, Nature 29 May 289].
See also: geosyncline

euhedral: [Greek *eus* = good, well + *hedra* = seat, base.] adj. Said of a mineral grain that is completely bounded by planar crystal faces characteristic of the mineral. Euhedral crystals have grown freely without interference by

adjacent grains during crystallization from a melt or fluid phase, or during recrystallization during metamorphism. The shape of the growing crystal is thus controlled by its own internal crystallographic structure. Term coined by petrologist Whitman Cross in 1906.
Cited usage: Euhedral, well faced, completely bounded by crystal planes, automorphic, idiomorphic [1906, P. Iddings, Rock Minerals ii. 64].
Synonyms: idiomorphic, automorphic
See also: anhedral, idiomorphic

eukaryote: [Greek *eus* = good, well + *karyon* = kernal, nut.] n. Any one of a major group of organisms having as its fundamental structural unit a cell that contains a membrane-bounded nucleus, and displays complex protoplasmic organization. First usage in English c. 1925.

eurypterid: [Greek *eurys* = broad, wide + *pteron* = wing.] n. One of a group of extinct segmented aquatic arthropods belonging to the subphylum Crustacea and the subclass Eurypteridae, which were abundant during the Paleozoic Era. The name alludes to a pair of broad, winglike swimming appendages that are the hindmost of a series of appendages attached to the cephalothorax. First usage in English c. 1870.
Cited usage: Contemporaneous with the Trilobites were the Eurypterids, which vary from one foot to five or six feet in length [1871, Hartwig, Subterr. W. ii. 125].

eustasy: [German *eustatisch* = eustasy; < Greek *eus* = well, good + *stasis* = standing, stoppage < *stare* = to stand.] n. A uniform worldwide change in sea level that affects the world ocean. Eustasy is the anglicized form first used c. 1945, being a back-formation of the original German term *eustatisch*, coined by Austrian geologist Eduard Suess c. 1900.
Cited usage: Such movements of the sea-level are called eustatic, and the phenomenon, glacial eustasy [1946, F. E. Zeuner, Dating the Past iii. 47].

eutaxitic: [Greek *eus* = good, well + *taxis* = arrangement.] adj. In igneous petrology, pertaining to the banded texture of volcanic rocks , such as welded tuffs or ignimbrites in which a planar fabric is defined by flattened and elongate pumice clasts (fiamme), usually 10-40 mm in length and set in a lighter matrix of flattened and sintered, ash-size, glassy shards. Term coined by German petrologists R. von Fritsch and W. Reiss in 1868.

eutectic: [Greek *eutektos* = easily melted < *eus* = good, well + *tektos* = melted < *tekein* = to melt.] adj. Said of a chemical system consisting of two or more solid phases and a liquid phase, all coexisting at a single (eutectic) temperature, the minimum melting temperature for the assemblage of solids. At this temperature, further addition of heat will change the proportion of solids to liquids, but will not change the temperature of the mixture until one of the constituents is entirely melted. Term coined by F. Guthrie in 1884, who applied it to the crystallization of granites.
See also: anatexis, peritectic

eutrophic: [Greek *eutrophos* = well-nourished < *eus* = good, well + *trophy* = food, nourishment < *trephein* = to nourish.] adj. Said of a lake or pond having waters rich in dissolved nutrients that promote a proliferation of plant life, especially algae. The effect is often seasonal, and the decaying organic matter reduces the dissolved oxygen content in the hypolimnion, resulting in depletion or extermination of aerobic species. First usage in English c. 1880.

euxinic: [Toponymous term after Latin and Greek *Pontus Euxinus*, the Black Sea (also called Euxine Sea).] n. Pertaining to the environment of a restricted basin where lack of circulation causes stagnant or anaerobic conditions and creation of toxic bottom waters. The characteristic sediments are black, organic-rich, and hydrogen- or sulfide-bearing, much as are found in the Black Sea and some fjords. In petrology, said of a rock facies that includes black shales and graphitic sediments. *Pontus Euxinus*, meaning 'Hospitable Sea' was the name given to the Black Sea by ancient Greek navigators. The Turks later named it Karadeniz, meaning Black Sea, due to its unusual dark color in comparison to the Mediterranean. *Cited usage*: Analyses show the presence in the Black Sea muds of a relatively high proportion of organic matter and various investigators have seen in this fact a possible source for the formation of materials analogous with the black shales of some geological formations. Van der Gracht proposes the term euxinic facies (Pontus Euxinus = Black Sea) for deposits of this type [1937, Bull. Amer. Assoc. Petroleum Geologists XXI. 1115].

evaporite: [Latin *evaporare* = to disperse in vapor < Greek *e* = out + *vaporis* = steam.] n. A chemically precipitated, nonclastic sedimentary rock that results from the evaporation of seawater or saline lakes. They are composed primarily of relatively soluble minerals, such as halite, nitratite, borax, gypsum, trona, dolomite. First usage in English c. 1920.
Cited usage: Salt deposits (evaporites), such as are left behind when salt lakes dry up, or when enclosed bodies of sea water are strongly evaporated [1965, A. Holmes, Princ. Physical Geol. 2nd ed. iv. 78].

everglade: [Toponymous term after the Everglades of the U.S. Southeast in Florida; < English *ever* = interminable + *glade* = open space in a forest.] n. A term used especially in the southern U.S. for a large tract of low marshy land, characterized by many branching waterways, mostly under water and covered in places with tall grass, the Florida Everglades, for example. Prior to the 18th century, the term glade had the distinct meaning of a light or sunny place, alluding to its possible derivation from the Old English *glad*, meaning sunny, joyous, bright. First usage in English c. 1820.

Cited usage: The back country presents a singular alternation of savannas, hammocks, lagoons, and grass ponds, called altogether the Everglades [1837, J. L. Williams, Florida 13].

evolution: [Latin *evolutio* = an unrolling, opening (as a book or scroll) < *evolvere* = to roll out, unfold < *e* = out + *volvere* = to roll.] n. In biology and paleontology, the gradual and permanent change in the genetic composition of organisms of successive generations or populations as a result of natural selection. The process results in the latest members of the succession differing significantly from the earliest, and ultimately developing into a new species. First usage in English c. 1620.

ex-: [Latin *ex* = out of, from, beyond.] Combining prefix meaning out of or from, as in extract, exfoliate, and exhale; also sometimes meaning upward, as in extol, or thoroughly, as in excruciate.

exfoliation: [Latin *exfoliare* = to strip off leaves < *ex* = out of, from + *folium* = leaf.] n. The process by which concentric sheets or plates of rock are successively separated from the exposed surface of a rock mass. At the largest scale, exfoliation joints or sheet joints are formed by the release of confining pressure on a once deeply buried rock of isotropic texture, such as granite or sandstone. As the rock is brought nearer to the surface by uplift and erosion, such joints bound exfoliation plates or sheets ranging in thickness from hundreds of meters down to tens of centimeters. This results in dome-shaped hills, such as Half Dome in Yosemite National Park and Uluru (Ayers Rock) in Australia. At the smaller scale of individual boulders, exfoliation results from the expansion of minerals exposed to near-surface chemical weathering. This creates spalling of concentric shells of rock ranging in thickness from several meters down to several

millimeters. The unweathered core stones formed by this process are spheroidally shaped rock masses. First usage in English c. 1675.
Synonyms: spalling, sheet jointing, sheeting, scaling, desquamation

exo-: [Greek *exo* = outside, out of.] Combining prefix meaning outside, outer, or external.

exogenous: [Greek *exo*= outside + *genesis* = origin, creation < *genesthai* = to be born, come into being; < IE *gen* = to give birth, beget.] adj. In volcanology, referring to a lava dome that has grown chiefly by extrusion of viscous flows onto the surface of the dome, or by projection of spines of lava through the outer crust of the dome. In contrast to endogenous dome. First usage in English c. 1825.
See also: endogenous

exoskarn: [Greek *exo* = outside + Swedish *skarn* = rubbish.] n. During contact metasomatism, a skarn formed by replacement of the country rock by the action of hydrothermal solutions. Exoskarns usually form in limestone or dolomite and surround the intrusion. In contrast to endoskarn.
See also: endoskarn

exoskeleton: [Greek *exo* = outside + *skeletos* = dried up, mummy.] n. An external covering that lies outside the body tissues of an animal, and forms a supporting and protective framework, such as in the carapace of crustaceans and trilobites, the integument of insects, and the shells of bivalves. First usage in English c. 1845.
See also: endoskeleton, carapace

exsolution: [Latin *ex* = out of, from + *solutus* < *solvere* = to loosen, dissolve.] n. In mineralogy, the process whereby an initially homogeneous crystalline phase separates at a particular temperature into two distinct crystalline phases of differing chemical composition. The two new phases form members of an isomorphous or solid solution series (e.g., the separation on cooling in the solid state of one homogeneous alkali feldspar into exsolution lamellae of orthoclase and albite). In chemistry and geochemistry, the process whereby a single homogeneous fluid phase, such as a gas or liquid, separates into two fluid phases of distinctly different composition (e.g., the exsolution of carbon dioxide gas from an aqueous liquid upon decrease in confining pressure).

externides: [Latin *externus* = outward.] n. The outer part of an orogenic belt nearest to the craton or foreland, commonly the site of a miogeosyncline during early stages of mountain building. Externides constitute this part of an orogenic belt that is later subjected to marginal deformation, such as folding and thrusting. Term coined by Austrian geologist Leopold Kober.

extinct: [Latin *ex* = out of, from + *stinguere* = to quench.] adj. In paleontology and biology, said of a species or higher taxon that has totally disappeared so that it no longer exists anywhere on Earth. In volcanology, said of a volcano that is no longer active and is expected to have no future eruptions. First usage in English c. 1425.

extinction: [Latin *ex* = out of, from + *stinguere* = to quench.] n. In evolution, the total and permanent disappearance of a species or higher taxon (as opposed to regional extermination). In optical mineralogy, the position in which a thin section of a birefringent mineral gives maximum darkness when viewed under crossed nicols with a polarizing microscope. First usage in English c. 1400.

extrusive: [Latin *ex* = out + *trudere* = to thrust.] adj. Said of an igneous rock that has erupted onto the surface of the Earth, including lava flows and pyroclastic materials. The term extrusive is not synonymous with volcanic, because some volcanic rocks solidify at shallow depth below the Earth's surface, and are therefore intrusive.
Synonyms: effusive
See also: igneous, intrusive, volcano, sill

exudation: [Latin *exsudare* < *ex* = out + *sudare* = to sweat.] adj. In glacial geology, used as an adjective pertaining to a spoon-shaped basin or depression on the ice surface at the head of an outlet glacier. First usage in English c. 1610.

fabric: [Latin *fabrica* = workshop, craft < *faber* = workman in metal, stone, or wood.] n. In structural geology, the complete spatial and geometrical arrangement of all components making up a deformed rock, encompassing texture, structure, and preferred orientation. In sedimentology, the orientation of discrete particles and cement that compose a sedimentary rock. In pedology, the physical nature of a soil determined by the spatial arrangement of particles and voids. First usage in English c. 1480.

facet: [French *facette* = little face; < Latin *facia* or *facies* = form, figure, appearance.] n. In gemology, any of the plane surfaces on a gemstone crafted by a lapidary. In sediment morphology, a nearly planar surface on one or more sides of a rock fragment produced by wind abrasion or glacial abrasion. In geomorphology, any planar surface intersecting the general slope of the land, produced by faulting or erosion, such as a triangular facet. The Latin *facies* may derive from either *facere*, meaning to make or *fa*, meaning to appear, shine. First usage in English c. 1620.
See also: facies

facies: [Latin *facies* = form, aspect, appearance.] n. In stratigraphy, the appearance, attributes, and characteristics of a rock unit, usually reflecting the conditions of its origin to distinguish the unit from adjacent and associated units. It can be considered as a mappable, areally restricted sedimentary unit, which is in the same lithostratigraphic body as other units, but which differs from other units in fossil content or lithology, each representing a different sedimentary facies. In petrology and ore deposits, different facies are represented by by distinctive rock types, broadly corresponding to different environments, or modes of origin, such as red-bed facies or black-shale facies. In paleoecology, the environment in which a rock was formed, such as eolian facies, volcanic facies, fluvial facies, proximal facies, or distal facies. In metamorphic petrology, rocks formed within certain temperature-pressure conditions, such as greenschist facies and granulite facies. In tectonics, rocks broadlydefined based on different tectono-stratigraphic environments, such as shelf facies, geosynclinal facies, and foreland facies. Term first used ingeology by Amanz Gressly in 1838, but since thattime the term has been widely applied and should be qualified when used. The Latin *facies* may derive from either *facere*, meaning to make, or *fa*, meaning to appear, shine. The first usage in English c. 1375.
Cited usage: Above all, there are two major facts which define everywhere the sum of the variables which I call facies or the aspects of a stratigraphic unit: one is that within a stratigraphic unit the occurrence of a specific lithology necessarily also requires the occurrence of a specific paleontological association; and the other is that a given paleontological association rigorously excludes those genera and species of fossils which are frequent in other facies [1838, Amanz Gressly, Observations geologiques sur le Jura Soleurois Ges. Ges. Naturw. 2: 1-112].

famatinite: [Toponymous term after *Sierra de Famatina*, the discovery locality in El Riojo Province, Argentina.] n. A gray to copper-red tetragonal mineral, copper antimony sulfide (Cu_3SbS_4). It is the high-temperature dimorph (above 320°C) of luzonite and shows extensive solid solution toward arsenic-bearing enargite.

family: [Latin *familia* = household < *famulus* = servant.] n. In biology and paleontology, a category or taxon in the hierarchy of biological classification developed by Carl Linnaeus that is ranked below order but above genus. In zoology, the name of the family characteristically ends in '*-idae*' (e.g., Hominidae), whereas in botany the ending is '*-aceae*' (e.g., Roseaceae). First use c. 1375.
Cited usage: An order of plants is composed of one or more families. The family usually represents a more natural unit than any of the higher categories [1951, G. H. M. Lawrence, Taxon. Vascular Plants iv. 46].

fan: [Old English *fann* < Latin *vannus* = fan, winnowing basket.] n. In geomorphology, a gently sloping, fan-shaped mass of alluvium, commonly deposited where there is a marked decrease in gradient of a stream. Alluvial fans generally originate from a relatively constricted upstream source and broaden out in large plain or valley lowlands, taking the shape of a flattened cone. In marine geology, a submarine fan is a cone-shaped deposit located seaward of river mouths and submarine canyons, and usually having a terrestrial sediment source. First usage in English before 900.

fanglomerate: [A portmanteau word from English *fan* + *(con)glomerate*.] n. A rock consisting of poorly sorted rock fragments deposited in an alluvial fan, variously cemented, and often consolidated into a solid mass. Term coined by U.C. Berkeley geologist Andrew C. Lawson in 1912.
Cited usage: Fanglomerate, a detrital rock at Battle Mountain, Nevada [1912, A. C. Lawson, Bulletin of Geol. Soc. Amer. XXIII. 72].

fascicular: [Latin *fasciculus* = little bundle < *fascis* = bundle, group.] adj. In mineralogy, said of an aggregate of acicular crystals. In biology and paleontology, said of a columella of cylindrical scleractinian corallites twisted together in ribbon-like shapes, or bound together by connecting tubules. The Italian *Fascisti* movement founded by Benito Mussolini in 1919 took their name and emblem from the same Latin root. The movement's emblem was the Latin *fasces*, meaning a bundle of sticks or rods bound around a projecting axe-head that was carried before an ancient Roman magistrate by an attendant as a symbol of authority and power.

Fata Morgana: [Eponymous term after Fata Morgana, legendary sister of King Arthur who had the magical power to create castles out of thin air..] n. A type of meteorological and optical phenomena in the general category called towering. It results in visions of large proportion, is typical in coastal areas, and is probably responsible for ancient mariners' reports of land where none exists.

fathom: [Old English *faethm* = the two arms outstretched, embrace, a measure of about 6 feet < Greek *petalos* = spreading, broad.] n. A unit of length equal to six feet, used mostly in nautical measure of depth, and found on many marine bathymetric maps. Its origin can be traced to the length covered by outstretched arms. The IE term *pet*, meaning to spread, is a common root for the Latin *patere*, meaning to be open; the Greek *petalos*, meaning spreading broad, is from meaning leaf, as well as the Old English *faaethm*. IE derivatives include: pace, expand, passport, spawn, and petasos (Greek broad-brimmed hat). First usage in English before 900.

fault: [Italian *falta* = fault; < Latin *fallita* = a failing < *fallere* = to fail, deceive, be wrong.] n. A fracture in the Earth along which there has been movement or displacement of the adjacent surfaces parallel to the plane of fracture. First usage in English c. 1275.

fauna: [Eponymous term after *Fauna* = female woodland deity in Roman mythology, sister of Faunus, later identified with the Greek god Pan.] n. A collective term applied to the entire living animal population or fossil assemblage of any particular region, environment, formation, or time span. Term made popular as a synonym for the animal kingdom by Carolus Linnaeus in the title of his 1746 work *Fauna Suecica* (*Animals of Sweden*), a companion volume to his 1745 *Flora Suecica*. Fauna is sometimes incorrectly used to include plant assemblages.

fayalite: [Toponymous name after *Fayal Island*, the discovery locality in the Azores.] n. A brown to black orthorhombic mineral of the olivine group, iron silicate (Fe_2SiO_4). Fayalite is isomorphous with forsterite (Mg_2SiO_4) and occurs chiefly in mafic and ultramafic igneous rocks. Term coined by J. F. Gmelin in 1840 after the discovery locality Fayal Island, comprising a single basaltic volcano with lavas containing olivine.
Alt. spelling: faialite
Synonyms: iron-olivine, iron-chrysolite, knebelite
See also: forsterite, olivine

fecal: [Latin *faeces* = plural of *faex* = dregs.] n. In paleontology and biology, usually referring to fecal pellets. This form of animal excrement is primarily from marine invertebrates, and is found intact in contemporary unconsolidated sediments or fossilized in sedimentary rocks. Fecal pellets are generally ovoid in form, less than mm in size, and smaller than coprolites, another type of excretory remain. First usage in English c. 1540.
Alt. spelling: faecal
See also: coprolite

feldspar: [German *feldspath* < *feld* = field + *spath* = spar, crystal.] n. The most widespread of any mineral group, making up over one-half of the Earth's crust. The general formula is (K, Na, Ca, Ba, NH_4)(Al, Si)$_4O_8$. Feldspar crystals are white to clear to translucent unless colored by impurities, such as iron, rubidium, or inclusions of hematite or ilmenite. Feldspars occur as components of all rock families. On chemical decomposition they yield a large proportion of the clay in soils. The erroneous spelling *felspar*, often used in Britain, was introduced by Kirwan based on the mistaken derivation from German *vels*, meaning rock, cliff. Spar is a general term for a number of minerals more or less lustrous in appearance and having pronounced cleavage. First usage in English c. 1755.
Cited usage: If the characters of this field-spar are accurately examined [1772, Cronstedt's Mineralogy App. 8].
Alt. spelling: felspar, fieldspar

feldspathoid: [German *feldspath* = feldspar + Greek *oeidos* = form.] n. A group of rock-forming minerals similar in chemical composition to the feldspars, but with lower silica content. The feldspathoids take the place of feldspars in igneous rocks that are undersaturated in silica, and therefore lacking in quartz. Examples include: leucite, nepheline, and sodalite. First usage in English c. 1895.

fell: [Old Norse *fjall* = hill, mountain.] n. A term used in Scotland and England for a bare, uncultivated hillside or mountain. First usage in English c.1325.
Synonyms: fell-field

felsenmeer: [German *felsen* = rock < *vels* = rock, cliff + *meer* = sea.] n. An expanse of angular, frost-heaved blocks developed over solid or weathered bedrock, alluvium, or colluvium. Felsenmeer have no apparent upslope source, such as a cliff or headland. They typically develop in gently sloping or flat terrain in polar regions or alpine climates on mountain slopes above the treeline. The etymology 'sea of rocks' is in allusion to the expanse of rocks all on the same level mimicking a rippling surface.

Cited usage: It is common to find at high altitudes accumulations of riven rocks known as felsenmeere [1954, W. D. Thornbury, Princ. Geomorphol. v. 6].
Alt. spelling: felsenmeere
Synonyms: block field, stone sea, blockmeer
Similar terms: stentorg
See also: stentorg

felsic: [Composite mnemonic term from English *fe(ldspar)* + *l(enad)* = feldspathoid + *si(ilica)* + *c*.] adj. Pertaining to igneous rocks containing abundant light-colored silicate minerals, such as feldspars, feldspathoids, quartz, and muscovite. Also, pertaining or to these minerals as a group. Common felsic rocks include granite and rhyolite. In contrast to mafic. Term coined by W. Cross, J. P. Iddings, L. V. Pirsson H.S. Washington in 1912.
Cited usage: We suggest the term felsic for the group of modal feldspars, feldspathoids, and quartz [1912, W. Cross et al, Jrnl. Geol. XX. 561].
See also: femic, mafic, lenad

femic: [Composite mnemonic term from English *fe(rric)* + *m(agnesium)* + *ic*.] adj. Pertaining to igneous rocks containing one or more of the dark-colored iron-, magnesium-, or calcium-rich minerals as major components. Also, pertaining to these minerals as a group. Term coined by W. Cross, J. P. Iddings, L. V. Pirsson H. S. Washington in 1902.
Cited usage: To express concisely the two groups of standard minerals and their chemical characters in part, the words sal and fem have been adopted. Fem indicates Group II, since its minerals are dominantly ferromagnesian. As adjectives to express these ideas the words salic and femic will be used [1902, W. Cross et al, Jrnl. Geol. X. 573].
See also: felsic, mafic

fen: [Old English *fenn* < Dutch *veen* = water-meadow, bog; < Old Norse *fen* = quagmire; < Gothic *fani* = mud.] n. A low-lying, boggy, water-saturated tract of land that is either wholly or partially submerged. Fen are characterized by reeds and decaying vegetation that may, under anoxic conditions, transform into peat. First usage in English before 900 CE.
See also: marsh, muskeg, bog

fenestra: [Latin *fenestra* = window; < Greek *phanein* = to show.] n. In paleontology and biology, a small opening in an invertebrate, such as an open pore space in a reticulate bryozoan. In sedimentology, a term used for shrinkage pore, prism crack, or microsheet crack, being a primary or penecontemporaneous gap in rock framework, larger than grain-supported interstices. Term coined in the sedimentary sense by G. E. Tebbutt, C. Conley, and D. W. Boyd in 1965. First usage in English c. 1825. Not to be confused with fenster.
See also: fenster

fenster: [German *fenster* = window; < Latin *fenestra* = window; < Greek *phanein* = to show.] n. In structural geology, an eroded area of a thrust sheet that exposes the younger rocks below. First usage in English c. 1920.
Cited usage: Occurrences, where the erosion of the older, overlying rocks has exposed the younger or underlying rocks, have generally been called fensters [1925, Bull. Virginia Geol. Surv. xxv. 45].

ferberite: [Eponymous term after German mineralogist Moritz Rudolph Ferber (1805-1875).] n. A black, monoclinic mineral, iron tungstate, $FeWO_4$. It is a member of the wolframite series and is isomorphous with manganese-bearing hübnerite. It occurs most commonly in pegmatite and high-temperature hydrothermal veins. Mineral discovered and named by Johan Friedrich August Breithaupt in 1863.

ferric: [Latin *ferrum* = iron + *-ic* = of the nature of.] adj. Pertaining to, or containing iron, especially iron in the trivalent state. First usage in English c. 1795.

ferrimolybdite: [Portmanteau word from *ferri(c)* + *molybd(enum)*.] n. A yellowish, orthorhombic mineral, iron molybdate hydrate, $Fe_2(MoO_4)_3 \cdot 8H_2O$. It is formed by oxidation of molybdenite and occurs usually as an earthy powder or encrustation, and sometimes as raditating fibrous crystals.
Synonyms: molybdic ochre

ferrosilite: [Portmanteau word from *ferro(us)* + *si(lica)*.] n. A black orthorhombic mineral in the orthopyroxene group, iron silicate ($FeSiO_3$). It is dimorphic with clinoferrosilite and forms an isomorphous series with enstatite. It is most commonly found in metamorphosed iron formations, or in tactite.
Synonyms: iron hypersthene

festoon: [French *feston* < Italian *festone* = decoration for a feast; < Latin *festa* = feast, festival.] adj. In stratigraphy, pertaining to a type of trough cross-bedding consisting of elongate scooplike structures that are filled by sets of thin laminae that crosscut each other, resulting in a curved or looped appearance. In volcanology, pertaining to pahoehoe flows displaying a ropy undulating surface that has been dragged into looping patterns. In geomorphology, used as a noun for the looping downfolded masses of soil that have moved by frost action during the slow viscous downslope flow of water-saturated surficial material (solifluction). The term relates to the shape of a string of garlands suspended in loops or curves between points. First usage in English c. 1675.

fetch: [Old English *feccean* < *fetian* = to fetch, bring, to marry; < Old Norse *feta* = to find one's way; < IE *ped* = foot.] n. In oceanography, the distance of open water over which the wind blows, generally in one direction. Along with wind intensity and duration, fetch is a controlling factor in building a wave system that, with no limiting conditions, can evolve into what is called a fully developed sea. First usage in English before 1000 CE.
Cited usage: In the Mediterranean, where the length of fetch is restricted, the highest waves reported are 4 to 5 m. [1950, P. H. Kuehnen, Marine Geol. i. 76].
Synonyms: fetch length, generating area

fiducial: [Latin *fiducia* = trust, confidence < *fidere* = to trust < *Fides* = Roman goddess of faith and trust.] adj. Said of an arbitrary line or point in a surveying, astronomical, or seismographic instrument, used as an index and assumed as a fixed basis of reference for comparison. With reference to time, said of a mark made on a record corresponding to an arbitrary reference or synchronization time.
Cited usage: The Line Fiduciall, because from this line proceeds the beginning of the degrees in the Circle [1644, NYE Gunnery].

filiform: [Latin *filare* = to spin < *filum* = thread + *forma* = form.] adj. In mineralogy, said of a habit of minerals that form exceedingly slender hairlike or threadlike crystals. In volcanology, referring to the form of threadlike lapilli known as Pele's hair. Etymologically related terms include: filament, fillet, defile, and profile. First usage in English c. 1755.

fiorite: [Italian *fiorito* = flowery < *fiorire* = to flower.] n. The white to grayish, low density, porous, opaline variety of silica forming a hydrothermal encrustation deposited from the waters of geysers and hot springs, often with a spiny habit. The type locality is Mount Santa Fiora, in Tuscany, Italy.
Synonyms: siliceous sinter, geyserite

firn: [German *firn* = old, last year's.] n. A granular material that is transitional between snow and glacial ice. Literally meaning last year's snow, firn is older and denser than snow, but has not yet transformed into glacial ice. Snow becomes firn in response to alternate thawing and

freezing during one summer melt season. Firn becomes glacial ice when its permeability to liquid water decreases to zero. The German root is cognate with Old English *fyrn*, meaning former, ancient. First usage in English c.1850.
Cited usage: The 'firn', or consolidated snow of the Alpine glaciers [1853, Elisha Kent Kane, U.S. Grinnell Expedition in Search of Sir John Franklin viii. 61].
Alt. spelling: firne
Synonyms: névé, firn snow

firth: [Scottish *firth* = lowlands.] n. A long narrow arm of the sea, especially in Scotland, variously being the seaward end of an estuary (e.g., Firth of Forth), a fjord (e.g., Firth of Lorne), or a strait (e.g., Pentland Firth). Term introduced into English from Scottish c. 1600.
Cited usage: A neck of sea possessing all the appearance of a navigable firth. [1839, W. Chambers, Tour of Holland].
Synonyms: frith

fissile: [Latin *fissilis* < *fissus* = cloven, cleft, divided < *findere* = to split, cleave.] adj. Said of a rock that is easily split along closely spaced and roughly planar surfaces that are parallel to bedding planes when found in shale, and parallel to foliation when found in phyllite or schist. In sedimentary rocks, an important distinction is that shale is fissile, but mudstone is nonfissile. Not to be confused with cleavage, which is the tendency of certain rocks and minerals to fracture along planes of weakness, not necessarily parallel to bedding planes. First usage in English c. 1660.

fissure: [Latin *fissura* = fissure < *fissus* = cloven, cleft, divided < *findere* = to split, cleave.] n. In geology, a fracture in rock along which there is a distinct separation and a narrow opening. In glacial geology, a fissure in a glacier synonymous with crevasse. In ore deposits, used as an adjective, as in fissure vein indicating mineral deposition that is open-space filling, as opposed to host-

rock replacement. In volcanology, used as an adjective, as in fissure eruption where lava erupts from an elongate fracture rather than a central vent. First usage in English c. 1400.

fix: [Latin *fixus* = fixed < *figere* = to fasten.] v. The act of locating a point in space, using various means of measurement, such as a compass, sextant, theodolite, or global positioning system (GPS). Also used as a noun for the location point itself. The term was first used c. 1375, with reference to the act of fixing one's eyes or mind upon an object. Etymologically related terms include: affix, fixture, prefix, and suffix.

fjall: [Old Norse *fell* or *fjall* = mountain, hill.] n. A Swedish term for a mountain rising above the timberline and having flat, undissected, or plateaulike areas.
Alt. spelling: fell, fel

fjard: [Swedish *fjard* = inlet, bay.] n. A Swedish term for a narrow inlet or bay formed by submergence of a glacial valley excavated in a lowland along a rocky coast. Fjards are in areas of low relief, and are often fringed by rocky, reef-forming islands or skerries that are submerged at high tide. A fjard is shorter, shallower, and broader than a fjord. It also lacks a shallow sill or lip at the mouth, and has relief on the order of several hundred meters. A similar feature even shallower than a fjard is called a ria.
Cited usage: Norway is a fiord country, while Sweden has few fiords but many fiards [1940, C. M. Rice, Dict. Geol. Terms 135/2].
Alt. spelling: fiard
See also: ria, fjord, skerry

fjeld: [Norwegian *field* = field.] n. A Norwegian term for an elevated, rugged, almost barren plateau above the timberline, covered with snow during the winter.
Similar terms: fjell, fjall
See also: fjall

fjord: [Old Norse *fjördh* = fjord.] n. A long, narrow, U-shaped, relatively deep inlet or arm of the sea between high rocky cliffs along a mountainous coast, and carved by glacial erosion. The glacial trough has been partially submerged due to rising sea level. Fjords are deeper than fjards and generally are more than several hundred meters deep. A Fjord has a shallow sill or lip of bedrock or sediment near its mouth, with depth increasing towards its head. Classic examples of fjords occur along the glaciated coasts of Norway, Greenland, Alaska, and Chile.
Alt. spelling: fiord
See also: fjard, ria, skerry

fladen: [German *fladen* = flat cake.] n. A pancake-shaped mass of glass mixed with rock and mineral fragments. A fladen exhibits flow structure and surface sculpturing that is apparently produced by aerodynamic forces as it moves through the atmosphere. Examples of fladen are found in the impact breccias (suevites) of the Ries basin meteorite impact structure in Germany.
Cited usage: Glass bombs (fladen) and suevites from the 15 Ma old Ries Crater and the North American strewnfield from 35 Ma illustrate the survivability of high-silica glasses [1999, P. H. Schultz, Generation and Dispersal of Impact Glasses..., Brown University, Providence, RI].
Similar term: pseudotachylite
See also: suevite

flagellum: [Latin *flagellum* = whip, lash.] n. A long, whiplike appendage that serves for locomotion through fluids, occurring singly or in groups, and common in many microorganisms, typically protozoans, algae, bacteria, and zoospores. First usage in English c. 1805.

flagstone: [Middle English *flagge* = piece of sod < Old Norse *flaga* = slab + Old English *sten* < German *stein* = stone; < Greek *stia* = pebble.] n. As a general rock description, a hard durable sandstone, usually micaceous and fine-grained with shale partings that cause it to split uniformly along bedding planes into thin slabs suitable for use in construction. In pedology, a relatively thin, flat rock fragment occurring in the soil, and having a length in the range of 15-38cm. Flagstone is often Paleozoic in age and found in large quantities on continental stable platforms. Frequently used in the plural in proper names, such as the Elland Flags of the Lower Coal Measure Series in England, and the Lingula Flags found in the lower Silurian or Cambrian system in Wales, and containing large numbers of Lingula fossils. First usage in English c. 1725.
Synonyms: cleftstone, grayband, slabstone, flag

flaser: [German *flaser* = dialect form of *flader* = streak, vein.] adj. In petrology, pertaining to a structure of a highly sheared metamorphic rock in which lenses of relatively unaltered parent rock represented by augen of quartz or feldspar are surrounded by sweeping aggregates of scaly mica or chlorite, giving the appearance of crude flow structure. With increased crushing and cataclasis, there can be a continuous transition between flaser structure and mylonite.
Cited usage: The terms flaser gabbro and flaser granite have long been applied by European petrologists to cataclasites in which granulated streaks and laminae swirl around and between streamlined eyes of undestroyed gabbro or granite [1954, H. Williams et al., Petrography XI. 204].
See also: mylonite

fleches d'amour: [French *fleches* = arrows + *d'amour* = of love.] n. Acicular, hairlike crystals of rutile (TiO_2) embedded in sagenitic quartz. The term is used loosely as a synonym for sagenitic quartz, and was formerly used for amethyst containing needles of goethite.
Synonyms: Cupid's darts, Venus' hairstone
See also: sagenite, rutile

fleckschiefer: [Old Norse *fleckr* = spot, streak; < German *fleck* = to stain, to soil + *schiefer* = schist, slate, or shale.] n. A type of spotted slate

characterized by minute flecks or spots of indeterminate material.
Synonyms: garbenschiefer, knotenschiefer

flexure: [Latin *flectere* = to bend.] n. In structural geology, the location of maximum bending or curvature, usually in the form of a line. As an adjective, used in many ways with regard to bending, folding, hinging, or slippage. In structural geology, used in such terms as: flexure fold, flexural-slip fold, and flexure-flow fold. In paleontology and biology, a collective term for the structures of the dorsal portion of a bivalve that function during opening and closing of the valves.
Synonyms: hinge

flint: [Old English *flint* = hard rock; < Greek *plinthos* = brick, tile.] n. In mineralogy, a massive, very hard, usually black or gray variety of chalcedony. It breaks with conchoidal fracture, sparks when struck with steel, and was used as one of the early neolithic tools for making fire. In sedimentology, flint has been widely used as a synonym for the dark-gray or black variety of chert. In southern England, the term is used for siliceous nodules of chert that commonly occur in Cretaceous chalk beds. First usage in English before 1000.
Synonyms: firestone, black chert, silex
See also: chalcedony, chert

flocculation: [Latin *floccus* = tuft of wool.] n. The process by which minute suspended particles of sedimentary grains or colloidal clay clump together into an open-structured mass. On contact with saltwater, suspended particles will flocculate and settle out as tiny lumps or clusters. First usage in English c. 1825.
Cited usage: The property, possessed by lime of flocculating and precipitating clay sediments [1877, Joseph Le Conte (co-founder, with John Muir, of the Sierra Club), Elements of Geology].

floe: [Norwegian *flo* = layer; < Old English *flöh* = piece; IE *plak* = to be flat.] n. A mass of floating sea ice that is neither glacial nor coastal (fast ice) in origin. Floes are larger than ice cakes and smaller than ice fields. In lateral extent, they range from about 20m to over 10km. They are usually fragments of oceanic salt ice, and are classified according to size (e.g., giant, vast, big, medium, and small). Etymologically related terms include: flake, flatfish, and flagstone.
Cited usage: The great stratified masses of salt ice are fragments broken from the edges of the perennial floes. We called them floe-bergs in order to distinguish them from, and express their kinship to, icebergs [1878, Edward L. Moss, Shores of the Polar Sea].

floetz: [German *floetz* = flat layer, stratum, seam.] n. An obsolete term for the stratified, flat-lying, fossiliferous rocks including the whole range of strata from the Devonian through the Tertiary. Term coined by Abraham Gottlob Werner in the 1790s. This was part of Werner's fourfold classification system based on four major time periods of Earth history: Primitive, Transitional, Floetz, and Aufgeschwemmte (swept together) or Alluvial. Werner thought Floetz represented emerging mountains exposed by a receding ocean.
Cited usage: A term applied by Werner to the Secondary strata, because they were 'flotz' or flat-lying, compared with the Primary and Transition rocks [1865, D. Page, Handbook of Geological Terms (2nd ed.)].
Alt. spelling: Floetz, flotz

fluorine: [Latin *fluor* = a flowing < *fluere* = to flow.] n. The most reactive nonmetallic element, a pale-yellow toxic gas (symbol F). Along with chlorine, bromine, iodine, and astatine, it makes up the halogen series on the periodic table. The element was discovered by K.W. Schele in 1771, but not named until 1813 by English chemist Sir Humphry Davy with reference to the fluxing action of the mineral fluorite, from which fluorine was derived.

Cited usage: When chlorine is present in place of fluorine it is called chlor-apatite, and when the reverse fluor-apatite [1882, James D. Dana, Manual of Mineralogy & Lithology 4th ed. 213].

fluorite: [Latin *fluere* = to flow; < IE *bhleu* = to swell, well up, overflow.] n. A colorless, purple, green , yellow, blue, or brown mineral, calcium fluorite (CaF$_2$). The now obsolete name *fluores lapides* (later replaced by fluorite) was coined by Georgius Agricola in 1546. He called it that because the mineral was used as a flux, and it significantly lowered the melting temperature of the mixtures of minerals in glass making and iron smelting. The word fluorescent derives from fluorite, as it was one of the first fluorescent minerals to be studied. Etymologically related terms include: fluorescent, fluid, fluvial, effluent, phlegm, reflux, solifluction, and influenza. First usage in English c. 1865.
Synonyms: fluorspar, fluor

flute: [Latin *flatus* = a blowing < *flare* = to blow; or, possibly, a composite word < French *flageolet* = wind instrument + *laut* = lute.] n. A primary sedimentary structure that is a scoop-shaped or lobate groove or depression, generally 2-10cm in length, caused by the scouring of turbulent sediment-laden water flowing over a muddy bottom. The depression has a steep up current end where the depth is greatest and the long axis is parallel to the current. In architecture, a groove or furrow in a shaft or column relating to the longitudi-nal cross section of a wind instrument, the flute. Term coined in geology by Maxson and Campbell in 1935 for marks scoured in boulders found in the Colorado River Valley, which have a shape similar to architectural flutes. In 1955, Crowell coined the term *flute cast* for the structure formed when sediment fills a flute. First usage in English c. 1375.

fluvial: [Latin *fluvius* = river < *fluere* = to flow; < IE *bhleu* = to swell, well up, overflow.] adj.

Pertaining to a river or rivers. In sedimentology, referring to deposits transported by or laid down by a stream. IE derivatives include: fluorescent, fluid, fluent, flume, effluent, confluent, and influenza.
See also: alluvium

flysch: [Swiss *flysch* < German *fliessen* = to flow, melt.] n. A marine sedimentary facies loosely characterized by a thick sequence of poorly fossiliferous, rhythmically bedded turbidites, usually including shales and graywackes. The flysch facies is often considered to be the sedi-mentary rocks that fill a trough adjacent to a rapidly rising mountain belt. The name was originally coined in the Swiss Alps for crumbling to fissile material that slides or flows. First usage in English c.1850.
See also: molasse, lamina

foehn: [German *fohn* = foehn; < Latin *Favonius* = the west wind < *favere* = to be favorable.] n. A warm, dry, katabatic wind descending the lee slopes of a mountain range, and warmed by compression as it descends. The type locality is the northern slope of the Alps, but local names are used elsewhere, such as chinook. First usage in English c.1860.
Synonyms: chinook
See also: chinook

foggara: [French *foggara* = canal; < Arabic *falaj* = water channel.] n. In Middle Eastern and North African desert regions, a human-constructed, gently sloping tunnel system for conveying water by intercepting groundwater at the foot of a mountain. Physically, it is composed of a series of shafts dug down to a horizontal conveyance tunnel at the aquifer level, usually located in the upper reaches of alluvial fans, but sometimes as deep as 100 meters.
Similar terms: qanat, qarez (karez), falaj
See also: qanat

foliation: [Latin *foliatus* = leafy <*folio* <*folium* = leaf.] n. In metamorphic petrology, the texture resulting from a planar arrangement of crystals produced during recrystallization. In structural geology, a planar arrangement of textural or structural features in any type of rock. In glacial geology, the planar or layered structure in glacial ice produced by plastic deformation. First usage in English c. 1620.

fondo: [Spanish *fondo* = bottom; < Latin *fundus* = bottom, foundation.] n. The environment of sedimentation that lies on the deep floor of a water body. It may be used alone or as a combining form, as in fondothem, the rock units deposited in a fondo environment. Term coined by John L. Rich in 1951.
Cited usage: The major Pennsylvanian-Permian sedimentary cycle assumed a topographic form which, in profile, resembles the classic delta, with topset (unda), foreset (clino), and bottomset (fondo) segments [1964, D.C. Van Siclin, Symposium on Cyclic Sedimentation, Kansas Geological Survey, Bulletin 169].

footwall: [Old English *fot* = foot; < Latin *ped* = foot.] n. The block of rock which underlies an inclined fault, vein, or mine working.

foraminifer: [Latin *foramen* = an opening < *forare* = to bore + *ferre* = to bring; < IE *bher* = to carry, to bear (as in children).] n. Any protozoan belonging to the order Foraminifera. They are characterized by a shell or test most commonly composed of calcite and usually perforated by pores called foramina, through which numerous pseudopods protrude. Most forams are marine, but some freshwater species are known, such as Gromia terricola. They range from the Cambrian to the present. Originally thought to be minute cephalopod molluscs, in 1826 Alcide d'Orbigny published the *Tableau méthodique de la classe des Céphalopodes* (*Classification Method of Cephalopods*), which recognized them as new species in the modern taxonomy.

Cited usage: In some places it [nummulitic limestone] is composed mainly of foraminifera [1882, Geikie, Texbook Geol. VI.].
See also: nummulite

forb: [Greek *phorb* = fodder < *pherbein* = to graze.] n. Any herbaceous, vascular plant other than a grass, sedge, or rush, with no persistent woody stem above ground and especially one growing in a field, meadow, or prairie and used as food for range animals. Term coined by F. E. Clements c. 1920.
Cited usage: ... the term was coined by the ecologist F. E. Clements, The earliest mention of 'forb' I've come across is in Dayton's "Glossary of Botanical Terms Commonly Used in Range Research" (USDA Misc. Pub. 110, 1931) [1967, F. J. Herman (US Forest Service), letter to M. M. Mathews, University of Chicago].

formation: [Latin *formatio* = book <*forma* = form, figure, model.] n. In geologic mapping, the basic cartographic unit being a body of sedimentary, metamorphic, or igneous rock having early recognizable boundaries that can be traced in the field. Formations are large enough to be represented on a geologic map. In stratigraphy, the fundamental lithostratigraphic unit that can be separated into discrete members, or combined with other formations into groups. It is a body of rock with identifiable lithic characteristics and is mappable at the Earth's surface or traceable underground. In geomorphology, a naturally formed topographic feature that is strikingly different from adjacent features, such as a conspicuous erosional landform. In speleology, formation is synonymous with speleothem. First usage in English c. 1400.

forsterite: [Eponymous term after English mineral collector Adolarius Jacob Forster (1739-1806).] n. A white to yellowish orthorhombic mineral in the olivine group, magnesium silicate, Mg_2SiO_4. It is isomorphous with fayalite, and

occurs chiefly in metamorphosed dolomites and skarns at the contact with basaltic dikes. Term coined by French mineralogist A. Levy in 1824.
Cited usage: I have chosen for it the name of forsterite [1824, Levy, Ann. Phil. Ser. ii. VII. 62].
See also: fayalite

fortnight: [Middle English *fourtenight* = back formation of *fourtene night* = fourteen nights; < Old English *fowertne* = fourteen; < IE *kwetwer* = four.] n. A period of 14 days. This unit relates to the time between opposite phases of the moon, e.g., full moon to new moon, or waxing moon to waning moon. It is useful because a fortnight is the time between successive spring tides (monthly strongest), or successive neap tides (monthly lowest). IE derivatives include: four, quarantine, squad, quarrel, and farthing (the fourth part of a penny).
See also: tide, neap, spring

fosse: [Latin *fossa* = ditch < *fodere* = to dig.] n. In general usage, a long, narrow waterway, moat or ditch. In glacial geology, a long, narrow trough formed at the edge of a retreating glacier along its confining valley wall, possibly caused by accelerated melting due to absorbed heat from the valley walls. First usage in English c. 1375.
Alt. spelling: foss
Synonyms: fosse lake

fossicker: [English *fossick* = a troublesome person < back-formation from *fossicking* = making trouble < *fossick* = to dig up, rummage.] n. An Australian term for a prospector or mineral collector, often searching abandoned mine workings in the hope of scavenging minerals overlooked by other miners. First usage in English c. 1850.
Cited usage: Steady old fossickers often get more than the first one who opens ground [1864, Rogers, New Rush II].

fossil: [Latin *fossilis* = dug up < *fossus* = past participle of *fodere* = to dig.] n. Any evidence of past life from prehistoric time (prior to the Holocene Epoch). Fossils are preserved in the Earth's crust and include: hard parts, impressions, or traces, such as burrows and tracks. Originally, the term was applied to any curious object obtained through digging, including crystals and stone tools, as well as remains of life. First usage in English c. 1560.
Cited usage: Those bodies that are generated in the earth called Fossilia [1563, W. Fulke, Meteors].

fossula: [Latin *fossula* = diminutive of *fossa* = ditch < *fodere* = to dig.] n. A prominent vacant space between septa of rugose corals, caused by failure of one or more septa to develop as rapidly as others. First usage in English c. 1840.
Synonyms: septal fossula

fracture: [Latin *frangere* = to break; < IE *bhreg* = to break.] n. A general term for any break in a rock resulting from mechanical failure due to stress, whether or not displacement has occurred. The term fracture includes cracks, joints, and faults. In mineralogy, the characteristic manner in which a mineral breaks other than along planes of cleavage. Examples of mineral fractures are: conchoidal, hackly, and fibrous. IE derivatives include: breccia, breach, rubble, and brioche (crumbly, breakable bread).

fragipan: [Latin *fragilis* < *frangere* = to break + German *phanna* = pan; < Latin *patina* < Greek *patane* = shallow pan, platter; < IE *pete* = to spread.] n. A dense subsurface soil horizon whose hardness and low permeability are chiefly due to compaction rather than high clay content (as in clay pan) or cementation (as in hard pan or duripan). Fragipan contains much silt and sand, but little clay, and appears indurated when dry, but moderately to weakly indurated when moist.
See also: duripan

framboidal: [French *framboise* = raspberry.] adj. In ore petrology, said of a texture displayed by pyrite aggregates in shale, characterized by minute spheroidal clusters of crystals resembling raspber-ries. Originally, the texture was thought to be caused by colloidal precipitation, but it is now considered to be due to bacterial processes.

frangite: [Latin *frangere* = to break; < IE *bhreg* = to break.] n. An obsolete term for all sedimentary rocks and their regionally metamorphosed equivalents, formed from the mechanical disinte-gration of igneous rocks, where the sediment has undergone little sorting or chemical decomposi-tion. Examples of sedimentary frangite are arkose and graywacke. Term coined by Bastin in 1909.

franklinite: [Toponymous term after Franklin, New Jersey, the discovery locality.] n. An iron-black cubic mineral in the spinel group, zinc iron oxide, $ZnFe_2O_4$. It commonly forms octahedral crystals and resembles magnetite, but is only weakly magnetic. Franklinite contains variable amounts of the element manganese as well. The town of Franklin was named after American inventor and statesman Benjamin Franklin (1706-1790). First usage in English c. 1815.

frazil: [French *fraisil* = coal cinders.] n. Ice crystals with a spicular or discoid habit formed in supercooled turbulent water, such as fast-flowing streams or churning seawater. Frazil forms a spongy, slushy mass immediately below waterfalls in winter. First usage in English c. 1885.
Synonyms: needle ice, frazil ice

fretting: [Old English *fretan* = to eat (possibly) < Italian *frettare* < Latin *fricare* = to rub.] v. A rarely used term for the wearing away of a rock surface by stream erosion or wave action. First usage in English c. 1375.
See also: fritting

friable: [Latin *friabilis* < *friare* = to crumble into small pieces.] adj. Said of a rock or mineral that is easily broken or pulverized, such as a poorly cemented sandstone.

Cited usage: Stone walls of matter moldring and friable, have stood two or three thousand years [1614, Sir Walter Raleigh, History of the World].

fritting: [Italian *fritta* < *friggere* = to fry; < Latin *frigere* = to roast, fry.] n. The process of partial melting of grains of quartz and other minerals so that each grain becomes enveloped by a sheath of glass. Fritting results from contact metamorphism, usually due to the intrusion of a mafic magma into a silicic country rock. The term was originally used in ceramics, referring to the fusion of material used in glazes and enamels. First usage in English c. 1660.

fruchtschiefer: [German *frucht* = fruit + *schiefer* = schist, slate, or shale.] n. A type of spotted slate with spots suggestive of grains of wheat.

frustule: [Latin *frustulum* = diminutive of *frustum* = piece broken off.] n. The siliceous cell wall of a diatom, comprised of two valves, the epivalve (upper), and the hypovalve (lower). The valves combine to form a shape like a pill box with one overlapping the other.
Cited usage: The frustules which are long and slender slip over each other, yet so always to adhere [1857, M. J. Berkeley, Intro. to Cryptogamic Botany].

fuchsite: [Eponymous term after mineralogist J. N. von Fuchs.] n. A bright-green, chromium-rich variety of muscovite, $K(Al,Cr)_3Si_3O_{10}(OH)_2$. Fuchsite contains up to 6% chromium, and occurs in hydrothermally altered serpentinite. It also occurs in association with chromium-rich rocks containing rubies.

fucoid: [Latin *fucus* = rock lichen, red dye, rouge; < Greek *phukos* = face painting cosmetic + *-oeides* = resembling, having the likeness of < *eidos* = form.] adj. In paleontology, an informal name applied to any trail or tunnel-like sedimentary structure identified as a trace fossil, but not ascribed to any particular genus. Fucoid fossils

include crustacean tracks, worm burrows, molluscan trails, impressions, such as those resembling algae, and even marks, such as rill marks made by tidal effects and waves. In biology, fucoid refers to any seaweed of the family Fucaceae, order Fucales, and class Phaeophyceae (brown algae). In antiquity, the term *phukos* referred to the cosmetic prepared from seaweed, which Greek women employed in painting their faces with the "red and white of nature." First usage in English c. 1835.
Synonyms: rockweed

fugacity: [Latin *fugax*= fleet < *fugere* = to flee.] n. In thermodynamics, a function expressed in units of pressure and a measure of the nonideality of a gas.

fulcrum: [Latin *fulcrum* = bedpost, supporting foot of couch < *fulcire* = to support.] n. In mechanics, the point of support for a lever in moving an object, used to gain mechanical advantage or position. In biology and paleontology, an anatomical structure that acts as a hinge or a point of support, such as the hinge of a bivalve or the bony skeleton of *Homo sapiens*. In coastal geology, the intersection of the end of a recurved spit with the next succeeding stage in development of a compound spit. First usage in English c.1670.

fulgurite: [Latin *fulgur* = lightning; < Greek *phlegian* = to burn; < IE *bhel* = to shine.] n. An irregular, glassy, tubular, or rodlike structure produced by lightning passing into and fusing dry sand or sandy soil, especially found on exposed mountaintops, or in dune areas of deserts and lakeshores. Fulgurites are also formed in existing rocks when lightning fuses the silica-rich portion. They may measure as much as 40cm in length and 5cm in diameter. In industry, the term was adopted by inventor Raoul Pictet for an explosive made from nitroglycerine mixed with farinaceous substances. IE derivatives include: flame, phlogo-

pite, phlogiston, flamboyant, flagrant, and flamingo. First usage in English c. 1830.
Cited usage: In sand or rock, where lightning has struck, it often forms long hollow tubes, known to the calmly discriminating geological intelligence as fulgurites [1884, Cornh. Mag. Nov. 526]. At Geneva a trial has been made in a quarry with the new explosive, 'fulgurite', under the direction of the inventor Raoul Pictet [1894, Daily News, Jan. 22].

fumarole: [Latin *fumarium* = smoke chamber < *fumus* = smoke.] n. A vent through which gases and vapors are emitted, usually from a volcanic region during a late stage of activity. Fumaroles are frequently described by the composition of their gases, such as a sulfur fumarole or solfatara. They may be distributed along fissures or scattered chaotically thorough a volcanic region. First usage in English c. 1810.
Cited usage: Fumeroles or small crevices in the cone through which hot vapours are disengaged [1830, Charles Lyell, Principles of Geology I. 342]. The Californian Geysers are rather fumaroles–an immense collection of vents from which hot air is emitted [1881, W. G. Marshall, Thro. Amer. xv. 315].
See also: solfatara

fungi: [Latin *fungus* = mushroom.] n. Any of diverse group of eukaryotic, heterotrophic organisms belonging to the kingdom Fungi, which lack chlorophyll and vascular tissue. Fungi vary from single-celled organisms to immense multiton masses of branched filamentous hyphae (mycelium) that often produce specialized fruiting bodies. The kingdom includes the yeasts, molds, smuts, and mushrooms. They are commonly microscopic and may be saprophytic, symbiotic, or parasitic on green plants. Many of the phyla in this kingdom appeared on Earth during, or possibly prior to, Late Precambrian time. *Fungi* and *spongia* are Latin cognates both presumed to derive from the Greek *sphongos*, meaning sponge.

furrow: [Old English *furh* = furrow; < Old Norse *for* = trench, drain; probably influenced by English *farrow* = young pig; < Latin *porcus* = swine.] n. A long, narrow linear depression caused by removal of surface material, as by a glacier or small stream course. In Africa, a natural or artificial watercourse used for irrigation. The etymological connection to pig is the inference that long furrows result from the incessant rooting of swine. First usage in English before 900.

fusain: [French *fusain* = Spindle tree, or the charcoal made from its wood; < Latin *fusus* = spindle.] n. A friable type of coal with a fibrous structure and silky luster, occurring in strands or patchy masses in bituminous seams. First usage in English c. 1865.
Cited usage: Fusain is a variety of coal, resembling wood-charcoal in appearance. Some stalks, the interior of which is composed of fusain, are covered with a bark which has been converted into coal [1883, Jrnl. Chem. Soc. XLIV. 941].
See also: clarain, durain, vitrain

fusible: [Latin *fus-* = melted, poured, cast < *fundere* = to pour.] adj. Capable of being melted. Not to be confused with shape-related terms, such as: fusil, fusulinid and fusiform. The confusion arises from the verb *fuse*, meaning to blend by melting together, and the noun *fuse*, meaning cord-shaped flammable device for igniting explosives. Etymologically related terms include: fusion, funnel, confuse, refuse, infuse, and fondue. First usage in English c. 1375.
Synonyms: fusile
See also: fusiform, fusion, fusulinid

fusiform: [Latin *fusus* = spindle + *forma* = form, figure, model.] adj. Shaped like a spindle (i.e., rounded and tapering to a point at each end from a swollen middle). In volcanology, used especially in description of volcanic bombs having this shape. First usage in English c.1745.
See also: fusulinid, fusion, fusible

fusion: [Latin *fus-* = melted, poured, cast < *fundere* = to pour.] n. In physics, the nuclear reaction combining two atoms whose nuclei fuse together to form a heavier atom, as in the fusion of two hydrogen atoms to form one helium atom with the release of tremendous energy in the process. Nuclear fusion is the energy-making process in stars, and in the hydrogen bomb. Elements up to the weight of iron are produced by fusion in common stars, whereas all elements heavier than iron are produced in supernovae. In petrology, the process of melting whereby a solid becomes liquid by heating. First usage in English c. 1550.

fusulinid: [Latin *fusus* = spindle + *ina* = diminutive suffix.] n. Any extinct protozoan belonging to the order Foraminifera, the suborder Fusulinia, and the family Fusulinidae. They are characterized by an elongate, multi-chambered test composed of calcite, commonly resembling a grain of rice. Fusulinids range from Ordovician to Triassic, and are valuable index fossils for Late Paleozoic dating. The fusulinids were single-celled organisms related to modern amoebas, but having complex shells easily preserved as fossils. Giants among protists, they attained a maximum length of one centimeter and, when abundant, formed sizable deposits of rock called "rice rock."

Ga: [Latin *giga annum* = one billion years; < Greek *gigas* = giant + Latin *annum* = year.] A chronologic abbreviation meaning billions of years before the present. The literal meaning of the term *giga annum* would be "one-billion years," but standard geologic usage of this abbreviation references it back in time from the present, therefore it now means one billion years before the present. *See also*: Ma, Ka

gabbro: [Toponymous name after *Gabbro*, a town in the Tuscany region of northern Italy.] n. An dark-colored, mafic, plutonic rock composed chiefly of calcic plagioclase, usually labradorite or bytownite, and pyroxene (augite). Gabbro is the plutonic equivalent of basalt. Term coined by German geologist Christian Leopold von Buch in 1810. Gabbros are commonly found intruded as ring complexes (e.g., Ardnamurchan and Skye in Scotland), large lopoliths (Bushveld complex in South Africa), or layered complexes (Skaergaard in eastern Greenland). *Similar terms*: norite, euphotide

Gaia hypothesis: [Eponymous term after Greek *Gaea* = goddess of the Earth.] n. An ecological theory that recognizes the Earth as a complete and complex web of interconnected, interrelated, and interdependent physical and biological systems. The Gaia hypothesis was formulated by James Ephraim Lovelock in 1979 in collaboration with microbiologist Lynn Margulis, but the term name was actually proposed by William Golding, author of *Lord of the Flies*. Thinking of the ancient Greek earth-mother goddess and of her dual character of caring supporter of beings who fit and ruthless annihilator of those who do not, Golding suggested the name Gaia hypothesis. *Cited usage*: Gaia is more a point of view than a theory. It is a manifestation of the organization of the planet [1973, Lynn Margulis].

galaxy: [Latin *galaxias* < Greek *galaktos kyklos* = the Milky Way < *galaktos* < *gala* = milk.] n. In astronomy, any one of billions of large systems of stars held together by mutual gravitation and separated from other similar star systems by vast regions of space. Each galaxy itself contains billions of stars, our own solar system is the called the Milky Way Galaxy. The etymology with reference to milk derives from the Greek myth recounting how Hera, tricked into breast-feeding Hercules, pulled her breast from his mouth when she realized her mistake and sprayed milk into the sky, forming the stars and heavens. *Via Lactea*, meaning "Road of Milk" is the Latin equivalent for the Milky Way. First usage in English c. 1375.

gale: [Norwegian *geil* = uproar, unrest, boiling < *galen* = mad, furious, bad; < Old Norse *galenn* = mad, frantic, bewitched < *gala* = to sing.] n. A wind having a sustained average velocity from 28-47 knots, being stronger than a breeze and milder than a storm. On the Beaufort Wind Scale, the term gale is further subdivided into: a) near gale (force 7, from 28-33 knots), b) gale (force 8, from 34-40 knots), and c) full gale (force 9, from 41-47 knots). At sea, a gale is distinguished by high waves, whitecaps, spindrift, reduced visibility, and confused seas (waves converging from multiple directions). The term gale is usually reserved for winds at sea; the same intensity winds on land are normally part of a storm or frontal system. First usage in English c. 1545.

galena: [Latin *galena* = lead ore.] n. A lead-gray cubic mineral, lead sulfide, PbS. Galena commonly occurs as cubic to octahedral crystals. It is isomorphous with selenium-bearing clausthalite and commonly shows solution toward silver. It is common in medium to low temperature hydrothermal veins, as well as in syngenetic sedimentary deposits associated with black shales. Galena is the chief ore mineral of lead as well as an important ore of silver. In antiquity, the term galena was applied to lead ore, and the 1st-century Roman historian Pliny used the term for the dross or waste material separated from molten lead during smelting of lead ore.
Cited usage: The ore is what is called potter's ore, or galena, and has a broad shining grain [1812, Henry Marie Brackenridge, Views of Louisiana 148].

gamete: [Greek *gamete* = wife, and *gametes* = husband < *gamein* = to marry < *gamos* = marriage.] n. A mature reproductive cell, as a sperm or egg, having the haploid number of chromosomes, and capable of fusing with another gamete of the opposite sex to produce a new organism. Etymologically related terms include monogamy and polygamy. First usage in English c. 1885.
See also: haploid

gangue: [German *gang* = a vein or lode of metal.] n. The valueless rock or mineral material in a mined ore that is separated from the desired ore minerals during processing and concentration. First usage in English c. 1805.
Cited usage: An earthy or stony substance which is termed the gangue or matrix of the metal or ore [1815, William Phillips, Outline of Min. & Geol.].
Similar terms: matrix
See also: matrix

ganister: [Toponymous term after Ganister Quarry in England.] n. A hard, fine-grained quartzose sandstone used in the manufacture of refractory bricks for furnace lining. It is composed of subangular quartz grains of silt to fine-sand size, cemented with secondary chalcedony. Ganister displays a splintery fracture with subconchoidal surfaces, like that found in the rock at Ganister Quarry, in Granny's Wood at Low Laithe, in the Lower Coal Measures of Yorkshire. The quarry operated until the 1960's; the siliceous sandstone found there is associated with coal seams. First usage in English c. 1810.

gap: [Middle English < Old Norse *gap* = chasm.] n. In geomorphology, used in the eastern and southern U.S., as in 'wind gap,' for a break or opening in a mountain ridge or short pass thorough a mountain range. Also used in such terms as water gap, for a ravine or gorge deeply cut by ongoing stream erosion through a mountain ridge or hilly impasse. First usage in English c. 1375.

garbenschiefer: [German *garben* = sheaf + *schiefer* = schist, slate, or shale.] n. A type of spotted slate characterized by concretionary spots whose shape resembles that of a caraway seed.

garnet: [French *grenat* = pomegranate-red; < Latin *granatum* = pomegranate, granular < *granatus* = seed.] n. Any of the minerals of the garnet group. They are in the cubic system and have the general formula $A_3B_2Si_3O_{12}$. The A site is occupied by Ca, Mg, Fe, or Mn; the B site is occupied by Al, Fe, or Cr. There are two major garnet groups: a) the aluminum-bearing pyralspite group, including pyrope, almandine, and spessartine; and b) the calcium-bearing ugrandite group, including uvarovite, grossularite, and andradite. Each of the two groups form isomorphous series between members. Garnets occur in a wide variety of colors with dark red as the most common. The mineral occurs widely as distinctive euhedral dodecahedral crystals in metamorphic rocks, especially in schist, gneiss, and eclogite. Garnets

also occur in many igneous rocks, especially in granite and pegmatite. Garnets are used both as gemstones and as abrasives. Term coined by German natural philosopher and Dominican friar Albertus Magnus in the 13th century in his treatise 'Mineralium' contained in his *Opera Omnia* (complete works). Magnus was St. Thomas Aquinas's teacher. Because it was similar in both sound and meaning, the term was likely influenced by the medieval Latin *granum*, meaning grain, or cochineal (a red-dye extract from the cochineal insect that Spanish explorers mistakenly thought came from seeds they called 'granos'). An etymologically related term is grenade. First usage in English c. 1275.
See also: ugrandite, pyralspite, pyrope, uvarovite, almandine, grossularite, spessartine

gas: [Greek *khaos* = chaos, empty space.] n. Along with solid and liquid, one of the three phases or states of matter commonly found on Earth (there are two other phases of matter, plasma and Bose-Einstein condensates, but they are not common on Earth). Like liquids, gases have no fixed shape, but unlike liquids, due to the relatively great space between molecules, gases are highly compressible, or will expand indefinitely to fill a container. In gases, the atoms or molecules have sufficient thermal energy to break away from their neighbors and move around freely. Term coined by Flemish chemist Jan Baptista van Helmont (1577–1644), and may have been suggested by Paracelsus' use of the word *chaos* for the element then considered to be the substance of spirits, such as gnomes. First usage in English c. 1640.

gastr-: [Greek *gaster* = stomach, belly.] Combining prefix meaning stomach or intestine, used in such terms as: gastric, gastrolith, gastropod, and gastronomic.

gastrolith: [Greek *gaster* = stomach, belly + *lithos* = rock, stone.] n. A rounded, often polished, stone from the stomach of some reptiles, especially dinosaurs, plesiosaurs, and crocodilians. Gastroliths are thought to be used for grinding up food as a digestive aid, but marine reptiles may have employed them for stability in aquatic environments. First usage in English c. 1850.
Cited usage: There are found at the sides of the stomach two lenticular calcareous masses, which are known as 'crabs-eyes' or gastroliths [1880, Huxley, Crayfish 29].
Synonyms: gizzard stone, stomach stone

gastropod: [Greek *gaster* = stomach, belly + *poda* = foot.] n. Any mollusc member belonging to the class Gastropoda, the largest class in the phylum Mollusca. Gastropods are characterized by: a) a foot upon which the rest of the body (visceral mass) sits, b) a well-developed head with eyes and antennae, and c) a nonchambered calcareous and sometimes spiralled shell. The class includes: snails, slugs, sea hares, sea slugs, limpets, and abalone. The name alludes to the main appendage for locomotion being also the organ for ingestion and digestion. They range from the Cambrian to the present. First usage in English c. 1825.
Cited usage: The gasteropods, including land-snails, sea-snails, whelks, limpets, and the like, are the types of the mollusca [1851, Woodward, Mollusca 97].
Alt. spelling: gasteropod

geanticline: [Greek *ge* = earth + *anti* = opposite, against + Latin *clinare* < Greek *klinein* = to slope, bend.] n. An upwarping of the Earth's crust having regional extent, as contrasted with a geosyncline. More specifically, a laterally compressed anticlinal structure that develops in geosynclinal sediments. First usage in English c. 1895.
Cited usage: A mountain range includes all the mountain ridges made over the area and border of a single geanticline [1895, J. D. Dana, Manual of Geology 4th ed. 389].

geest: [German *geest* = dry sand; < Frisian *gamac* = barren.] n. Alluvial material not of recent origin that is lying on the surface. Specifically, the sandy region of a beach, or the glacial deposits along the North Sea coast in Germany.

gel: [Latin *gelare* = to freeze.] n. A translucent to transparent apparently homogeneous substance, generally elastic and jellylike, and formed as a colloidal dispersion or network of a solid with either a liquid or gas. A gel is a more congealed form of a sol. First usage in English c. 1895.

gelifluction: [Latin *gelare* = to freeze + *fluctus* = a flowing < *fluctuere* = to flow.] n. The progressive and gradual lateral flow of soil material under periglacial conditions in a region underlain by frozen ground.
Alt. spelling: gelifluxion
Synonyms: congelifluction
See also: solifluction

gem: [Middle English *gemme* = gem, jewel; < Latin *gemma* = flower bud, jewel < *gen-* = to produce.] n. A cut and polished stone used as ornamentation, generally having the beauty, rarity, size, and durability to warrant a high intrinsic value. Also said of rough stones, such as diamond that have the necessary shape, purity, and color that allow it to be suitable for cutting and polishing in the making of jewelry. First usage in English before 900.
Similar terms: jewel

gemstone: [Middle English *gemme* = gem, jewel; < Latin *gemma* = flower bud, jewel < *gen-* = to produce + Old English *stan* = stone; < IE *stai* = stone.] n. Any mineral, rock, or other natural material that is usually cut and polished, and has sufficient beauty and durability for use as an ornament. Examples of biologically derived gemstones include: pearl, amber, jet, ivory, and coral.

genus: [Latin *genus* = kind, sort ; < Greek *genea* = race, generation < *genos* = race, kind < *gonos* = birth, offspring; < IE *gen* = to give birth.] n. In biology and paleontology, a category or taxon in the hierarchy of classification developed by Swedish biologist Carl Linnaeus (1707-1778). Genus lies below family but above species. In the Linnaean system, the generic (genus) and specific (species) names are always in Latin or resemble Latin construction and pronunciation. Together they form the proper scientific name of any organism. The genus name stands first and is capitalized, followed by the species name in lower case, as in *Homo sapiens*, *Dendraster sullivani* (a type of sand dollar), or *Ptyocephalus yersini* (a species of trilobite). First usage in English c.1550. *Cited usage*: Genus is a general word, the which is spoken of many that differ in their kind [1551, Sir Thomas Wilson, The Rule of Reason Conteyning the Arte of Logike].

geo: [Old Norse *gja* = chasm.] n. A term used in northern Scotland for a long, deep, narrow coastal inlet bordered by rocky cliffs, and formed by marine erosion along a line of weakness, such as a joint or fault. Not to be confused with the prefix *geo-*.

geo-: [Greek *ge* = Earth; eponymous term after *Gaea* = goddess of the Earth.] Combining prefix meaning the planet Earth. In Greek mythology, out of Chaos the goddess Gaea was born. She gave birth to, then married her son Uranus. Together they gave birth to the Titans and the Cyclopes.

geocentric: [Greek *ge* = arth + *centrum* = center; < Greek *kentron* = center of a circle < *kentein* = to prick.] adj. In astronomy, pertaining to the Earth as a central point, as in the geocentric theory of the universe. In geophysics, pertaining to the center of mass of the Earth in defining coordinate systems, such as geocentric latitude. First usage in English c. 1685.

geode: [Latin *geodes* < Greek *geodes* = Earthlike < *ge* = Earth.] n. A hollow or partially hollow, subspherical rock mass, generally separable by weathering from the rock in which it occurs. It is characterized by a thin outermost layer of chalcedony, and by a cavity that is partly filled by an inner drusy lining of inward-projecting crystals. Geodes are most commonly composed of quartz or calcite deposited from solution on the cavity walls. These inner crystals usually differ in composition from that of the enclosing rock. First usage in English c. 1675.
See also: druse

geodesic: [Greek *ge-* = Earth + *daiesthai* = to divide.] adj. Said of or pertaining to geodesy, used in such terms as: geodesic curve, geodesic line, and geodesic dome. The term geodesic was made famous by the American designer and architect, R. Buckminster Fuller, in his popularization of the geodesic dome, based on the principles of geodetic construction, which combine the structural advantages of the sphere (enclosing the most space within the least surface, and being strongest against internal pressure) and the tetrahedron (enclosing the least space with the most surface, having the greatest stiffness against external pressure).

geodesy: [Greek *ge-* = Earth + *daiesthai* = to divide.] n. The branch of applied mathematics dealing with the size and shape of the Earth, and the measurement of large tracts or regions of the Earth's surface, thus the adjective *geodetic*. First usage in English c. 1565.

geoecology: [Greek *ge-* = Earth + *oikos* = house, dwelling + *logos* = word, speech, discourse, reason < *legein* = to speak, to say; IE *leg* = to collect.] n. The branch of geology dealing with the application of geology to the study of terrestrial ecosystems. It particularly emphasizes the physical controls, such as bedrock, soil, hydrology, and landforms, on the restriction of populations of organisms to certain regions, known as edaphic endemism. Also, defined as a branch of ecology restricted to high-elevation alpine environments.

geognosy: [Greek *ge* = Earth + *gnosis* = knowledge.] n. An obsolete term, now replaced by the term geology; geognosy was the 18th-century science of the Earth, particularly in reference to the origin, distribution, and sequencing of minerals and rocks in the Earth's crust. First usage in English c. 1790.
Cited usage: Werner directed his attention to what he termed 'geognosy,' or the natural position of minerals in particular rocks, together with the grouping of those rocks, their geographical distribution, and various relations [1830, Lyell, Principles of Geology I. 55].

geography: [Greek *ge* = Earth + Latin *graphicus* < Greek *graphikos* < *graphe* = writing < *graphein* = to write; < IE *gerbh* = to scratch.] n. The description of the Earth's surface, treating of its form and physical features, and its natural and political divisions, as well as the climate, natural resources, population, etc., of the various regions so defined. Term coined by Eratosthenes, head of the library at Alexandria during the 3rd century BCE. Eratosthenes was the first person to closely calculate the circumference of the Earth.

geoid: [Greek *ge* = Earth + *-oid* = combining suffix meaning resembling < Greek *eidos* = shape, form; < IE *weid* = to see.] n. A theoretically continuous surface nearly identical with the terrestrial spheroid, but at every point perpendicular to the direction of gravity; used as a reference datum for geodetic levelling, astronomical observations, and navigation. Term coined by German mathematician Johann Benedict Listing in 1872.

geology: [Greek *ge* = Earth + *logos* = word, speech, discourse, reason < *legein* = to speak, to say; IE *leg* = to collect.] n. The scientific study of the Earth, its physical features, the materials of which the planet is composed, and the dynamic processes acting on these materials. It also includes the history of the planet and its life forms since its origin. The word geology was used, perhaps for the first time, in the mid-14th century by Richard de Bury, bishop of Durham. He described it as the "science of earthly things", which he applied to the study of earthly law as distinguished from the arts and other sciences which were divine law, or matters concerned with the works of God. Jean-André de Luc used the term in a stricter scientific context in 1778. A year later, Horace Bénédict de Saussure (1740-1799) introduced the term into scientific nomenclature with the publication of the first of four volumes of his monumental work, *Voyages dans les Alpes* (*Travels in the Alps*), a compilation of more than 30 years of geologic studies.
Cited usage: Geology is divided into the following subordinate branches; (i) Geography, which treats of the Earth or Land; (ii) Hydrography, which treats of Water; (iii) Phytography (iv) Zoography [1735, Benjamin Martin, Philosophical Grammar].

geomorphology: [Greek *ge* = Earth + *morphos* = shape, form + *logos* = discourse, knowledge.] n. The science, based in both geology and geography, dealing with the origin, classification, description, and evolution of the landforms of the Earth's surface, and their relationships to underlying geologic structures. It also encompasses the history of geologic changes as evidenced by these surface features. First usage in English c. 1890.

geopetal: [Greek *ge* = Earth + *petalon* = leaf, thin plate, lamina < *petalos* = to outspread < *pet* = to spread.] adj. Pertaining to any rock feature that helps determine top from bottom at time of formation of the rock, such as cross-bedding, flute marks, and flame structures in sedimentary rocks.

geostrophic: [Greek *ge* = earth + *strophe* = turn < *strophein* = to turn.] adj. Said of or relating to the pseudo-force known as the Coriolis force, especially as applied to a wind or a current of water in which there is a balance between the Coriolis force and the horizontal pressure gradient.

geosuture: [Greek *ge* = Earth + Latin *sutura* = seam, suture < *suere* = to sew.] n. As used by J. Tuzo Wilson, a region of the Earth's crust where two areas of continental crust have collided. As used by Robert S. Dietz, a boundary zone between contrasting tectonic units of the crust, in many places marked by a fault that extends entirely through the Earth's crust.

geosyncline: [Greek *ge* = Earth + *syn* = together, alike + *klinein* = to bend.] n. A term proposed by American geologist James Hall in 1859 for a large-scale, either elongate or basinlike, active downwarp of the Earth's crust. Sedimentary and volcanic rocks were thought to accumulate in geosynclines up to thicknesses in the thousands of meters. With the development of plate tectonics theory in the 1970s, it was recognized that no contemporary examples of the proposed geosynclines actually exist. Most geologists today consider that the assemblages of rocks, originallythought to form in a geosyncline, actually represent assemblages that formed separately and were later joined by plate collisions.
Cited usage: The geosyncline was a late refinement from 1873 in response to Hall's 1857-1859 "theory of mountains with the mountains left out" (according to Dana) [1997, Robert H. Dott, Jr., James Dwight Dana's Old Tectonics Global Contraction Under Divine Direction, Amer. Jrnl. of Sci. vol. 297 p.283].
See also: geanticline

geothermal: [Greek *ge* = Earth + *thermos* = hot < *thermotes* = heat.] adj. Said of or pertaining to the internal heat of the Earth. First usage in English c. 1875.

geyser: [Toponymous term after Icelandic *Geysir*, a hot spring in southwest Iceland < *geysir* = gusher < *geysa* = to gush; < IE *gheu* = to pour.] n. A periodically erupting hydrothermal feature that intermittently ejects a fountainlike column of water and steam into the air. Active natural geysers occur in five major locations worldwide, these clusters ranked by number are: a) Yellowstone National Park, Wyoming, U.S. (513), b) Kamchatka Peninsula, Russia (200), c) Worth Island, New Zealand (51), d) Chile (46), and e) Iceland (25). Natural geysers are apparently restricted to areas of rhyolitic volcanic rocks where restricted channelways develop in hydro-thermal systems. Geysir is near the active volcano Hekla in Iceland. The term was first used in the generic sense in 1847 by the German chemist Robert Wilhelm von Bunsen (1811-1899). Bunsen analyzed the gases in Icelandic geysers and gave a physical explanation for their eruption.

ghat: [Hindi *ghat* = in India, stairway leading down to a landing on the water.] n. A term used in India for a mountain pass or a path leading down from a mountain. Also used for a wide set of steps descending to a river and used for bathing as in the bathing ghats of Benares. The term was originally erroneously by applied by Europeans to various mountain ranges in India dissected by rivers. The Sahyadri, Nilgiri, Annamalai, and Cardamom Hills refer to the Western Ghats, whereas the Mahendra Girir Hills refer to the Eastern Ghats.

gibbsite: [Eponymous term after mineralogist Colonel George Gibbs (1776-1833).] n. A white to grayish monoclinic mineral aluminum hydroxide, $Al(OH)_3$. It is polymorphous with bayerite, doyleite, and nordstrandite. Gibbsite is the principal constituent of bauxite. Colonel Gibbs donated his collection of over 12,000 mineral specimens to Yale University. First usage in English c. 1820.
Synonyms: hydrargillite

gilgai: [Kamilaroi (Australian aboriginal) *ghilgai* = a saucerlike depression forming a natural reservoir for rainwater.] n. A microrelief that occurs in heavy clay soils with high coefficients of expansion and contraction. This feature is especially prevalent in vertisols that repeatedly change moisture content, thus causing irregular surfaces to develop. First usage in English c. 1895.

gipfelflur: [German *gipfelflur* = summit plain < *gipfel* = summit, acme, peak + *flur* = lea, tract of cultivable land.] n. Prior to plate tectonics theory, an assumption that the world's highest mountain peaks were of similar height due to the balance between isostatic limitations and uplift compared to the erosion rate balance. In modern usage, often used as an adjective to describe a region of peaks, ridges, and crests all at about the same elevation. *Cited usage*: The north face of Nanga Parbat has several kinds of erosion-surface remnants, as well as gipfelflur ridge crests at ~5000 m altitude, together with approximately equally spaced valleys with similarly inclined slopes [2003, John Shroder & Michael P. Bishop, Erosion Surfaces and Gipfelflur in the Western Himalayas, AAG Annual Meeting Publication].

girasol: [Italian *girasole* = sunflower, opal < *girare* = to turn + *sole* = sun.] adj. Said of the effect produced by light transmission from any gemstone that exhibits a gleaming billowy area of light that appears to float as the stone is moved, such as in sapphire, chrysoberyl, moonstone, and fire opal. As a noun, used as a name for gemstones having this effect, specifically, a variety of fire-opal with brilliant flame-like yellow, orange, and red colors. First usage in English c. 1585.

glacier: [French *glace* = ice; < Latin *glacis* = ice < *glaciare* = to freeze.] n. A large consolidated mass of ice formed, at least in part on land, by compaction and recrystallization of snow which survives from year to year. Glaciers move slowly downslope under the force of gravity, or radially outward from centers of accumulation due to the stress of their own weight. The three main types of glaciers are: a) alpine or mountain glaciers which occur at high elevations, b) continental glaciers or ice sheets of continental size, and c) ice shelves which float on the ocean but are fed in part by glaciers formed on land.
Cited usage: Glacier breeze is a cold breeze, blowing down the course of a glacier, which owes its origin to the cooling of the air in contact with the ice [1930, Meteorol. Gloss. (2ed.)].

glacio-: [Latin *glacis* = ice.] Combining prefix meaning ice, used in glacial-related terms such as, glaciology, glacio-fluvial, and glaciokarst.

glance: [Middle English *glauncen* < *glenten* = to shine; < Dutch *glans* < German *glanz* = brightness, luster.] n. A old mining term for any of various minerals having a lustre that indicates its metallic nature. Examples of glance are: copper glance (chalcocite), lead glance (galena), iron glance (specular hematite), silver glance (argentite), and antimony glance (stibnite). First usage in English c. 1800.

glass: [Old English *glaes* < German *glas* = glass.] n. An amorphous or noncrystalline solid, usually formed by the rapid cooling of molten material. The atomic arrangement of glass is disordered and close to that of a liquid, being intermediate between the orderly arrangement of a crystalline solid and the complete disorder of a gas. At room temperature, glasses are metastable and are often called supercooled liquids, but they continue to display most physical properties of a solid, such as definite shape.

glauconite: [German *glaukonit* < Greek *glaukos* = bluish-green.] n. A dull green, earthy or granular, monoclinic mineral of the mica group, hydroxy potassium iron magnesium aluminum silicate, $(K,Na)(Fe,Mg,Al)(Al,Si,P)_4O_{10}(OH)_2$. Glauconite occurs abundantly in greensand, and is the most common authigenic sedimentary iron silicate. It is often regarded as an iron-rich analogue of illite. It is found in marine sedimentary rocks from the Cambrian to the present, and is an indicator of slow sedimentation rates in marine basins starved of detrital material. Named in 1827 by Keferstein (1784-1866).
Cited usage: Marls and sands, often containing much green earth called glauconite [1865, Lyell, Elements of Geology (6 ed.)].

glaucophane: [Greek *glaukos* = bluish-green + *phanesthai* = to appear.] n. A grayish-blue, bluish-black, or blue monoclinic mineral of the amphibole group, hydroxy sodium magnesium iron aluminum silicate, $Na_2(Mg,Fe)_3Al_2Si_8O_{22}(OH)_2$. It has a fibrous to prismatic habit, is isomorphous with iron-rich ferroglaucophane, and occurs principally in metamorphic rocks, such as blue schist and eclogite. Glaucophane is formed by recrystallization at low temperature and high pressure in a subduction zone. The protolith for glaucophane schist is often considered to be sodium-rich spilite (albitized altered basalt). First usage in English c. 1845.

glen: [Welsh *glyn* < Gaelic *gleann* = mountain valley.] n. A usually narrow mountain valley, often containing a stream or lake. The term is especially applied to the narrow glaciated mountain valleys in Scotland and Ireland, and is narrower and more steep-sided than a strath. First used in English by Edmund Spenser (1552-1599) in *Shepherd's Calendar* (1579) and *The Faerie Queene* (1590).
Cited usage: There in a gloomy hollow glen she found A little cottage, built of stickes and reedes [1590, E. Spenser, Faerie Queene Book 1].
Alt. spelling: glyn
See also: strath

gley: [Russian (Ukrainian) *glei* = clayey earth, sticky bluish clay.] adj. Said of a soil with gray to bluish-gray mottles that have developed under conditions of poor drainage and intermittent water saturation. The colors are due to reduction of iron, manganese, and other elements under anaerobic conditions. Also, a term not used in geology for a soil mottled with brownish oxidized patches due to uneven degrees of dryness. First usage in English c. 1925.
Alt. spelling: glei
Synonyms: G horizon, gley horizon

globigerinid: [French *globe* < Latin *globus* = a ball, sphere, round body + *gerere* = to carry, to bear.] n. Any planktonic foraminifer belonging to the superfamily Globigerinacea. They are characterized by a perforated calcite test. Globigerinids range from Jurassic to present and are a common constituent of deep-sea pelagic ooze sediment (generally found in depths below 1000m). First usage in English c. 1845.
Cited usage: There stomachs were full of globigerina, of which forminiferous creatures…the oozy bed of the ocean at that vast depth was found to be exclusively composed [1863, Sir Charles Lyell, The Geological Evidence of the Antiquity of Man xiv. 268].

glomeroporphyritic: [Latin *glomeratus* = past participle of *glomerare* = to wind into a ball < *glomus* = ball + Greek *porphyrites* = purple, hard, crystalline rock quarried in Egypt < *porphyros* = purple < *porphyra* = the purple-whelk.] adj. Said of the texture of an igneous rock containing clusters of crystal of the same mineral.

glory hole: [Scandinavian *glaury* < *glar* = to make muddy + Old English *hol* = hole, a hollow place; < IE *kel* = to conceal, to cover.] n. In mining, a large open pit from which ore has been extracted through a system of passageways beneath the ore body. First usage in English c.1835.

glossopterid: [Greek *glossa* = tongue + *pteris* = fern.] n. The informal name for fossil flora of the gymnosperm genus Glossopteris and its allies. Fossils of glossopterids were common in the Permian period throughout the supercontinent of Pangea, and their subsequent separation in the Mesozoic provided important evidence for migration of the continents. Term coined by A. Brongniart in 1822. An etymologically related term is glossolalia (speaking in tongues < Greek *glossa*, meaning tongue + *lalia*, meaning speaking).

gneiss: [German *gneist* = spark.] n. A foliated metamorphic rock that has bands or lenticles of granular minerals, such as quartz or feldspar, alternating with bands of flaky, tabular, or elongate prismatic minerals, such as biotite or hornblende. Varieties are distinguished by texture (e.g., augen gneiss), characteristic minerals (e.g., hornblende gneiss), or general composition (e.g., granite gneiss). The etymological connection to spark comes from the sparkling appearance of the aligned minerals. An alternate etymology states that gneiss is a term long used by the miners of the Harz Mountains to designate the country rock in which mineral veins occur; it is believed to be a word of Slavonic origin meaning rotted or decomposed.
See also: lenticle

gob: [Middle English *gobbe* < *gobet* < Old French *gobe* = a mouthful.] n. In general usage, a mass or lump. In mining, waste or barren material that has been used to fill the space where coal has been removed from an underground mine. First usage in English c. 1375.
Synonyms: goaf

goethite: [Eponymous term after Johann Wolfgang von Goethe of Germany (1749-1832).] n. A yellowish-brown to dark brown orthorhombic mineral, hydroxy iron oxide, FeO(OH). It is one of the most common products of weathering of iron-bearing minerals. Goethite occurs with radiating,

stalactitic, botryoidal, and massive habits, Goethite is polymorphous with lepidocrocite, *gamma*-FeO(OH) and akaganeite, *beta*-FeO(OH,Cl). The term was coined in 1806 by German mineralogist Johann Georg Lenz to commemorate Goethe, a philosopher, literary giant, scientist, and avid mineral collector. Goethe established the first system of weather stations, and made the one of the earliest systematic classifications of minerals.
Alt. spelling: göethite

gold: [German *geld* < Teutonic *ghelto* = gold; < Sanskrit *jyal* < IE *ghel* = to shine.] n. A soft, yellow, corrosion-resistant isometric native metallic element (symbol Au). It commonly forms natural alloys with silver and copper, and is widely found in hydrothermal veins and alluvial deposits. It is the most malleable and ductile metal, and can be easily distinguished from pyrite, chalcopyrite, and weathered biotite by its high specific gravity and sectility. It is estimated that all the world's gold purified to 24-carat would only make a cube 20-meters square. The symbol for gold, Au, comes from its Latin name *aurum*. IE derivatives include: glass, glitter, glisten, and gleam. First usage in English before 900.

Gondwana: [Sanskrit *gonda* = fleshy navel + *vana* = forest.] n. A term coined in 1885 by Austrian geologist Eduard Suess (1831-1914) for the Paleozoic supercontinent of the southern hemisphere that joined with Laurasia to form Pangea. According to the theory of plate tectonics, Pangea broke up into India, Australia, Antarctica, Africa, and South America. Gonda comes from the term *Gond*, a member of a Dravidian people, many of them jungle-dwellers of central India, who display a profoundly fleshy navel. When Alfred Wegener proposed the idea of continental drift, Gondwana was used to designate similar rock systems with the same characteristic fossil flora found in other countries. Eduard Suess

coined the term Gondwana in 1885 in his book *Das Antlitz der Erde* (*Face of the Earth*) after the name given to an extensive series of rocks in India, chiefly sandstone and shales of fluvial origin, ranging in age from Pennsylvanian through Jurassic.

goniometer: [Greek *gonios* = angle + French *metre* = meter; < Greek *metron* = measure, poetic meter; < IE *me* = to measure.] n. An instrument used in crystallography for measuring the angles between crystal faces. An etymologically related term is goniatite (a cephalopod having angular sutures). First usage in English c. 1765.

gorge: [French *gorge* = throat; < Latin *gurges* = whirlpool, abyss.] n. A narrow, deep valley with nearly vertical, rocky walls. A gorge is smaller than a canyon and more steep-sided than a ravine. First usage in English c. 1350.

gossan: [Cornish *gos* = blood.] n. The surficial weathered capping overlying a deposit of primary iron-bearing sulfide, such as pyrite. It is formed by the oxidation of sulfides, the leaching out of most metals and sulfur, and the deposition of iron-hydroxides, such as goethite and limonite. The result is a rust-colored outcrop resembling dried blood. In mining, an exposed, oxidized portion of a mineral vein, especially a rust-colored outcrop of iron ore. Many terms from the Cornish dialect spread to mining areas throughout Europe because of the productive tin mines of Cornwall in SW England, which were even known to ancient Greek traders.
Alt. spelling: gozzen, gozzan
Synonyms: iron hat, chapeau de fer

gouge: [French *gouge* < Old Prussian *goja* = gouge; < Welsh *gylf* = beak; < Late Latin *gubia* = rostrum; < Celtic *gulban* = beak, boring tool.] n. In ore deposits, a relatively thin layer or selvage of soft, earthy, fault-comminuted rock material along the wall of a vein. In structural geology, the crushed rock material that has accumulated between or along the walls of a fault. Gouge is so named because a miner is able to gouge out the unconsolidated material, facilitating mining of the valuable vain material. First usage in English c. 1325.

graben: [German *graben* = ditch, trench < graban = to dig.] n. An elongate, basinlike, structural land form caused by down-dropping of a block of the Earth's crust, and bounded by faults along its long sides. It is a structure which is common in rift valleys and regions experiencing crustal extension. First used with a geological sense in 1883 by Austrian geologist Eduard Suess in *Das Antlitz der Erde* (*Face of the Earth*).
Cited usage: This left a great open Rift Valley (or, to use Prof. Suess's term, a 'Graben') [1896, J. W. Gregory, Great Rift Valley xii. 220].
Similar terms: trough

grade: [Latin *gradus* = step; < *grady* < IE *ghredh-* = to walk, go.] n. In engineering, the angle of inclination with respect to the horizontal of some engineering structure, such as a road. In ore deposits, the percentage of ore mineral or valuable element in an ore body. In metamorphic petrology, the intensity or rank or metamorphism indicating in general the pressure-temperature environment in which metamorphism took place. In geomorphology of streams, the condition of balance between erosion and deposition, used in describing a stream that has reached grade. In evolution, the common level of development attained independently by separate but related lineages of organisms. Etymologically related terms include: prograde, retrograde, transgressive, regressive, gradual, egress, congress, and centigrade. First usage in English c. 1510.
Similar terms: slope, gradient

granite: [Italian *granito* = grainy < *granire* = to make grainy; < Latin *granum* = grain.] n. A coarse-grained igneous plutonic rock dominated by light-colored minerals, of which the feldspars and quartz are the primary constituents, and potassium feldspar exceeds plagioclase. Additional minerals include ferromagnesian silicates and muscovite. Granite and granodiorite together make up the majority of all plutonic rocks. Granite is the approximate plutonic equivalent of rhyolite. Etymologically related terms include: garnet, kernel, gram, grenade, and pomegranate. First usage in English c. 1645.

granitoid: [Italian *granito* = grainy < *granire* = to make grainy; < Latin *granum* = grain + Greek *oidos* = resembling.] n. A rock resembling granite in texture and color, having visible crystals and a light-colored, mottled appearance.
Similar terms: granitic

grano-: [Italian *granito* < past participle of *granire* = to make grainy; < Latin *granum* = grain.] Combining prefix meaning granitelike or grainy textured.

granoblastic: [Latin *granum* = grain + Greek *blastos* = bud, seed, germ.] adj. Referring to the texture of a nonfoliated metamorphic rock containing roughly equidimensional crystals with well-sutured boundaries. Rocks with a granoblastic texture include: quartzite, marble, hornfels, and charnockite.
See also: homeoblastic

granodiorite: [Latin *granum* = grain + Greek *diorizein* = to distinguish < *dia* = apart, between + *horizein* = to divide, limit.] n. A coarse-grained igneous plutonic rock that is intermediate in composition between granite and diorite, especially one in which there is at least twice as much plagioclase as alkali feldspar. Granodiorite is the approximate plutonic equivalent of dacite. Term coined in 1893 by Swedish ore deposit geologist Valdemar Lindgren (1860-1939).

granophyre: [Latin *granum* = grain + French *phyre* = combing suffix meaning porphyry.] n. A relatively fine to medium-grained, generally hypabyssal igneous intrusive rock of granitic composition displaying a porphyritic texture, and a groundmass with a micrographic intergrowth of quartz and feldspar. Term coined in 1867 by German mining engineer and petrologist Hermann Vogelsang (1838-1874).
Cited usage: Vogelsang has proposed to classify this (porphyritic) type in three groups: 1st, Grano-phyre, where the ground-mass is a microscopic crystalline mixture of the component minerals, with a sparing development of an imperfectly individual-ized magma; 2nd, Felsophyre; and 3rd, Vitrophyre [1882, Geikie, Textbook Geol. Iii 90].

graphic: [Latin *graphicus* < Greek *graphikos* < *graphe* = writing < *graphein* = to write; < IE *gerbh* = to scratch.] adj. Pertaining to the texture of an igneous rock, such as graphic granite, in which quartz and potassium feldspar form a regular intergrowth in a manner controlled by their crystal structure. The quartz forms elongate rods with wedge-shaped cross sections producing the effect of cuneiform writing on a background of feldspars. Less commonly, other pairs of minerals, such as ilmenite and pyroxene form graphic intergrowths. First usage in English c. 1835.

graphite: [Latin *graphicus* < Greek *graphikos* < *graphe* = writing < *graphein* = to write; < IE *gerbh* = to scratch + Greek *-ites* < *-lite* = pertaining to rocks < *lithos* = rock, stone.] n. An opaque, lustrous, iron-black to steel-gray, hexagonal mineral and native form of carbon (symbol C). It is one of the softest minerals (H = 1-2) and is poly-morphouswith diamond, chaoite, and lonsdaleite. It is common in metamorphic rocks. The term was coined in German as *graphit* by Abraham Gottlob Werner in 1789, alluding to its use in pencils. IE derivatives include graffiti, paragraph, epigraphy, topography, and seismograph.
See also: carbon, diamond

graptolite: [Greek *graptos* = written, painted or marked with letters < *graphikos* < *graphe* = writing < *graphein* = to write; < IE *gerbh* = to scratch + *-lite* = pertaining to rocks < *lithos* = rock, stone.] n. Any colonial marine fossil organism belonging to the class Graptolithina, ranging from the Late Cambrian to the Early Mississippian Period. They are characterized by cup or tube-shaped individuals (theca) arranged along one or more branches (stipes) to form a colony. The chitinous exoskeleton is highly resistant, and commonly occurs in black shales where it provides an important index fossil for the Early Paleozoic Era. The name alludes to the resemblance of graptolite fossils to markings with a slate pencil. First usage in English c. 1835.
Cited usage: The Florentine, or ruin marble, the dendritical ramifications on many limestones, and the moss-like forms in agates, were ranked (by Linnaeus) as Graptolites [1838, Penny Cyclopedia. XI. 363/1].

graticule: [Middle English *grata* = a grating; < Latin *graticula* < *craticula* = gridiron < *cratis* = wicker work, intertwined branches.] n. In cartography, the network of lines representing meridians of longitude and parallels of latitude on a map or chart. Even though the etymology refers to gridiron, graticule is not synonymous with grid. A grid consists of a series of regular-ly-spaced parallel lines set at right angles to each other, whereas a graticule can display a wider variety of line orientations, depending on the map projection used. First usage in English c. 1885.
Similar terms: grid

gravel: [French *gravele* = diminutive of *grave* = pebbly shore, sea shore, coarse sand; < Celtic *grouan* = gravel.] n. An unconsolidated natural accumulation of rounded rock fragments, predomi-nantly larger than sand (diameter more than 2mm), such as granules, pebbles, cobbles, and boulders. Gravel is the unconsolidated equivalent of

conglomerate. In engineering geology, the term is confined to water-worn rock fragments ranging in size from a diameter of 1.87 to 3.0 inches. The term is also commonly used in industry for a mix of sand, clay, and small water-worn stones much used for laying roads and paths. The term gravel is probably cognate with the Old English *greot*, meaning grit. First usage in English c. 1275.
Cited usage: During the gradual rise of a large area several kinds of superficial gravel must be formed [1833, Lyell, Principles of Geology III. 146].
See also: sand, clay, silt

gravity: [French *gravité* = heaviness; < Latin *gravis* = heavy; < IE *gwer-* = heavy.] n. The natural force of attraction between objects having mass. According to Newtonian physics, gravitational force is directly proportional to the product of the masses in consideration, but is inversely proportional to the square of the distance between those masses. First used in the literary figurative sense, the scientific usage came into English in the 17th century. Etymologically related terms include: grief, grave, and guru (heavy, venerable).
See also: barysphere

graywacke: [German *grauwacke* < *grau* = gray + *wacke* = large stone; < Old High German *waggo* or *wacko* = boulder rolling on a riverbed < *wag* = to move about; < IE *wegh* = to go, transport.] n. An old rock name with a diversity of usage, but now generally applied to a dark-gray, firmly indurated, coarse-grained sandstone, consisting of poorly sorted, angular to subangular grains of feldspar, quartz, and dark rock fragments embedded in a compact clayey matrix. Graywacke generally represents an environment in which the processes of erosion, transport, and deposition operated at a fast rate. A typical environment for formation of graywacke is in an orogenic belt. Graywackes are typically interbedded with marine shales, and are believe to be deposited by submarine turbidity currents. They frequently show sole marks and graded bedding. First usage in English c. 1810. An IE derivative is wiggle.
Cited usage: The fundamental rock of the Eifel is an ancient secondary sandstone and shale, to which the obscure and vague appellation of 'graywacke' has been given [1833, Lyell, Principles of Geology. III. 194].

greisen: [German *greiszen* = to split.] n. A rock composed chiefly of quartz, mica (muscovite or lepidolite), and topaz, formed by high-temperature hydrothermal or pneumatolytic alteration of granitic rock. Common accessory minerals include tourmaline, fluorite, rutile, cassiterite, and wolframite. Greisen is a common rock associated with the ancient tin mines of Cornwall, England. The terms greisenization and greisening are used in reference to the process of high-temperature alteration to form greisen. Term possibly influenced by German *greis*, meaning gray, alluding to the gray hue greisen may take on during alteration. First usage in English c. 1875.

grenz: [Back-formation from German *grenzhorizont* = recurrence horizon < *grenz* = boundary, barrier.] n. A horizon in coal beds resulting from a temporary cessation in the accumulation of organic matter, frequently marked by a layer of clay or sand.
See also: back-formation

grid: [English *grid* = back-formation from gridiron; < Latin *graticula* < *craticula* = gridiron < *cratis* = wicker work, intertwined branches.] n. A pattern of regularly spaced horizontal and vertical lines set at right angles, used as a reference for locating points on a map, chart, aerial photograph, or optical device. In contrast to a graticule, the lines of a grid always intersect at right angles, while the lines of a graticule may or may not be at right angles. First usage in English c. 1835.
See also: graticule

grike: [From a Northern dialect of Britain, possibly Scottish.] n. A term used in Scotland for any crack, slit, or fissure in rock; or a ravine in a hillside. In geomorphology, a solution fissure in limestone that separates clints. Contemporary usage has "clint and grike" topography used as a compound term, much like basin and range, where clint is the ridge and grike is the solution fissure of a karst region.
Alt. spelling: gryke
Synonyms: kluftkarren, scailps
See also: clint

griotte: [French *griotte* = a shiny, slightly acidic, reddish-black cherry; the morello cherry.] n. A French quarryman's term for a red-hued limestone or marble, often displaying purple, white, or brown streaks and spots.

grit: [Old English *greot* = sand; < Old Norse *grjot* = pebble, boulder.] n. A coarse-grained sandstone, especially one composed of angular clasts. Also applied to any sedimentary rock that looks or feels gritty on account of the angularity of its grains, the more angular the clasts, the rougher the feel. First usage in English before 1000.

groin: [Old French *groin* = snout.] n. A long, narrow jetty, usually constructed of large rocks, timbers, or pieces of steel that project into the sea perpendicular to the shoreline, either to protect the shore from erosion by currents and waves, and/or to trap sand carried by the longshore current in order to build up or retain a beach. First usage in English c. 1375.
Cited usage: Since the Point of Dungeness has advanced, forming a great natural groin, it intercepts the shingle which formerly travelled eastward, and was accumulated by artificial groins at Hythe [1872, Lyell, Principles of Geology. I. ii. xx. 533].
Alt. spelling: groyne

grossularite: [New Latin *grossularia* = gooseberry bush; < French *groseille* = gooseberry.] n. A usually pale-green of the garnet group, but also found as colorless, yellowish, orange, brown, or red cubic minerals, calcium aluminum silicate, $Ca_3Al_2(SiO_4)_3$. It forms an isomorphous series with the other calcium garnets of the "ugrandite" group, uvarovite and andradite. Grossularite is commonly found in skarns formed by contact metamorphism of impure limestone. The name alludes to the resemblance of pale-green specimens to the gooseberry (Ribes grossularia). First usage in English c. 1845.
Cited usage: Pale green garnets are not invariably grossularite [1868, James D. Dana, A System of Mineralogy 267].
Alt. spelling: grossularite, grossularia
Synonyms: gooseberry stone, gooseberry garnet
See also: ugrandite, pyralspite, pyrope, uvarovite, andradite, almandine, garnet

grotto: [Italian *grotta* < Latin *crypta* = vault, chamber, subterranean passage; < Greek *kryptein* = to hide.] n. A small cave, or a small room in a cave, often supplied with a natural spring and serving as a place for retreat, recreation, or reverie. Dante Alighieri (1265-1321), author of *The Divine Comedy*, is credited with first using the modern spelling *grotto*.

grus: [German *grus* = grit, fine gravel, debris.] n. The fragmental products of in situ granular disintegration of granitic rocks, especially when accumulating below an exposure of saprolite. Grus consists primarily of residual grains of minerals that are resistant to chemical weathering, such as quartz and potassium feldspar.
Alt. spelling: gruss, grush

guano: [Spanish *guano* < Quechua *huanu* = dung.] n. A phosphate or nitrate deposit formed by accumulation of the excrement of seabirds, found in great abundance on some seacoasts; sometimes also found in significant quantities in

caves from bat droppings. Some famous guano deposits are found on islands of the West Indies (Guano Islands), and of the Eastern Pacific (Chincha and other islands near Peru). In 1857, the entrepreneur Peter Duncan staked a claim on the Caribbean island of Navassa to mine guano. He based his claim on the little-known Guano Act of 1856, which states that any uninhabited island containing guano, and not under control by any other nation, may be claimed by a U. S. citizen. First usage in English c. 1600.
Cited usage: It is called Guano, not because it is the Dung of Sea-fowls (as many would have it understood), but because of its admirable virtue in making ploughed ground fertile [1669, Edward Montagu the Earl of Sandwich, translation of Albaro Alonso Barba's El Arte de Los Metales (The Art of Metals)].

gulch: [Echoic term from Middle English *gulchen* = to drink greedily, to sink in (of land).] n. A term used especially in the Western U.S. for a narrow, deep, steep-sided ravine, especially one formed by water erosion. A gulch is larger than a gully. In California and the Yukon, the term is often used in names of stream valleys containing placer gold, such as French Gulch in Northern California and Gambler Gulch in the Yukon. First usage in English c. 1830.
Cited usage: From the Peak to the sea shore, the earth is cut into gullies. The settlers call these ravines gultches [1832, Augustus Earle, A Narrative of 9 Months Residence in N.Z. In 1827].

gulf: [Greek *kolpos* = bosom, gulf.] n. A relatively large part of a sea or ocean partially enclosed by land in a relatively large, broad bight of the coast, and often connected to the sea via a strait. The distinction between a gulf and a bay is somewhat ambiguous, but is usually determined by size, or by its already given name, rather than by any particular shape. Gulfs are usually larger than bays (e.g., the Gulf of Mexico and the Gulf of Alaska), however, the Hudson Bay in Canada is larger than both the Gulf of Finland and the Gulf of Tehuantepec, Mexico.

gully: [Middle English *golet* = throat; < French *goule* < Latin *gula* = throat.] n. A very small and narrow channel or ravine eroded by water in unconsolidated sediment or soil. A gully is only occupied by running water after a rain or snow-melt. In size, a gully is an erosion channel smaller than a gulch, but large enough that it cannot be traversed by wheeled vehicles. Etymologically related terms include gullet and gullible. First usage in English c. 1535.
Alt. spelling: gulley

gumbo: [Louisiana French *gombo* = stew thickened with okra; < Bantu (Angola) *kingombo* = okra < *ki-* = a common Bantu vocalizing prefix + *ngombo* = okra.] n. A term used locally in the U.S. for a clay-rich soil that becomes sticky and plastic when wet. Forms an unusually sticky mud when wet. This term originated in the Mississippi Valley. The related term gumbotil refers to the clay-rich B horizon of a mature soil developed on glacial till, which becomes very sticky and plastic when wet. First usage in English c. 1800.
Cited usage: The soil here is largely mixed with a kind of blue clay, locally known as 'gumbo' [1891, Cecil Roberts, Adrift in America 27].
Synonyms: gumbotil

guyot: [Eponymous term after Arnold Henri Guyot (1807–1884), Swiss-born American geologist and geographer.] n. A flat-topped, seamount whose platform-shaped summit is attributed to wave erosion. Cooling and contraction of the ocean crust results in the gradual subsidence of the wave-cut platform to depths of hundreds of meters below sea level. Term coined in 1946 by the oceanographer and navy captain Harry Hess, in recognition of Guyot, Princeton's first geology professor. After his tour of duty, Hess became head of Princeton's geology department in 1950.

Cited usage: In discussing these submerged flat-topped peaks which rise from the normal ocean floor, the writer will henceforth call them 'guyots' after the 19th century geographer, Arnold Guyot [1946, H. H. Hess, Amer. Jrnl. Sci. CCXLIV. 772]. These 'tablemounts' (or 'guyots', if one prefers to use the non-descriptive term of the discoverer, Prof. H. Hess of Princeton) must be deep-drowned ancient islands [1959, New Scientist 1 Jan. 14/3].

gymnosperm: [Greek *gymnos* = naked + *sperma* = seed.] n. A vascular plant whose seeds are commonly in cones, but are never enclosed in an ovary. Major groups include conifers and cycads. Gymnosperms first appeared in the Devonian Period. First use c. 1825.

gypsum: [Greek *gypsos* = chalk, gypsum.] n. A commonly colorless to white, but also variously tinted, monoclinic mineral, hydrated calcium sulfate, $CaSO_4 \cdot 2H_2O$. It occurs in many varieties: a) transparent tabular to bladed crystals (selenite), b) massive fine-grained (alabaster), c) fibrous (satin-spar), d) flexible filaments protruding from a cave wall (cave cotton), and e) an impure earthy surface efflorescence containing clay and sand (gypsite). Gypsum is the most common sulfate, and is frequently associated with halite and anhydrite in evaporites. The term gypsum is probably cognate with Hebrew *gephes*, meaning plaster, alluding to the principle use for the mineral in manufacture of plaster of Paris. First use c. 1645.

gyre: [Latin *gyrus* < Greek *gyros* = ring, circle.] n. Any large-scale, circular ocean current, at least one being found in each of the world's ocean basins. The largest gyres are centered in subtropical high-pressure regions. Water movement is driven by a combination of poleward convective flow of warm surface water, prevailing winds, and deflection by the Coriolis effect (caused by the Earth's rotation). Gyres are generated mainly by the winds that are in frictional contact with the ocean surface. They generally rotate clockwise in the Northern Hemisphere and counterclockwise in the Southern Hemisphere.

gyroid: [Greek *gyros* = ring + *oidos* = resembling.] n. An isometric crystal form with 24 faces, that may be right- or left-handed, and therefore displays rotational symmetry in a spiral.

gyttja: [Swedish *gyttja* = mud, ooze.] n. A dark-brown to black freshwater mud that contains much organic matter, and is characterized by a significant quantity of silica-rich diatomaceous shells. Gyttja is found in marshes, lakes, and low-energy aquatic marine environments that are rich in nutrients and oxygen. Term coined in a technical sense by H. A. von Post in 1862.
Cited usage: In the shallow lakes and enclosed bays of the sea there began to be formed and still is in course of formation a deposit known by the name gyttja, characterized by the diatomaceous shells it contains [1887, Encycl. Britannica XXII. 740/1].

habit: [Latin *habitus* = appearance, condition, demeanor, dress < *habere* = to have, to hold, possess; IE *ghabh* = to give or receive.] n. In general, the mode or condition in which one exists or exhibits oneself, often expressed externally as outward appearance. In crystallography, the characteristic external crystal form or morphology of single crystals, or the characteristic overall appearance of aggregates of many smaller crystals. Examples of single crystals include: cube, octahedron, and rhombohedron; examples of aggregates include: acicular, dendritic, botryoidal, and massive. First usage in Old French c. 1225. IE derivatives include: habitat, inhabit, prohibit, and avoirdupois.

hachure: [Middle French *hacheu* = to cut up < *hache* = ax.] n. One of a series of short parallel lines drawn on a topographic map perpendicular to and intersecting the contour lines to show relief. Hachures are used particularlyas inward pointing "ticks" trending downslope from a depression contour. First usage in English c. 1855.

hade: [English dialect *head, hade* = to slope, incline; < uncertain origin.] n. In structural geology, especially in mining, the complement angle of the dip. The angle of inclination measured from the vertical of any inclined planar feature, such as a vein or fault. First usage in English c. 1680.

Cited usage: The side on which the Plim Line will fall is called the Hadeing-side; and according to the Hadeing of this the other flys off, and that we call the Hanging-side [1747, Wm. Hooson, Miner's Dict.]. The hade, slope, or inclination of the vein is chiefly estimated by miners from the lower side [1811, John Pinkerton, Petralogy (sic), or A Treatise on Rocks II. 578].

Hadean: [Eponymous term after Greek *Hades* = Homeric god, ruler of the underworld.] adj. In the geologic time scale, generally used in reference to the first eon of Earth history, being the time span from the initial accretion of the Earth (4.6Ga) to the oldest known Earth rocks (3.96Ga). The additional eons of geologic time are: Archean (3.96-2.5Ga), Proterozoic (2.5Ga-542Ma), and Phanerozoic (542Ma to present). The term was coined by geologist Preston Cloud in 1972 for lunar rocks of this age because the early rock record on the Moon is better preserved than on Earth. There is controversy between the International Union of Geological Sciences (IUGS) and the International Commission on Stratigraphy on whether to call the Hadean an era or an eon. If the Precambrian is considered an eon, then the Hadean, Archaean, and Proterozoic are all eras.
Synonyms: Rockless Era
Similar terms: Priscoan Period
See also: Archean, Proterozoic, Cryptozoic, Precambrian

haldenhang: [German *haldenhang* = below talus slope < *halde* = acclivity, upward slope, heap + *hang* = inclination, slope, leaning.] n. A slope of unconsolidated debris that is less steep than a gravity slope and is found near the base of areas in relief. Term coined by German geomorphologist Walther Penck in 1924.
Synonyms: wash slope

halite: [Greek *hals, halos* = salt, sea < IE *sal* = salt.] n. A colorless, white, or rarely blue isometric

mineral, sodium chloride (NaCl), commonly occurring as cubic crystals or massive compact aggregates. Deposits often result from evaporation of shallow inland seas and pluvial or playa lakes. Introduced as a mineral term by Glocker in 1847. The term rock salt correctly applies to polycrystalline aggregates of halite. In the related terms: halocline, halophyte, haloxene, and halophilic, the prefix halo- implies high salinity of various salts, not simply halite. IE derivatives include: sea, saline, salary, salad, and salami.
Synonyms: common salt
Similar terms: rock salt

halo-: [Modern Latin *halites* = the mineral sodium chloride (NaCl); < Greek *hals, halos* = salt, sea < IE *sal* = salt.] Combining prefix meaning salty or saline, often with reference to the sea. IE derivatives include: salt, silt, salsa, salami, and salmagundi.
Cited usage: They probably belong to the same quasi-marine, or what I shall in future call the Halolimnic group [1898, J.E.S. Moore, Proc. R. Soc. LXII].

halogen: [Greek *hals* = salt + *genesis* = origin, creation < *genesthai* = to be born, come into being; < IE *gen* = to give birth, beget.] n. Any of a group of five chemically related nonmetallic elements in Group VII-A of the periodic table, including fluorine, chlorine, bromine, iodine, and astatine. They combine with metals to make ionically bonded salts, thus inspiring the name. The term was coined in 1825 by Jons Jakob Berzelius (1779-1848) for the elements fluorine, chlorine, and iodine (bromine was not discovered until 1826 and astatine in 1940).

halse: [Teutonic *holsoz* < *kolsos* < Latin *collum* = neck.] n. A narrower and lower part of a line of hills, joining two heights; a pass or col.
Alt. spelling: hals, hause
Synonyms: col
See also: col

hammada: [Arabic *hammada* = hot stony plain.] n. An extensive, flat, upland desert surface that has been swept free of sand and finer sediment by the wind, and thus consists of either bare bedrock or bedrock with a thin veneer of wind-abraded pebbles. It is characteristic of plateaus in the Sahara, Gobi, and Australian deserts.
Cited usage: Aghadez is situated òn a hamadah, or lofty plateau of sandstone and granite formation [1853 J. Richardson, Narrative of Mission to Central Africa performed in the years 1850-51 II. iv. 60].
Alt. spelling: hammadah, hammadat, hamada, hamadet
Synonyms: nejd, rock desert

hanging wall: [Old English *heng* < Old Norse *hengja* = to hang (influenced by English *hade* = angle from vertical of a mineable vein).] n. The block of rock above an inclined fault, vein, or mineworking. In the early days of English mining, a plumb line was dropped in a mineshaft from above. The angle of inclination from the vertical defined the *hade* of the mine shaft. The plumb bob rested therefore, on the "hadeing-side"; the opposite side then became known as the "hanging-side", because that is the side from which the plumb line was hung. First usage in English c. 1775.
Cited usage: "You can walk up the footwall, and you can hang your lamp from the hanging wall, and if things get tough after hanging your lamp, you can hang yourself from the hanging wall" [2002, David Mustart, oral comm.].
Synonyms: hanging side, hanger
See also: footwall, hade

haplo-: [Greek *haplos* = single, simple.] Combining prefix meaning single or simple, as in: haplogranite, haploid, haplophyre, haplophase and haplozoon. This prefix is often erroneously used to mean 'one-half' instead of 'single,' due to its sound, its spelling, and the fact that *haploid* gamete cells are often thought to have one-half the number of chromosomes of a 'normal' cell. Actually, the 'normal' cells are diploid, defined as having twice as many chromosomes as the single set found in gametes.

haplogranite: [Greek *haplos* = single, simple + Italian *granire* = to make grainy; < Latin *granum* = grain.] n. In experimental petrology, generally used in referring to the chemical system albite-orthoclase-quartz-water, investigated as a simple analogue for natural granites. First usage in English c. 1950.

hapteron: [Greek *haptein* = to fasten.] n. A specialized rootlike projection formed most commonly by differentiated algal cells, which adhere to objects and serves to hold most macro-algae in place. Because most species of macro-algae live in high-energy tidal and subtidal environments, the hapteron (holdfast), along with the stipe (stem) and theca (cell covering), is exceptionally strong. Haptera are also found in various types of fungi (birds nest fungus) and lichen (fruticose lichen). First usage in English c. 1890. *Cited usage:* It creeps over the rocks to which it adheres by hairs or by exogenous projections known as haptera, which secrete a cement from their discoid tips [1967, C. D. Sculthorpe, The Biology of Aquatic Vascular Plants v.111]. *Alt. spelling:* haptera (pl.) *Synonyms:* holdfast

hardebank: n. A term originating in South African diamond mines for unaltered kimberlite below the zone of surface weathering, called blue ground.

hartschiefer: [German *hart* = tough, calloused + *schiefer* = slate, shale or schist.] n. A fine-grained, dense, banded, usually siliceous, metamorphic rock formed by recrystallization of ultramylonite in which any trace of original megascopic crystals has been obliterated. Term coined by Swedish geologist P. Quensel in 1916.

haru ichiban: [Japanese *haru* = springtime + *ichiban* = first.] n. Strong, warm, southerly winds blowing from the Pacific Ocean that are generated by low pressure air masses in the Sea of Japan. The low pressure masses originate in China during early spring before moving eastward. The first of these winds is called haru ichiban.

harzburgite: [Toponymous name after town of Harzburg, Germany (formerly in Saxony).] n. A rock of the peridotite group consisting primarily of olivine and orthopyroxene. Term coined by German petrologist Karl Heinrich Rosenbusch in 1887. *Cited usage:* The typical norite has 23-49% of silica, the well-known 'schillerfels' or bastite-serpentine-rock (harzburgite of Rosenbusch) [1890, Mineral. Mag. IX. 41].

hectare: [French *hecto* < Greek *hekaton* = 100 + French *are* = a square of which each side measures ten meters; < Latin *area* = open space, courtyard < *arere* (possibly) = to be dry.] n. In the metric system, a unit of area equal to $10,000m^2$ (2.47 acres), or 100 *ares* (an *are* = $100m^2$). First usage in English c. 1805.

hedenbergite: [Eponymous term after Swedish chemist Ludwig Hedenberg.] n. A black to dark green monoclinic mineral of the clinopyroxene group, calcium iron magnesium silicate, $Ca(Fe,Mg)Si_2O_6$. It forms an isomorphous series with diopside mineral, commonly occurring in skarns. First usage in English c. 1820.

hedreocraton: [Greek *hedron* = seat, base, foundation + *kratos* = power.] n. A stable continental craton characterized by a complete absence of volcanos, and only rare occurrences of earthquakes. It includes both the continental shield and platform. *Synonyms:* craton *See also:* craton

helicitic: [Latin *helix* < Greek *helix* = anything with spiral form < *helissein* = to twist; < IE *wei* = to turn, roll.] adj. Referring to a metamorphic rock texture consisting of contorted and curved strings of inclusions cutting through later-formed porphyroblasts. The inclusions are relicts of original bedding in the protolith. This texture can be seen as schistosity bands that cut through later formed crystals.
Alt. spelling: helizitic
See also: poikiloblastic

helico-: [Latin *helix* < Greek *helix* = anything with spiral form; < IE *wel* = to turn, roll.] Combining prefix meaning anything of spiral form, used in terms such as: helicopter, helico-conical, and *helicobacter pylori* (the corkscrew-shaped bacterium that causes stomach ulcers). IE derivatives, with reference to curved, enclosing objects include: valley, whelk, wallow, volume vulva and platyhelminthes.

helictite: [Portmanteau word from *helic(o)* + *(stalic)tite* Latin *helix* < Greek *helix* = anything with spiral form < *helissein* = to twist; < IE *wei* = to turn, roll + Greek *stalaktos* = dripping.] n. A tube-like speleothem which may be curved, spiral-shaped, or branching and twig-like. Helictites grow at the distal end by crystallization of calcium carbonate from an aqueous solution emerging from a minute central opening. First usage in English c.1882.
Cited usage: The term 'helictite' has been suggested as appropriate to these contorted growths [1882. H.C. Hovey, Celebrated Amer. Caverns xi. 186].
See also: stalactite

helio-: [Greek *helios* = sun.] Combining prefix pertaining to the Sun, as in helium or heliotrope.

helium: [Greek *helios* = sun.] n. A colorless, odorless inert gaseous element (symbol He). The English astronomer Sir Joseph Norman Lockyer

coined the term in 1868 on observing the solar spectrum of a previously undiscovered element (wavelength = 587.6nm), which he named helium. It was later discovered on Earth by the Scottish chemist Sir William Ramsey in 1895.

hematite: [Latin *haematites* = blood stone; < Greek *haimatites* = bloodlike stone <*haima* = blood + -*ites* < -*lite* = pertaining to rocks < *lithos* = rock, stone.] n. A widespread trigonal mineral, iron oxide, Fe_2O_3. Hematite is dimorphic with maghemite, and is the principal ore of iron. It commonly occurs as both reddish-brown earthy masses and as submetallic steel-gray to black crystals. It also sometimes occurs in botryoidal masses, and as polycrystalline aggregates, such as the bladed hematite roses. All forms have a consistent red streak, common to this sesquioxide-state of iron. Known since antiquity, the large deposits of hematite originally found in the Mediterranean area were thought to be accumulations of blood that leached into the ground following ancient battles. First usage in English c. 1540.
Alt. spelling: haematite
Synonyms: red ocher, oligist iron, bloodstone, sanguine lodestone
See also: iron, sesquioxide

hemera: [Greek *hemera* = day.] n. In stratigraphy, the span of geological time during which any particular taxon was at its greatest abundance, or at its apex of evolution (acme state). Term coined by the English paleontologist S. S. Buckman in 1893. The etymology stems from the observation that species stay at the peak of their success for a relative moment in geologic history.
Cited usage: The shortest geological time-division is a hemera: that is, the time during which a particular species had dominant existence [1893, S. S. Buckman, Q. Jrnl. Geol. Soc. XLIX].

hemi-: [Greek *hemi-* = half.] Combining prefix meaning one-half, used in terms such as: hemi-

sphere, hemitropous (half turning), and hemisome (1/2 the body of an animal, especially on one side of an axis of symmetry). Words formed from Latin generally use the corresponding prefix *semi-*, but there are instances of hybridism, where roots from the two languages are combined into one term, as in hemicerebrum.
Synonyms: semi-
See also: semi-, hemichordate, hybrid

hemichordate: [Greek *hemi* = half + Latin *chorda* < Greek *khorde* = gut, string, cord.] n., adj. Any of various wormlike marine animals of the phylum Hemichordata, having a primitive notochord and gill slits. The sense being that the hemichordates have a less-developed notochord compared to that of the chordates. First usage in English c. 1880.
See also: hemi-, chordate

hemihedral: [Greek *hemi* = half + *hedra* = seat, base.] adj. In crystallography, having half the number of equivalent faces of the holohedral form in the same crystal system. A tetrahedron is the hemihedral form corresponding to the holohedral octahedron. First usage in English c.1835.
See also: holohedral

hemimorphite: [Greek *hemi* = half + *morphos* = shape, form.] n. A white to pale green, blue, or yellow orthorhombic mineral, hydrated hydroxy zinc silicate, $Zn_4Si_2O_7(OH)_2 \cdot H_2O$. It is an ore of zinc and is found commonly in the oxidized zone of zinc sulfide deposits. The name was inspired by its crystal morphology, where different faces appear at each end of the prism. In the gem trade, the name is used incorrectly as a synonym for the similar appearing zinc carbonate, smithsonite. First usage in English c. 1870.
Synonyms: calamine, galmei

hemipelagic: [Greek *hemi* = half + *pelagikos* < *pelagos* = the sea.] adj. Said of deep-sea sedi-

ment, usually accumulated near the continental margin, in which debris of terrigenous, volcanogenic, or neritic origin makes up more than 25% of the coarse fraction; as contrasted to eupelagic sediment where the continentally derived fraction is less than 25%.
See also: terrigenous, neritic, eupelagic

Hercynian: [Latin *Hercynia* (sc. silva) = Greek *Orkynios drymos* = Hercynian forest; < (possibly) Old Celtic *Perkunya* < IE *perq* = oak, oak forest, wooded mountain.] adj. Referring to the late Paleozoic orogeny of Europe extending through the Carboniferous and Permian Periods, roughly comparing in time to the Allegheny orogeny in the Eastern U.S. Also used in reference to the Harz Mountains, located between the Weser and Elbe Rivers of northern Germany. The term was used by the ancient writers in reference to the wooded mountains of Middle Germany. Julius Caesar described it as the northern forest which could be traversed by a fast runner in nine days. The word was first used in geology as the German *hercynisch* by von Buch; in 1887 Marcel Bertrand first used the term in reference to the major European orogeny, a term that largely replaced the *variscisch und armoricanisch* of Suess, and later became common usage in English.
Cited usage: The 'Hercynian system' of Bertrand includes a long range of dislocated Devonian and Carboniferous rocks extending from Brittany to the Vosges and Ardennes, and beyond along the Black Forest, the Harz to Bohemia [1895, James D. Dana, Manual of Geology 4th ed. IV. iii. 734].

hermatolith: [Greek *herm* = reef + *lithos* = rock.] n. A synonym for reef rock, and thus a massive unstratified rock composed of calcareous exoskeletal remains of reef-building organisms, often with intermingled carbonate sand and siliciclastic material, all cemented by calcium carbonate.
Synonyms: reef rock

hervidero: [Spanish *hervidero* = hard-boiled, like an egg < *hervir* = to boil.] n. A synonym for mud volcano, and thus a conical accumulation of mud and rock formed by ejection of volcanic or petroliferous gases. The term has been used for a mud cone, unrelated to volcanism, such as accumulation of clastic material associated with surface eruptions of earthquake triggered debris. In Spanish-speaking countries hervidero is also a common term for bubbling hot springs, such as the Hervidero in Guanoco, Venezuela. The synonym macaluba is a toponymous term after Macaluba Volcano, in Sicily. The term salinelle is a toponymous synonym after Le Salinelle, mud volcanoes common along the foot of Mt. Etna, Sicily.
Synonyms: macaluba, salinelle, mud cone, mud volcano
See also: macaluba, salinelle

hetero-: [Greek *hetero* = the other of two, other, different.] Combining prefix meaning different, other, or sometimes opposite. Often used in contrast to *homo-*, and sometimes in contrast to *auto-, iso-, ortho-*, and *syn-*. Used in terms such as heterozygote, heterodont, and heterosexual.
See also: homo-, ortho-, iso-, syn-

heteroblastic: [Greek *hetero* = different + *blastos* = sprout, shoot, germ.] adj. In petrology, pertaining to a crystalloblastic texture composed of grains of two or more distinct sizes. Term coined by German petrologist Friedrich Johann Karl Becke in 1903 to distinguish textures distinct from homeoblastic.
See also: homeoblastic

heterochronous homeomorphy: [Greek *hetero* = different + *kronos* = time; Greek *homeo* = same + *morphos* = shape, form.] n. The trait of an organism that resembles an organism from a prior geologic age, also called atavism (from Latin *atavus*, a great-grandfather's grandfather). Used in contrast to isochronous homeomorphy.
Cited usage: Isochronous and heterochronous homeomorphy state whether the homeomorphous species lived in the same or at different times [1913, Q. Jrnl. Geol. Soc. LXIX. 166].

heterochthonous: [Greek *hetero* = different + *khthonos* = earth, soil, land, country; < IE *dhghem* = earth.] adj. In stratigraphy, said of a transported rock, fossil, or sedimentary particle that was not formed in the place where it now occurs. It differs from the similarly defined *allochthonous* in the sense that it refers to individual fragments only. IE derivatives include: humus, homo-, autochthonous, and chernozem. First usage in English c. 1890.
See also: allochthonous, autochthonous, chthonic, hetero-

heterogeneous: [Greek *hetero* = different + *genea* = race, generation < *genos* = kind, race < *gonos* = birth, offspring; < IE *gen* = to give birth, beget.] adj. In general English usage, referring to something composed of parts of different kinds. In experimental petrology, referring to equilibrium in a system of more than one phase where identification of the phases present, and their bulk composition is primary; in contrast to homogeneous equilibrium, where reactions of species within a single phase are investigated. First usage in English c. 1620. IE derivatives include: genus, gender, genius, germ, nature, connate, native, pregnant, and naïve.

heteromorphism: [Greek *hetero* = different + *morphos* = shape, form.] n. In paleontology and biology, an organism that displays a form that differs form the norm, such as an ammonoid that deviates from the normal planispiral form of coiling. Also, organisms that exhibit different forms in different stages of development, as with insects having larval and pupal stages.
See also: isomorphism, polymorphism

heteropic: [Greek *hetero* = different + *topos* = place.] adj. In stratigraphy, said of sedimentary rocks that are contemporaneous but that differ in lithology, and replace one another laterally.

heterotactic: [Greek *hetero* = different, other + *taxis* = order, arrangement < *tassein* = to arrange.] adj. In stratigraphy, said of sedimentary strata of abnormal or irregular arrangement, such as in two contemporaneous stratigraphic columns in which the coal beds were deposited at widely separate times.
Alt. spelling: heterotactous, heterotaxic

heterotrophic: [Greek *hetero* = different, other + *trophe* = food, nourishment < *trophein* = to feed.] adj. Said of any organism that cannot manufacture its own food by photosynthesis or chemosynthesis, in contrast to an autotroph. Instead of synthesizing food from inorganic material, heterotrophs must utilize already formed organic material to synthesize essential compounds for sustenance. First usage in English c. 1895.
Synonyms: zootrophic

hiatus: [Latin *hiatus* = gap, opening < *hiare* = to gape.] n. An interruption in the continuity of the geologic record, such as the absence of strata that would normally be expected, but were either eroded or never deposited. Also, a lapse in time, such as the time value of an episode of nondeposition or erosion represented by an unconformity.
Cited usage: Well developed paleosols and continental redbeds sandwiched between shallow marine sediments with clear evidence for erosional hiatus [2002, Joe Meert, Geology and the Great Flood].
Similar terms: unconformity, lacuna
See also: lacuna

hill: [Old English *hyll* < Middle Dutch *hille* < Latin *collis* < Greek *kolonos* = hill.] n. Elevation of the Earth's surface rising rather prominently above the level of the surrounding land, usually having a rounded, rather than a peaked, outline. Hills generally have a height from the base to summit of less than 300m, but the distinction between hill and mountain is arbitrary, for example, in Great Britain, elevations under about 600m (2,000ft) are generally called hills. An etymologically related term is column. First usage in English before 1000.
Similar terms: mountain
See also: mountain

hinterland: [German *hinter* = behind + *land* = land.] n. In general English usage, the remote or less-developed parts of a region; the back country. In tectonics, an area bordering or within an orogenic belt on the internal side and away from the direction of overfolding and thrusting. It is the site of eventual plutonism in the orogenic belt. In geomorphology, the low-lying region behind the coast. First usage in English c. 1885.
Synonyms: backland
Similar terms: internides

histogram: [Greek *histos* = beam, mast, tissue, web + *gramma* = something written < *graphein* = to write.] n. A bar graph representing a frequency distribution in which vertical bars are used to indicate the relationship between variables. The width of the bars equates to the class intervals (e.g., sediment size), and the height of the bars equates to the frequency of occurrence within the class (e.g., number of particles). Term coined by Karl Pearson in 1895. There are several possible etymological explanations: a) each vertical bar represents a mastlike shape, b) each bar represents a 'cell' in the web or tissue of the graph, and c) the graph represents some history of an occurrence (Greek *histor*, meaning knowing, learned, wise man, judge).

history: [Middle English *historie* < Greek *historia* = learning, knowing by inquiry < *histor* = knowing, learned, wise man, judge.] n. The branch of knowledge dealing with past events. As an adjective, used in historical geology as the branch of geology concerned with the evolution of the Earth, and its life forms from its origin to the present. First usage in English c. 1375.

histosol: [Greek *histois* = tissue, web + *sol* = soil.] n. A worldwide soil type rich in organic matter such as peat, especially prevalent in wet, poorly drained areas. In soil taxonomy of the U.S.D.A., one of the twelve soil orders. First usage in English c. 1970.

hoarfrost: [Middle English *hor* < Old English *har* < Old Norse *harr* = grey with age (as in hair); < Old High German *her* = old + Old English *frost* = frost; < IE *preus* = to freeze, burn.] n. A thin veneer of ice crystals formed on objects in water-saturated air. The ice crystals form a drusy coating on the cold objects, thus resembling gray hair. The distinction between hoarfrost and rime is that hoarfrost is a crystalline deposit of ice formed by the sublimation of water vapor, while rime is a deposit of rough ice crystals formed by the rapid freezing of supercooled droplets of fog when they are brought by air currents into contact with a surface below the freezing temperature. First usage in English c. 1275
Alt. spelling: hoar-frost
See also: rime

hogback: [English *hog* < Old English *hogg* = pig, boar + Old English *boec* = back.] n. A long, sharp-crested ridge with steep flanks of nearly equal slope on both sides, formed by the outcropping of steeply-inclined strata, or intrusive dikes that are particularly resistant to erosion and resembling in outline the back of a hog. First usage in English c. 1660.
Alt. spelling: hog-back

Synonyms: stone wall, hog's back
Similar terms: cuesta, dike wall

hollow: [Middle English *holwe* < Old English *holh* = hole, burrow; < German *hulwe* = pool, puddle, slough.] n. A depression, or low tract of land surrounded by hills or mountains, such as a sink, cirque, cavern, blowout, basin, or small valley. First usage in English before 900.
Similar terms: cirque

holo-: [Greek *holos* = whole, entire.] Combining prefix meaning whole, wholly, or completely; often used in contrast to mero-, hemi-, or semi-. Used in such terms as holography, holoethnic and holocaust (holy burnt).

holoblast: [Greek *holos* = whole, entire + *blastos* = germ, embryo, seed.] n. A crystal that is newly and completely formed during metamorphism. Term first used in geology by Bruno Sander in 1951.

Holocene: [Greek *holos* = whole, entire + *kainos* = new, recent.] adj. Pertaining to the most recent geologic epoch from the end of the Pleistocene, approximately 10,000 years ago, to the present time. The Pleistocene and Holocene Epochs together make up the Quaternary Period. Term coined by P. Gervais in 1860, who defined the base of the Holocene Epoch as evidenced by the first use of neolithic implements, corresponding to the end of the Younger Dryas cold period. Paleontologists currently define the base of the Holocene Epoch by the abundant appearance of the coccolithophore *Emiliania huxleyi* in marine sediments.
See also: Pleistocene, Quaternary

holocrystalline: [Greek *holos* = whole, entire + Latin *crystallum* = crystal ice; < Greek *krystallos* = clear ice, rock crystal.] adj. Referring to the texture of an igneous rock composed entirely of crystals and containing no glass.

holohedral: [Greek *holos* = whole, entire + *hedra* = seat, base.] adj. In crystallography, pertaining to that crystal class having the highest degree of symmetry for that crystal system. Also said of a crystal form that is a member of the holohedral class, and displays all the required faces. First usage in English c. 1835.
Cited usage: Hemihedral forms may be derived from a holohedral form, as the tetrahedron is from the octahedron [1855, Wm. A. Miller, Elements of Chemistry, p.103].
Synonyms: holosymmetric
See also: hemihedral

holohyaline: [Greek *holos* = whole, entire + *hyalos* = glass.] n. An igneous rock composed entirely of glass, such as obsidian or pumice. Also used to describe the texture of shock-metamorphosed impactite melt rock or matrix, such as an impactite breccia.
See also: obsidian, pumice

holotype: [Greek *holos* = whole, entire + *typos* = impression, figure, type < *typtein* = to beat, strike.] n. In taxonomy, the single specimen that is chosen as the type specimen in the original description of a new species. First usage in English c. 1895.
See also: lectotype

homeo-: [Greek *homoios* < *homos* = same, similar, alike.] Combining prefix meaning same, similar, or alike; used in terms such as: homeostasis, homeocrystalline (homogranular), and homeopathy.
See also: homo-

homeoblastic: [Greek *homoios* < *homos* = same, similar, alike + *blasto-* = seed, germ, sprout.] adj. In petrology, referring to the texture of a metamorphic rock in which the essential minerals are all of similar size. Depending on the habit of the minerals involved, this texture may also be more specifically called granoblastic, lepidoblastic, nematoblastic, or fibroblastic. Term coined by Friedrich Johann Karl Becke in 1903.
See also: lepidoblastic, granoblastic

homeomorphism: [Greek *homoio* < *homos* = same, similar, alike + *morphos* = shape, form.] n. In mineralogy, the characteristic of crystals having similar form and habit, but different chemical compositions. In paleontology and biology, the characteristic of individual organisms to bear a close resemblance to one another while having different ancestors.

homo-: [Greek *homos* = same, alike, or similar.] Combining prefix meaning same, alike, or similar. As a noun, borrowed directly from Latin *homo*, meaning man, and used as the genera name for species of humans, e.g., *Homo erectus* ("upright man"), *Homo habilis* ("handy man"), Homo *neanderthalensis* ("Neanderthal Valley man"), and *Homo sapiens* ("wise man"). The prefix *homo* has a slightly different nuance of meaning from *homeo*, but for all geologic interpretations they can be considered synonymous. Often used in contrast to *hetero-*.
See also: homeo-

homocline: [Greek *homos* = same, similar, alike + *klinein* = to slope, bend.] n. In structural geology and stratigraphy, the general term for a series of strata having the same dip.
Cited usage: For convenience the word 'homocline' will here be used as a general name for any block of bedded rocks all dipping in the same direction. A 'homocline' may be a monocline, an isocline, a tilted fault-block, or one limb of anticline or syncline [1915, R. A. Daly, in Geol. Survey Canada Mem. lxviii. 53].

homogeneous: [Greek *homos* = same, similar, alike + *genea* = race, generation < *genos* = race, kind < *gonos* = birth, offspring; < IE *gen* = to give

birth, beget.] adj. In general English usage, referring to a whole that is composed of individual parts that are identical in composition, or of the same fundamental kind, nature, or character. In experimental petrology, referring to equilibrium reactions between species within a single phase; as contrasted with heterogeneous equilibrium between different phases. The term can be confused with the similar term *homogenous*, probably because of the similarity in meaning to the verb *homogenize*, meaning to mix particles of different compositions into a single whole of uniformly dispersed composition. Reference books are known to perpetuate the confusion; for example, the 1961 edition of Webster's Dictionary states that homogenous and homogeneous are synonymous. First usage in English c. 1640.
See also: homogenous, heterogeneous

homogenous: [Greek *homogeneia* = community of origin < *homogenes* = of the same race, family, kind < *homos* = same, similar, alike + *genea* = race, generation < *genos* = race, kind < *gonos* = birth, offspring; < IE *gen* = to give birth, beget.] adj. In paleontology and biology, referring to similarity but not identity between parts of different organisms as a result of evolution from a common ancestor. These parts, although related in appearance are often dissimilar in function. In general English usage, the term is considered an alteration of homogeneous. This is due to the related verb *homogenize*, meaning to unite into a single whole of uniform composition, can create confusion with the term homogeneous. First usage in English c. 1865.
Cited usage: Structures which are genetically related in so far as they have a single representative in a common ancestor, maybe called homogenous. We may trace an homogeny between them and speak of one as the homogen of the other. Thus the fore limbs of Mammalia, Sauropsida, Batrachia, and Fishes, may be called homogenous [1870, Sir Edwin Ray Lankester, Annals of Natural History VI. 36].
Synonyms: homologous
See also: homogeneous, heterogeneous

homoiothermic: [Greek *homos* = same, similar, alike + *therme* = heat < *thermos* = hot.] adj. Said of an animal that maintains its internal body temperature at a relatively constant value by using metabolic processes to counteract fluctuations in the temperature of the environment. Birds and mammals are homoiotherms. First usage in English c. 1890.
Alt. spelling: homeothermic
Synonyms: endothermic, warm-blooded
See also: endothermic, poikilothermic, ectothermic

homologous: [Greek *homologos* = agreeing < *homos* = same, similar, alike + *logos* = word, proportion.] adj. In paleontology and biology, referring to parts of different organisms that are similar in appearance and structure, but not necessarily in function. This similarity is a result of evolution from a common ancestor. In structural geology, said of faults in separated areas that are in the same relative position and of the same type and structure. In stratigraphy, pertaining to strata that can be correlated, and are of the same general lithologic character. Such strata can occupy the same relative structural positions in the stratigraphic column. First usage in English c.1655.

homolographic: [Greek *homologos* = agreeing < *homos* = same, similar, alike + *logos* = word, speech, discourse, reason < *legein* = to speak, to say; IE *leg* = to collect + Latin *graphicus* < Greek *graphikos* < *graphe* = writing < *graphein* = to write; < IE *gerbh* = to scratch.] adj. In cartography, pertaining to a method of projection in which equal areas on the Earth's surface are represented by proportionally equal areas on the map or chart. First usage in English c. 1860.

Cited usage: The problem proposed by Babinet, and solved by Cauchy, of the homolographic (or, as I prefer to call it, the equigraphic) projection of maps; that is of the construction of maps in which all areas shall be correctly given [1866, Proctor, Handbook of Stars 22].

homotaxial: [Greek *homos* = same, similar, alike + *taxis* = arrangement < *tassein* = to arrange.] adj. In stratigraphy, pertaining to a similarity of arrangement of strata, or of fossil assemblages, that are in different locations, but which are not necessarily contemporaneous. The term is derived from homotaxis, proposed by Huxley in 1862. It emphasizes the point that similarity in succession does not prove age equivalence of comparable units.
Synonyms: homotactic

homothetic: [Greek *homos* = same, similar, alike + *thetikos* = to place, to set < *tithenai* = to lay down.] adj. In geomorphology, referring to land forms that are similar in shape but may differ in size. Also used in geometry to describe a type of symmetry in figures contained within the same plane, such as the two sides of a bow tie. First usage in English c. 1875.
Synonyms: synthetic

hoodoo: [Louisiana French *hoodoo* < West African *hoodoo* = one who practices voodoo < *voodoo* < Ewe *vodu* = voodoo.] n. A column, pillar, or pinnacle of layered sedimentary or volcaniclastic rock, usually of fantastic, grotesque, or eccentric shape. They are produced by differential weathering and erosion, often with a resistant cap rock at the apex. Good examples of sedimentary hoodoos are found in Bryce Canyon, Utah and Cappadocia, Turkey; examples of volcanic hoodoos are found around Crater Lake, Oregon. The term is used primarily in North America, and is inspired by the resemblance of spirits and animals that populate the voodoo pantheon. It arrived when slaves were brought to the Americas from Africa. The Ewe language (pronounced 'eh-way' or 'ev-way') is the etymological root of the term, and the primary language spoken in West Africa, near the Volta and Mono Rivers (Ghana and Togo). Today, voodoo is practiced chiefly in Brazil and Haiti, combining the syncretized elements of Roman Catholic rituals with the spiritual beliefs and practices of West Africa, in which deified ancestors communicate with living practitioners. In Colorado, U.S. hoodoos are also called tepees due to their similar shape to the stereotypical Native American dwelling.
Cited usage: I done hoodooed the hoodoo man [c.1900, Cotton-eyed Jack, Blues song].
Synonyms: demoiselle, rock pillar, fairy chimney, tepee
See also: demoiselle

horizon: [Latin *horizon* < Greek *horizon* = limiting, bounding < *horizein* = to limit < *horos* = boundary.] n. In general usage, the line that forms the apparent boundary between Earth and Sky. In stratigraphy, a thin distinctive stratum or interface between strata that is useful in correlation, such as the location of a particular fossil or mineral concentration that serves to identify the stratum with a specific event or time period. In pedology, any of the distinctive layers in a soil profile with characteristic physical or chemical properties, and designated by capital letters, such as A, B, C, or R. In archaeology, a discrete regional, cultural period or level of cultural development during which the influence of a specified culture spread over a defined area, and is marked by some easily recognizable criterium or trait. First usage in English c. 1545.
Cited usage: Each [species] is most abundant in one horizon, and becomes gradually less frequent in the beds above and below [1856, Samuel Woodward, A Manual of Mollusca iii. 411].

horn: [Old English *horn* < German *horn* < Latin *cornu* < Greek *keras* = horn.] n. In geomorphology, a body of land or water shaped like a horn, or the pointed ends of a dune or beach cusp. In glacial geology, a high, sharp mountain peak bounded by steep head-walls of three or more cirques cut back into the mountain by headward erosion of the glaciers. During formation, the peak itself was not covered by ice, but stood above it. Relative terms include: hornblende, ceratopsian, cladoceran, rhinoceros, triceratops, and cerebrum. First usage prior to 900.
Synonyms: pyramidal peak, cirque mountain, Matterhorn

hornblende: [German *horn* = horn + *blenden* = to deceive, blind..] n. A black to dark-green to brown monoclinic mineral, hydroxy calcium sodium magnesium iron aluminum silicate, $(Ca,Na)_2(Mg,Fe)_4(Al,Fe)(Al,Si)_8O_{22}(OH)_2$. Hornblende is the most abundant mineral of the amphibole group, and is common in igneous rocks, such as: granite, granodiorite, diorite, syenite, and andesite, as well as in metamorphic rocks, such as gneiss and schist. The name was used by German miners to refer to any dark, shiny, bladed mineral found with metalliferous ores but containing no valuable metal. Finding no commercial use for the mineral, the miners named it for its deceptiveness to the untrained eye. Sphalerite, commonly called simply *blende*, has a similar reputation. First usage in English c. 1765.
Cited usage: The great weight of the stone called hornblende made the miners at first imagine it contained some metal, but finding none except iron they called it blind [1796, Richard Kirwan, Elements of Mineralogy 2nd ed. I. 215].
Similar terms: amphibole
See also: amphibole, tremolite, asbestos

hornfels: [German *horn* = horn + Middle English *fel* < Old Norse *fjall* = mountain; < German *fels* = rock, cliff.] n. A dark, fine-grained metamorphic rock composed of equidimensional grains without preferred orientation and formed by recrystallization of fine-grained siliceous or argillaceous sediments by contact metamorphism. Hornfels are tough, flintlike rocks that tend to resist weathering and erosion, and are thus often exposed in prominent outcrops and cliffs.
Cited usage: We use the term hornfels instead of its etymological equivalent, hornstone, because in many cases the rocks termed hornfels are distinctly crystalline [1888, J. J. H. Teall, British Petrography xii. 374].

hornito: [Spanish *hornito* = diminutive of *horno* < Latin *furnus* = oven.] n. A small, beehive or oven-shaped mound of spatter usually formed on the solidified crust of a pahoehoe flow and formed by globs of lava ejected through an opening in the roof of an underlying lava tube. The gas responsible for the ejection of spatter is often created by the lava flowing over marshy ground.
Cited usage: The small conical mounds at Jorullo called 'hornitos' or ovens [1830, Lyell, Principles of Geology. I. 378].
Synonyms: driblet cone
Similar terms: monticule

horse: [Middle English *hors* < Teutonic *horso* = horse; < pre-Teutonic *kurs* < Latin *currere* = to run.] n. In structural geology, a large rock mass that has been caught between the walls of a fault, or a coherent unsheared rock mass between two branches of a fault. In ore deposits, a miner's term for a mass of barren rock that occurs within a vein. First usage prior to 900.

horse latitudes: [Translation (possibly) from Spanish *Golfo de las Yeguas* = Mares' Sea.] n. The belts of calm seas and light winds at approximately 30-35° N & S latitudes that border the trade-wind belts. Possibly named because extended calms on the high seas necessitated eating the livestock, or throwing them overboard, as feed supplies

dwindled. First usage in English c. 1770.
Cited usage: The latitudes where these calms chiefly reign, are named the horse-latitudes by mariners because they are fatal to horses and other cattle which are transported (to America) [1777, Johann G. A. Forster, *A Voyage Round the World, in His Britannic Majesty's Sloop Resolution*, Commanded by James Cook II. 581].

horseback: [Old English *hors* = horse + Middle English *bak* < Old English *baec* = back.] n. In glacial geology, a term used especially in Maine for an esker, kame, or other low, sharp-crested ridge of sand or gravel, resembling the back of a horse in shape.

horst: [German *horst* = heap, mass, cluster.] n. An elongate block of the Earth's crust which has been uplifted relative to adjacent blocks and is bounded by faults on its long sides. Term coined by Eduard Suess in 1883 in his book *Das Antlitz der Erde* (*Face of the Earth*).
Cited usage: If the outer borders of two fields of subsidence approach each other so that a ridge is left between them, on both sides of which the two areas of depression descend more or less in the form of steps, then we have what we shall distinguish, making use again of a common mining word, as a horst [1904, H. Sollas, translation of Suess's Face of the Earth I.].
See also: graben

how: [Old Norse *haug* = mound, cairn; cognate with Teutonic *hauh* = high.] n. A small hill, mound, or hillock in a valley; the term is frequently used in northern England, as in Great How, Brant How, and How Hill. Not to be confused with the Scottish homonym *howe*, meaning hollow or depression in the Earth's surface.
Cited usage: Round burial mounds are the commonest objects of antiquity met with in the field. They are called by different names in different parts of the country barrow, low, howe, cairn [1963, Field Archaeology, 4th ed., 45].
Alt. spelling: howe

hoya: [Spanish *hoyo* = large hole, cavity, pit, grave.] n. A stream bed, valley, or basin in a rugged mountainous area. Along with the masculine form *hoyo*, *hoya* is often used in proper place names, such as the *Hoya de Quito* in Ecuador.
Cited usage: The Gran Hoyo de los Pozos (Great Valley of the Potholes) was only one of several large enclosed valleys [1962, OUCC Proceedings 1, The Oxford University Expedition to Northern Spain].
Alt. spelling: hoyo

huebnerite: [Eponymous term after 19th-century German metallurgist Adolph Hübner.] n. A reddish-brown to black monoclinic mineral of the wolframite series, manganese iron tungstate, $(Mn,Fe)WO_4$. It is isomorphous with ferberite, and is a minor ore of tungsten.
Alt. spelling: hübnerite

huerfano: [Spanish *huerfano* = orphan.] n. An elevated mass of older rock completely surrounded by younger, usually sedimentary, rocks of different lithology from the protruding hill or mountain. The term is used primarily in the U.S. Southwest.

hum: [Toponymous term after Croatian town of Hum < Serbo-Croation *hum* = hill.] n. An isolated hill in a karst region, surrounded by a plane that is usually covered with alluvium.
Cited usage: Residual limestone masses or hummocks rising above polje floors are called 'hums' [1937, Wooldridge & Morgan, The Physical Basis Geography: An Outline of Geomorph. xix 289].
Synonyms: haystack, mogote, pepino, karst tower
See also: mogote

hummock: [German *humpel* or (possibly) *hummel* = a small height, a hump + English *-ock* = combining suffix used to form diminutives.] n. A rounded knoll, mound, hillock, or other small

elevated tract of land. Term coined in a geologic sense by Richardson in 1808, after its use by marine navigators to describe distant billows on the sea that appeared like hills and mounds. Both hummock and hump are probably derived from the German *humpel*, because it has been shown that the English hump appears in the language 140 years later than hummock. First usage in English c. 1550.
Cited usage: To these may be compared the stratified basaltic hummocks so profusely scattered over our area. Navigators use the word hummock to express circular and elevated mounts, appearing at a distance; I adopt the word from them [1808, Richardson in Phil. Trans. XCVIII. 218].
Similar terms: hillock

humus: [Latin *humus* = soil, mold, ground, earth; < IE *ghthem* = earth, ground.] n. The dark-brown or black material resulting from the slow decomposition of organic matter present in soil. Such humified organic material is fairly stable, and is at such an advanced level of pedogenic transformation that the original source material may not be identifiable. IE derivatives include: human, homicide, bonhomie, exhume, humility, autochthonous, chameleon, chamomile, and chernozem. First usage in English c. 1795.
Cited usage: That stratum called humus, which serves as a basis to the vegetable kingdom [1796, H. Hunter, translatation of St.Pierre's Studies of Nature (1799) I. 474]
Synonyms: humified organic matter

hurricane: [Spanish *huracan* < Taino (Caribbean) *hurákan* = violent storm.] n. A severe tropical cyclone originating in the equatorial regions of the Atlantic Ocean and the Caribbean Sea, or the eastern regions of the Pacific Ocean. Hurricanes are defined by the Beaufort scale as having winds with a speed greater than 64 knots (74mph/119kmph).

hyalite: [Greek *hyalos* = glass.] n. A colorless variety of common opal that may be as transparent as glass or translucent and milky. Hyalite occurs commonly as botryoidal crusts lining cavities and fractures, or as globular concretions resembling drops of melted glass. First usage in English c. 1790.
Synonyms: water opal, Müller's glass

hyalo-: [Greek *hyalos* = glass.] Combining prefix meaning glassy in appearance or composed of glass. The prefix is often added to names of rocks with a glassy texture, such as the volcanic hyalobasalt, and the metamorphic hyalomylonite.

hyaloclastite: [Greek *hyalos* = glass + *klastos* = broken.] n. A volcanic breccia formed by lava flowing or intruding into water, ice, or water-saturated sediment. The lava shatters into small angular fragments of glass, which, subsequently, often undergo devitrification.

hyalocrystalline: [Greek *hyalos* = glass + *krystallos* = ice.] adj. Pertaining to a texture of a porphyritic volcanic rock in which the proportions of phenocrysts and glass ground mass are in approximately equal amounts.

hyalophitic: [Greek *hyalos* = glass + *ophis* = serpent.] adj. Pertaining to the texture of a porphyritic volcanic rock in which the last-formed intersticial material (mesostasis) is glassy and makes up approximately 10-50% of the rock, a proportion intermediate between hyalopilitic and hyalocrystalline.

hyalopilitic: [Greek *hyalos* = glass + *pilos* = felt.] adj. Pertaining to the texture of a porphyritic volcanic rock with needlelike feldspar laths, that are embedded in a glassy ground-mass. The phenocrysts comprise less than approximately 10% of the total mass. Introduced in German as *hyalopilitisch* by Harry Rosenbusch in 1887.

hybrid: [Latin *hibrida* = a cross-bred animal (mongrel), offspring of a tame sow and a wild boar.] n. In petrology, said of an igneous rock whose chemical composition is the result of assimilation. In biology, used as a noun for offspring of two animals or plants of different genera, species, variety, or breed.
See also: hybrid (in Appendix 1)

hydrate: [Greek *hydor* = water.] n. A solid compound formed by the union of molecules of water with other molecules or atoms. The water molecules are combined in a definite ratio as an integral part of the crystal. Not to be confused with hydroxide, which is a compound composed of a element or radical with one or more hydroxyl- (OH-) based groups. The terms are often used in combination (e.g., hydrated hydroxy phosphate).
See also: hydroxide, hydrogen, oxygen

hydraulic: [Greek *hydraulos* = water organ < *hydor* = water + *aulos* = pipe, flute.] adj. In general usage, pertaining to water or other liquids in motion. The term was first incorporated it into engineering by Roman authors, such as Vitruvius. First usage in English c. 1625.

hydric: [Greek *hydros* = water.] adj. Of, characterized by, or adapted to an extremely moist habitat. In comparison to mesic and xeric.
See also: mesic, xeric

hydro-: [Greek *hydor* = water; IE *wed* = water.] Combining prefix meaning water. In addition to the prefix *hydro-*, the Greek noun *hydor* is the root for terms like hydrant and the biological genus *hydra*, based on the nine-headed, snakelike mythological monster. Other IE derivatives include otter, vodka, and whiskey.
See also: water

hydrogen: [Greek *hydor* = water + *genesis* = origin, creation < *genesthai* = to be born, come into being; < IE *gen* = to give birth, beget.] n. A colorless, odorless, flammable gaseous element with an atomic number of 1 (symbol H). Hydrogen is the lightest of all elements. It combines with oxygen to form water, thus inspiring its name "progenitor of water". Discovered by Henry Cavendish in 1766, but not named until 1787 by French chemists Louis Bernard Guyton de Morveau, Antoine Laurent Lavoisier, and others. They collaborated on a publication titled <u>Méthode de Nomenclature Chimique</u> (*The Method of Chemical Nomenclature*). Prior to its being recognized as a discrete element, hydrogen was called "phlogiston" or "inflammable air."
See also: deuterium, tritium

hydrology: [Greek *hydor* = water + *logos* = word, speech, discourse, reason < *legein* = to speak, to say; IE *leg* = to collect.] n. The science dealing with the occurrence, distribution, circulation, and properties of water, on and under the Earth's surface, and in the atmosphere. First usage in English c.1760.
Cited usage: Hydrology is that part of natural history which examines and explains the nature and properties of water in general [1796, Charles Hutton, A Mathematical and Philosophical Dictionary].

hydrolysis: [Greek *hydor* = water + *lysis* = dissolving < *luein* = to dissolve.] n. A chemical decomposition reaction in which a bond is broken and a compound is split into new components by reaction with water. Examples of hydrolysis include the dissociation of a dissolved salt, or the catalytic conversion of starch to glucose. In geology, it commonly implies reaction between an aqueous solution and a silicate mineral, such as hydrolysis of potassium feldspar to form a clay mineral. First usage in English c. 1875.

hydrometer: [Greek *hydor* = water + *metron* = to measure.] n. An instrument for measuring the specific gravity of a liquid, such as seawater, acids, or molten metals. The term relates to the Greek root *hydor*, because the specific gravity of any substance is a unitless comparison of its density to the density of water. Historical evidence alludes to the hydrometer being originally invented by Hypatia, the last librarian of the great library at Alexandria, but was not widely used until Robert Boyle rediscovered it in 1675.

hydrophilic: [Greek *hydor* = water + *philos* = friend, loving.] adj. Having a strong affinity for water; readily absorbing it, dissolving in it, or being miscible with it. Hydrophilic molecules are typically polarized and capable of hydrogen bonding. First usage in English c.1900. The term hydrophobic, coined by George Weinberg, is used in contrast to hydrophilic.
See also: hydrophobic

hydrophobic: [Greek *hydor* = water + *phobos* = fear, dread.] adj. In chemistry; having little or no affinity for water; used in reference to substances that resist entering into solution or absorption with water. In medicine, hydrophobia is the abnormal fear of water. It is also another name for disease rabies, because, when contracted, people lose their ability to drink.
See also: hydrophilic

hydrothermal: [Greek *hydor* = water + *therme* = heat < *thermos* = warm, hot; < IE *gher* = to warm, heat.] adj. Pertaining to hot aqueous solutions, the action of such solutions, or the products of this action, such as a mineral deposit precipitated from a hydrothermal solution. Also used to designate a family of rocks formed by crystallization from heated aqueous solutions, both on or beneath the Earth's surface. IE derivatives include: burn, brandy, brimstone, brand, furnace, fornicate, and ghee (heated and clarified butter). First usage in English c. 1845.

Cited usage: Along with igneous, metamorphic, and sedimentary, 'hydrothermal' should be considered as a fourth major rock family [1990, David Mustart, oral communication].

hydroxide: [Greek *hydor* = water + French *oxide* < Greek *oxus* = sharp, acid.] n. A chemical compound containing the hydroxyl group (OH), such as potassium hydroxide (KOH) and ammonium hydroxide (NH_4OH). In mineralogy, designating a mineral class in which a metallic element or radical is linked with the hydroxyl (OH) ion, such as goethite, FeO(OH). Not to be confused with hydrate, which is a solid compound formed by the union of molecules of water with other molecules or atoms., such as hydrate of lime. First used in 1787 by Antoine Lavoisier, G. de Morveau, and others in their work *Méthode du Nomenclature Chimique* (*Method of Chemical Nomenclature*).
See also: hydrogen, oxygen

hygrometer: [Greek *hygros* = wet, moist < *hydor* = water + *metron* = to measure.] n. An instrument made to measure the water vapor content (humidity) of the atmosphere. There are various types of hygrometers used; they are named based on the method of measurement used, such as: the hair hygrometer, the electro-carbon hygrometer, and the infrared hygrometer. It was invented by French physicist Guillaume Amontons in 1687. First usage in English c.1675.

hypabyssal: [German *hypabissisch* = hypabyssal; < Greek *hypo* = under, beneath + *abyssos* = bottomless, the deep.] adj. Pertaining to an igneous intrusion, or to rock of that intrusion whose depth of solidification is intermediate between that of plutonic (abyssal) and surface extrusive lavas. The term includes such features as: dikes, sills, and laccoliths. Igneous rocks of such intrusions display textures intermediate between coarse-grained plutonic rocks and fine-grained extrusive volcanic rocks. The distinction is

not considered relevant by some petrologists because the term 'volcanic' is generally considered to include both extrusive lavas and hypabyssal intrusions. The etymology can be misleading if taken literally from the ancient Greek as 'below the bottomless deep,' especially when considering that a sill or laccolith may be just a few meters below the surface. But, modern interpretation of the prefix *hypo-* also means 'less than normal,' so hypabyssal intrusions can mean ones that form at less than the normal lithification depth for plutonic rocks. Term coined by Brogger in 1896.
Cited usage: Brogger has insisted on the necessity of a division intermediate between the plutonic and the volcanic, which he terms 'hypabyssal' [1896, Science Progress IV. 476].

hyper-: [Greek *hyper* = over, beyond.] Combining prefix meaning 'over' in the sense of excess, exaggeration, or more than normal, as in hyperbole, hyperbola, hypercritical, and hypersensitive.
See also: hypersthene

hypersaline: [Greek *hyper* = over, beyond + Latin *sal* = salt.] adj. Said of aqueous solutions that are excessively salty. In oceanography, said of water with a salinity substantially greater than that of normal seawater. In most cases, the term is used for solutions supersaturated with respect to halite (NaCl).

hypersthene: [Greek *hyper* = over, beyond + *sthenos* = strength.] n. A greenish-black to dark brown orthorhombic mineral of the orthopyroxene group, magnesium iron silicate, $(Mg, Fe)SiO_3$. It is isomorphous with enstatite and ferrosilite. Hypersthene often has a bronze to greenish-brown play of color called schiller, which results from preferred arrangement of minute inclusions. It is an essential constituent of the plutonic rock norite. Term coined by René-Just Haüy in 1803 due to its superior

hardness as compared with hornblende, a mineral with which it is often confused.
See also: hornblende, hyper-, enstatite, gabbro

hypidiomorphic: [Greek *hypo* = under, beneath + *idios* = self, peculiar, separate, distinct + *morphos* = shape, form.] adj. Referring to the texture of an igneous rock in which the individual crystals are subhedral, meaning each is bounded only in part by crystal faces. Term coined by Harry Rosenbusch in 1887 because the crystal faces are less than or not quite euhedral.
Cited usage: The order being first plagioclase in more or less idiomorphic lath-shaped individuals, lying in all positions, then augite generally allotriomorphic, sometimes hypidiomorphic [1888, A. C. Lawson, Amer. Geologist Apr. 204].
Synonyms: subautomorphic, subidiomorphic
See also: euhedral, anhedral

hypo-: [Greek *hypo* = beneath, under, below.] Combining prefix meaning a) below, beneath, or under, as in hypodermic or hypothermia; b) less than normal or deficient, as in hypabyssal or hypoesthesia; and c) in the lowest state of oxidation, as in hypoxanthine or hypochlorous acid. Used in contrast to *epi-* or *hyper-*.

hypocenter: [Greek *hypo* = beneath, under, below + *kentron* = needle, point, center point in drawing a circle.] n. The focus of an earthquake, being the point of initial rupture where strain energy is first converted into elastic wave energy. First usage in English c. 1900.
See also: epicenter

hypogene: [Greek *hypo* = below, beneath + *genesis* = origin, creation < *genesthai* = to be born, come into being; < IE *gen* = to give birth, beget.] adj. In general geologic usage, said of a process and its resultant features that form beneath the Earth's surface, as opposed to epigene. In ore deposits, said of mineral deposits formed by ascending solutions, as opposed to supergene. Term coined by Charles Lyell in 1833 in reference to the idea that the Earth had not yet given birth to the rocks it had created.

Cited usage: We propose the term 'hypogene' a word implying the theory that granite and gneiss are both nether-formed rocks, or rocks which have not assumed their present form and structure at the surface [1833, Lyell, Principles of Geology III. 374].

See also: epigenetic, supergene

hypolimnion: [Greek *hypo* = below, beneath + *limno* = lake, marsh.] n. In a thermally stratified lake, the denser and lowermost layer of water which, in contrast to higher layers, is generally colder, stagnant, and oxygen-depleted. From top to bottom, the usual layering of a thermally stratified lake is: epilimnion, metalimnion (thermocline), and hypolimnion.

Cited usage: In summer months the upper layers warm up more quickly than the lower regions and a sharp division in temperature–a thermocline–is formed. The reservoir or lake becomes divided into a lower, anaerobic cool layer or hypolimnion and an upper, warm aerobic epilimnion [1971, Nature 26 Feb. 596/1].

hypopycnal: [Greek *hypo* = below, beneath + *pyknos* = dense, thick.] adj. Said of water that is less dense than the body of water it enters, frequently said of freshwater from glacial melt or rainwater runoff that enters the sea and remains floating on the surface.

hypsometric: [Greek *hypsos* = height, top + *metron* = to measure.] adj. Said of a map or curve showing the area or proportion of the Earth's surface above any given elevation or depth, thus the comparative altitude of different places on the Earth's surface. In geology, referring to a cumulative-frequency profile representing the relative areas of the Earth's surface above and below a reference datum, usually sea level. First usage in English c. 1565.

Cited usage: If the whole Earth be divided in to square kilometers and these are arranged in a series according to their height above sea-level, the well-known hypsometric curve of the Earth's surface is obtained [1924, J. G. A. Skerl, translation of Alfred Lothar Wegener's Origin of Continents and Oceans ii. 28].

Synonyms: hypsographic

hysteresis: [Greek *hysteresis* = coming short, deficiency < *hysterein* = to lag behind, come late < *hysteros* = late, coming behind.] n. The lagging of an effect behind its cause, or the phenomenon by which changes in a property lag behind changes in its causative agent, as when the change in magnetism of a body lags behind changes in the magnetic field. In structural geology, a lag in the return of an elastically deformed body to its original shape after the load has been removed. In geophysics, the property exhibited by a rock when its magnetization is nonreversible (e.g., the hysteresis or elastic rebound of rock during an earthquake is the time required to come to rest after the stress has dissipated). Term coined by J. A. Ewing c. 1882.

Cited usage: When there are two qualities M and N such that cyclic variation of N cause cyclic variations of M, then if the changes of M lag behind those of N, we may say that there is hysteresis in the relation of M to N [1885, J. A. Ewing, Phil. Trans. 176 p. 524].

Iapetus: [Eponymous term after *Iapetus* = The Titan from Greek mythology, son of Uranus and Gaia, and father of Atlas, for whom the Atlantic Ocean is named.] adj. Referring to the ocean that existed prior to formation of Pangea, the supercontinent created when ancestral Europe and Africa collided with ancestral North America during the Late Paleozoic.
Synonyms: proto-atlantic

ice: [Old English *is* < German *eis* < Old Norse *iss* = ice.] n. Formed in nature by the freezing of liquid water, ice is a colorless to pale-blue to greenish-blue hexagonal mineral, hydrogen oxide, H_2O. It is commonly white, due to included gas bubbles. More than nine polymorphs of ice are known with remarkable similarities to the polymorphs of silica. First usage in English before c. 900.

ichnofossil: [Greek *ikhnos* = footprint, track + Latin *fossilis* = dug up < *fossus* = past participle of *fodere* = to dig.] n. A trace fossil, being a sedimentary structure representing the fossilized track, trace, burrow, tube, boring, or tunnel resulting from the life activities of an animal, apart from the growth of the organism.

ichnology: [Greek *ikhnos* = footprint, track + *logos* = word, speech, discourse, reason < *legein* = to speak, to say; IE *leg* = to collect.] n. The branch of paleontology dedicated to the study of trace fossils, such as tracks, trails, burrows, imprints, or other traces of animal activity. Term coined by F. T. Buckland in 1844.
Cited usage: Ichnology, as a science, began with "Dr. E. A. Hitchcock" [1864, Proc. Amer. Phil. Soc. IX. 445].

ichor: [Greek *ichor* = in Greek mythology, the ethereal fluid that flows in the veins of the gods.] n. An obsolete term for a theorized residual magmatic fluid, thought to be responsible for the process of granitization, in which sedimentary or metamorphic rocks are transformed to granitic rocks without melting. Evidence from experimental petrology has generally rejected the significance of such solid-state granitization. Term coined by J. J. Sederholm in 1926.
Cited usage: The writer proposes to introduce, instead of the word granitic juices, the term granitic ichor, preliminarily with no more strictly defined signification than that possessed by the word juice. It will soon be possible to give to the term a stricter definition [1926, J. J. Sederholm, Bulletin de la Commission Géologique de Finlande XII. LXXVII. 89].
Similar terms: mineralizer

iddingsite: [Eponymous term after American petrologist Joseph Paxton Iddings (1857-1920).] n. A reddish-brown mixture of silicates and oxides formed by the alteration of olivine. It forms rust-colored patches and rims around olivine crystals that are contained in mafic and ultramafic igneous rocks.

idio-: [Greek *idio-* = self, peculiar, separate, distinct.] Combining prefix meaning distinct, especially relating to the self or individual as opposed to the group; used in such terms as idiomorphic, idioblastic, and idiolect.

idioblastic: [Greek *idio* = self, peculiar, separate, distinct + *blastos* = sprout, shoot, germ.] adj. In

petrology, said of a mineral formed by recrystallization in a metamorphic rock having distinct crystal faces. The 'idioblastic series' refers to a ranking of metamorphic minerals expressing their relative ability to develop distinct crystal faces when competing with each other for growth space. Term coined by Friedrich Johann Karl Becke in 1903.
Cited usage: Most of the hornblende existing in the amphibolite...is clearly secondary, and from its idiomorphic forms would be called 'recrystallized'.... For such cases Prof. F. Becke has proposed the term idioblastic [1908, Q. Jrnl. Geol. Soc. LXIV. 482].

idiomorphic: [Greek *idio* = self, peculiar, separate, distinct + *morphos* = shape, form.] adj. In mineralogy, referring to euhedral crystals. In petrology, referring to a texture of igneous rocks that are characterized by euhedral crystals. Term coined by Rosenbusch in 1887.
Cited usage: An idiomorphic mineral is bounded by crystal planes. [1888, W. S. Bayley, Amer. Naturalist Mar. 208].
Synonyms: euhedral, automorphic
See also: euhedral

idocrase: [Greek *eidos* = form, figure + *krasis* = mixture, combination < *keranunai* = to mix.] n. An obsolete mineral name, replaced by vesuvianite. Term coined by René-Just Haüy in 1796.
Synonyms: vesuvianite
See also: vesuvianite

igneous: [Latin *igneus* = of fire < *ignis* = fire < *ignire* = to set on fire < Sanskrit *agni* < IE *egni* = fire.] adj. Referring to or said of a rock or mineral formed by solidification from molten or partially molten material. Igneous is also used to designate one of the four families into which rocks are divided. The other three families are: sedimentary, metamorphic, and hydrothermal. The two sub-families of igneous rocks are volcanic and plutonic. The term igneous alludes to the god Agni, who was the personification of the sacrificial fire,

being one of the three chief gods described in the Rig Veda, the ancient Hindu body of sacred knowledge that includes "Divine Hymns" to prominent deities. He was a prototype of other torchbearers in mythology, such as Lucifer and Prometheus. Etymologically related terms include ignition and ignite.

ignimbrite: [Latin *ignis* = fire < *ignire* = to set on fire < *ignis* < Sanskrit *agni* < IE *egni* = fire + *imbris* = shower of rain, storm cloud.] n. A fine-grain igneous volcanic rock composed of pyroclastic material, formed by the deposition and consolidation of ash flows. The term originally implied dense welding, as in welded tuffs, but is now applied to nonwelded ash-flow tuffs or sillar. Any pyroclastic rock, typically a welded tuff or a nonwelded sillar, deposited from or formed by the settling of a nuée ardente. Term coined in 1935 by New Zealand geologist P. Marshall.
Cited usage: This is matched in volume only by the large welded tuff sheets of the central North Island of New Zealand, the type locality for ignimbrite [1970, Nature 12 Sept. 1125/1].
Synonyms: flood tuff
See also: sillar, tuff

illite: [Toponymous term after *Illinois*, a state in the U.S. Midwest..] n. A general term for a group of three-layer micas and clays, common in argillaceous sediments, especially of marine origin. Illite is intermediate in structure and composition between muscovite and montmorillonite. Illite has less potassium, but more water than muscovite; it has more potassium than montmorillonite. In social history, the Illites were members of a confederation of Algonquian Indian tribes formerly inhabiting an area in and around the U.S. state of Illinois. Term coined in 1937 by R. E. Grim and coworkers, in recognition of the state of Illinois, where study of clay minerals has been encouraged, and where the type locality for illite in the Maquoketa Shale is found.

Cited usage: There remains only the alternative of giving a new name to the mica occurring in argillaceous sediments, and the term illite, taken from the State of Illinois, is here proposed. It is not proposed as a specific mineral name, but as a general term for the clay mineral constituent of argillaceous sediments belonging to the mica group [1937, R. E. Grim et al., Amer. Mineralogist XXII. 816].

illuviation: [Latin *il-, in-* = in, into, on, upon + *luvia* = washing < *luere* = to wash.] n. In pedology, the process of accumulation of soluble salts or suspended material in a lower soil horizon, that was transported by percolating water from an upper horizon by the process of leaching or eluviation. As the adjective illuvial, said of the lower horizon of accumulation which is usually a B horizon. First usage in English c. 1920.
Cited usage: Three main horizons are generally recognized, the A or eluvial horizon, the B or illuvial horizon, and the C horizon, which consists of the parent material [1924, Geol. Mag. LXI. 451].
See also: alluvium, diluvium, eluviation, proluvium

ilmenite: [Toponymous term after the discovery locality, Lake Ilmen, in the Ilmen Mountains in the southern Urals of Russia..] n. An iron-black, opaque, trigonal mineral, iron titanium oxide, $FeTiO_3$. It is isomorphous with manganese-bearing pyrophanite, and magnesium-bearing geikielite. Ilmenite shows complete solid solution with hematite above 950°C. Ilmenite is the chief ore mineral of titanium. It occurs in large masses formed by magmatic segregation in an anorthosite plutons, is a common accessory mineral in mafic igneous rocks, and is concentrated with other dense minerals in beach sands. First usage in English c.1825.
Cited usage: Ilmenite, or titanic iron (Fe Ti)$_2$O$_3$, an ore in which one of the iron molecules of hematite is replaced by the metal titanium [1827,

Edinburgh New Philosophical Jrnl. III. 187].
Synonyms: titanic iron ore

im-: [Variant form < Latin *in* = in, within, with intensive force, against.] Combining prefix meaning against, not, unable, in, within, or with intensive force. In the negative sense meaning against, it is used in words such as, immature, immiscible, impotent, imbecile, impasse, and impermeable. In the sense meaning within, or with intensive force, used in words such as impregnate, immerse, and impose. In English usage, the prefix *im-* is used before *b, m,* or *p*, whereas its base form is used before most other letters.
Synonyms: in-
See also: in-

imbricate: [Latin *imbricatus* = covered with roof tiles < *imbrex* = roof tile < *imber* = rain.] adj. In general usage, to be arranged with regular overlapping edges. In sedimentology, referring to a structure characterized by flattened pebbles or cobbles, all tilted in the same direction and generally dipping upstream. In structural geology, said of a series of closely spaced overlapping thrust faults, all verging in the same direction, dipping toward the source of stress, and located above a decollement or detachment fault. In paleontology and biology, said of the scales of many reptiles and fish, or the scales of a botanical cone or bud. In optical mineralogy, the scaly microscopic texture of certain minerals such as tridymite. First usage in English c. 1665.
See also: schuppen

immiscible: [Latin *im* < *in* = against, not + *miscere* = to mix; < IE *meig* = to mix.] adj. Said of two or more phases incapable of completely dissolving in one another. The term is usually used in reference to liquids, such as oil and water, but also refers to solid phases, such as albite and

orthoclase, that are miscible at high temperatures, but become immiscible and separate from each other at lower temperatures. First usage in English c. 1670.
See also: phase

impactite: [Latin *im* < *in* = against, not + *pangere* = to fasten, to drive in.] n. A vesicular glassy to finely crystalline material created by fusion of the target rock by the heat of impact of a large meteorite. Impactite is found in and around the resulting crater. It usually consists of individual bodies of glass mixed with rock fragments and traces of meteoritic material. The term impactite has been used incorrectly for any shock-metamorphosed rock. Term coined by H. B. Stenzel c. 1935, then director of the geology museum at the University of Texas.
Cited usage: Spencer's meteorite splash origin is valid for the formation of certain glasses. Glasses of this type will be distinguished in general from most of those now included under tektites. These meteorite splashes should be given a distinctive name, such as impactites. This name was suggested by Dr. H. B. Stenzel [1940, V. E. Barnes, N. Amer. Tektites in Univ.Texas Publ. #3945, p. 558].
See also: tektite

impermeable: [Latin *im* < *in* = against + *permere* = to penetrate < *per* = through + *meare* = to pass; < IE *mei* = to change.] adj. Said of a rock layer, sediment, or soil that does not permit the passage of pressurized water or other fluid. First usage in English c. 1695.
Synonyms: impervious

impregnated: [Latin *in-* = in, into + *praegnatus* = pregnant.] adj. Said of a dispersed or disseminated deposit of a usually metalliferous or mineral ore that is younger than the host rock.
Similar terms: disseminated
See also: epigenetic

impression: [Latin *im* < *in* = in, within, toward + *prestare* = to furnish, lend, give < *praest* = it is near, at hand.] n. An indentation, concavity, or shallow mold made in a soft sediment surface by a usually harder structure, such as a fossil shell or leaf. Also, the circular pit made by hail or rain drops. The term is also used for such trace fossils as: footprints, tracks, and burrows. The original French meaning of this term was to hire for military service by advance payment. First usage in English c. 1350.
Synonyms: imprint, shallow mold

in-: [Latin *in* = in, within, with intensive force, against.] Combining prefix having two main, but nearly opposite, meanings; one meaning is negative, used in such terms as: inorganic, insoluble, infertile, intangible and inarticulate; the other meaning is as an intensive used for emphasis in such terms as: inflame, incense, and intense. This dual-meaning combining prefix is potentially very confusing. It is both homophonic and homographic, therefore it sounds and is spelled the same way, but has distinctly different meanings, in this case opposites. The textbook case of the confusion that can arise with opposing homophones occurred with the prefix *in-*. In the 1950's, commercial fuel trucks used to have warning signs that read *inflammable*, meaning that the contents were explosively flammable (the intensive form), but too many people took it for the negative form, thinking it meant "not flammable." To the disgust of some stalwart linguists, traditional English was sacrificed for the sake of safety, a new word was coined, and the warning signs were changed to read *flammable*. As a result of this change, and in an effort to be completely clear, the antonym *nonflammable* was soon coined and put into use on containers that may have looked dangerous, but actually contained safe materials.
Synonyms: im-
See also: im-

inarticulate: [Latin *in* = not + *artus* = joint; < Greek *arthron* = joint, article.] n. Any brachiopod belonging to the class Inarticulata, characterized by calcareous or chitinophosphate valves.

incandescent: [Latin combining suffix means intensive + *candescence* = to become bright.] adj. Said of lava flow or ash flow that is glowing due to its high temperature. First usage in English c. 1790

inch: [Latin *uncia* = twelfth part.] n. In linear measurement, one-twelfth of a foot. The unit of weight *ounce* also comes from the Latin *uncia* because in Troy measure (used in gemology), there are 12 ounces in a pound (as opposed to 16 per pound in avoirdupois measurement).

incidence: [Latin *in* = in, within + *cadere* = to fall.] adj. Said of the angle formed by the intersection of a ray and a line. In the physical world it usually involves a ray of energy (e.g., light, radio waves) reflecting off of, or refracting into, a surface.

inclination: [Latin *in* = in, toward + *clinare* = to lean; < IE *klei* = to lean.] n. The slope attitude of any structure or body. Etymologically related terms include: monoclinic, triclinic, climate, pericline, anticline, syncline, climax, and clitoris (Greek *kleitoris*, meaning small inclined hill).

inclusion: [Latin *includere* = to enclose < *in* = in, within + *claudere* = to close.] n. A relatively small piece of older rock contained within an igneous rock with which it may or may not be compositionally related.
Similar terms: enclave, enclosure
See also: xenolith, autolith

incongruent: [Latin *in* = not + *congruent* = to agree.] adj. In general use a term meaning inconsistent or disagreeing. In petrology and chemistry said of melting in which one solid phase (such as

orthoclase) reacts on melting to produce a new solid phase (such as leucite) and a liquid differing in composition from the original solid (such as melt richer in silica than orthoclase). First usage in English c. 1530.
See also: congruent

incrustation: [Latin *in* = on + *crusta* = crust; < Greek *kruos* = icy; < IE *kreus* = to begin to freeze.] n. The coating of a rock surface with a thin layer of usually crystalline minerals. Etymologically related terms include: crystal, crystallo-, crustacean, and crouton.
Alt. spelling: encrustation
Similar terms: druse, drusy
See also: druse

index: [Latin *index* = forefinger.] n. A fundamental aspect, quality, or quantity that acts as a marker, guide, or indicator in referencing or measuring. In geology, used in compound terms, such as a marker bed in stratigraphy, or as a reflection marker in seismology. This term is used to represent markers because the index finger is used for pointing to and marking an object.

indigenous: [Latin *indigena* = native to an area < *indu* = in, within + *genea* = race, generation < *genos* = race, kind < *gonos* = bith, offspring; < IE *gen* = to give birth, beget.] adj. In paleontology and biology, said of an organism originating in a specific locale, a mineral or resource formed 'in situ,' or a stream that lies wholly within its drainage basin.
See also: autochthonous

induration: [Latin *in* = intensive prefix + *durare* = to harden < *durus* = hard.] n. The hardening of a rock by heat, pressure, or the action of a cementing material.
Cited usage: Solid deposits of indurated sandstone [1848, Hugh Miller, The First Impressions of England and Its People].

inert: [Latin *in* = not + *ars* = skill.] adj. In chemistry, said of elements such as the inert gases that have little or no ability to react, and occur uncombined in the atmosphere. The inert gases are: helium, neon, argon, krypton, xenon, and radon. First usage in English c. 1645.

inertia: [Latin *inertia* = idleness < *in* = not + *ars* = skill.] n. In physics, the tendency of a body to maintain its state, whether in motion or at rest, unless acted on by an external force.

influent: [Latin *in* = in + *fluere* = to flow.] n. A surface stream that flows into a larger body of water (as opposed to outward-flowing effluent). In ecology, an organism which affects the ecological balance of a plant or animal community.
Cited usage: F. E. Clements coined the term influent to cover those organisms which have important relations in the biotic balance and interaction [1926, V. E. Shelford, Ecology VII. 389].
See also: biome, effluent

infra-: [Latin *infra* = below, beneath, within.] Combining prefix meaning below or beneath, as in infrared, infralittoral, and infraglacial; also meaning within, as in infrastructure and infra-corpus (within the body). Usage similar to the prefix *sub-*, and in contrast to *supra-*.

infralittoral: [Latin *infra-* = below, beneath, under.] adj. Pertaining to the zone or region of the sea below the littoral region; the lower part of the intertidal zone that is exposed to air only during the very lowest spring tides.

inlier: [Latin *in* = within + Middle English *lien* = that which lies < Old English *licgan* to lie.] n. An area or formation of older rocks completely surrounded by younger layers, as opposed to outlier. Term coined by Drew c. 1860.

Cited usage: Inlier, a term introduced by Mr. Drew, of the Geological Survey, to express the converse of 'outlier'. 'It means a space occupied by one formation which is completely surrounded by another that rests upon it' [1859, David Page, Handbk. Geol. Terms 256].
See also: outlier

inosilicate: [Greek *inos* = muscle, fibre, nerve, strength + Latin *silex* or *silic* = hard stone, flint.] n. The structural group of silicate minerals containing silicon-oxygen tetrahedra (SiO_4) linearly linked in single or double chains by the sharing of oxygen atoms. Formerly called metasilicate.
See also: cyclosilicate, nesosilicate, phyllosilicate, sorosilicate, tectosilicate

inselberg: [German *inselberg* = island mountain < *insel* = island + *berg* = mountain; < IE *bergh* = high.] n. An isolated hill or mountain that rises abruptly from its extensive lowland surroundings, typically from a plain in a hot, dry region, such as the deserts of southern Africa and Arabia. Inselbergs are generally barren and rocky, but usually partially buried in erosional debris that accumulates on the gentle slopes at the bases of alluvial fans. Term coined by W. Bornhardt. IE derivatives include, iceberg and bourgeois.
Cited usage: Two important features of Triassic geography were the hills of Charnwood and the Mendips. These were inselbergs rising above the general level of the Triassic landscape but gradually buried as the deposits accumulated [1969, Bennison & Wright, The Geological History of the British Isles xii. 272].
Synonyms: island mountain
See also: monadnock

insequent: [Latin *in* = not + *(con)sequens* = consequent < *com-* < *cum* = intensive used for emphasis + *sequere* = to follow.] adj. Said of a stream, valley, or drainage pattern having a course or form that appears haphazard, exhibiting no apparent relation to the form, slope, or structure of the land. The main reason for this is that the primary features are largely controlled by headward erosion on horizontally stratified and homogeneous rock. This results in erosion without pattern. Term coined by W. M. Davis in 1897. *Cited usage*: Then the side streams, growing headwards, are accidentally located; and streams of this class have been called autogenetic by McGee. Insequent may prove to be a more satisfactory name for such streams, as it is of the same etymological family as consequent, subsequent and obsequent. As insequent has proved serviceable in my lectures during the past winter, it is now submitted for trial by others [1897, W. M. Davis, Science 2 July 24/1].

in situ: [Latin *in* = in, within + *situ* < *situs* = place.] adj. Latin borrowed phrase meaning 'in its original place' or 'in position.'

insolation: [Latin *in* = in, on, upon, towards, against + *sol* = sun.] n. The combined direct solar and cosmic radiation reaching a given celestial body, such as the Earth or the Moon. *See also*: insulation

insulation: [Latin *insulatus* = made into an island < *insula* = island.] n. Material used to create an isolated condition, such as to prevent or reduce the transfer or leakage of heat, electricity, and sound. First usage in English c. 1795. *See also*: insolation

inter-: [Latin *inter-* = between, among, amid, in between, in the midst.] Combining prefix meaning between, as in intertidal, interglacial, and interfacial; also meaning amid or among, as in interference, intercalate, and intercept.

intercalation: [Latin *inter* = between + *calare* = to proclaim; < IE *kel* = to shout.] n.

An atom, molecule, or substance that enters between the layers of the crystal lattice of another substance. In stratigraphy, thin layers of material that exist between thicker layers of a different character, such as beds of shale or ash layers that are intercalated in a body of sandstone. In volcanology, thin lava sheets between sedimentary strata. In paleontology, a distinct fossil horizon between fossil zones of different character. In mineralogy, said of the occurrence of lamella of one mineral in another in such a way that the lamellae are oriented in planes related to the crystal structure of the host mineral, such as intercalations of albite in perthite. Originally, to intercalate was to insert a day into the calendar, hence, the need to "proclaim" it. First usage in English c. 1610.

interfluve: [Latin *inter* = between, among, in the midst + *fluvius* = river.] n. The area between rivers, especially the relatively undissected upland region between the valleys of adjacent streams flowing in the same general direction. Term coined prior to 1900 as a back-formation from interfluvial. *Synonyms*: interstream area

intergranular: [Latin *inter* = between + *granatus* = seed.] adj. In igneous petrology, said of the texture of an igneous rock in which augite occurs as an aggregation of grains in the interstices between feldspar laths. In glacial geology, referring to movement of ice grains that have rotated and slid over each other, contributing to flow near the surface of the glacier. First usage in English c. 1930.

interlobate: [Latin *inter* = between + *lobos* = pod, lobe.] adj. In glacial geology, said of glacial deposits, such as till or outwash lying between two glacial lobes. Term coined in 1881 by American geologist Thomas Crowder Chamberlin (1843-1928). *Cited usage*: A peculiar morainic type to which the term intermediate or interlobate

moraines will be applied [1881, T. C. Chamberlin, Rep. U.S.G.S. 313].
Synonyms: intermediate moraine

internides: [Latin *inter-* = between, among, amid, in between + *nidus* = nest.] n. The internal part of an orogenic belt, farthest away from the craton, thought to be the site of a eugeosyncline in its early stages, and later subjected to folding and plutonism. Term coined by Kober.

intersertal: [Latin *inter* = between, in the midst + *serere* = to set, put, place, insert.] adj. In igneous petrology, referring to the texture of a porphyritic igneous rock that has groundmass composed of glassy material occupying the interstices between unoriented feldspar laths. In thin section, platy or tabular crystals often appear lath-shaped.
Cited usage: Among the basic melaphyres there occurs a texture analogous to the ophitic, namely the intersertal [1916, A. Johannsen, translation of Weinschenk's Fund. Princ. Petrology].

interstade: [Latin *inter* = between + *status* = stood < *stare* = to stand.] n. A warmer episode of a glacial periodic stage marked by temporary glacial retreat or still stand. Etymologically related terms include: stadium, status, and stage.

interstice: [Latin *interstitium* = to stand or put between < *inter* = between + *sistere* = to cause to stand, to set up.] n. A void, opening, crevice, space, or pore in the Earth, as in a rock or a mass of soil. As the adjective interstitial, said of minerals that fill voids in a host rock. Also, said of atoms that occupy defects within a crystal lattice. First usage in English c. 1600.

interval: [Latin *inter* = between + *vallum* = rampart, palisade.] n. In general usage, a space between two objects or points; also, the amount of time between two specified instants or events. In stratigraphy, an interval is the body of rock between two stratigraphic markers. In chronostratigraphy, a polarity interval is the fundamental unit of polarity classification, referring to rock formed during a specific polarity episode. In glacial geology, an interval is an informal subdivision of an interstade. In cartography, an interval is the space between two adjacent contour lines. First usage in English c.1275.

intra-: [Latin *intra-* = on the inside, within.] Combining prefix meaning inside or within, used in terms such as intrazonal, intramuscular, and intrafacies (subordinate facies occurring within a major facies). Used in contrast to the prefix *extra-*. Not to be confused with the prefixes *inter-*, meaning between, amid; or *infra-*, meaning below, within.

intratelluric: [Latin *intra* = within + *tellus* = earth.] adj. Occurring, taking place, or formed in the interior of the Earth. In igneous petrology, said of phenocrysts that formed at depth in a magma prior to extrusion of the molten rock as lava. Term coined by Harry Rosenbusch, initially in German as *intratellurisch*, then quickly anglicized. First usage in English c. 1885.
Cited usage: After their slow development in the magma during an intra-telluric period [1889, Nature 17 Jan. 273/2].

intrusion: [Latin *intrudere* = to intrude, to thrust in < *in* = in + *trudere* = to thrust.] adj. In igneous petrology, the process of emplacement of magma into a pre-existing host rock. Also, the igneous rock mass so emplaced. In sedimentology, the forced injection of a relatively large amount of plastic sediment, such as: mud, clay, chalk, salt, or liquefied sand. In hydrology, generally referring to displacement of fresh surface or ground water by influx of salt water. Used as an adjective in terms such as: intrusion breccia, intrusion displacement, intrusive tuff, and intrusive vein. First usage in English c. 1275.

Cited usage: By this intrusion of the petrifying particles, this substance also becomes hard [1665, Robert Hooke, Micrographia xvii. 109].

intumescence: [Latin *intumescere* = to swell up < *in* = prefix used for emphasis + *tumere* = to swell.] n. The property shown by certain minerals to swell or froth when heated due to the release of gas. Kernite is an example, being a hydrated sodium borate that releases water and swells when heated. First usage in English c. 1665.

invar: [French *invar(iable)* = trademark for human-made metallic alloy, invariable.] n. An alloy of iron or steel (approx. 64%) plus nickel (approx. 36%), having a very small coefficient of thermal expansion (one twenty-fifth that of steel). It is used in the construction of surveying instruments, such as level rods, tapes, and pendulums. The material was invented, and the term coined by Dr. C. E. Guillaume in 1897. Invar was originally copyrighted as a proprietary name and trademark, but is now a common term.
Cited usage: The more recent discovery of the nickel-steel alloy, *Invar*, by Dr. C. E. Guillaume has, however, to a considerable extent revolutionized compensated pendulums [1928, J. E. Haswell, Horology iii 29].

inver: [Manx *inver* from Gaelic *inbhir* = mouth of the river < *in* = in + *beir* = to carry.] n. A place where a river flows into the sea; often used in location names, such as Inverness, Scotland and Inverness, California (the Gaelic equivalent is *Inbhirnis*).

invertebrate: [Latin *in* = not, without + *vertebra* = spinal joint.] n. Any animal belonging to the phyllogenetic subkingdom Invertebrata. Taxonomically, the term is also defined by exclusion, being any animal not classified within the phylum Chordata and subphylum Vertebrata. Invertebrates lack a backbone, are ectothermic (cold-blooded), and are by far the most numerous grouping of animals on Earth, comprising over 2 million species (98% of known species). Invertebrate phyla include: Arthropoda, Mollusca, Cnidaria, Porifera, Protozoa, and Echinodermata; classes within these phyla include: Insecta, Crustacea, Decapoda, Asteroidea, Bivalvia, and Copepoda. The classification of vertebrate and invertebrate animals was originally conceived by Lamarck, but the term invertebrate does not appear in his *Système des Animaux sans Vertèbres* of 1801. He does use the term in his *Philosophie Zoologique* of 1809; however, by this time the term had already been in use by Cuvier and Duméril for four years. First usage in English c. 1805.
Cited usage: Invertebrate animals are divided by Lamarck into two great groups, which he calls 'animaux apathiques', and 'animaux sensibles' [1838, Penny Cyclopedia. XII. 488/1].

iodine: [Greek *ioeides* = violet-colored < *ion* = violet + *oeides* = resembling.] n. A grayish-black non metallic halogen element (symbol I). Iodine sublimes to a dense violet vapor when heated. The element was discovered in 1811 by French chemist Bernard Courtois, and the name was coined in 1814 by British chemist Sir Humphrey Davy.

ion: [Greek *ion* = something that goes < *ienai* = to go.] n. Any individual atom or molecule having a net electric charge. If the charge is positive, it is called a cation, and if negative, an anion. Term coined by British physicist Michael Faraday in 1834.
Cited usage: Finally, I require a term to express those bodies which can pass to the electrodes. I propose to distinguish these bodies by calling those anions which go to the anode of the decomposing body; and those passing to the cathode, cations; and when I have occasion to speak of these together, I shall call them ions [1834, Faraday, Phil. Trans. R. Soc. CXXIV. 79].
See also: anode, cathode

iron: [Old English *isarn* = iron < *iss* = ice + *aes* = metal; < Anglo-Saxon *iren* = iron < German *eisen* = holy or strong metal; < Celtic *isarnon* = powerful, holy; < IE *eis* = strong, passion.] n. A silvery to silver-white, heavy, malleable, ductile, magnetic, metallic element (symbol Fe). It occurs rarely in the native state on Earth, as it is readily oxidized (generally into the +2 ferrous or +3 ferric state). Native iron is, however, common in meteorites. Iron is the most widely used metal, and occurs in various oxides, sulfides, silicates, and other minerals. Iron's symbol *Fe*, the prefix *ferr-*, and the words ferrous and ferric all derive from the Latin root *ferrum*, meaning iron; < Greek *hieros*, meaning filled with the divine, holy. IE derivatives include: estrogen, irate, and hierarchy. Due to the many variant forms of the word, the term iron did not gain universal acceptance in English until the early 1600's. First usage in English before 900.

isinglass: [Dutch *huisenblas* = sturgeon's bladder < *hausen* or *hus* = sturgeon + *blas* = bladder.] n. Muscovite in thin transparent sheets. The original isinglass is from a transparent, almost pure gelatinous substance prepared from the air bladder of the sturgeon and certain other fishes, and used as an adhesive and a clarifying agent. Due to the transparency of this substance, conventional spelling of this term evolved from *huisenblas* to *huisenglas*, and finally to *isinglass*. This, in turn, resulted in a folk etymology relating the root back to the word glass. A plentiful source of exceptionally large and transparent crystals of isinglass was discovered in Russia near Moscow. These crystals became known as Muscovy Glass, which was later named muscovite. First usage in English c. 1540.

island: [Middle English *ilond* < Old English *egland* = island < *ea* = watery place, pertaining to water; < IE *akwa* = water + Old English *land* < IE derivatives include: *lendh* = open land.] n. A general term for a body of land that is smaller than a continent, and is completely surrounded by water. Islands are found in the world ocean, in seas, lakes, and streams. The ten largest islands of the world are: a) Greenland (2,175,600 sq km/840,000 sq mi); b) New Guinea (808,510 sq km/312,170 sq mi); c) Borneo (751,100 sq km/290,000 sq mi); d) Madagascar (587,041 sq km/226,658 sq mi); e) Baffin Island (507,451 sq km/195,928 sq mi); f) Sumatra 473,605 sq km/182,860 sq mi); g) Honshu (230,455 sq km/88,979 sq mi); h) Great Britain (229,870 sq km/88,753 sq mi); i) Ellesmere Island (196,236 sq km/75,767 sq mi); and j) Victoria Island (217,291 sq km/83,897 sq mi). First usage in English before 900.

iso-: [Greek *isos* = equal.] Combining prefix meaning equal, used primarily in scientific terms of Greek origin, such as isotope, isocline, isochron, isomorph, and isotherm. The Latin equivalent prefix is *equi-*, used in terms such as: equidistant, equigranular, and equinox.

isobar: [Greek *isos* = equal + *baros* = weight < *barys* = heavy.] A line on a map, chart, or graph connecting points of equal barometric pressure at a given time. Generically, lines connecting points of equal value are called isopleths, isolines, or isograms. First usage in English c. 1695.
Synonyms: isogonic line
Similar terms: isopiestic
See also: isopiestic

isobath: [Greek *isos* = equal + *bathos* = depth.] n. In oceanography and limnology, a line on a map or chart that connects points of equal water depth. In hydrology, an imaginary line that connects points at the same vertical distance relative to a aquifer or the water table. First usage in English c. 1885.
Synonyms: depth contour, bathymetric contour

isochore: [Greek *isos* = equal + *khorein* = to spread about < *khoros* = place, room.] n. In geochemistry, a line connecting points that

represent shapes of equal volume. In stratigraphy, drilled holes of equal size used for showing the thickness of a stratigraphic unit. In petroleum geology a line on a map showing the oil-water contact, used for determining reservoir volume.
See also: isopach

isocline: [Greek *isoklinos* = equally balanced < *isos* = equal + *klinein* = to bend, slope, slant.] n. An anticline or syncline so tightly folded that the rock beds of the opposing limbs are nearly parallel. *Cited usage*: Where a series of strata has been so folded and inverted that its reduplicated members appear to dip regularly in one direction, the structure is termed isoclinal [1882, Geikie, Textbook Geol. 503].

isogon: [Greek *isos* = equal + *gonos* = angled < *gonia* = angle; < IE *genu* = knee.] n. In meteorology, a line of equal or constant wind direction. In geology and mining, a synonym for isogonic line, being an isomagnetic line connecting points of equal magnetic declination. In mathematics, a polygon having all corners at the same angle; an equiangular polygon. IE derivatives include: orthogonal, diagonal, and genuflect.

isograd: [Greek *isos* = equal + *gradus* = to step, walk.] n. A line on a map joining points where metamorphic recrystallization has taken place at about the same temperature and pressure. Such a line often identifies the boundary between two contiguous metamorphic facies or zones of metamorphic grade. Each of these zones is defined by the appearance of specific index minerals (e.g., garnet isograd, staurolite isograd). As the adjective isograde, isogradal, or isogradic, referring to rocks that are at the same grade of metamorphism.

isogyre: [Greek *isos* = equal + *gyros* = ring, circle.] n. In optical mineralogy, a black or shaded part of an interference figure that may look like one of the four branches of a black cross for

uniaxical minerals, and a block parabola for biaxical minerals. It is produced by extinction, and indicates the emergence of those components of light with equal vibration directions.

isohaline: [Greek *isos* = equal + *halinos* = salty.] n. A line on a map or chart that connects points of equal salinity in the ocean or other body of salt water. As an adjective, referring to a constant salinity throughout a system or body of water. First usage in English c. 1900

isohyet: [Greek *isos* = equal + *hyetos* = rain.] n. A line drawn on a map connecting points of equal precipitation. First usage in English c. 1895.

isohypse: [Greek *isos* = equal + *hypsos* = height.] n. A line drawn on a map joining points of equal height or elevation, usually in reference to a constant geopotential datum surface. *Synonyms*: contour

isomer: [Greek *isos* = equal + *meros* = part.] n. In organic chemistry, a compound, ion, or radical that has the same chemical composition as one or more other compounds, ions, or radicals, but differs in structural arrangement. The term isomer is distinct from the term polymorph, in that polymorphs are variations of a crystalline solid whereas isomers are variations of a noncrystalline substance in solution. In meteorology a line connecting points of average precipitation. In nuclear physics, nuclides having the same number of protons and neutrons but differing in energy levels. First usage in English c. 1835.

isometric: [Greek *isos* = equal + *metron* = measure, poetic meter; < IE *me* = to measure.] adj. Of the six crystal systems, referring to the one characterized by three equal-length axes at right angles to one another. The other five crystal systems are: tetragonal, orthorhombic, hexagonal, monoclinic, and triclinic. First usage in English c. 1835.

isomorph: [Greek *isos* = equal + *morphos* = shape, form.] n. In mineralogy and crystallography, a mineral or synthetic compound which has the same crystal structure, but different chemical composition as another mineral or compound. Both minerals are so closely related that they form members of an isomorphous or solid solution series. Ionic substitution is possible between such isomorphs; for example, the end members of the olivine series, forsterite and fayalite. Etymologically, the term isomorph is derived by back-formation from isomorphous. First usage in English c. 1860.
Cited usage: Isomorph, a substance which has the same crystalline form with another [1864, Webster's Dictionary].
See also: heteromorphism, polymorph, convergence

isopach: [Greek *isos* = equal + *pakhus* = thick < *pachos* = thickness.] n. A line drawn on a map joining points of equal true thickness of a particular stratum or group of stratigraphic units. Isopachs are usually drawn on an isopach map, made specifically for the purpose of representing cross-sectional information on a plan view map. First usage in English c. 1915.
Synonyms: thickness contour
See also: isochore

isopiestic: [Greek *isos* = equal + *piestos* = compressible < *piezein* = to compress, press.] adj. In hydrology referring to a line drawn on a map or diagram that joins points of equal water pressure (pressure head) of ground water in a subsurface aquifer. The flow of groundwater is perpendicular to these lines. In general, the terms refers to lines of equal pressure. This meaning is further constrained in engineering , refering to a process taking place at constant pressure. First usage in English c. 1870.
Synonyms: equipotential, isopotential, piezometric
See also: isobar

isopleth: [Greek *isoplethes* = equal in number < *isos* = equal + *plethos* = multitude, quantity.] n. A general term for a line on a map connecting points of equal value that represent a unique variable, such as elevation, temperature, pressure, or salinity. In geochemistry and experimental petrology, an isocompositional section portaryed on a phase diagram having constant composition, but variable pressure and temperature. In meteorology, a line drawn on a graph joining points of equal magnitude, such as: rainfall, temperature, or barometric pressure. First usage in English c.1905.
Cited usage: Isobars are isopleths of pressure, isotherms are isopleths of temperature, and so on [1911, N. Shaw, Forecasting Weather 4].
See also: choropleth

isopod: [Greek *isos* = equal + *pedos* = foot.] n. Any marine, freshwater or terrestrial crustacean belonging to the order Isopoda, characterized by seven pairs of legs adapted for crawling. In general, isopods have a compressed body, and sessile eyes, but lack a protective carapace. Examples of isopods are: wood lice, several species of crab parasites, and numerous aquatic bottom-dwellers. First usage in English c. 1830.

isopycnic: [Greek *isos* = equal + *pyknos* = dense, close.] n. A line drawn on a map connecting all points of equal density. In astrophysics, a line connecting points of equal or constant density with respect to space or time. First usage in English c. 1885.
Synomyn: isosteric

isostasy: [Greek *isos* = equal + *stasis* = setting, weighing, standing.] n. In geology, the tendency toward a state of gravitational equilibrium of the Earth's lithosphere, which is hydrostatically supported by the denser asthenosphere. Isostatic adjustments of the lithosphere can result in either downwarping or upwarping. Downwarped lithosphere occurs as a result of crustal loading,

such as by accumulation of glacial ice, sediment, lava flows, or bodies of water. Upwarped lithosphere (isostatic rebound) occurs from removal of crustal loading, such as by melting of glaciers, erosion of high ground, and evaporation of water bodies. The dynamic balance occurs by compensation to changes in heat flow, in melting and lithification rates, and by the processes of erosion and accumulation. Term coined by American geologist Clarence E. Dutton in 1889.
Cited usage: For this condition of equilibrium of figure, to which gravitation tends to reduce a planetary body, irrespective of whether it is homogeneous or not, I propose the name isostasy. [1889, C.E. Dutton, Bull. Philos. Soc. Washington XI. 53]. It must be clearly realized that isostasy is only a state of balance; it is not a force or a geological agent [1944, A. Holmes, Princ. Physical Geol. iii. 34].

isotope: [Greek *isos* = equal + *topos* = place.] n. A variety of one particular chemical element, which is distinguished from the other varieties of the same element by having a different number of neutrons in the nucleus, while maintaining the same number of protons. It thus has a different atomic mass, but shares the same atomic number. Term coined in 1912 by English chemist and Nobel laureate Frederick Soddy (1877-1956), so called because isotopes occupy the same place in the periodic table.
Cited usage: When the arithmetical sum (of particles in the nucleus) is different... gives what I call 'isotopes' or 'isotopic elements', because they occupy the same place in the periodic table. They are chemically identical, and save only as regards the relatively few physical properties which depend upon atomic mass directly, physically identical also [1913, F. Soddy, Nature 4 Dec. 400-1].

isotropic: [Greek *isos* = equal + *tropos* = turning.] adj. In general use, said of a substance whose properties are the same in all directions. In crystallography, said of a crystal whose physical properties are the same in all crystallographic directions (e.g., a crystal within which light travels at the same speed in any direction). Amorphous substances, such as glass and minerals in the isometric system (halite, garnet, fluorite), are isotropic. First usage in English c. 1860.

isthmus: [Greek *isthmos* = neck, neck of land (usually between two seas), narrow passage.] n. A relatively long and narrow stretch of land which is bounded on each side by water. An isthmus forms the connection between two larger land masses. Examples include the Isthmus of Panama and the Isthmus of Suez. When the Greeks associated this term to physical geography, they were thinking specifically of the Isthmus of Corinth, which connects Peloponnesus with northern Greece. It also served to separate the Gulf of Corinth to the northwest, from the Saronic Gulf to the southeast. First usage in English c. 1550.
Alt. spelling: isthmi (pl.)

itabirite: [Toponymous term after *Itabira*, a city in Minas Gerais, Brazil, the type locality for this rock.] n. A laminated metamorphosed oxide-facies of banded iron formation (BIF), where visible grains of quartz and the iron are present in the form of thin layers of iron oxides, such as hematite, magnetite, and martite.
See also: BIF

ivory: [Middle English *ivurie* < Latin *ebur* = ivory; < Coptic *ebou* = elephant.] n. The hard, white, cream-colored, or yellowish-white, fine-grained substance composed of dentin (mostly the mineral apatite), and comprising the main part of the tusks of the elephant, mammoth (fossil ivory), hippopotamus, walrus, and narwhal. First usage in English c. 1275.

jack: [Middle English *jahke* = generic form for addressing any male, especially a social inferior; < French *Jaques* < Latin *jacobus* < Greek *Iakobos* = male peasant of lower orders.] n. In mining, a term commonly used for apparently worthless material accompanying more valuable ore. In coal mining, applied to carbonaceous shale interbedded with coal, or volatile rich coal interbedded with shale. Also, a large ironstone nodule occurring in the coal beds. In metal mining, used for dark iron-rich sphalerite (black jack), which miners considered as worthless material occurring with valuable silver-bearing galena. First usage in English c. 1375

jackrock: [Welsh *jack* = mining term for a worthless shale parting between coal beds + English *rock*.] n. In mining, a regional term used in the Appalachian Mountains of the U.S. for shale interbedded between coal seams, and acting as a parting to separate the beds. Also, a device consisting of nails welded together in such a way that a sharp point always points upward. During heated labor disputes in the mining industry, jackrocks were spread across roadways to puncture tires, and occasionally used in hand-to-hand combat. At the height of the U.S. labor strikes during the early 20th century, police measured the jackrocks used during disputes in pounds rather than by count. The Illinois Criminal Code still has laws prohibiting their use and possession. The following is an excerpt from that code (720 ILCS 5/21-1.4, Sec. 21-1.4): "a person who knowingly sells, gives away, manufactures, purchases, or possesses a jackrock or who knowingly places, tosses, or throws a jackrock on public or private property commits a Class A misdemeanor. This Section does not apply to the possession, transfer, or use of jackrocks by any law enforcement officer in the course of his or her official duties." Also, the jackrock became an energetic dance from the Boogie-Woogie era of the 1940s.
Cited usage: The Fire Clay coal is commonly separated into two distinct layers or benches by a flint-clay and shale parting called the "jackrock parting" by miners [1999, Stephen F. Greb & John K. Hiett, Geology of the Fire Clay Coal in Part of the Eastern Kentucky Coal Field].

Jacob's staff: [Eponymous term after *Saint James (Jacobus)*, the saint whose symbols in religious art are a pilgrim's staff and a scallop shell.] n. A single straight staff or pole used as a surveying tool, particularly for measuring the thickness of sedimentary beds in a stratigraphic section. It is usually distinctly marked in some units of linear measurement so that it can be easily read at a distance. The original Jacob's staff, or cross-staff, was a single pole device invented by mathematician Levi ben Gerson (1288-1344). The pole was marked in degrees, and the altitude of the stars could be determined by using a sliding wooden panel on the rod. Sailors made such extensive use of these staffs measuring the height of the sun above the horizon in celestial navigation, that many of them went blind in one eye. Thus, the common caricature of the sailor wearing an eye patch was based on this fact. First usage in English c. 1545.
Cited usage: Jacobs Staff, a Pilgrims staff, so called from those who go on pilgrimage to the city of St Jago, or St James Compostella in Spain [1656, Thomas Blount, Glossographia of the Harder Words Found in Whatsoever Language].
Synonyms: cross-staff

jade: [French *(le) jade* < alteration of *(l)'ejade* < Spanish *(piedra de) ijada* = flank stone, loin stone, colic stone; < Latin *ilium* = flank.] n. As most commonly used, jade is the name for a hard, extremely tough, compact, monomineralic rock consisting of either: a) nephrite, an amphibole, and a double-chain silicate, or b) jadeite, a pyroxene, and a single-chain silicate. Nephrite varies in color from dark or deep green to grayish-white, whereas jadeite ranges to light green to purple, blue, or white. Both types of jade are especially tough and therefore resist breakage. But, they are not extremely hard, which makes them ideal for sculpture, particularly because jade often occurs in large masses. The Spanish Conquistadors of the 1500's noticed that indigenous Meso-Americans wore jade amulets as a mystical protection against kidney dysfunction. The Spanish named the green stone *piedra de ijada*, meaning stone of the loins. French was the official language of European diplomacy and letters, so *piedra de ijada* became *pierre de l'ejade*, which then became masculinized to *le jade*. This transformation of the French feminine form into the masculine was an error made when the word was as yet unfamiliar to Europeans. Scientists then Latinized the term to *nephriticus*, which means kidney. Thus the mineral came to be called nephrite jade. Burma (now Myanmar) opened its borders to trade in the 1700s, and became known worldwide as a source for Burmese jade. At first, this mineral from Burma was called nephrite due to its similarity to the amulets from South America, but Burmese 'nephrite' had an imperial color, then was later recognized as being composed of the distinctly different mineral, jadeite. The term jade has also been applied to various other hard, compact aggregates of certain green minerals, such as: vesuvianite (California jade), tuxtlite (omphacite), and green-dyed calcite (Mexican jade). In addition, the term has been applied to green varieties of the minerals: sillimanite, pectolite, garnet, and serpentine. First usage in English c. 1375.

Cited usage: A kinde of greene stones, which the Spaniards call Piedras Hijadas, and we use for spleen stones [1595, Sir Walter Raleigh, The Discovery of Guiana 24]. Under the name of 'oceanic jade' M. Damour has described a fibrous variety found in New Caledonia and in the Marquesas Islands differing from ordinary nephrite in the proportion of lime and magnesia which it contains. If this oceanic jade be recognized as a distinct variety, the ordinary nephrite may be distinguished as 'oriental jade' [1881, F. W. Rudler, Encycl. Britannica XIII. 540/1].
Similar terms: nephrite, jadeite
See also: nephrite, jadeite

jadeite: [French *(le) jade* < alteration of *(l)'ejade* < Spanish *(piedra de) ijada* = flank stone, loin stone, colic stone; < Latin *ilium* = flank.] n. A white to apple-green to emerald-green to blue or violet, monoclinic mineral, sodium aluminum iron silicate, $Na(Al,Fe)Si_2O_6$. It is found in metamorphic rocks formed at high pressures, and as a member of the clinopyroxene group it rarely occurs as isolated crystals, but is usually formed in compact fibrous aggregates that are extremely tough, and are recognized as one of the forms of the gem, jade. Jadeite is the rarest, and therefore the most desirable form of jade. The best and most abundant sources are found in Myanmar (formerly Burma). Nephrite, the other mineral commonly referred to as jade, is a gem quality form of tremolite and actinolite, a double-chained inosilicate of lime and magnesia. Nephrite, due to its greater abundance, has become the more popular variety, often worn by people who believe in its health-inducing properties. They are known to wear it constantly as a preventative for renal dysfunction. First usage in English c. 1860.
See also: jade, nephrite

jasper: [Greek *iaspis* = jasper, chalcedony; < (perhaps) a blend of Akkadian *yapu* = chalcedony and *apû* = jasper.] n. A red, yellow, or brown,

opaque cryptocrystalline variety of chalcedony (SiO_2), colored by inclusion of iron oxide impurities, and commonly making up the sedimentary rock, chert. The ancient Romans adopted the Greek term *iaspis* and used it for any bright-colored chalcedony except carnelian. First usage in English c. 1325.
Synonyms: jasperite, jaspis
Similar terms: jasperoid

jebel: [Arabic *jabal* = hill, mountain, mountain chain.] n. A hill, mountain, or mountain range, especially one in North Africa or the Middle East. The term is commonly used in place names, such as: a) *Jebel Usdum*, an erosional pillar of salt near the Dead Sea, which may have inspired the Biblical story of Lot's wife, b) *Jebel Musa*, the promontory in North Africa at the eastern end of the Strait of Gibraltar, marking the entrance to the Mediterranean Sea. Gibraltar is the promontory on the European side of the Strait of Gibraltar, its name being derived from the French form, *djebel*. In 711C.E., Tarik, leader of the Saracens, named Gibraltar *Jebel el Tarik*, meaning Mountain of Tarik.
Alt. spelling: djebel, gebel, jabal, gibal (pl.)

jetty: [Old French *jetee* = the action of throwing, a projecting part of a building < *jeter* = to throw.] n. An engineered breakwater, groin, or seawall extending out from shore into a body of water, designed to direct and confine a current or the tides, to protect a harbor, or to prevent sediment accumulation in a navigation channel. Also, used as a term for a wharf or landing pier. First usage in English c. 1400.

jheel: [Hindi *jheel* = swamp, lake.] n. A backwater pool, marsh, or lake, especially as applied to the backwaters of the Ganges flood plain, which remain from prior inundation.
Alt. spelling: jhil

john: [English *john* = back-formation from *bluejohn*; < French *bleu-jaune* = blue-yellow.] n. An English term for a massive fibrous or columnar variety of fluorite, which is frequently banded, colored blue or purple alternating to yellow, and referred to as bluejohn. Also called Derbyshire spar, in reference to the Derbyshire mines, caves, and caverns in northwestern England, the type locality for the bluejohn variety of fluorite. This variety has long been considered as gem quality in Europe. During the eighteenth century, Derbyshire fluorite was exported to France for jewelry making. French craftsmen referred to it as *bleu-jaune*, meaning blue-yellow, in allusion to its banded and alternating-colored appearance. When samples of this mineral returned to England as jewelry, the French lapidary term was Anglicized to 'bluejohn.' First usage in English c. 1770.
Cited usage: Known in London by the name of the Derbyshire drop. But in the spot it is called Blue John [1782, Wm. Gilpin, Observations on Cumberland and Westmorland].
Alt. spelling: Blue John

joint: [French *joindre* = to join; < IE *yeug* = to join.] n. In structural geology, a fracture or crack in a rock mass with no evidence of displacement (in contrast to a fault). Joints are often found in parallel groups called joint sets. In general usage, the place at which two things or parts are joined or fitted together; a junction or articulation. Not to be confused with fault. IE derivatives include: conjugate, yoke, adjust, junta, and yoga (from Sanskrit *yoga*, meaning union). First usage in English c. 1275.
See also: fault

jokulhlaup: [Icelandic *jokull* = icicle, glacier < *jaki* = piece of ice + *hlaup* = flood burst, avalanche, run.] n. An Icelandic term for a glacial outburst flood, especially when an ice dam impounding a glacial lake breaks. A massive jokulhlaup drained ancient glacial Lake

Missoula and created the Channeled Scablands in
the U.S. Pacific Northwest.
Cited usage: Icelandic jokulhlaups can originate
from the failure of ice-dammed lakes or from
subglacial volcanic and/or geothermal activity
[1940, Sigurdur Thorarinsson].
Alt. spelling: jokulhaup
Synonyms: glacial outburst flood

jokull: [Icelandic *jokull* = icicle, glacier < *jaki*
= piece of ice.] n. A term used in Iceland for a
glacier or an ice sheet.
Cited usage: The fire is generally contained in
these mountains covered with ice, or, as they are
called in the country, jokuls [1777, Dr. Uno Von
Troil, Letters on Iceland,1772 p. 233].
Alt. spelling: yokul, jokul
See also: jokulhlaup

Jurassic: [Toponymous term after the Jura
Mountains, the German name for the range lying
between France and Switzerland; < Gaulish *jura* =
forest.] adj. The middle of the three periods of the
Mesozoic Era, from 199.6-145.5Ma, noted for its
abundance of dinosaurs, and its unique forests of
ginkgo and cycad. Term coined in 1795 by
Alexander von Humboldt, for the stratigraphic
system of sedimentary rocks found in the Jura
Mountains between France and Switzerland.
They are characterized by oolitic limestone.
Gaulish was one of the primary Celtic languages
of Europe in ancient times. Its similarity to Latin
allowed for its rapid assimilation during the
Roman expansion period.
Cited usage: Sedimentary formations...as
modern as the Jurassic or Oolite formations [1833,
Sir Charles Lyell, Principles of Geology III. 372].

kalsilite: [Portmanteau word from Latin *kal(ium)* + English *sil(ica)*.] n. A white to gray hexagonal mineral in the feldspathoid group, potassium aluminum silicate, $KAlSiO_4$. Kalsilite is isostructural with nepheline, with which it forms a complete solid solution series above 1000°C. At lower temperatures, kalsilite exsolves from nepheline forming finely intergrown exsolution lamellae, analogous to that in the alkali feldspars. Kalsilite is polymorphous with kaliophilite, panunzite, and trikalsilite. It occurs in silica-undersaturated lavas and nepheline-bearing plutonic rocks. Kalsilite is also formed as an alteration product of blast furnace brick when it reacts with potash in cement kilns (the bricks also react with soda to form albite).

kalutoganiq: [Inupiat (Kovakmiut) *ganiq* = ice crystals (that form on objects).] n. Kalutoganiq are arrowhead-shaped snow drifts that form on top of wind-hardened tundra snow (upsik). Sometimes called barkhans due to their similarity to dunes found in the Sahara Desert.
Synonyms: barchan, snow drift

kame: [Middle English *camb* < *comb* = a low ridge; < IE *gembh* = tooth, nail.] n. A low mound, hummock, or ridge of stratified sand and gravel of glaciofluvial or glaciolacustrine origin, whose precise mode of formation is uncertain. Kame are found at the lower end of the glacial valleys in Scotland, their type location. IE derivatives include comb, unkempt, and gem. First usage in English c. 1860.
Similar terms: esker, osar

kämmererite: [Eponymous term after Prussian mine surveyor August Alexander Kämmerer (1789-1858).] n. A pale- to deep-purple to reddish chromium-bearing variety of clinochlore in the chlorite group, a monoclinic mineral, hydroxy magnesium aluminum chromium silicate, $Mg_5(Al,Cr)_2Si_3O_{10}(OH)_8$. It occurs with chromite in alpine serpentinites.
Alt. spelling: kaemmererite

kandite: [Industrial mnemonic term from English *ka*(olinite) + *n*(acrite)+ *di*(ckite).] n. A name for the kaolin group of clay minerals, including kaolinite, nacrite, dickite, and halloysite.
See also: kaolin

Kansan: [Toponymous name after the U.S. State of Kansas, named for the Kansa group of Native Americans originally inhabiting the Great Plains area; < Sioux *kansa* = south-wind people.] adj. A term used in North America for the second-oldest of the four classical glacial advances of the Pleistocene Epoch. These advances, separated by glacial retreats of various duration, were originally recognized in glacial deposits on the continent, and named after the U.S. states where prominent moraines were found. The four stages are: a) Nebraskan (2.1-1.7Ma), b) Kansan (1.3-0.9Ma), c) Illinoian (0.7-0.3Ma), and d) Wisconsinan (75-10Ka). Recent work has found evidence of many more glacial advances than the classical four.

kaolin: [Toponymous term after *Kao-ling*, a mountain in Jianxi Province in northwest China; < *kao* = high + *ling* = hill.] n. A group of structurally related, two-layered white clay minerals, including kaolinite, nacrite, dickite and anauxite. The kaolin minerals have the general formula,

$Al_2Si_2O_5(OH)_4$ and are usually formed by the in situ decomposition of alkali feldspar and mica during weathering or hydrothermal alteration. In contrast to other clay minerals, such as montmorillonite and illite, the kaolin group absorbs less water and thus its members tend to be less plastic when wet, and undergo less shrinkage when drying. Because of the lack of color-inducing chromophores, kaolin remains white on firing and is used in the manufacture of porcelain, ceramics, refractories, and paper. The term originally applied only to one mineral, kaolinite, but now includes at least four. Although kaolin is thought to have been introduced to Europe in 1712 by Pere Francois d'Entrecolles in a discussion about the fabrication of porcelain, it had actually been incorporated in the making of porcelain a few years earlier by Johann Bottger at the Meissen Manufactory near Dresden, Germany. Bottger had a reputation as an alchemist and was held in servitude for six years by Frederick Augustus the Strong of Saxony on the order to convert base metals to gold. He failed to accomplish this demand, but did formulate the process for making porcelain. Together with Augustus, they set up the Meissen Manufactory and succeeded in making Dresden Porcelain the most popular form of fine china made in Europe (a veritable gold mine for them).
Cited usage: Decomposed white felspar, or kaolin, produced from the granite rocks of Cornwall [1813, Robert Bakewell, Intro. to Geology Illustrative to the General Structure of the Earth].
Alt. spelling: kaoline
Synonyms: porcelain clay, China clay, bolus alba
See also: kandite

kaolinite: [Toponymous term after *Kao-ling*, a mountain in Jianxi Province in northwest China; < *kao* = high + *ling* = hill.] n. A white, triclinic clay mineral of the kaolin group, hydroxy aluminum silicate, $Al_2Si_2O_5(OH)_4$. It is polymorphous with dickite, halloysite, and nacrite. It is usually derived by alteration of alkali feldspars and micas. First usage in English c. 1865.

kar: [German *karen* = plural of *kar* = cirque.] n. A relatively steep, horseshoe-shaped depression occurring at the upper end of a mountain valley, and formed by erosional plucking at the head of a glacier.
Cited usage: Above the shoulders the valley slopes are far from being regular; often they form cirque-like niches, at the bottoms of which little tarns occur. These are the Kar of the Alps, the 'corries' of Scotland [1905, Jrnl. Geol. XIII. 2].
Synonyms: cirque (French), cwm (Wales), corrie (Scotland), kaar (Holland)
See also: cirque, cwm

karat: [Arabic *qirat* = pod, husk, weight of four grains, < Greek *keration* = carob seed.] n. A unit of measure for the proportion of pure gold in an alloy, equal to 1 part in 24. Pure gold is, therefore, 24 karat, and gold that is 50% pure is 12 karat. The abbreviation is *k* or *kt*. First usage in English c. 1555.
Cited usage: In U.S. usage "karat" designates the proportion of fine gold in an alloy, whereas "carat" is applied to the weight of a stone [1945, Jewelers' Dict.].
Similar terms: carat
See also: carat

karren: [German *karren* = trolley, cart, barrows.] n. A general term for solution grooves ranging in width from a few millimeters to more than a meter, commonly occurring in karst topography. Karren are commonly separated by sharp, knifelike ridges, and may originate under a soil cover or while exposed at ground surface. A karrenfield is a karst area where such furrows, fissures, or grikes are abundant. The etymology refers to the grooves, ruts, or tracks left in the ground by wheeled vehicles.

Cited usage: The chief features of such limestone regions are those known as karren, dolinen, blind valleys, and poljen [1894, Geogr. Jrnl. III. 322].
Synonyms: lapies, lapiaz, solution grooves

karst: [Toponymous term after the Kars or Krs region of the former Yugoslavia; < German *Karst* < Serbo-Croatian *Kras* = limestone plateau near Trieste Bay in western Slovenia.] n. A type of topography formed primarily by chemical dissolution on sedimentary terrains of limestone and rock gypsum, as well as other soluble rocks. Karst is characterized by numerous sinkholes, caves, solution valleys, and underground drainages. The term evolved from Serbo-Croatian *Krs* to the German *Karst* when the name was applied to the type locality in the Dinaric Alps of western Slovenia near Trieste Bay. First usage in English c. 1900.
Cited usage: Some prominent features of Serbian karst are: largest uvalas and poljes, major karst springs, longest underground coursers in Serbian karst, and longest and deepest caves in Serbia [1999, Predrag Djurovic, Speleological Atlas of Serbia].
See also: polje, uvala, doline

kata-: [Greek *kata-* = down, downward; < IE *kat-* = down.] Combining prefix meaning down, used in terms such as katazonal and katabatic. *Cata* is the equivalent Latin prefix, but has a wider range of meaning. Used in contrast to the prefix 'ana-.'
See also: katabatic, katazonal, cata-, ana-, catadromous

katabatic: [Greek *kata* = down, downward + *batis* = going, walk + *-ic* = pertaining to.] adj. In meteorology, said of a wind blowing down from an elevated region, especially when caused by increased density of air induced by surface cooling during the night. This term relates to the 'katabasis,' a massive military retreat from the historical narrative *The Anabasis* by Xenophon (444-354 BCE), the Greek historian who described a military campaign in Asia, the advance (anabasis) being led by Cyrus the Younger, and the subsequent massive retreat (katabasis) being a loosely organized democratic decision by the masses.
Cited usage: Where the valley is long and rather steep, the down-flowing air current may attain the velocity of a gale and become a veritable aerial torrent. This drainage flow is known as the mountain breeze, or mountain wind; also canyon wind, katabatic wind, and gravity wind. [1920, W. J. Humphreys Physics of Air vii. 111].
See also: anabatic

katamorphism: [Greek *kata* = down + *morphos* = shape, form.] n. An obsolete term for the chemical decomposition of minerals that takes place at or near the Earth's surface. This process involves complex minerals that are broken down to simpler ones through oxidation, reduction, solution, and hydration.
Cited usage: The zone of katamorphism may be defined as the zone in which the alterations of rocks result in the production of simple compounds from more complex ones [1904 C. R. Van Hise in Monogr. U.S.G.S. XLVII. 43].

katatectic: [Greek *kata* = down + *tekton* = builder.] adj. Said of a sedimentary layer that is created by accumulation and compaction of relatively insoluble residue, such as gypsum or anhydrite, as a result of preferential solution of halite in an evaporite sequence of a salt dome. Term coined in 1933 by American geologist M. I. Goldman, in reference to the layers building in a downward direction.

katazonal: [Greek *kata* = down + Latin *zona* < Greek *zone* = girdle, celestial region < *zonnenai* = to gird.] adj. In igneous petrology, according to the classification formulated by American geolo-

gist Arthur F. Buddington in 1959, referring to the lowermost depth zone of igneous intrusions, where migmatites and other products of ultra-metamorphism occur. In metamorphic petrology, according to the classification of German petrologist Ulrich Grabermann in 1904, the lowermost depth zone of metamorphism, where modern usage emphasizes temperature and pressure conditions of high metamorphic grade. These conditions produce such rocks as: amphibolite, granulite, eclogite, and high-grade schist and gneiss.
Alt. spelling: catazonal

kay: [Spanish *cayo* = shoal, rock, barrier-reef.] n. Small, flat, marine island formed from coral-reef material or sand. The term is commonly used in describing the low-lying, sparsely vegetated islands off the coast of southern Florida. Originally the same word as *quay*, meaning wharf, reinforced bank; after colonization of the West Indies however, the Spanish version *cayo* adopted a meaning more relevant to the islands of the Caribbean, rather than the established ports of the European mainland. The term was no doubt influenced by the native Caribbean language Taino, as cayo in that language now means "a pass between islands."
Alt. spelling: key

Keewatin: [Toponymous term after a district in the Northwest Territories of Canada.] adj. A term primarily used in Canada, denoting the oldest division of the Archean Eon in North America, and the rocks found in the Canadian Shield formed during this time.
Cited usage: The most appropriate name for the series that suggests itself to me is 'Keewatin', the Indian name for the North-west, or the North-west wind, which has been applied to the district within which the rocks occur [1886 A. C. Lawson in Annual Report Geol. Survey Canada 1885 I. 14].

kelyphytic: [Greek *kelyphos* = rind, shell.] adj. In petrology, said of a rim or peripheral zone of pyroxene or amphibole around olivine crystals, or around garnet crystals, as a result of reaction between the central crystal and the surrounding melt or fluid. Term coined in 1882 by Austrian mineralogist Albrecht Schraub (1837-1897). An etymologically related term is reaction rim.

keratophyre: [Greek *keras* = horn, horny substance + *phyre* = contraction of porphyry.] n. In igneous petrology, a term applied to intermediate and silicic (salic) volcanic rocks that are characterized by the presence of albite, epidote, and calcite. Keratophyres generally result from postmagmatic sodium metasomatism. Some keratophyres contain sodic orthoclase, amphiboles, and pyroxenes, are commonly associated with spilites, and are often interbedded with marine sediments. Term coined in 1874 by German geologist C. W. Von Gümbel, in his monograph *Die Paläolithischen Eruptivgesteine des Fichtelgebirges* (*The Paleolithic Volcanic Rocks of the Fichtel Mountains*). The name alludes to the similarity in appearance of this volcanic rock to hornstone, and obsolete term for chert, flint, or hornfels that are compact, tough, and splintery.
Cited usage: Microscopical examination and chemical analysis show that these rocks consist, in part at least, of soda-felsites or keratophyres. The keratophyres (so named from their resemblance to hornstone) were first described by Gümbel [1889, Geol. Mag. Feb. 71].

kernite: [Toponymous term after Kern County, California.] n. A colorless to white monoclinic mineral, hydrated sodium borate, $Na_2B_4O_7 \cdot 4H_2O$. It is an important ore of boron. At its discovery locality, formed as a result of recrystallization and dehydration of borax. Kernite forms large crystals, often one meter long. The discovery locality was the Kramer deposit near the town of Boron in Kern County, California.

kerogen: [Greek *keros* = wax + *genesis* = origin, creation < *genesthai* = to be born, come into being; < IE *gen* = to give birth, beget.] n. The bituminous insoluble organic material found fossilized in oil shales and other sedimentary rocks. Kerogen can be converted by heating and distillation into other more volatile petroleum products. Term coined in 1906 by Scottish chemist Alexander Crum Brown (1838-1922), who also was the first to use structural formulas in organic chemistry.
Cited usage: We are indebted to Professor Crum Brown, F.R.S., for suggesting the term Kerogen to express the carbonaceous matter in shale that gives rise to crude oil in distillation [1906, D. R. Steuart in H. M. Cadell's Oil-Shales of Lothians iii. 142].

kettle: [Old English *cetel* < Latin *catinus* = large bowl.] A steep-sided, bowl-shaped depression, commonly without surface drainage, located in outwash or other glacial-drift deposits. Kettles are formed by the melting of a large detached block of stagnant glacial ice. Such ice blocks become completely or partially buried in the glacial drift. The resulting kettles can be tens of meters deep and up to 13 kilometers in diameter. The Kettle Moraine area in Wisconsin and Walden Pond in Massachusetts are examples. First usage in English c. 1880.
Synonyms: kettle hole

khadar: [Toponymous term after *Khadar*, an agricultural region in India; < Hindi *khadar* = silted river bank.] n. A term used in India for a low-lying area or alluvial plain that is prone to overbank flooding by a river. Along with Bhangar and Nardak, Khadar is one of three agricultural regions in the Yamuna River Valley of north-central India. In Hindi, the term khadar, meaning a salted river bank, is also used for a hand-spun, open-weave cotton cloth (also called a khadi). They are made primarily in the khadar regions of India, because much cotton is grown there. Use of the khadar was encouraged by Mahatma Ghandi in a successful effort to break the British monopoly on the manufacture of cotton.
See also: bhangar

khamsin: [Arabic *rih-al-hamsin* = wind of the 50 days < *hamsin* = fifty < *hams* = five.] n. A generally southerly hot wind from the Sahara that blows across Egypt from late March to early May. First usage in English c. 1680.

khor: [Arabic *khawr* = dry wash, wadi.] n. A term used in northern Africa for a ravine or watercourse, especially one that is dry.
Synonyms: arroyo, dry wash, wadi, koris
See also: wadi, arroyo, koris

kieselguhr: [German *kiesel* = flint <*kies* = fine gravel, pebble + *guhr* = earthy deposit, ferment < *gähren* = to ferment.] n. A fine-grained, powdery biologic sediment composed of the siliceous remains of diatoms. After lithification, kieselguhr forms the sedimentary rock diatomite. With further diagenesis, it is converted to chert. Term coined c. 1870 by German microscopist Christian Gottfried Ehrenberg (1795-1896).
Synonyms: diatomaceous earth
See also: diatom

kill: [Dutch *kil* = riverbed, channel.] n. A stream, creek, or tributary river. The term is used in names of places originally settled by the Dutch, especially places in Delaware and New York State, such as the Catskill Mountains, Fresh Kill, Dutch Kill, and Westkill. First usage in English c. 1665.

kilo-: [Greek *khilioi* = one thousand.] A combining prefix referring to amounts 1,000 times the base unit, used largely for weights and measures in the metric system, as in kilogram and kilometer, but also in other types of units, such as kilohertz and kilotons.
See also: meter

kimberlite: [Toponymous term after Kimberley, a town founded in 1871 on the Northern Cape, South Africa, known primarily as a diamond-mining center.] n. A porphyritic alkalic peridotite containing phenocrysts of phlogopite (commonly chloritized) and olivine (commonly serpentinized), and sometimes chromium pyrope. Kimberlite is the main host rock for diamonds. It is usually brecci-ated, and displays a fine-grained groundmass of calcite, olivine, and phlogopite, with the accessory minerals serpentine, chlorite, and ilmenite. Kimberlite typically occurs in pipe-shaped intrusions having diameters of several hundred meters. These pipes originate in the Earth's mantle at depths greater than one hundred kilometers. Term coined in 1886 by American geologist Henry Carvill Lewis.
Cited usage: There appears to be no named rock-type having at once the composition and structure of the Kimberley rock. ... It is now proposed to name the rock Kimberlite... Kimberlite is a rock sui generis, dissimilar to any other known species [1887, H. C. Lewis, Papers on the Diamond].
See also: diatreme

kingdom: [Old English *cyning* = king; < IE *gen* = beget, to give birth + Old English *dom* = jurisdic-tion.] n. In the hierarchical phylogenetic classifi-cation system devised by Carolus Linnaeus (Carl von Linne, 1707-1778), the term kingdom was the category of highest rank. In 1977, Carl Woese proposed an even higher rank then kingdom, called domain. He created a three-domain system comprised of Eukaryota, Bacteria, and Archaea. In this revision of Linnaean classifica-tion, the domain Eukaryota includes the four kingdoms: Protista, Fungi, Plantae, and Animalia; the domain Bacteria includes most prokaryotic organisms; and the domain Archaea includes organisms living in extreme environ-ments, usually (but not exclusively) small, prokaryotic species of ancient heritage.

kipuka: [Hawaiian *kipuka* = a variation of *puka* = hole, opening.] n. An isolated islandlike area of land completely surrounded by one or more younger lava flows. Because they are surrounded by recent flows, kipukas are often areas of mature vegetation that become isolated refuges for rare species of plants and animals. The related term puka shell gets its name because of the hole worn into its center by wave action on sandy beaches.

klint: [Danish *klint* = cliff.] n. A term used in Scandinavia for a tall (often several hundred meters high), vertical rock wall or sea cliff, espe-cially a steep cliff along the shores of the Baltic Sea. Not to be confused with clint.
See also: clint

klippe: [German *klippe* = rock protruding from the sea floor; < Dutch *klippe* < Swedish *klippa* = cliff, to cut.] n. An isolated rock unit that is an erosional remnant or outlier of a nappe or thrust sheet.
Cited usage: These [structures], called Klippen, are abrupt pyramidal masses, the beds in the upper part being not only older than those in the lower, but also 'contorted, fractured, crushed, and mixed up', while the newer are comparatively undis-turbed [1902, Encycl. Britannica XXV. 333/2].

kloof: [Dutch *kloof* = cleft.] n. A term used in South Africa for a deep rugged gorge, ravine, or other steep-sided valley, or for a mountain pass.

kluft: [German *kluft* = fissure, cleft, chasm, abyss.] n. A German term for a fissure, joint, or cleft in a rock mass, especially as used in the Swiss Alps. Klufts are usually open fissures where large quartz crystals form from hydrothermal solutions.

knickpoint: [German *knick* = bend + *punkt* = point.] n. Any interruption or break in slope of the ground surface, especially an abrupt change in the longitudinal profile of a stream, resulting from glacial erosion, outcropping of a resistant bed, or uplift. *Cited usage*: Since erosion progresses headward when a region is uplifted by a uniform tilting movement, the longitudinal profiles of those streams will record a knickpoint at the headward limit of quickened erosion [1924, Bull. Geol. Soc. Amer. XXXV. 638].

knock: [Gaelic *cnoc* = knoll, rounded hill.] n. A term used in Ireland and Britain for a hill.

knocker: [Old English *cnocian* = probably onomatopoetic, of echoic origin. Also, Gaelic *cnoc* = knoll, rounded hill.] n. In petrology, a colloquial field term denoting a resistant, rounded monolith up to several hundred meters across. A knocker stands out from the surrounding mélange terrain. The term is used for both tectonic and exotic blocks. In mining, a goblin or dwarf is said to live beneath the surface who directs miners to ore by knocking on the mine surfaces. Mythical stories of these quixotic creatures originated with the Saxons from central Europe, who worked the mines in southwest England. Knockers are among the most persistent of mining superstitions, still believed in and heard from by miners in Wales and England. First usage in English c. 1375.
Cited usage: A spirit or goblin imagined to dwell in mines, and to indicate the presence of ore by knocking [1888, Pall Mall G. 20 Apr. 11/2 Cardiff].

knoll: [Dutch *knol* = clod, ball, turnip; also German *knollen* = clod, lump, knot, tuber.] n. In geomorphology, a small, low, rounded hill or mound. Also, the rounded top of a hill or mountain. In marine geology, a moundlike hill occurring on the seafloor that is less than 1,000m high. The rounded top of a hill or mountain, whether on continental terrain or the seafloor. First usage in English before 900.
Alt. spelling: knowe, knowle

komatiite: [Toponymous term after the Komati River in the Barberton Mountain Land of Transvaal, South Africa.] n. In igneous petrology, used most widely for volcanic rocks of ultramafic composition (both lavas and near-surface intrusives). Komatiites display spinifex texture, having tabular crystals of olivine up to one meter long. They are considered to have been extruded in a totally liquid state but they have never been observed erupting, so their volcanic nature has necessarily been deduced from the properties and textures of known outcrops. The evidence suggests that the lavas erupted at extremely high temperatures, around 1600°C. Except for one known occurrence that is about 80 million years old, all komatiites are at least three billion years old, and are found as lava flows and shallow sills within Archaean terrains. Compositionally, they represent the volcanic equivalent of peridotite, but they are unique in containing significant amounts of nickel sulfide. Since 1977, the term has been extended to include not only ultramafic volcanic rocks, but also the associated cumulates and more silicic differentiates. The term was originally coined in 1969 by Richard P. and Morris J. Viljoen.
Cited usage: All of these peridotitic rocks have a distinctive and unusual chemical composition. They are unlike any class of peridotitic or picritic rock known and a new name, peridotitic komatiite,..has been proposed [1969, Viljoen Viljoen, Geological Society of South Africa Special Publication v 2]
See also: tholeiitic

kona: [Toponymous term after the *Konako 'ne* district on the leeward (westward) side of the island of Hawaii.] n. A term used in Hawaii for the leeward side of the islands, the side not facing the prevailing trade winds. Also used for a southwesterly winter wind in Hawaii, which blows in the opposite direction to the prevailing trade winds, and is often strong, bringing rain. First usage in English c. 1860.

kopje: [South African Dutch *kopje* = small hill < *kop* = hill, head + *-je* = diminutive suffix.] n. A term used in South Africa for a small but prominent hill occurring in the veld, sometimes reaching 30 meters in height. Kopjes are usually isolated erosional remnants or inselbergs composed of granitic rock. They are partially scrub-covered, making them a favorite resting place for lions. First usage in English c.1880.
Alt. spelling: koppie
See also: veld, inselberg

koris: [Wolof (language spoken in Senegal) *kori* = a dry valley, or wadi.] n. A usually dry valley cut by intermittent watercourses. The term is common in North Africa, and is comparable to the term arroyo used in the U.S. Southwest.
Synonyms: arroyo, dry wash, wadi, khor
See also: wadi, khor, arroyo

kratogen: [Greek *kratos* = strength + *genesis* = origin, creation < *genesthai* = to be born, come into being; < IE *gen* = to give birth, beget.] n. An obsolete term for an area of a continent that has resisted deformation over a long period of time. Term proposed in 1921 by Austrian geologist Leopold Kober (1883-1970) in *Der Bau der Erde* (*The Building of the Earth*).
Cited usage: The 'Kratogens', once the area of the most ancient geosynclines, may, after they are peneplained, be widely flooded by epeiric seas [1923 Bull. Geol. Soc. Amer. XXXIV. 210].
Synonyms: craton

kum: [Turkish *kum* = sand.] n. A term applied to the sandy deserts of Central Asia, such as Kizil Kum (Red Sand) of Kazakhstan.

kuroko: [Japanese *kuroi* = black + *oko* = ore.] n. In ore deposits, referring to massive base-metal sulfide deposits, especially those of copper, lead, and zinc, originally recognized in Japan. Kuroko are volcanogenic, strata-bound deposits of Tertiary age that are primarily found precipitated on the sea floor near fumaroles, or at hot springs located on the flanks of submarine dacite domes. In places, kuroko deposits are associated with massive pyrite and gypsum deposits.

kurtosis: [Greek *kurtosis* = a convexity < *kurtos* = bulging, convex.] n. In statistics, a graphical representation of a frequency distribution showing the degree of flatness or peakedness of the curve, especially with respect to the concentration of values near the mean as compared with the normal distribution. First usage in English c.1900.
Cited usage: The kurtosis is useful in determining if a frequency distribution differs from the normal error curve. The kurtosis of a normal distribution is equal to 3; smaller values than 3 indicate a flatter distribution than the normal (a platykurtic distribution), while values above 3 indicate a more sharply peaked distribution than the normal (a leptokurtic distribution) [1952, W.O. Gore, Statistical Methods for Chem. Exper. 216].

kyanite: [Greek *kyanos* = azure, dark-blue.] n. A blue or greenish, triclinic mineral, aluminum silicate, Al_2SiO_5. It is trimorphic with sillimanite and andalusite, and occurs as long, thin, bladed crystals in high-grade schists and gneisses. Term coined by Abraham Gottlob Werner in 1789, as a variant spelling of the original name cyanite, alluding to its cyan-blue color.
Alt. spelling: cyanite
Synonyms: disthene, sappare
See also: andalusite, sillimanite

kyle: [Gaelic *caol* = narrow.] n. A narrow channel between two islands, or between an island and the mainland. Also, a sound or strait. The place name Kyle is one of the three principle regions of Ayrshire, England. These regions are defined agriculturally into: a) Carrick, a wild hilly area for grazing, b) Cunningham, a rich dairyland, and c) Kyle, a strong corn-growing soil. First usage in

English c. 1545.
Cited usage: Ane right dangerous kyle or stream
[1549, Sir Donald Monro, A Description of the
Western Isles of Scotland, called Hybrides].

kyr: [Turkmenian *kyr* = stony hard ground.] n.
Central Asian term for a flat terrain, plateau, or top
of a hill or mountain, especially in areas of stony
hard ground, such as desert pavement.

labile: [Middle English *labil* = forgetful, wandering; < Latin *labilis* = apt to slip < labi = to slip.] adj. Said of rocks and minerals that are mechanically or chemically unstable and are likely to change, such as a labile sandstone containing abundant fragments of feldspar and rock particles that are easily decomposed. The term is used in contrast to its antonym, stabile, which means what it almost spells, stable. First usage in English c.1425.

labradorite: [Toponymous term after Labrador, Canada, where it was first discovered in large masses.] n. A dark-gray, triclinic mineral of the plagioclase feldspar solid solution series, with a composition ranging from $Ab_{50}An_{50}$ to $Ab_{30}An_{70}$. It commonly shows a play of colors in blue to green, resulting from extremely fine exsolution lamellae. It is common in mafic igneous rocks, such as gabbro and norite. First usage in English c. 1810.

laccolith: [Greek *lakkos* = pond, reservoir + *lithos* = rock, stone.] n. A concordant igneous intrusion, having a flat floor and a convex roof, and formed when the overlying sediments were forced to bulge upward, resulting in an overall plano-convex form. A laccolith is roughly circular in plan, less than 10km in diameter, up to several hundred meters thick, and is thought to have a feeder dike at its thickest point. Originally, the term was named laccolite in 1877 by American geologist G. K. Gilbert in his report on the Henry Mountains,

but was later replaced by the term laccolith. *Cited usage:* A special type of sill is one where the lava swells out to form a lens-shaped mass which according to its particular form is known as a laccolite (with a flat base) or a phacolite (with a curved base) [1946, L. D. Stamp, Britain's Struct. ix. 80]. Laccoliths, which are closely related to sills but which were formed from a magma too viscous to spread far, may form local dome-like features when exposed by erosion [1960, B. W. Sparks, Geomorphol. vii. 151].
Synonyms: laccolite
See also: lopolith

lacuna: [Latin *lacuna* = pit, hole, hollow, gap < *lacus* = lake, pool, basin, tub.] n. In stratigraphy, a chronostratigraphic term representing a gap in the stratigraphic record, specifically the missing interval at an unconformity. A lacuna results from either nondeposition (hiatus) or erosion. In paleontology, used for a pore or opening in the structure of some invertebrates, such as bryozoans and brachiopods. First usage in English c.1660.
See also: hiatus

lacustrine: [Latin *lacus* = lake, pond, basin, tub.] adj. In general usage, pertaining to lakes. In limnology, said of sediment and rock strata that originated by deposition at the bottom of lakes and of the flora and fauna resident in a lake. First usage in English c.1825.
Cited usage: The lacustrine and alluvial deposits of Italy [1830, Lyell, Principles of Geology i. iii. 49].

lade: [Middle English *laden* < Old English *hladen* = to load, heap, draw water.] n. The mouth of a stream or river. Etymologically related terms include ladle and laden. First usage in English before 900.

lag: [Norwegian *lagga* = go slowly, last, delay.] adj. In sedimentology, said of a residual accumulation of usually hard-rock fragments that remain on the surface after a finer material has been preferentially removed by water or wind. In volcanology, said of deposits from pyroclastic flows, such as lag breccias, that are composed of clasts too large to be transported for long distances, so they accumulate close to the eruptive vent. In structural geology, said of a type of overthrust fault where differential movement in the hanging wall block leaves the upper part of the geologic cross section at some distance behind the lower part of the section. First usage in English c. 1510.

lagg: [Swedish *lagg* = edge of a bog or marshland, river bank.] n. The depressed margin of a raised, dome-shaped bog forming a natural ditch. Because laggs are frequently full of water, they look a lot like moats.
Cited usage: Raised bog is a type confined to the area of a wet basin. The surface becomes convex so that the feeble drainage channels (lagg) that run across the basin are pushed to one or both sides of the peat dome [1968 R. F. Daubenmire Plant Communities iii. 149].
Synonyms: bog moat

lagoon: [French *lagune* < Italian & Spanish *laguna* < Latin *lacuna* = pit, hole, hollow, gap < *lacus* = lake, pool, basin, tub.] n. In oceanography a shallow stretch of sea water or brackish water that is partly or completely separated from the ocean by a narrow, elongate strip of land, such as a coral reef, barrier island, sand bank, or spit. The term is commonly used in the islands of the South Pacific, due to the preponderance of lagoons enclosed within coral atolls, and to the European occupation of Oceania in the 19th century. First usage in English c. 1610.
Cited usage: Lagoons along the sea-margin are for the most part shallow and narrow, running parallel with the coast, from which they are separated by a strip of low land formed of sand, gravel, or other loose material. [1877, A. Geikie,

Elem. Lessons Physical Geogr. iv. 271]
Synonyms: lagune, laguna
See also: atoll

lahar: [Javanese *lahar* = mudflow < back formation of *berlahar* = to expel lava.] n. A usually massive mudflow of volcaniclastic material occurring on the flank of a volcano. Lahars are often triggered by sudden melting of snow or ice at volcanic summits, or by torrential rain falling on ash-covered slopes. The term was adopted by volcanologists during studies of Indonesia's volcanoes in the 1920s, and became more widely used during eruptions of Mt. Merapi on Java in the 1930's. First usage in English c. 1925.
Cited usage: Lahars follow mainly channels already existing, filling them temporarily to the brim with rushing torrents but leaving them empty again and eventually depositing their loads of debris on low ground many miles beyond [1944, C. A. Cotton, Volcanoes xiii. 247].

lake: [Latin *lacus* = lake, pond, basin, tub.] n. Any inland body of standing water, either fresh or saline, that occupies a depression of appreciable size. A lake is larger than a pond, and is deep enough to prevent terrestrial vegetation from becoming established across the expanse of water. First usage in English before 1000.
Cited usage: The whole assemblage must terminate somewhere: where they reach the boundary of the original lake-basin [1833, Lyell, Principles of Geology III. 9].
Alt. spelling: lac, lago, loch, lough

lamella: [Latin *lamella* = small thin plate < (diminutive of) *lamina* = thin plate.] n. In mineralogy, one of the thin, lensoid-to-platelike units formed by exsolution in alkali feldspar (exsolution lamellae), by polysynthetic growth twinning in plagioclase (twin lamellae), or by deformation twinning in calcite (also twin lamellae). In biology and paleontology, a thin scale, plate, or layer of

body tissue, as in the gills of a bivalve mollusk, or around the minute vascular canals in bone. As the adjective lamellar, the term is frequently used to indicate an arrangement of thin layers stacked like the leaves of a book, such as the crystal habit of the mica minerals. First usage in English c. 1675.
Alt. spelling: lamellae (pl.)
See also: lamina

lamellibranch: [Latin *lamella* = small thin plate < *lamina* = thin plate + *branchia* = gills.] n. Any benthic bivalve mollusk belonging to the class Bivalvia (also called Pelecypoda or Lamellibranchia), including clams, scallops, oysters, and whelks. First usage in English c. 1850.
Synonyms: pelecypod
See also: pelecypod

lamina: [Latin *lamina* = thin plate.] n. In stratigraphy, the thinnest recognizable unit layer of original deposition in a sediment or sedimentary rock. Laminae are less than 1 cm in thickness, and differ from other layers in color, particle size, or composition. Several laminae may constitute a bed. In paleontology, a thin platelike or sheetlike structure in an organism, such as the layered structure deposited daily in some corals, and the layers of aragonite and calcite in the shells of bivalves. As the adjectives laminar and laminated, used in such terms as: laminar flow, laminar layer, and laminated texture. First usage in English c. 1655.
Cited usage: The finer beds of clay or sand will all be arranged in thicker or thinner layers or laminae [1872, Nicholson, Palaeont. 6].
Alt. spelling: laminae (pl.)
See also: lamella, flysch

lamprophyre: [German *lamprophyr* < Greek *lampros* = clear < *lampein* = to shine + French -*phyre* = combining suffix representing porphyry.] n. A dark-colored, porphyritic, hypabyssal, intrusive igneous rock found in dikes and sills.

Lamprophyres are characterized by panidiomorphic texture, and a high percentage of mafic minerals forming phenocrysts, such as: biotite, hornblende, and various pyroxenes. They are commonly associated with carbonatites. Term coined in 1874 by C. W. Gumbel, for certain dark rocks that contain phenocrysts of biotite and hornblende, usually in dikes of Paleozoic age. The German petrologist Harry Rosenbusch broadened the term to include a wide variety of hypabyssal rocks containing ferromagnesian phenocrysts. The lamprophyre group includes: minette, kersantite, alnöite, vogesite, fourchite, camptonite, and monchiquite.
Cited usage: Lamprophyres are for the most part melanocratic to mesotype rocks. Nearly all of the true lamprophyres are characterized by a definitely alkaline aspect, indicated by the presence of abundant biotite, alkali feldspar, soda pyriboles, nepheline, or analcite, combined with a low silica content [1959, W. W. Moorhouse, Study of Rocks in Thin Section xvii. 325].

lapidary: [Old French *lapidaire* < Latin *lapidarius* = stonecutter < *lapid-* < *lapis* = stone.] n. The art of cutting and polishing colored or precious stones other than diamond. Also, a term for a worker who cuts and polishes semiprecious and precious gems. First usage in English c. 1350.

lapidofacies: [Latin *lapis* = stone + *facia* < *facies* = form, figure, appearance < (perhaps) *facere* = to make, or < *fa-* = to appear, shine.] n. Facies that are related to diagenesis.

lapilli: [Latin *lapilli* = little stones, plural of *lapillus* = diminutive of *lapis* = stone.] n. Pyroclastic material between 2-64mm in diameter that is intermediate in size between volcanic ash and bombs. Lapilli may be either solid or partially molten when the particles first come to rest. Common shapes of lapilli include: spheroid, teardrop, dumbbell, and button-shaped. Examples

of lapilli are: cinders, Pele's tears, and Pele's hair (filiform lapilli). First usage in English c.1745.
Cited usage: Ashes and lapilli of the size of nuts [were projected] as far as 40 miles [1875, Lyell, Principles of Geology II. ii. xxvi. 18].
Alt. spelling: lapillus (sing.)
See also: Pele's hair, Pele's tears

lapis lazuli: [Latin *lapis* = stone + *lazuli* < *lazulum* = azure blue; < Arabic *al-lazward* = blue color, lapis lazuli.] n. An azure-blue rock composed of a granular aggregate of the minerals lazurite, calcite, and pyrite. Lapis lazuli may also contain sodalite, hauyne, and other minerals. It has been prized since ancient times as a semi-precious gemstone and as a pigment. It is probably the original sapphire of the ancients. The term lapis is also used in the fine arts for the deep-blue and vividly colored pigments made from lapis lazuli. First usage in English c.1375.

lapse: [Middle English *lapsen* = to deviate from the normal; < Latin *lapsus* = a slip or fall < *labilis* = apt to slip < labi = to slip.] v. Referring to the rate at which some atmospheric property, usually temperature, decreases with altitude. The average is about 0.6°C per 100 meters, or about 4°F per 1,000 feet.

Laramide: [Eponymous term after the Laramie Formation of Wyoming and Colorado.] adj. Referring to the orogeny and time of deformation that is recorded in the rock record of the Cordilleran Mountain belt of the Western U.S. The Laramide orogeny had several phases extending from the late Cretaceous period through the Paleocene epoch of the Tertiary period. Intrusives and associated ore deposits emplaced about this time are commonly called Laramide (e.g., the Boulder Batholith of Montana, and many porphyry copper deposits of Arizona, Utah, and New Mexico).

latent: [Latin *latere* = to be hidden.] adj. In general usage, referring to characteristics that are present but not apparent or visible. In chemistry, referring to the excess heat released or absorbed by a material during a phase change, such as the latent heat of crystallization, fusion, or vaporization. First usage in English c. 1615.

laterite: [Latin *later* = brick + *ite* = belonging to.] n. An older term for a highly weathered, red subsoil that develops in humid tropical to subtropical regions, as well as in forested temperate regions. Laterites are nearly devoid of mobile bases (e.g., K, Na, Ca, Mg, etc.), as well as primary silicates and quartz. They are, however, rich in secondary oxides and hydroxides of iron and aluminum. Due to severe leaching, lateritic soils are depleted in nutrients, which, at first, seems contradictory, considering the rapid growth rate of plants in tropical rainforests. However, the source of organic material that sustains high growth rates is the rapidly decomposed leaf litter and other organic material on the forest floor. Exposure of the lateritic subsoil, due to natural or human-induced removal of the forest, to repeated wetting and drying of laterite causes it to become irreversibly hardened. In this state, it can be cut and used for bricks. First usage in English c. 1805.
Cited usage: The word 'laterite' is used by us for the sesquioxide-rich, highly weathered clayey materials that change irreversibly to concretions, hardpans, or crusts [1949, Publ. Inst. Nat. pour l'Étude Agronomie du Congo No. 46. 7].

latite: [Toponymous term after *Latium*, a region near the Tiber River, annexed to Italy in 1870.] n. A porphyritic, volcanic igneous rock containing phenocrysts of plagioclase and potassium feldspar, with little or no quartz embedded in a groundmass that is finely crystalline to glassy. Latites are the volcanic equivalent of monzonite and, as such, belong to the calc-alkaline magma series. With increased alkali feldspar, latites grade

into trachyte; with increased plagioclase, they grade into andesite or basalt. Term coined in 1898 by American geologist F. L. Ransome.
Synonyms: trachyandesite, trachybasalt

latitude: [Middle English = geographical latitude; < Old French = width; < Latin *latitudo* = width, geographical latitude < *latus* = wide.] n. In cartography, the angular distance north or south of the Earth's equator, measured in degrees along a meridian between 0° at the equator and 90° at the poles. A degree of latitude on the Earth's surface measures 60 nautical miles (69 statute miles). In general, on a sphere, the elevation of a point measured from the plane of the equator. Nicole Oresme (1320-1382) was perhaps the first person to use the terms latitude (the abscissa) and longitude (the ordinate) with reference to a mappable Cartesian grid system superimposed on a sphere. First usage in English c. 1375.

lattice: [Middle English *latis* < German *latte* = lath.] n. In crystallography, the three-dimensional geometric arrangement of points in space that represent the translational periodicity of a crystal structure. The lattice points represent locations where the atoms, molecules, or ions of a crystal occur. The term alludes to laths, being linear strips of material, usually arranged in an orderly and rectilinear manner, forming a latticework or repeating pattern of framework and space. First usage in English c. 1375.

Laurasia: [Portmanteau word from Modern Latin *Laur(entia)*, referring to the Canadian Shield and vicinity drained by the Saint Lawrence River + *(Eur)asia*, referring to Europe and Asia.] n. The Paleozoic supercontinent of the northern hemisphere that joined with Gondwana to form Pangea. Pangea eventually broke up in the Mesozoic, forming North America, Greenland, Europe, and most of Asia north of the Himalayas. The term was coined in 1928 by

Swiss geologist Riedolf Staub (1890-1961), in *Der Bewegungsmech Anismus der Erde* (*The Mechanics of the Agitation of the Earth*).
Cited usage: The Atlantic and Indian Oceans originated from the break up of the Gondwanan and Laurasian continents [1973, Nature 1 June 278/ 2].
See also: Gondwana, Pangea, protocontinent, supercontinent

Laurentian: [Latin *Laurentius* = Lawrence.] n. A name for the granites and orogenies of Precambrian age found in the Laurentian Highlands of the Canadian shield, north of the St. Lawrence River (radiometrically dated at about 1Ga). This term is widely misapplied, as it is often used to describe older Archean granites lying northwest of Lake Superior. Term coined in 1863 by Canadian geologist William Edmund Logan (1798-1875), in his monograph 'The Geology of Canada'. Logan was the founder of the Geological Survey of Canada.
Cited usage: The Laurentian rocks are the oldest formations at present known in the world [1863, A. C. Ramsay, Phys. Geog. V.].

lava: [Italian *lava* = a stream caused suddenly by rain; < Latin *lavare* = to wash (possibly influenced by *labes* < *labi* = to fall, slide).] n. A general term for molten rock on the Earth's surface that has extruded from a volcano or volcanic vent. The term is also used for the rock formed by solidification of molten lava on the Earth's surface. The people of Naples, Italy (Neapolitans), living in the vicinity of Mount Vesuvius, gave the term lava its modern meaning. They borrowed the Italian word *lava*, then meaning a stream caused suddenly by rain, and applied it to the streams of molten rock flowing down the slopes of the volcano. Mt. Vesuvius is the only active volcano on the European mainland. Its most famous eruption occurred in 79 CE, when it buried the cities of Pompeii and Herculaneum, killing the Roman

explorer and statesman Pliny the Elder.
Cited usage: This lava is a very hard substance, like stone, of a slate colour [1750, Phil. Trans. XLVII].

lawsonite: [Eponymous term after Andrew Cowper Lawson (1861-1952), Professor of Geology, University of California, Berkeley.] n. A colorless, to bluish-gray or pale-blue, orthorhombic mineral, hydrated hydroxy calcium aluminum silicate, $CaAl_2Si_2O_7(OH)_2 \cdot H_2O$. It is a common constituent of glaucophane schist and other metamorphic rocks, formed at high pressures and low temperatures in subduction zones.

leach: [Old English *lece* = muddy stream < *leccan* = to moisten.] v. To selectively remove, while in solution or suspension, by action of a percolating liquid, usually water. Leaching, or lixiviation, is an essential process in the formation of soils, and in the origin of certain types of ore deposits. First usage in English c. 1450.
Synonyms: lixiviation

lead: [Old English *lead* = heavy metal; < German *lot* = weight, plummet, sounding-lead, solder; < IE *ploud* = to flow.] n. A soft, dense, malleable, ductile, silvery to bluish-gray, isometric mineral and metallic element (symbol Pb). It has a relatively low melting temperature and is the heaviest of the base metals. It occurs rarely in the native state, and is extracted chiefly from the sulfide mineral galena (PbS). The symbol Pb comes from the Latin *plumbum*, meaning lead, and is the root of words such as: plumber, plum-bob, and plumbago. The term lead is often misapplied to the mineral graphite, as in 'lead' pencils. The element was known as early as 3500 BCE. First usage in English before 900.

league: [Middle English *lege* < Old French *liue* < Latin *leuga* = a measure of distance equal to 1.5 Roman miles.] n. A variously defined linear unit of measured distance used in different countries, and ranging in length from 2.4 to 4.6 miles. In England, for example, a land league is equal to 3 statute miles, whereas a marine league is equal to 3 nautical miles. A league is also defined as a unit of land area equal to one square league. First usage in English c. 1375.

lechatelierite: [Eponymous term after chemist Henri LeChatelier (1850-1936), who discovered LeChatelier's rule regarding disturbance of equilibrium.] n. Naturally fused amorphous silica occurring in fulgurites, tektites, and compact craters as a result of the melting of quartz from the heat of a lightning strike or meteorite impact. Lechatelierite differs from obsidian in being composed of pure silica. Term coined in 1915 by French mineralogist Alfred Lacroix (1863-1948). First usage in English c. 1915.

lectotype: [Greek *lektos* = chosen < *legein* = to choose; < IE *leg* = to collect + Greek *typos* = figure, impression.] n. In taxonomy, the single specimen selected as the type specimen for a particular species or subspecies when the original author did not designate the holotype for the new found species. IE derivatives include: eclogite, lexicon, and dyslexia.
Cited usage: A lectotype is a specimen or other element selected from the original material to serve as the nomenclatural type, when the holotype was not designated at the time of publication, or when the holotype is missing [1951, G. H. M. Lawrence, Taxon. Vascular Plants ix. 204].
See also: holotype, eclogite, lignite

lee: [Old English *hleo* < Old Norse *hle* = shelter, warmth < *hlyja* = to protect.] adj. Said of the sheltered side of any object facing away from the wind, or the side opposite to where the wind first hits an object, such as the side of a landmass (island, hill, sand dune), a ship, or a building. Also, in glaciology, said of the side of a hill, knob, or

other protrusion facing away from an advancing glacier. First usage in English before 900.
Cited usage: Rob Roy's cave under the Lee of Ben Lomond [1847, D.G. Mitchell, Fresh Gleanings 223].

legend: [Middle English *legende* = written account < Latin *legenda* = what is read < *legere* = to read.] n. In cartography, an explanatory table or list of the symbols and cartographic conventions appearing on a map or chart. In general usage, a story handed by tradition from earlier times, popularly accepted as historical. First usage in English (generally) c. 1325; in cartography, c. 1900.

lekolith: [Greek *lekos* = dish + *lithos* = rock, stone.] n. A mass of extrusive igneous rock that is equant in plan view, has a diameter greater than its depth, a nearly level upper surface, and a lower surface determined by the contour of the basin filled (e.g., a mass formed by a solidified lava lake). Term coined in 1968 by American volcanologist Robert Roy Coats.

lenad: [Portmanteau word from the feldspathoid minerals *le(ucite)* and *ne(phaline)*, + *ad*.] n. A mnemonic group name for the feldspathoid minerals.
See also: leucite, nepheline, feldspathoid

lens: [Latin *lens* = lentil *lentil* = a type of legume, doubly convex.] n. In geology, a rock unit bounded by converging surfaces, at least one of which is a curved surface. A lens is thick in the middle and thins out toward the edges. A lens may be plano-convex or double-convex. The term derives from the Latin word for lentil, a legume with a double-convex seed. This shape is in contrast to a double-concave shape, which is thin in the center and thickens towards the outer edge; especially in optics, a translucent material of this shape designed to refract light is also called a lens. First usage in English c. 1690.

Cited usage: A glass spherically convex on both sides (usually called a Lens) [1704, Isaac Newton, Opticks I].
Synonyms: lentil
See also: lenticle

lenticle: [Latin *lens* = lentil *lentil* = a type of legume, doubly convex.] n. A lens-shaped rock or stratum, regardless of size. Not to be confused with the botanical term lenticel.
See also: lens

lepidoblastic: [Greek *lepido* < *lepis* = scale + *blasto* = germ, sprout, bud.] n. In petrology, referring to the texture of foliated metamorphic rock, where the essential minerals have a flaky, scaly, or micaceous habit, and are all about the same size.
See also: homeoblastic

lepidocrocite: [Greek *lepido* < *lepis* = scale + *krokis* = fiber, thread.] n. A ruby-red to reddish-brown, orthorhombic mineral, hydroxy iron oxide, $FeO(OH)$. It is polymorphous with goethite and akaganéite. The name was coined in 1813 by German mineralogist J. C. Ullmann, alluding to the mineral's feathery to scaly habit. First usage in English c. 1820.

lepidolite: [Greek *lepido* < *lepis* = scale + *lithos* = rock, stone.] n. A lilac, yellow, rose, or gray monoclinic mineral of the mica group, hydroxy potassium lithium aluminum silicate, $K(Li,Al)_3(Si,Al)_4O_{10}(F,OH)_2$. It is found in pegmatites, and is an ore of lithium. Term coined in 1792 by German chemist M. H. Klaproth, alluding to the scaly appearance of many specimens. The etymology refers to the scalelike or platy habit of the micaceous minerals.
Cited usage: A violet variety [of common mica] occurring in small scales, has been distinguished by the name lepidolite [1837, James D. Dana, A System of Mineralogy 264].

lepto-: [Greek *leptos* = fine, thin, delicate < *lepein* = to peel.] Combining prefix meaning thin, used in such terms as: leptochlorite, leptopel, and leptothermal.

leptothermal: [Greek *leptos* = fine, thin, delicate < *lepein* = to peel + *therme* = heat.] adj. In ore deposits, said of hydrothermal mineral deposits that are formed at temperatures and depth conditions intermediate between epithermal and mesothermal. The term has generally been discarded due to the difficulty of distinguishing such deposits.

leucite: [Greek *leukos* = white.] n. A white to gray mineral of the feldspathoid group, potassium aluminium silicate ($KAlSi_2O_6$). It is usually found as vitreous trapezohedral crystals in volcanic rocks, especially in lavas from Mt. Vesuvius, Italy. Term coined in 1791 by Abraham Gottlob Werner.
Cited usage: Many of the older lavas yield agates, leucite and other precious minerals [1876, Page, Advd. Texbook. of Geol. vii. 146].
Synonyms: white garnet, Vesuvian garnet

leuco-: [Greek *leukos* = white.] Combining prefix meaning white, used in such terms as: leucocratic, leucophyre, leucite, and leucoxene.
See also: leucocratic

leucocratic: [Greek *leukos* = white + *kratos* = rule, sway, authority.] adj. Said of light-colored igneous rocks that are relatively poor in mafic minerals, usually less than 30-35% of all minerals.

levee: [French *levee* < *lever* = to rise.] n. A long, low, broad ridge or embankment of sand and coarse sediment, built by a stream along both sides of its channel, especially during overbank flooding. Such embankments are also called natural levees. The term can also refer to various similar forms of natural embankments, created by other means, such as mud flows, lava flows, along submarine channels, or along the borders of deep-sea fans. Levees are also human-made, as in an embankment built along a stream or arm of the sea to prevent flooding (the levee system of the Mississippi River maintained by the Army Corps of Engineers, or parts of the California Aqueduct). First usage in English c. 1715.
Cited usage: The town (New Orleans) is secured from the inundations of the river by a raised bank, generally called the Levée [1770, P. Pittman, European Settlement of Mississippi].

lichen: [Greek *leikhein* = to lick; < IE *leigh* = to lick.] n. Any complex organism of the subdivision Lichenes, composed of a fungus and an alga living in a mutualistic symbiotic relationship. Most lichen are in the order Ascomycetes, whereas a few are in Basidiomycetes. Lichens are also classified by growth form, the most common being crustose, a form that attaches very tightly to its substrate, such as a: rock, tree, or building. Other growth forms include: foliose, fruticose, and squamulose. Lichens may be mistaken for weathered or stained rock when viewed at a distance. IE derivatives include: lick, and lingula.

lido: [Toponymous term after Italian *Lido*, a small island chain near Venice; < Latin *litus* = shore.] n. Originally, the name of the chain or spit of barrier islands that protects the lagoon of Venice, Italy. The term is now also used generally for such a spit enclosing a swimming lagoon or fashionable beach resort area. The Italian Lido, like Monaco, is an international playground. First usage in English c. 1925.
Cited usage: The broad sandspit or lido separating the lagoons from the sea. [1934, Discovery Aug. 215].
Synonyms: plage

liesegang: [Eponymous term after German chemist Raphael Eduard Liesegang (1869-1947).] adj. In geology, referring to concentric rings or parallel bands caused by rhythmic precipitation of crystalline or amorphous compounds at discrete intervals in a porous rock. Liesegang rings form due to diffusion of one or more dissolved substances through the fluid-saturated rock. Such bands can be produced in the laboratory by diffusion of two soluble substances from opposite ends of a gel column, resulting in rhythmic precipitates at discrete intervals in the column. *Cited usage*: The concentric rings in the 'common gall stones' of inflammatory origin are a manifestation of the liesegang phenomenon [1932, Jrnl. of Physical Chem.].

life: [Old English *lif* = life, person, body; < IE *leip* = to stick, adhere, fat.] n. A complex chemical assemblage that is capable of capturing and transforming energy into matter or another form of energy and is capable of creating a flow of this energy to evolve areas of greater order and adaptability. On a cosmic level, life is, therefore, a zone of relatively low entropy. IE derivatives include the prefix lipo- (fat) and liver (formerly believed to be the blood-producing organ).
Cited usage: He of life ewat [c.1100, Beowulf, line 2471]. As long as there's Life there's Hope [1697 Collier, Immor. Stage 288].

ligament: [Latin *ligamentum* = bandage < *ligare* = to bind, tie.] n. In paleontology, and biology, a tough structure of connecting tissue in an animal. In vertebrates, it is the fibrous band of tissue connecting skeletal elements. In invertebrates, such as bivalves, it is a band of elastic tissue connecting the valves of the shell. First usage in English c. 1400.

ligand: [Latin *ligandus* < *ligare* = to bind, tie.] n. In geochemistry, an ion or functional group (usually negatively charged) that is bonded to the central atom or ion (usually a metal and positively charged). An array of ligands around a center (usually a cation) is called a coordination compound or metal complex. Etymologically related terms include ligament and ligature. First usage in English c. 1945.

lignin: [Latin *lignum* = wood, firewood.] n. A complex and amorphous organic substance that, together with cellulose, forms the chief constituent of wood. First usage in English c. 1820.

lignite: [Latin *lignum* = wood, firewood; < IE *leg* = to collect, gather.] n. A soft, brownish-black coal in which the alteration of plant material has proceeded further than in peat, but not as far as in a sub-bituminous coal. First usage in English c. 1805. IE derivatives include: eclogite, lectotype, homologous, logic, logarithm and legal.
Synonyms: brown coal
See also: eclogite, lectotype

liman: [Russian *liman* < Greek *limen* = harbor.] n. Originally applied to the salt-marshes at the mouth of the Dnieper river in Russia, now a general term for a shallow, muddy, and salty lagoon at the mouth of a river, especially one that is protected from the open sea by a barrier bar or spit. First usage in English c. 1855.
Cited usage: Liman, a shallow narrow lagoon, at the mouth of rivers, where salt is made [1858, Simmonds, Dict. of Trade].

lime: [Old English *lim* = lime; < Dutch *lijm* < German *leim* < Latin *limus* = mud slime.] n. A white, caustic, isometric mineral, calcium oxide, CaO. It is a very rare mineral found only in some volcanic areas, such as on Mount Vesuvius. The term is also used for the industrial material, calcium oxide (also called quicklime), produced by thermal decomposition of calcium carbonate (calcining), thus removing carbon dioxide in a reversible reaction. The term has been used incorrectly by miners and drillers in referring to limestone or calcite. It is also frequently used as a short form for the common noncaustic hydrated lime. IE derivatives include: loam, slimy, and schlep. First usage in English before 900.
See also: calcine

limnology: [Greek *limne* = lake, marsh + *logos* = word, speech, discourse, reason < *legein* = to speak, to say; IE *leg* = to collect.] n. The study of the physical, chemical, geological, and biological aspects of lakes, pools, ponds, and other bodies of inland waters, especially those having measurable residence times. First usage in English c. 1890. *Cited usage*: Friedrich Simony became the founder of the special branch of science termed by Forel, Limnology, or the scientific study of lakes [1893 Geogr. Jrnl. I. 353].

limonite: [Greek *leimon* = meadow.] n. A general field term for a dark-brown to yellow-brown mixture of amorphous and crystalline hydroxy iron oxides, including goethite and lepidocrocite, along with possible hematite and various amounts of absorbed water. It is a common secondary material formed by oxidation (weathering) of iron-bearing minerals. Limonite may form as a biogenic precipitate in bogs, lakes, or springs. The term was coined in 1813 by German mineralogist Johann Friedrich Ludwig Hausmann (1782-1859). At first, the meaning was restricted to bog-iron ore, and prior to 1813, it was known in Germany as *weisenerz*, meaning meadow-ore, alluding to its formation in bogs and swamps.
Cited usage: Limonite occurs in stalactitic, mammillated, pisolitic, or earthy, conditions [1879, Rutley, Study Rocks x. 156].

lineament: [Latin *linea* = line, string, cord < *linum* = flax, thread, linen; < IE *lino* = flax.] n. A linear geologic feature of regional extent, that is believed to reveal something about the crustal structure. Examples of lineaments are: fault lines, aligned volcanoes, and linear stream courses. First usage in English c. 1425. IE derivatives include: line, lineage, linseed, linen, lint, and lingerie.

lingulid: [Latin *lingula* = little tongue < *lingua* = tongue.] n. Any of the inarticulate brachiopods belonging to the family Lingulidae of the order Lingulida and the class Lingulata. Lingulids are characterized by an elongate oval to spatula-like outline, and a biconvex shell. They range from Lower Cambrian to Recent. The inarticulate brachiopod genus Lingula is the oldest unchanged animal known, the first occurrence being early Cambrian, possibly older than 500Ma. Lingula is primarily an Indo-Pacific genus. It is harvested for human consumption in Japan and Australia. The term lingulid refers to the tongue-like shape of the shell.
Synonyms: lamp shell

linguoid: [Greek *lingua* = tongue + *ordos* = resembling.] adj. Said of an aqueous ripple mark generated by a current, characterized by a tongue-shaped outline, or by having a barchan shape whose horns point into the current. It is usually developed best on shallow stream bottoms. Term coined in 1919 by American paleontologist Walter Hermann Bucher (1888-1965).

linn: [Gaelic *linne* = pool.] n. A term used chiefly in Scotland and England for a pool of water, especially a deep one below a falls. Also, a torrent, cascade or waterfall. Dublin, the capital of the Republic of Ireland, is derived from *dubh*, meaning black and *linne*, meaning pool.

Linnaean: [Eponymous name after Carolus Linnaeus (1707-78), Swedish botanist who developed the modern classification system for plants and animals.] adj. Referring to the system of taxonomic classification and binomial nomenclature for living organisms that was developed by Carolus Linnaeus (who Latinized his name from Carl von Linné).

linsey: [English *linsey-woolsey* = A coarse, woven fabric of wool and linen.] n. A sedimentary rock composed of alternating, interbedded, striped shales and streaky-banded sandstones. So named because the interbedding resembles the weave of linsey-woolsey fabric, and because the type variety is found in Lancashire, England, where linsey-woolsey is commonly used.

liparite: [Toponymous term after the Lipari Islands, northeast of Sicily; < Greek *liparos* = shining.] n. A generally obsolete synonym of rhyolite, used by some German and Russian authors. The term was originally coined in 1847 by C.F. Glocker, but his 'liparite' was actually fluorite. German geologist J. Roth gave the term its modern meaning in 1861.
Synonyms: rhyolite
See also: rhyolite

lithification: [Greek *lithos* = rock, stone.] n. The process by which unconsolidated sediment is converted to coherent solid rock by some combination of cementation, compaction, and dehydration. Lithification may occur soon after or long after deposition of the sediment. First usage in English c. 1870.
Synonyms: induration

lithium: [Greek *lithion* = diminutive of *lithos* = rock, stone.] n. A soft, silvery, highly reactive, metallic element (symbol Li). Lithium has an atomic number of 3, and is directly below hydrogen on the periodic table. The element was discovered in 1817 by Swedish chemist Johann August Arfvedson, who identified it in the mineral petalite, $LiAlSi_4O_{10}$. The term was coined a year later by Swedish chemist Jons Jacob Berzelius (1779-1848), because lithium was first discovered in a mineral, distinguishing it from previously discovered alkalis, such as potassium and sodium, which were separated from plant material of vegetable origin.
Cited usage: A substance is separated, which may be called 'lithium', the term 'lithia' being applied to its oxide [1851, Richardson, Geology v.81].

litho-: [Greek *lithos* = rock, stone.] Combining prefix meaning rock or stone, used in such terms as: lithology, lithofacies, lithoclast, and lithogenesis.
Synonyms: -lith, -lite
See also: -lith, -lite

lithoid: [Greek *lithos* = rock, stone + *oidos* = resembling.] n. Said of material that has rocklike characteristics, used in such terms as lithoid tufa, a gray, compact, bedded tufa that is older and more stonelike than the overlying thinolitic tufa and dendroid tufa.

lithology: [Greek *lithos* = rock, stone + *logos* = word, speech, discourse, reason < *legein* = to speak, to say; IE *leg* = to collect.] n. The description of rocks in hand specimen and in outcrop, including such characteristics as: mineralogic composition, grain size, and color. First usage in English c. 1715.

lithophile: [Greek *lithos* = rock, stone + *philos* = loving.] adj. Said of an element that is concentrated in the silicate rather than the sulfide (chalcophile) or metal (siderophile) phases of meteorites. These elements concentrate primarily in the Earth's crust, rather than the mantle or core, as described by Victor Moritz Goldschmidt in his three-fold system for dividing Earth's solid elements. Goldschmidt's fundamental laws of geochemistry, which include his tripartite division of Earth's elements, appear in his series of monographs titled Geochemische Verteilungsgesetze der Elemente (*Geochemical Laws of Distribution of the Elements*). First usage in English c. 1920.
See also: chalcophile, siderophile

lithophysae: [Greek *litho* = rock, stone + *physa* = bellows.] n. Hollow spherulites, shaped like a bubble, and having concentric shells composed of fine crystalline quartz, alkali feldspar, and other minerals. Lithophysae are found in silicic volcanic

rocks, especially rhyolite, obsidian, and perlite. The term alludes to the hollow interior, likened to a bellows.
Alt. spelling: lithophysa (sing.)
Synonyms: stone-bubble
Similar terms: thunderegg

lithosome: [Greek *lithos* = rock, stone + *soma* = body.] n. A largely homogeneous mass of rock penetrated by tongues of rock from adjacent masses having a different lithology.
Synonyms: magnafacies
See also: magnafacies

littoral: [Latin *littoralis* or *littus* = shore.] adj. Said of a biogeographic zone or related processes that are located or taking place near the shore of the ocean, sea, or a lake. The term is generally defined as the intertidal depth zone between high and low water. Due to tidal variation, the areal extent of the littoral zone ranges widely. In some places it is defined by tidal flux, whereas in others by species extent. First usage in English c. 1655. An etymologically related term is *lido*.
Cited usage: There were then also littoral formations in progress, such as are indicated by the English Crag [1830, Lyell, Principles of Geology I. 151].
Synonyms: intertidal zone

lixiviate: [Latin *lixivium* = made into lye < *lix* = ashes, lye + *luv* = to wash.] v. To wash out soluble matter from a larger mass, usually by percolation. First usage in English c. 1645.
See also: leach

llano: [Spanish *llano* = plain; < Latin *planus* = level; < IE *pel* = flat.] n. A level, treeless plain or steppe in the northern parts of South America. IE derivatives include feldspar, field, planet, and polka.
Cited usage: Peru is divided into three parts, which they call Llanos, Sierras, and Andes. The

Llanos or Plaines on the Sea-coast have ten leagues in bredth [1613, Purchus His Pilgrimes 873].

loam: [Old English *lam* = clay.] n. An easily worked, fertile soil composed of approximately equal amounts of clay, silt, and sand. An etymologically related term is lime < Latin *limus*, meaning mud.

lobe: [Greek *lobos* = lobe of the ear, of the liver or kidney, or a pod of leguminous plants.] n. In glacial geology, a large marginal projection from the body of a continental ice sheet, or a rounded tonguelike projection of glacial drift. In biology and paleontology, parts of organisms with the shape of a rounded protuberance. First usage in English c. 1520.
Cited usage: The moraines can be traced around continuously from one lobe to another [1889, Nature 3 Oct. 558].

loch: [Scottish and Gaelic *loch* = lake.] n. In the British Isles, a lake. Also, a narrow arm of the sea, generally resulting from glacial erosion. First usage in English c. 1375.

lode: [Middle English *lode* = way, load < Old English *lad* = way.] n. In ore deposits, a metal-bearing vein of ore that fills a fissure in a consolidated rock formation, as opposed to an alluvial placer deposit. The alternate spelling was used in naming the famous gold mining area of Lead in the Black Hills of South Dakota. First usage in English before 900.
Alt. spelling: lead
Synonyms: vein
See also: placer

loess: [German *loess* < *losch* = loose; < IE *leu* = to loosen, divide.] n. A widespread, generally homogenous, buff to yellowish-brown to gray, blanket deposit of fine-grained, slightly coherent, calcareous windblown silt or clay. Loess was transported in large quantities from deserts, glacial outwash plains, and alluvial valleys during the

Pleistocene epoch. A characteristic feature of loess is its ability to stand in vertical faces. Loess is widespread in Eastern China, North-central Europe, and in the U.S. in the Pacific Northwest and Mississippian Valley. First usage in English c. 1830.
Cited usage: There is a remarkable alluvium filled with land-shells of recent species which we may refer to the newer Pliocene era. This deposit is provincially termed 'Loess' [1833, Lyell, Principles of Geology III. 151].
Alt. spelling: löess

loess kindchen: [German *loess* < *losch* = loose + *kindchen* = small children < *kindern* = children.] n. A compound nodule or concretion of calcium carbonate found within a loess deposit and resembling the head of a child or a potato. It is often hollow, but may contain a loose stone.

loipon: [Greek *loypon* = remaining, the rest, residue.] n. A residual surface layer that remains after intense chemical weathering has broken down the more reactive component. Examples are bauxite deposits, gossans over ore bodies, and duricrust. Term coined by Shrock in 1947.
Alt. spelling: bauxite, gossan, duricrust

loma: [Spanish *loma* = hillock < *lomo* = back, loin, ridge.] n. Term used in the U.S. Southwest for a broad, gentle, elongated hill. Frequently used in place names, such as: Loma Prieta, Loma Linda, or Yucca Loma. First usage in English c. 1845.

longitude: [Latin *longitudo* = length < *longus* = long.] n. An imaginary great circle on the surface of a sphere passing through both poles at right angles to the equator. On Earth, longitude is expressed as the angular distance on the Earth's surface measured parallel to the equator from a given meridian to any other meridian. By convention, longitudes on Earth are referenced to the Prime Meridian, presently located at Greenwich,

England, site of the Royal observatory. It is usually expressed in degrees, minutes, and seconds; 15° of arc being equivalent to 1 hour. Formerly each nation took its own capital or principal observatory as the standard meridian from which longitudes were measured. By the turn of the 20th century, most nations had adopted Greenwich as the Prime Meridian. France, the last major holdout, signed the agreement to accept the Greenwich meridian as standard at the International Meridian Conference in Washington D.C. in 1884 (at the behest of the Canadians, who pointed out that 72% of the world's floating commerce already operated on Greenwich time). Not until after World War I however, did France abandon the meridian of the Paris Observatory as its standard for all nautical and astronomical purposes. First usage in English c. 1375.
Cited usage: Mr. Williams had made many ingenious advances towards a discovery of the longitude [1791, Boswell, Johnson an. 1755].
See also: latitude

longwall: [English description of a style of mining process, named after the long length of mining tunnels.] n. A highly efficient underground mining process in which a panel or block of coal, generally 700 feet wide and often over a mile long, is completely extracted using heavy machinery that functions like an auger. The working area is protected by a movable hydraulic roof support system. After mining out the coal, the supportsare removed and the roof allowed to collapse. Longwall mining accounts for 45% of the underground coal production. The remainder is by the older, traditional room and pillar technique. Most mining in England has been of the longwall advancing method, in which a long straight wall of coal is exposed and gradually cut away, advancing from the entrance shaft to the outer limits of the mineable coal bed. First usage in English c. 1835.

lophophore: [Greek *lophos* = crest of a helmet + *phoros* = bearing < *pherein* = to carry.] n. A circular or horseshoe-shaped ciliated organ surrounding the mouth of brachiopods, bryozoans, and wormlike phoronids that is used to gather food. Term coined in 1850 by G. J. Allman.
Cited usage: The sort of disc or stage which surrounds the mouth and bears the tentacula, I have called Lophophore [1850, Allman, Report for The British Association for the Advancement of Science 307].

lopolith: [Greek *lopas* = basin, a flat earthen dish + *lithos* = rock, stone.] n. A large, plano-complex, funnel-shaped, or lenticular, igneous intrusion that is sunken on both the roof and floor due to the sagging of the underlying rock. Term coined in 1918 by geologist F. F. Grout.
Cited usage: Professor John Barrell has suggested that as igneous forms they deserve a distinct name. The name proposed by the writer is 'lopolith' [1918, F. F. Grout, Amer. Jrnl. Sci. CXCVI. 518].
See also: laccolith

loran: [English acronym from the initial letters of *lo(ng)* + *ra(nge)* + *n(avigation)*.] n. A long-range navigation or position-fixing system that employs the difference in the times of arrival of pulsed radio signals from different stations in order to triangulate a position.

lottal: [Adaptation of English *lot'll* = a word from a poem by Richard Armour.] n. A field term for the aqueous clayey mixtures formed by mass wasting down hillslopes. Term coined in 1962 by geologist L.C. King largely as a jest.
Cited usage: Shake and shake / The catsup bottle. / None will come, / And then a lot'll [Richard Armour (1906-1989)].

louderback: [Eponymous term after American geologist G. D. Louderback (1874-1957).] n. A cap of old lava on a tilted fault-block, used as evidence of block faulting in the Basin and Range Province of the U.S. Term coined in 1930 by W.M. Davis.
Cited usage: It seems to me highly appropriate that the lava sheets, which were thus spread unconformably on the Powell surface of the worn-down King mountains, and which now cover the back slopes of the tilted Gilbert blocks, should be called Louderbacks, after their discoverer [1930, W. M. Davis, Bull. Geol. Soc. Amer. XLI. 299].

luminescence: [Latin *lumen* = an opening, light < *lux* = light; < IE *leuk* = light, brightness.] n. The emission of light that does not derive energy from the temperature of the emitting body, as in phosphorescence, fluorescence, and bioluminescence. Luminescence is caused by a variety of effects, such as: a) changes in chemistry, biochemistry, or crystallography, b) motions of subatomic particles, or c) radiation-induced excitation. Etymologically related terms include: luster, lunar, lunatic, and Lucifer.

luster: [Latin *lustrare* = to illumine < *lux* = light; < IE *leuk* = light, brightness.] n. The manner in which light reflects from the surface of a mineral, described in terms of its quality and intensity. Terms such as metallic, adamantine, silky, and vitreous describe the quality, whereas bright, shiny, and dull refer to intensity.

lutite: [Latin *lutum* = mud.] n. A general term for fine-grained sedimentary rocks making up shales and mudstones. Lutites are composed of silt or smaller-sized particles. Term coined in 1904 by A.W. Grabau. The analogous Greek term is pelite.
Cited usage: The third texture (that of 'rock flour or impalpable powder'), finally, may be designated by the term lutaceous, (from lutum, mud), and for consolidated rocks of this type the term lutyte may be used, irrespective of chemical composition. Pelyte has been used

particularly for argillaceous rocks of the group [1904, A. W. Grabau, Amer. Geologist XXXIII. 242].
Alt. spelling: lutyte
Synonyms: pelite
See also: pelite, arenite, rudite

lyell: [Eponymous term after Scottish geologist and writer Charles Lyell (1797-1875).] n. An ice-rafted block of rock transported by an iceberg, then deposited at some distance from the pick-up location. Term coined in 1961 by Hamelin, in recognition of the geologist Sir Charles Lyell, who first published the book *Principles of Geology* in 1830. Lyell continued to publish new revisions until he died in 1875. He was a determined proponent of uniformitarianism, but insisted that catastrophic events, even those naturally produced, were insignificant in Earth history.
Synonyms: dropstone

lynchet: [Middle English *lynch* = an unploughed strip serving as a boundary between fields; < Old English *hlinc* = a ridge or bank.] n. A bank of earth that moves and collects downslope from a disturbed surface area, such as a ploughed field, terraced hillside, or developed ridge. The term shares the same root with the term links, used to describe a golf course.

lysocline: [Greek *lusis* = a loosening < *luein* = to loosen; < IE *leu* = to loosen, divide, cut up + Greek *klinein* = to slope, bend.] n. The ocean depth at which the dissolving rate of calcium carbonate just exceeds its combined rate of deposition and precipitation, resulting in a net loss of calcite. The term is not synonymous with the carbonate compensation depth (CCD), because the lysocline marks the depth in the ocean where calcium carbonate begins to dissolve; however, the calcium carbonate compensation depth is the depth where calcium depth becomes completely dissolved. Some IE derivatives include: loess, catalysis, loosen, solve, soluble, and solution.

Ma: [Greek *megas* = great + Latin *annum* = year.] An abbreviation for mega-annum, meaning one million years before the present. Many authors prefer to use *mya*, meaning "million years ago" because Ma technically means simply, one million years, not necessarily in the past. This term is an example of an etymological hybrid, being a combination of roots from Greek and Latin.
Synonyms: million years
See also: Ga, Ka

maar: [German *maar* = crater lake < *mariska* = water-logged land (cognate with West Frisian *mar* = lake; < Old Frisian *mere* = sea); < Latin *mara* = standing water, lake < *mare* = sea; < IE *mori* = body of water.] n. In general, any low-relief crater formed by explosive volcanism and eruption of ash with no accompanying lava flows. The crater does not lie within a cone, but is surrounded by a tuff ring or crater ring. Maars are usually occupied by lakes, such as at the type locality in the Eifel district of Germany. Not to be confused with mar. IE derivatives include: marsh, marine, meerschaum, cormorant, and maré (craters of the Moon). First usage in English c. 1825.
Cited usage: The craters of this country (Prussia) have nearly, without exception become bodies of water, or maare, as they are called by the natives [1826, Edinburgh Jrnl. 5 152]. Maars are relatively flat-floored craters of explosion at vents that are either coneless or else provided with inconspicuous cones [1933, R. A. Daly, Igneous Rocks & Depths of Earth viii. 161].
See also: mar

macaluba: [Toponymous term after Macaluba, a mud volcano in Sicily.] n. A conical accumulation of mud and rock ejected by volcanic gases or volcanically heated ground water.
Synonyms: mud volcano
See also: hervidero, salinelle

maceral: [Latin *macerare* = to make soft, weaken, steep (like tea); < Greek *mattein* = to knead; < IE *mag* = to knead, fit.] n. Any one of the fundamental organic constituents that comprise coal. Maceral are most easily distinguished in polished or thin sections. The names of the three principle maceral groups, all ending with the suffix '-inite,' are: vitrinite, inertinite, and liptinite (exinite). There are also three microlithotype groups: the monomacerals (vitrite, liptite, and inertite), bimacerals (clarite, durite, and vitrinertite), and the trimacerals (duroclarite, vitrinertoliptite, and clarodurite). In hand specimen, the main macrolithotypes, all ending in the suffix '-ain,' are: vitrain, fusain, clarain, and durain. By crude analogy, macerals are to coal as distinct minerals are to inorganic rock. Term coined in 1935 by M.C. Stopes. IE derivatives include: magma, mass, masonry, maquillage, and macerate.
Cited usage: To construct an acceptable petrological classification, therefore, a prime need became obvious, viz., a word to cover all petrological units seen in microscopic sections of coals, as distinct from the visible units seen in hand specimens. I now propose the new word 'Maceral' (from the Latin, macerare, to macerate). The word 'macerals' will, I hope, be accepted as a pleasantly sounding parallel to the word 'minerals', conveying the suggestions of the fundamental difference between them. ...The concept behind the word 'macerals' is that the complex of biological units

represented by a forest tree which crashed into a watery swamp and there partly decomposed and was macerated in the process of coal formation, did not in that process become uniform throughout but still retains delimited regions optically differing under the microscope [1935, Dr. Marie Carmichael Stopes, Fuel in Sci. & Pract. XIV. 11/1].
See also: vitrain

macro-: [Greek *makros* = large, long.] Combining prefix meaning large, used in such terms as: macrofossil, macrocrystalline, and macroscopic. Used in contrast to the prefix *micro-*.
See also: micro

maculose: [Latin *macula* = spot, blemish.] adj. In petrology, said of contact metamorphic rocks that have a spotted or knotted appearance, such as spotted slates.

madrepore: [Italian *madrepora* < madre = mother; < Latin *mater* = mother + Italian *pora* = tufa, pore; < Latin *porus* < Greek *poros* = calcareous stone, stalactite.] n. Any true or stony coral of the order Madreporaria, the main builders of coral reef islands in tropical seas. They are genetically related to Actinaria, the order of common sea anemones. Madrepore are found in all the world oceans, but only those in the tropics form coral reefs. First usage in English c. 1750.
See also: coral

madreporite: [Italian *madrepora* < madre = mother; < Latin *mater* = mother + Italian *pora* < *poro* = tufa, pore; < Latin *porus* = passageway.] n. A perforated plate-like structure in most echinoderms that forms the intake for their water-vascular systems. The term may also be derived from the Greek *poros*, meaning calcareous stone or stalactite. First usage in English c. 1800.

maelstrom: [Dutch *malen* = to grind, whirl; < IE *mel* = to crush, grind + Dutch *stroom* = stream; < IE *sreu* = to flow.] n. Rapid, whirling, and confused currents, the strongest ones being created by strong wind-generated waves moving against an opposing tidal current, often with resultant rips, eddies, and whirlpools. In coastal areas, shape of the surrounding topography can also be a significant factor in their form, intensity, and duration. Whereas whirlpools in the open seas, such as the Sargasso Sea, are giant suctionless eddies, strong down-ward spiraling vortical whirlpools (maelstrom) occur in places such as: Charybdis in the Strait of Messina between mainland Italy and the island of Sicily; the Maelstrom, or Moskenstraumen, in the Lofoten Islands off Norway; and the Whirlpool Rapids below Niagara Falls. Another famous location is the Correyvrecken Maelstrom at Jura in the Hebrides, Scotland. There George Orwell wrote his novel *1984* in a farm house overlooking it. He and his son nearly lost their lives while crossing, having miscalculated the time of the strongest tide. IE derivatives of the root *mel* include: mold, molar, millet, mallet, malleable, maul, and pall-mall; derivatives of *sreu* include: rhyolite, diarrhea, hemorrhoid, and rhythm. First usage in English c. 1555.

mafic: [A portmanteau word from *ma*(gnesium) + *f*(err)*ic*.] adj. Referring to the group of dark-colored magnesium- and iron-bearing minerals that occur in igneous rocks. Also, said of an igneous rock, such as gabbro or basalt, composed chiefly of one or more dark-colored, ferromagnesian minerals. Term coined in 1902 by W. Cross, J.P. Iddings, L.V. Pirsson, H. S. Washington. The group worked together for three years on their system of "chemicomineralogical" classification of igneous rocks. It is curious that they never published after 1902 and also always credited the entire group for the discoveries made and the nomenclature coined.

Cited usage: We suggest the term mafic for the group of modal ferromagnesian minerals of all kinds [1902, W. Cross et al., Jrnl. of Geology]. *See also*: femic, felsic, salic

magma: [Middle English *magma* = sediment, dregs; < Greek *massein* or *mag-* = to knead.] n. Naturally occurring molten rock at some depth beneath the Earth's surface. Magma may contain suspended crystals and rock fragments, as well as vapor bubbles. Such molten material is capable of intrusion or extrusion, and upon solidification becomes igneous rock. First usage in English c. 1425.

magna-: [Latin *magna* = feminine form of *magnus* = great.] Combining prefix meaning great, used in such terms as: magnificent, magnum, and Magna Carta. *See also*: magnafacies

magnafacies: [Latin *magna* = feminine form of *magnus* = great + Latin *facia* < *facies* = form, figure, appearance < (possibly) *facere* = to make, or < (possibly) *fa-* = to appear, shine.] n. A major, continuous, homogeneous belt of sedimentary deposits having similar lithologic and paleontologic characteristics that reflect a distinct depositional environment, but that extend across time boundaries. Term coined in 1934 by Kenneth Edward Caster (1908-1992). *Synonyms*: lithosome *See also*: lithosome

magnesite: [Middle English *magnesia alba* = white magnesia, main ingredient of the Philosopher's Stone; < Greek *Magnesia lithos* = the lodestone < *Magnesia* = region in Thessaly of Asia Minor.] n. A white to grayish-yellow, or brown trigonal mineral, magnesium carbonate, $MgCO_3$. Magnesite is isomorphous with iron-bearing siderite. It is most commonly found as earthy masses, or in veins resulting from hydrothermal alteration of magnesium-rich rocks, such as dolostone or serpentinite.

magnesium: [Middle English *magnesia alba* = white magnesia, main ingredient of the mythical Philosopher's Stone; < Greek *Magnesia lithos* = the stone of Magnesia < *Magnesia* = region in Thessaly of Asia Minor.] n. A low-density, ductile, silver-white metallic element (symbol Mg). Magnesium does not occur in the native state, but is widely distributed in combination with other elements, chiefly as silicates, carbonates, and chlorides. It occurs in many minerals, such as: olivine, augite, hornblende, biotite, talc, dolomite, and kainite. In 1808, Sir Humphry Davy isolated the metal and called it *magnium* to avoid confusion with the previously discovered manganese. Despite his proposal, the term magnesium persisted. Derived from the medieval name *Magnesia lithos*, magnesium was the source for the ancient medicinal ingredient now known as 'milk of magnesia.' *See also*: manganese

magnetite: [Latin *magnes* = magnet; < Greek *Magnetes lithos* < *Magnesia lithos* = Magnesian stone.] n. A black, opaque, strongly-magnetic, isometric mineral of the spinel group, iron oxide, Fe_3O_4. It is isomorphous with magnesium-bearing magnesioferrite and manganese-bearing jacobsite. Magnetite is an important ore of iron, and is widely distributed in rocks of all types. It occurs commonly as octahedral crystals, as well as granular masses. Magnetite, which acts as a magnet itself (rather than simply being attracted to a magnet) is called lodestone. It was known to the ancient Greeks, such as Thales of Miletus (c. 625-547 BCE), who called this mineral '*magnet*,' because it was found in the land of the Magnetes in Thessaly, Greece. Pliny the Elder (23-79 CE), citing Nicander, called the mineral '*magnes*,' after a shepherd named *Magnes*, who found that the rock on Mount Ida attracted the iron nails in his shoes and the iron ferrule of his staff. First usage in English c. 1850. *See also*: magnesia

magnitude: [Latin *magnitudo* = greatness < *magnus* = great + *tudo* = suffix indicating an abstract noun.] n. In seismology, a measure of the size of an earthquake in terms of the total energy released as measured using a seismograph. Originally defined in 1935 by seismologist C.F. Richter as the logarithm to the base 10 of the amplitude in microns of the largest trace deflection on a standard torsion seismograph by an earthquake at a distance of 100km from the epicenter.

makhtesh: [Hebrew *makhtesh* = mortar.] n. A term used in Israel for a craterlike depression created by water erosion of a structural dome. Typically, the depression is surrounded by steep walls of resistant rocks made of limestone and dolomite, whereas the bottom of the basin consists of friable sandstone. Makhteshim frequently are sites of desert oases, having a wadi at their base (seasonal watercourse). In the Central Negev Hills located in the Eastern Sinai Peninsula of Israel, desert erosion has created Makhtesh Ramon, the largest known makhtesh.
Alt. spelling: makhteshim (pl.)

malachite: [Greek *molochitis lithos* = mallow stone < *malokhe* = mallow; < Semitic *mallua* = a salt-marsh plant < *mela* = salt.] n. A bright-green, monoclinic mineral, hydroxy copper carbonate, $Cu_2CO_3(OH)_2$. It is an ore of copper and a common secondary mineral. Malachite forms in the oxidized zone above primary deposits of copper sulfides. The term alludes to the resemblance of the color of the mineral to that of the leaves of the mallow plant. First usage in English c. 1375.

malacolite: [Greek *malakos* = soft + *lithos* = rock, stone.] n. An obsolete synonym of diopside. The term originally designated a light-colored, pale-green, or yellow, translucent variety of diopside from Sweden.

malacostracan: [Greek *malakostrakos* = soft-shelled < *malakos* = soft + *ostrakon* = shell.] n. Any crustacean belonging to the class Malacostraca (subphylum Crustacea), characterized by compound eyes and a thorax with eight subdivisions. Malacostracans be marine, freshwater, or terrestrial, such as those found in the orders: Decapoda (lobsters, shrimp, crabs), Amphipoda (whale lice, beach fleas), and Isopoda (pill bugs). First usage in English c. 1830.

malleable: [Latin *malleare* = to hammer < *malleus* = hammer; < IE *mel* = crush, grind.] adj. Capable of being shaped or formed without fracture, as by hammering or by application of pressure. First usage in English c.1375. IE derivatives include: maelstrom, mill, mallet, millet.

malpais: [Spanish *malo* = bad + *pais* = country, region.] n. A term used in the U.S. Southwest and Mexico for an extensive area of rough, barren lava flows, often called badlands. First usage in English c. 1840.
Cited usage: Others are volcanic, locally known as 'mallapy' (mal pais) [1942, Thomas Kearney & Robert Peebles, Flowering Plants and Ferns of Arizona 17].

maltha: [Greek *maltha* = mixture of wax and pitch.] n. A soft, sticky, petroleum-rich variety of native asphalt or bitumen. Maltha was formerly also used to describe a kind of mortar or water-proofing agent made by mixing liquid asphalt (pitch) with wax, fat, or sand, and other ingredients. First usage in English c. 1400.
Synonyms: brea, mineral tar, earth pitch
See also: bitumen, ozocerite

mamlahah: [Arabic *mamlahah* = salty inland lake.] n. A term used in the Arabian Peninsula for an interior salt-encrusted playa. The *mamlahah*, though often simply classed as a dry lake, differs from other types of interior basins in North Africa,

such as: *khabra*, having no salts or evaporites, but filled entirely with silt, or *sabkha*, which are almost exclusively coastal.
See also: sabkha

mammal: [Latin *mamma* = breast.] n. Any of various warm-blooded vertebrate animals of the class Mammalia, characterized by a covering of hair on the skin, bearing live young, and milk-producing mammary glands for nourishing the young.

mammillary: [Latin *mamma* = breast.] adj. In mineralogy, referring to mineral aggregates, such as malachite and limonite, that form smooth and rounded masses resembling portions of spheres or breasts. First usage in English c. 1610

manganese: [Italian *manganese* = a corruption from Middle English *magnesia alba* = white magnesia, main ingredient of the mythical Philosopher's Stone; < Greek *Magnesia lithos* = the Magnesian Stone < *Magnesia* = region in Thessaly of Asia Minor.] n. A hard, grayish-white to silver, brittle metallic element (symbol Mn). Manganese does not occur in the native state, but is widely distributed in minerals, such as pyrolusite and rhodochrosite, and in nodules on the ocean floor. The term is a corruption of *magnesia*, which is the root of the term magnesium. Whereas the element magnesium was isolated from *magnesia alba*, manganese was actually isolated from the manganese oxide *magnesia nigra* (MnO_2 or pyrolusite). This oxide of manganese has been known since early times. The element was first isolated by Johan Gottlieb Gahn in 1774, and was coined in the same year by Carl Wilhelm Scheele.
See also: magnesium

mangrove: [Portuguese *mangue* = mangrove; < Taino *mangi* = mangrove + Old English *graf* = grove.] n. Any tropical tree or shrub of the genus Rhizophora, which are chiefly low trees growing in

marshes and tidal flats down to the low water mark. Mangroves are characterized by large masses of interlacing roots above ground, which intercept mud and weeds, and thus cause the land to encroach on the sea. First usage in English c. 1610.

mantle: [Old English *mentel* = cloak; < Latin *mantellum* = cloak.] n. In geochemical descriptions of the Earth's structure based on composition, the zone between the crust and the core. In geophysical descriptions of the Earth's structure based on tensile strength, the mantle is divided into outer and inner parts located between the asthenosphere and the outer core. In either case, the mantle is approximately 2,900km thick. In geomorphology, the general term for the blanket of unconsolidated material, known as the regolith, that covers bedrock. In paleontology and biology, the lobe of tissue in a mollusc or brachiopod bearing the shell-secreting glands. First usage in English before 900.

map: [Latin *mappa* = tablecloth, napkin (possibly of Punic origin).] n. A graphic representation of a region of the Earth or celestial sphere, usually inscribed on a plane surface. The original *mappae* were pieces of material on which maps were drawn. First usage in English c. 1375.

mar: [Swedish *mar* = blocked bay (cognate with West Frisian *mar* = lake; < Old Frisian *mera* = to hinder).] n. A bay or creek whose entrance is blocked by a tidal bar so that the water is nearly fresh. Not to be confused with maar.
Alt. spelling: marer (pl.)
See also: maar

marais: [French *marais* = swamp, marsh.] n. A term meaning swamp, adopted for use in the Southern U.S. The most famous marais is the Marais of central Paris. This district, now the 3rd and 4th arrondissements, was originally a swamp;

drained by Christian monks in the thirteenth century, the resultant land became the '*Rive Droit*,' or Right Bank of the Seine River. First usage in English c. 1790.
Synonyms: swamp

marble: [Greek *marmaros* = shining stone < *marmairein* = to sparkle.] n. A metamorphic rock formed by recrystallization of limestone or dolostone. Marble is often irregularly colored by impurities, for example: red by hematite, yellow by limonite, green by serpentine, and blue by diopside. It weathers well in arid climates, but breaks down quickly in acidic humid conditions. Marble is often erroneously used as a collective term for any building stone that takes a polish, such as: limestone, alabaster, granite and serpentine.
Cited usage: In altered dolomites the magnesian silicates form before the lime combines, and we have not only serpentine marble but tremolite marble, diopside marble, phlogocite marble, talcose marble, and even garnet marble [1940, Frank Fitch Grout, J. F. Kemp's Handbook of Rocks].

marcasite: [Toponymous name after Markhashi, the Arabic name for a region on the Eastern Iranian Plateau, presumably an early source of the material.] n. A brass-yellow to grayish orthorhombic mineral, iron sulfide, FeS_2. It is dimorphous with pyrite, but is less stable chemically and usually has a paler color. Marcasite often occurs in sedimentary rocks, such as coal, in the form of nodules with a radiating and fibrous structure. Marcasite has referred to as both a form of fool's gold and as a lapidary stone, but these specimens of iron sulfite having a gold or silver luster are probably the minerals chalcopyrite and pyrite. First usage in English c.1400.
Synonyms: white pyrites, cockscomb pyrites, lamellar pyrites

maré: [Latin *mare* = sea.] n. Any of the large, dark, plainlike areas on the surface of the Moon that have fewer craters and are at lower elevation than the surrounding terra or highlands. Maré usually have circular outlines, such as Maré Imbrium, and were initially created as large asteroid impact craters, later filled with basalt flows. The name was first applied by Galileo Galilei (1564-1642), who, upon first seeing them through his telescope, believed that lunar features were seas. The term is also used for a large dark area on Mars of uncertain origin.
Alt. spelling: maria (pl.)

marigraph: [Latin *mare* = sea + *graphicus* < Greek *graphikos* < *graphe* = writing < *graphein* = to write; < IE *gerbh* = to scratch.] n. An instrument used for registering and recording the height of the tide at any instant. Such an instrument is usually activated by a float in a pipe connected to the sea through a small hole that filters out shorter waves. First usage in English c. 1865.

marine: [Latin *marinus* = of the sea < *mare* = sea.] adj. Of or relating to the sea. Said of oceans, seas, bays, estuaries, fjords, and other bodies of salt water contiguous with the oceans. In sedimentology, referring to sediments deposited in sea water; as contrasted to non-marine sediments deposited in other environments, such as rivers, deserts, and lakes (including salt lakes). First usage in English c. 1350.

mariposite: [Toponymous term after the county of Mariposa, California.] n. In mineralogy, a bright-green, chromium-rich, micaceous variety of phengite, $K(Al,Cr)_2(Al,Si)_4O_{10}(OH)_2$. In petrology, mariposite is considered a rock composed of a mixture of dolomite, milky quartz, and a bright-green, chromium-rich variety of muscovite called fuchsite. Mariposite is formed by hydrothermal alteration of chromite-bearing serpentinite.

marl: [Middle English *marle* < Latin *marga* = marl.] n. A term loosely applied to a variety of sedimentary substances, but most often referring to a grayish, crumbly, sediment composed of a mixture of calcium carbonate and clay. Specifically, marl is defined as being 35-65% base clay and 56-35% carbonate. This sediment mixture accumulates in relatively low-energy fresh water environments, and, less often, in marine environments. As marlstone, the term is used generally for rock consisting of consolidated marl; specifically, it is an argillaceous, iron-rich, oolitic limestone that lies between the Upper and Lower Lias Formations of England. Marl is used as a fertilizer in acid soils deficient in lime. First usage in English c.1350.

Mars: [Eponymous name after Mars, the Roman god of war.] n. The fourth planet in order from the sun. Mars has two moons, Deimos and Phobos. The surface of Mars is not as tectonically mobile as Earth's, therefore, it has developed both the largest canyon system and the largest volcanoes known in the solar system. The tectonic-generated trough *Valles Marineris* is Mars largest canyon-like structure, while *Olympus Mons* is the largest volcano in the solar system. Mars has long captivated the collective imagination of people, as it appears to have much in common with Earth. Recent discovery of water in the form of subsurface ice lends credibility to the belief Mars may once have sustained life as we know it.
See also: canal, shergottite

marsh: [Old English *mersc* = marsh; < Teutonic *mari* = sea, lake; < IE *mori* = body of water.] n. A tract of low-lying, boggy and water-saturated land, characterized by shallow interconnected ponds and thick hydrophilic vegetation (reeds, willows, etc.). may be entirely freshwater or brackish, if they are in estuaries where fresh and salt water mix. They are generally less wetand not as acidic as bogs, and therefore don't accumulate appreciable peat deposits. First usage in English before 900. IE derivatives include: maar, mermaid, meerschaum, morass, and cormorant.
See also: bog, muskeg, fen, marais

martite: [Toponymous term after the planet Mars.] n. Hematite that occurs as iron-black octahedral crystals (sesquioxide in isometric form). These crystals are pseudomorphs after magnetite. The planet Mars was named after the Roman god of war, also known as Ares in Greek mythology. Medieval alchemists associated the metal iron with Mars.
Syn: alchemical iron
See also: Mars

mascon: [Portmanteau word from English *mas(s)* + *con(centration)*.] n. In planetary geology, a term for a high-density lunar mass located below a ringed maré. Mascons create anomalous gravity increases generally in excess of 100 milligals at an elevation of 100 kilometers. Discovered by gravitational analysis of orbiting satellites that were experiencing difficulty maintaining stable orbits around the Moon, mascons are probably caused either by the presence of thick basalt flows that flooded lunar basins, or by uplift of high-density mantle material when ancient impacts removed lighter surface rock, allowing denser rock from the mantle to bulge upwards. First usage in English c.1965.

massif: [French *massif* = large mountain mass, massive; < Latin *massa* = mass.] n. In structural geology and physical geology, a large portion of an orogenic belt, or a compact group of connected mountains forming an independent portion of a range. A massif is generally more rigid than the surrounding rock units, and is often separated from a larger mountain range by a major river basin. Massifs may be protruding bodies of older basement rocks, or of younger igneous intrusions. Some notable examples are: the Massif Central of the Auvergne region in south-central France, the Massif Montgris in Spain, the Annapurna Massif in Nepal, the Vinson Massif in Antarctica, and the Atlantis Massif, part of the Mid-Atlantic Ridge in the North Atlantic Ocean. The term perhaps originated in French as an architectural one denoting a large, massive structure. It was first recorded in this sense in 1546, whereas, it wasn't used in relation to 'high ground' until 1796. In mountaineering, a massif is often used to identify the main mass of an individual mountain. First usage in English c. 1520.

matrix: [Late Latin *matrix* = womb; < Latin *matricis* = female animal kept for breeding < *mater* = mother; < IE *mater* = mother.] n. In petrology, the usually fine-grained material that fills the interstices between larger clasts in a sedimentary rock, the material that surrounds phenocrysts in an igneous rock. In paleontology, the rock material in which a fossil is embedded. In mineralogy and gemology, the rock material surrounding a crystal or natural gemstone. In mathematics, a two-dimensional rectangular grid of numeric of algebraic quantities arranged in columns and rows, and subject to mathematical operations. First usage in English c.1350. IE derivatives include: material, matter, madrepore, maternity, and metropolis.

mbuga: [Swahili < *mbuga* = seasonal swamp, savannah; < Nyamwezi *mbuga* = seasonal swamp or wetland.] n. A term used in Southeast Africa for the black sediment that is rich in clay and organic material, resulting from desiccation of a playa lake or temporary swamp, often found in the alternating wet/dry climate of the savannah region.
Cited usage: In a certain sense the South African 'vlei' is similar to the East African 'mbuga' [1955, Patrick Alfred Buxton, Nat. Hist. Tsetse Flies ix. 271].

meadow: [Old English *mead* = meadow; < IE *me* = to cut down grass or grain with a sickle or scythe.] n. A tract of grassland, either in its natural state or used as pasture or for growing hay. First usage in English before 1000. An IE derivative is mow.

mean: [Latin *medius* = middle.] n. The arithmetic average, derived by adding up all values and dividing the resultant sum by the number of discrete values. First usage in English c. 1325.

meander: [Toponymous term after the Maiandros River, the Greek name for a river in Phrygia, noted for its windings (now known as the Menderes River in Southwest Turkey).] n. One of a series of freely developing sinuous curves, bends, loops, or windings of a stream course. Meanders are produced by a mature stream swinging from side to side as it flows across its flood plain. In English, the word meander originally meant intricacies or confusing, and was first used in this sense in 1576, as derived from the Latin *maeandere*, which came from Greek *maiandros*, the meaning of which originally came from the Maiandros River in ancient Lycia now a region of mountainous and wild landscapes in Turkey. The first recorded use in English in reference to winding stream courses was not until 1612.
Cited usage: And they say that lawsuits are brought against the god Maiandros for altering the boundaries of the countries on his banks, that is, when the projecting elbows of land are swept away by him; and that when he is convicted the fines are paid from the tolls collected at the ferries. [Strabo, 12.8.19 (French translation of Strabo's *Geography* by Laporte du Theil, Korais, and Letronne was undertaken between 1805-1819 at the request of the emperor Napoleon)].

measures: [Middle English *mesure* < Latin *mensura* = to measure.] n. In geology, a group or series of sedimentary rocks having common characteristics, specifically applied as 'coal measures.' The term alludes to the old mining practice of distinguishing different coal seams by their measured thickness.

medano: [Spanish *medano* = sand bank, sand dune; < Portuguese *meda* = conical haystack, sand dune.] n. A Spanish term for a shifting, conical, steep-sided, crescent-shaped sand dune (barchan), especially one occurring along the coast of Peru and Chile. Notable medanos outside South America include: El Medano in Tenerife, Playa al Medano in Cabo San Lucas, Mexico, and Medano Creek in Great Sand Dunes, Colorado.

median: [Latin *medius* = middle.] n. In statistics, of a data set arranged in order by rank, the middle value in a distribution, above and below which lie an equal number of values. In paleontology and biology the center line dividing mirrored halves of bilaterally symmetrical organisms; also, relating to or directed toward the middle. Also, a line that joins a vertex of a triangle to the midpoint of the opposite side. First usage in English c. 1540.

Mediterranean: [Latin *mediterraneus* = inland < *medius* = middle + *terra* = land; < IE *ters* = to dry.] adj. In geography, referring specifically to the sea surrounded by North Africa, Europe, the Middle East, and Asia. Also, said of epicontinental seas having a narrow opening to the ocean. In meteorology, said of climates similar to those found in the Mediterranean region, characterized by two seasons of nearly equal length, a mild, rainy winter season, and a hot, dry summer season.

meerschaum: [German *meer* = sea + *schaum* = foam.] n. A massive variety of the magnesian clay mineral sepiolite, formed by weathering of serpentinite in alluvial deposits, as well as originating from hydrothermal solutions. Meerschaum was originally thought to be petrified sea froth and was collected on various seashores. Most modern meerschaum is mined in Tanzania and Turkey. When first dug it lathers like soap, and is used for cleaning by the Tartars. It is used in ornamental carvings and in fashioning pipe-smoking bowls. First usage in English c. 1780.
Syn: sepolite, Soap of Tartars
See also: sepolite

mega-: [Greek *megas* = great.] Combining prefix meaning great, used in such terms as: megabreccia, megaclast, megaripple, megalodon, and megabyte.
See also: micro-, macro-

meio-: [Greek *meion* = less, lesser; < IE *mei* = small.] Combining prefix meaning less, used in such terms as: meiofauna, meiosis, and meionite. The term is often used to signify an intermediate value, position, or state between mega- and micro-. IE derivatives include Miocene, Menshevik, and minestrone (from Latin *minister*, meaning servant).
See also: mega-, micro-, meso-

meizoseismal: [Greek *meizo* = greater + *seizmos* = earthquake.] adj. Pertaining to the maximum destructive force in an earthquake.

mélange: [French *mélange* = a mixture < *meler* = to mix.] n. A body of sheer rock mappable at 1:24,000 or smaller, characterized by a lack of internal continuity of contacts or strata, and by inclusion of rock fragments of all sizes. A mélange contains rocks that are both exotic and native, and are embedded in a fragmental matrix of finer-grained material. A sedimentary mélange with a chaotic nature is called an olistostrome. Tectonic mélanges are thought to form in subduction zones at shallow depth. In mining, an assortment of mixed-sized diamonds, each weighing more than 1/4 carat; the gems of a mélange are larger on average than those of a melee. Term coined in 1919 by British geologist Edward Greenly.
See also: olistostrome

melano-: [Greek *melanos* < *melas* = black.] Combining prefix meaning dark-colored or black, used in such terms as: melanite, melanocratic, melanin, and melanoma. The prefix *mela-* has the same root and is used in such terms as melagranite. This prefix is often used in cases where the specimen is darker than normally expected, for example, melagranite is used in contrast to leucogranite, from the Greek *leuco-*, meaning white. In biology, used unhyphenated, as a noun for an individual having excessive melanin, in contrast to albino.
See also: leuco-

melanterite: [Greek *melanteria* = black pigment, native copperas < *melano-* = black < *melainein* = to blacken + *-ter* = suffix meaning causative agent.] n. A greenish to yellowish to white monoclinic mineral, hydrated iron sulfate, $FeSO_4 \cdot 7H_2O$. It usually forms from the weathering and decomposition of iron sulfides, especially marcasite. It is one of only a few water-soluble sulfates. Melanterite has an astringent, metallic, and often bitter-sweet taste. It commonly forms an efflorescence on mine walls and timbers, as well as on the surface of rocks bearing marcasite or pyrite. Epsomite is an analogous magnesium sulfate. Copperas was a name given from ancient times to the protosulfates of copper (blue), iron (green), and zinc (white). Its etymology has been variously associated with the Greek *kalkanthos*, meaning 'flower of copper,' and the Latin *aqua cuprosa*, meaning 'water of copper.' In modern English usage, copperas refers exclusively to the green iron sulfate ($FeSO_4$).
Synonyms: copper, green vitriol, iron vitriol

melee: [French *melee* = mixture < *mesler* = to mix, mingle.] n. In general use, a jumble or a confused fight or struggle. In gemology, a collective term for a small, round-cut diamond weighing less than 0.25 carat (1 grain), and used commonly in jewelry. In mining this size of diamond has the most frequent occurrence, and therefore makes up the bulk of all diamonds mined. There is discrepancy among gemologists over the maximum size for the melee, ranging from a low of .15 carat to a high of 1 carat. The minimum size, however, is set at point .01 carat because that is a the size of a 1-point diamond (based on the fact that 1 carat equals 100 points by weight). First usage in English c. 1910.
See also: carat

mendip: [Toponymous term after the Mendip Hills in Somerset, England.] n. A buried hill exposed by the cutting of a valley. Also, a hill on an emergent coastal plain that was formerly an offshore island or sea stack.

mengwacke: [German *meng* = mix + *wacke* = large stone; < Old High German *waggo* or *wacko* = boulder rolling on a riverbed < *wag* = to move about; < IE *wegh* = to go, transport.] n. An obsolete term for a lithic sandstone (wacke) having a large amount of easily decomposed argillaceous material (33-90%). Term coined in 1934 by German petrologist Georg Fischer.
See also: graywacke

menhir: [French *menhir* = longstone; < Breton *men* = stone + *hir* = long.] n. In archeology, an upright monumental stone, standing either alone or with others, often in an alignment. The term is used most commonly around the areas of Cornwall, England, and Brittany, France. In modern Breton, the word for menhir is actually *peulvan*. The largest surviving menhir is at Locmariaquer, Brittany. It is called the *Grand Menhir Briseé* ("Great Broken Menhir") and once stood about 20 meters high. It now lies broken in four pieces, but when intact, it would have weighed around 330 tons and may have been the heaviest object ever moved by humans without powered machinery. First usage in English c.1835.

meniscus: [Greek *meniskos* = crescent, diminutive of *mene* = moon.] n. In general usage, a crescent-shaped body. In physics, the concave or convex surface of a column of liquid, its curvature being caused by surface tension. In biology and paleontology, a crescent-shaped cartilaginous disk that serves to cushion the ends of bones meeting at a joint. First usage in English c. 1690.

mephitic: [Latin *mefitis* = exhalation of sulfured water or gas < *Mephitis* = goddess of exhalations.] adj. Said of a noxious or poisonous exhalation from the earth. Also, said of any offensive smell, especially resulting from atmospheric pollution, as in mephitic air. First usage in English c. 1620.

mer de glace: [French *mer de glace* = sea of ice.] n. A general term for the large glaciers of the Pleistocene Epoch. Also, used specifically for the largest modern alpine glacier on the Mont Blanc massif in France.

mercury: [Eponymous term after Roman God *Mercurius*, the messenger to the gods, as well as the god of commerce, travel, and thievery.] n. The planet closest to the sun. Also, a silvery-white, toxic, extremely dense metallic element, the only metal which is liquid at room temperature (symbol Hg). It occurs in the native state as minute fluid globules associated with other Mercury minerals, such as cinnabar. Although mercury occurs as a liquid, it is considered a mineral, because, when frozen, it crystallizes in the hexagonal system. Mercury combines with most metals to form amalgams. Mercury has been known since antiquity, being found in Egyptian graves dating back to 1500 BCE. In the 6th century, alchemists considered mercury to be the basis of all metals, and closely related to gold, the metal associated with the Sun. Because mercury was closely related to gold, alchemists assigned this element to the planet closest to the Sun, Mercury. Owing to its long history, mercury has names in over a hundred languages. In German and English for example, it was called *quicksilver*, due to its fluidity and mirrorlike sheen. The Middle Eastern translation is "living silver." In Japanese and Chinese, the characters *sui*, meaning water, and *gin*, meaning silver, are used together in describing this element. The elemental symbol for mercury, Hg, is derived from the Greek *hydrargyrum*, meaning silver water, which itself derives from *hydros*, meaning water, and *arguros* meaning silver. First usage in English c. 1325. *Synonyms*: hydrargyrum, quicksilver

meridian: [French *meridien* = of midday; < Old French *meridiane* = southern; < Latin *meridianus* = midday, southern < *medidie* = at midday < *mediei* < *medius* = middle + *dies* = day; < IE *dyeu* = to shine.] n. An imaginary great circle on the Earth's surface passing through the North and South geographic poles. A meridian runs perpendicular to the equator, and connects all points of equal longitude around the sphere. All meridians, or lines of longitude, are great circles, but, only one line of latitude, the equator, is a great circle. First usage in English c. 1375. IE derivatives include: day, divine, and diary.

mero-: [Greek *mero-* = part, fraction.] Combining prefix meaning part or fraction, used in such terms as: meroplankton (organisms spending part of their lives as plankton), merohedral (a crystal lacking all its normal or expected faces), and meroblastic (a partially cleaved ovum forming an embryo).
See also: holo-

meroplankton: [Greek *mero* = part, fraction + *planktos* = wandering < *plazein* = to turn aside.] n. Any of various organisms that spend part of their life cycle as plankton, usually in the larval or egg stages. Used in contrast to holoplankton. First usage in English c. 1905

mesa: [Spanish *mesa* = table.] n. An isolated, elevated, nearly flat-topped landform bounded by steep erosional scarps on all sides, and capped by resistant and nearly horizontal sedimentary layers or lava flows. A mesa is similar to a butte, but with a more extensive summit plateau. Mesas are common topographical features of the U.S. Southwest. First usage in English c. 1755.
Alt. spelling: meseta, mesita, mesilla
Similar terms: butte

mesic: [Greek *mesos* = middle + *ic* = pertaining to.] adj. Said of a habitat receiving a moderate amount of moisture, as compared to xeric (little moisture) and hydric (abundant moisture). First usage in English c. 1925.
Cited usage: The curves of frequency show the relative abundance of lichens in oak-dominated woods and their rarity in mesic climax forests [1967, M. E. Hale Jr., The Biology of Lichens vii. 91].
See also: xeric, hydric

meso-: [Greek *mesos* = middle.] Combining prefix meaning middle, used in such terms as: meso-American, mesoderm, mesofauna, Mesozoic, and Mesopotamia. Often, when the prefix is used directly before a vowel the 'o' is dropped, as in mesaxonic (having the axis of symmetry run through the central digit), mesencephalon, and mesic.

Mesozoic: [Greek *meso* = middle + *zoikos* = pertaining to animals < *zoa* = plural of *zoon* = animal, living being.] adj. Designating the second of three eras in the Phanerozoic eon. The Mesozoic era (251Ma—65.5Ma) is divided into three periods: the Triassic, Jurassic, and Cretaceous. Term coined by John Phillips in 1840, who popularized the phrases, 'Age of Fishes' (Paleozoic), 'Age of Reptiles' (Mesozoic), and 'Age of Mammals' (Cenozoic) for the three eras of Phanerozoic eon.

Cited usage: Corresponding terms (as Palaeozoic, Mesozoic, and Kainozoic) may be made, nor will these necessarily require change upon every new discovery [1840, John Phillips, Penny Cyclopedia. XVII. 154].

mesozonal: [Greek *mezos* = middle + Latin *zona* < Greek *zone* = girdle < *zonnai* = to gird.] n. In igneous petrology, according to the classification system of American geologist Arthur F. Buddington, the intermediate depth zone between epizonal and catazonal. In metamorphic petrology, according to the classification system of German petrologist Ulrecht Grubermann, the intermediate depth zone of metamorphism, where modern usage emphasizes temperatures and pressures of moderate metamorphic grade. This zone is characterized by the amphibolite facies, composed of minerals, such as epidote, almandine, staurolite, kyanite, and sillimanite.
See also: catazonal, epizonal, meso-

meta-: [Greek *meta* = beyond, change (of place, order, condition, or nature), sharing, after.] Combining prefix meaning change, beyond, sharing, or after. When meaning change, used in such terms as metamorphism or metabolism; when meaning beyond, used in such terms as metaphysics and metazoan; when meaning sharing, used in such terms as metacenter; when meaning after, used in such terms as metacinnabar, as well as more than 40 other mineral names.
See also: metasomatism, metabolism, metamorphism

metabolism: [Greek *meta* = change + *ballein* = to throw.] n. The physical and chemical processes occurring within a living cell or organism that are necessary for the maintenance of life. First usage in English c. 1875.

metacinnabar: [Greek *meta* = after + Latin *cinnabaris* < Greek *kinnabari* = cinnabar; <

Arabic (possibly) *zinjafr* = dragon's blood.] n. A black isometric mineral, mercury sulfide, HgS. It is dimorphous with cinnabar and an ore of mercury.
Alt. spelling: metacinnibar
Synonyms: metacinnabarite

metacryst: [Greek *meta* = beside, change + *krustallos* = crystal < *krumos* = icy, cold; < IE *kreus* = starting to freeze, form a crust.] n. A large crystal formed in a metamorphic rock by recrystallization, such as a schist containing garnet or staurolite crystals.
Synonyms: porphyroblast

metal: [Latin *metallum* = metal, quarry; < Greek *metallon* = metal, mine, ore.] n. Any of the elements forming positively charged ions (electropositive). Metals usually have a shiny surface, are relatively easy to melt, alloy, and fuse, and are generally good conductors of heat and electricity. They are usually ductile, and can therefore be hammered into thin sheets, or drawn into wires. Typically metals form salts with nonmetals, oxides with oxygen, and alloys with other metals. By the 16th century, the term metal also acquired a figurative meaning, being "the stuff one is made of," or "one's character." There was no difference in spelling between the literal and figurative senses until about 1700, when the spelling mettle, originally just an alternate spelling of metal, was established to designate the meaning of 'fortitude.' There are numerous examples of similar word pairs that diverged in meaning over time. Examples are: trump/triumph and through/thorough. First usage in English c. 1275.

metamict: [Greek *meta* = change + *miktos* = mixed, blended.] adj. Said of a mineral containing radioactive elements in which radiation damage has resulted in various degrees of disruption of the crystal lattice, in addition to elimination of the internal ordered arrangement of atoms. However, the external morphology of the crystal is retained. Examples occur in zircon and thorite. Term coined in 1893 by Danish scientist W.C. Brögger.

metamorphism: [Greek *metamorphoun* = to transform < *meta* = change + *morphos* = shape, form.] n. The process of recrystallization of solid rocks in the Earth's crust, in response to changes in the existing physical and chemical conditions, primarily temperature, pressure, and bulk composition, but excluding diagenesis, lithification, and weathering. Metamorphism occurs in solid rocks, and excludes the igneous processes of melting and solidification. Metamorphic temperature are between 100°C and 600°C, when rocks begin to melt. Term coined in 1833 by Scottish geologist Charles Lyell.

metasomatism: [Greek *meta* = change + *soma* = body.] n. The process by which the chemical composition of a rock is changed, generally by interaction with hot fluids introduced from an external source. The process requires almost simultaneous solution of an old mineral, and replacement by a new mineral. The process generally occurs at constant volume, with little disturbance of textural or structural features. Metasomatism most commonly occurs at the contacts of an igneous pluton, where limestone or dolomite are replaced by calc-silicate bearing skarn. An etymologically related term is the name of the rock relating to this process, metasomatite. First usage in English c. 1885.
Synonyms: metasomatite

metastasy: [Greek *metastasis* = a changing < *meta* = change + *stasis* = setting, standing, weighing < *histanai* = to cause to stand, to place; < IE *sta* = to stand.] n. In tectonics, the name for lateral adjustments of the Earth's crust to attain gravitational equilibrium, as opposed to vertical adjustments (isostasy). Term coined in 1958 by Gussow. A similar, but now obsolete term, metastasis,

was proposed in 1886 by Bonney, and referred to the petrologic change in a rock that produces subtle changes in form, such as the recrystallization of limestone, or the devitrification of obsidian. IE derivatives include: enstatite, staurolite, stasis, eustasy, isostasty, prostitute, destitute, stoic, stamen, and stud.
Similar terms: metastasis
See also: isostasy, eustasy

metazoan: [Greek *meta* = after, beyond + *zoa* = plural of *zoon* = animal, living being.] n. An informal term for all multicellular animals. It is now largely obsolete, because it has no taxonomic significance. Term coined in 1874 by Ernst Haeckel, at a time when single-celled protozoans were considered animals. A functional remnant of his original definition is that these multicellular animals form a central cavity during embryonic development, and that the cells around this central cavity are first arranged in two layers, the ectoderm and endoderm.
Alt. spelling: metazoon
See also: protozoa, ecology

meteor: [Greek *meteoron* = a thing in the air < *meteoros* = raised into the air, lofty < *meta* = beyond + *eor* = to raise.] n. The visible transient streak of light resulting from the entry into the Earth's atmosphere of a solid extraterrestrial object and made luminous as a result of friction. A less accepted use of the term (replaced by meteoroid) refers to the object itself. Not to be confused with meteorite, being an extraterrestrial object that has reached the Earth's surface. Prior to any scientific understanding of the phenomena, atmospheric phenomena were formerly classified as: a) aerial or airy meteors (winds), b) aqueous or watery meteors (rain, snow, hail, dew, etc.), c) luminous or radiant meteors (the aurora, rainbow, halo, etc.), and d) igneous or fiery meteors (lightning, shooting stars, etc.). First usage in English c. 1575.
See also: meteorite, meteoroid

meteoric: [Middle English *metheour* = atmospheric phenomenon; < Old French *meteore* < Latin *meteorum* < Greek *meteoron* = a thing in the air < *meta* = beyond + *eor* = to raise < *aeirein* = to lift up; < IE *wer* = to raise, lift, or hold suspended.] adj. In meteoritics and planetary geology, said of the tiny particles (diameters up to 100 microns) released from melting meteoroids in the Earth's atmosphere, as in meteoric dust. In hydrology, said of water of recent atmospheric origin, as in meteoric water. First usage in English c.1630. IE derivatives include: air, aura, aria, malaria, and aorta.
See also: meteor, meteorite, meteoroid

meteorite: [Greek *meteoron* = a thing in the air + *-ite* < *lite* < *lithos* = rock, stone.] n. A stony or metallic extraterrestrial object that is large enough to survive its passage through the Earth's atmosphere, and reach its surface without being completely vaporized by intense frictional heating. Most meteorites originate in the asteroid belt, but some are fragments of other planets, such as Mars. First usage in English c. 1820.
See also: meteor, meteoric, meteoroid

meteoroid: [Greek *meteoron* = a thing in the air < *meteoros* = raised into the air, lofty < *meta* = beyond + *eor* = to raise + *oidos* = resembling.] n. Any of the small fragments of extraterrestrial material moving in interplanetary space that are distinguished by their small size, but are considerably larger than an atom. When such a fragment enters the Earth's atmosphere, it is heated to luminosity, and is seen as a meteor. First usage in English c. 1860.
See also: meteor, meteorite, meteoric

meter: [French *metre* = meter; < Greek *metron* = measure, poetic meter; < IE *me* = to measure.] n. The international standard unit of length, approximately equivalent to 39.37 inches. The metric system was originally conceived in 1202 by the Italian mathematician Fibonacci, who simply called it the decimal system. The original meter was chosen as the length of a pendulum with a period of two seconds. In 1790, after the French Revolution, the French Academy of Sciences proposed that a new system of measurement, the metric system (based on Fibonacci's decimal system) be established. In this system, the meter was redefined as one ten-millionth of the distance from the north pole to the equator, as measured along the meridian running through Paris. However, this measure was impractical, and was short by 0.2 millimeters because researchers miscalculated the flattening of the earth due to its rotation. Therefore, from 1889 to 1960 the meter was defined as the length of an actual object, being the distance between two lines scribed in a platinum-iridium bar housed at the International Bureau of Weights and Measures near Paris. After years of discussion, it was decided in 1983 to redefine it as the distance traveled by light in a vacuum in 1/299,792,458 of a second. Some IE derivatives are: dimension, diameter, and semester.

methane: [Greek *methu* = wine + *hule* = wood, substance.] n. A colorless and odorless flammable gas, being the principle constituent of commercial natural gas, and the simplest single-bonded hydrocarbon (CH_4). It is found associated with crude oil; it also emanates from coal seams, stagnant pools, and volcanoes. When mixed with seven or eight parts of air, it forms a violent explosive. The term methane is formed by back-formation from *methylene*, meaning wood wine, from the Greek *methy*, meaning wine, and *uln*, meaning wood. First usage in English c. 1865.

miarolitic: [German *miarolitisch* < Italian *miarolo* = a granite containing cavities + *lithos* = rock, stone.] adj. Referring to plutonic igneous rocks having irregular cavities into which well-formed crystals project. Miarolitic cavities usually result from evolution of a high-temperature aqueous fluid, late in the history of solidification of a magma. The cavities range in diameter from less than 1mm to several meters.
Cited usage: Vacant spaces are apt to occur, into which project the sharp angles of well-formed crystals. This miarolitic or drusy structure is more or less marked in some granites (e.g. the Mourne Mts in Ireland) [1895, Alfred Harker, Petrology for Students ii. 31].
Alt. spelling: myrolitic

mica: [Latin *mica* = grain, crumb (possibly derived from Latin *micare* = to flash).] n. Any of a group of complex phyllosilicates, common in igneous and metamorphic rocks, and characterized by perfect basal cleavage that is prone to splitting into very thin elastic sheets or laminae. The mica group includes: muscovite, biotite, lepidolite, phlogopite, zinnwaldite, roscoelite, paragonite, sericite. Micas range from colorless to silvery-white, from purple to brown, and from yellow to green or black. They crystallize as pseudohexagonal prisms that are actually monoclinic. First usage in English c. 1705.
Cited usage: Pinchbeck mica, iron pyrites, and titanite of iron occur as accidental constituents [1835, H. D. & T. Thomson's Records of General Sci. II. 445].

micelle: [Latin *mica* = grain.] n. In chemistry, an electrically charged particle formed by an aggregate of molecules or polymeric ions, and occurring in certain colloidal suspensions, such as in the common surfactant soaps and detergents. Surfactants are chemical amphipathic compounds, meaning that they contain both hydrophobic and hydrophilic ends. First usage in English c.1880.
Alt. spelling: micellae (pl.)

micrite: [English *micro-* < Greek *mikros* = small + *ite* = pertaining to rock.] n. In current usage, the term is used in a descriptive sense for a limestone formed by lithification of very fine-grained calcareous ooze (less then four microns in

223

diameter) and formed from either chemical or biological sources. The term was defined in 1959 by R.L. Folk, as a limestone having a matrix finer than sparite. There are a variety of limestone classifications based on composition. When a rock is composed of nearly pure calcareous ooze, it is classified simply as micrite, but if allochems (e.g., oolites, fossil shells) are abundant within the rock, it takes on the name of the dominant component, such as oomicrite and fossiliferous micrite.
Cited usage: Type III limestones, almost entirely ooze, are designated simply as 'micrite', without any allochem prefix [1959, Robert Louis Folk, Bull. Amer. Assoc. Petroleum Geologists XLIII. 17].
See also: sparite

micro-: [Greek *mikros* = small.] Combining prefix meaning small, used chiefly in scientific terms, such as microscope and microorganism. In measurement, meaning one-millionth, as in microliter and micrometer (micron).
See also: micron

microcline: [Greek *mikro* = micro- + *klinein* = to lean.] n. A white to gray to salmon to brick-red or green triclinic mineral of the alkali feldspar group, potassium aluminum silicate, $KAlSi_3O_8$. It is trimorphous with orthoclase and sanidine, being stable at lower temperatures, and is a common rock-forming mineral of granitic rocks and pegmatites. Microcline is often more commonly known as the semiprecious varieties amazonite and perthite. Amonzonite is a deep-green variety suitable for craving and polishing, whereas perthite is striped, veined, or zebra patterned, produced from lamellar intergrowths inside the crystal. The etymology of microcline derives from the fact that the cleavage angle is not exactly equal to 90°, therefore leaning slightly from the vertical. First usage in English c. 1845.
See also: orthoclase, sanidine

micron: [Greek *mikro* = small.] n. A millionth of a meter, denoted by the symbol *µm*, a composite of the Greek letter mu and the English letter m. In the International System of Units (SI units) the word has been replaced by micrometer. First usage in English c. 1880.

midden: [Middle English *midding* = midden; < Danish *mog* = muck + *dynge* = heap; < Old Norse *dyngja* = manure pile.] n. A mound or deposit of refuse, containing shells, animal bones, pottery shards, tools, and ash; middens usually indicate the site of a human settlement. First usage in English c. 1325.

migma: [Latin *miscere* = to mix; < Greek *migma* = mixture < *mignunai* = to mix; < IE *meig* = to mix.] n. An obsolete term for a mixture of magma and solid rock material that has been emplaced as discontinuous lenses in high-grade metamorphic rocks to form the composite rock, migmatite. Migma was considered to originate by partial melting during ultrametamorphism. The material defined originally as migma has now been included as magma. IE derivatives include: melange, medley, and promiscuous.
Cited usage: Between these extremes is a concept of a pore fluid, or migma, generated by the differential melting or partial fusion of the root portions of mountains [1974, Encycl. Britannica Micropaedia VI. 880/1].

migmatite: [Latin *miscere* = to mix; < Greek *migma* = mixture < *mignunai* = to mix; < IE *meig* = to mix + Greek *lithos* = rock, stone.] n. A term for a composite igneous/metamorphic rock formed by emplacement of light-colored quartzofeldspathic igneous rock into foliated schists and gneiss. Migmatites may appear layered, or the felsic portion may occur as pods or form a network of cross-cutting dikes. The igneous portion of the migmatite is generally considered to originate from partial melting, or from anatexis of the metamorphic host rock. Term coined in 1907 by J. J. Sederholm. *Cited usage*: For the gneisses here in question, characteristic of which are two elements of different genetic value, ...the author proposes the name of migmatites [1907, J. J. Sederholm, Bulletin de la Commission Géologique de Finlande 5 XXIII.110].

mil: [Latin *millesimus* = one thousandth < *mille* = thousand.] n. In measurement, the unit equal to one-thousandth of the whole. The term can be confusing because it is fundamentally unitless, especially because it has been ascribed different values in different parts of the world. For example, in North America it refers to one-thousandth of an inch, whereas, in Europe it is equal to one-thousandth of a meter (one millimeter). Etymologically related terms include: mile, million, millennium, millipede, and millenary. First usage in English c. 1720.

mile: [Latin *mille* = thousand.] n. A unit of linear distance measurement equal to 1,760 yards, 5,280 feet, or 1,610 meters. The term is derived from the Latin *mille passum*, meaning a thousand paces of a Roman soldier marching in formation. This distance was roughly equal to one statute mile. The term 'pace' is confusing, because it can mean either a single step from one foot to the other (military pace, about 2½ ft.), or stepping from one foot to where the same foot lands again (two single steps, a geometrical pace, or about 5 ft.). First usage in English before 1000.

milieu: [French *milieu* = middle, medium; < Latin *medius* = middle + *lieu* = place.] n. In paleontology and stratigraphy, a term for a particular sedimentary environment, characterized by a distinctive biotic assemblage (e.g., a deepwater facies characterized by chert-producing radiolarian ooze). First usage in English c. 1800.

millerite: [Eponymous name after English mineralogist William Hallowes Miller (1801-1880).] n. A brass-yellow to bronze-yellow hexagonal mineral, nickel sulfide, NiS. It generally occurs as fine hairlike or capillary crystals of extreme delicacy. It forms chiefly from low-temperature hydrothermal solutions in vugs and geodes in limestone or ironstone. Miller first studied the crystals of this mineral, as well as many others; he popularized the system of indexing crystal faces, now known as Miller indices.
Synonyms: capillary pyrites, nickel pyrites, hair pyrites

milli-: [Latin *mille* = thousand.] Combining prefix meaning one thousand, or one-thousandth. In the sense of one-thousandth, the prefix is used in such terms as: milligram, milliliter, millihenry, milliampere, and millicurie. In the sense of one thousand, used in such terms as: millennium, millipede, and millifold (a thousandfold increase).

mima mound: [Toponymous term after the Mima Prairie in western Washington State.] n. A term used in the U.S. Northwest for low, circular to oval, dome-shaped accumulations of unstratified gravely-silt and soil material built upon glacial outwash. The basal diameter varies from 1 meter to more than 30 meters, with heights up to 2 meters. Various theories of origin for mima mounds have been proposed, including: a) vestigial hillocks of a former forest that are resistant to water erosion due to root-anchoring, b) mounds resulting from seismic disruption of silty sediments in a relatively rigid planar substratum, and c) mounds formed by pocket gophers.
Alt. spelling: Mima mound
Synonyms: hog wallow (California), pimple mound (Texas), prairie mound

mimetic: [Greek *mimesis* = mimicry, copy <*mimos* = imitator, mime.] adj. In general usage, referring to anything that exhibits imitation or mimicry. In petrology, referring to a metamorphic texture that reproduces or mimics any pre-existing anisotropic rock fabric, such as bedding, schistosity, or sheared structure. In mineralogy, referring to a twinned or malformed crystal that appears to have higher symmetry than it actually possesses. First usage in English c. 1630.

mine: [(possibly) Latin *mina* = an excavation < *minare* = to excavate; or (possibly) Celtic *meini* = ore, metal.] n. An open-pit or underground excavation made with the intent of extracting mineral deposits. In England, under ancient mining laws no longer enforced, a royal mine is any mine which yields more ore than will cover the cost of working. By law, all such mines were liable to be claimed as the property of the Crown. First usage in English c. 1300.
See also: ore

mineral: [Latin *minare* = to excavate, make underground passages, to undermine, to lead or drive.] n. A naturally occurring, homogeneous, crystallographically continuous, inorganic (no C-H bonding) solid substance having a definite, but not necessarily fixed, chemical composition that can be represented by a formula. A mineral has a characteristic crystalline structure. In a nongeologic context, it has two distinct senses: a) any substance that is nonbiologic (neither animal nor vegetable), or b) as a general term for an element, such as calcium, iron, potassium, sodium, or zinc, that is essential to nutrition. There are eight mono-elemental minerals known as the native element minerals, they are: a) graphite and diamond, both different minerals of carbon, b) copper, silver, and gold, all in the same transitional metal group on the periodic table, and c) platinum, iron, and sulfur. (The semimetals arsenic and bismuth are sometimes also included in the list of native element minerals). First usage in English c. 1400.
See also: ice, mineraloid

mineraloid: [English *mineral* < Latin *minare* = to excavate, make underground passages, to undermine, to lead or drive + Greek *oidos* = resembling.] n. A naturally occurring inorganic solid that is either amorphous or that lacks long-range orderly arrangement of atoms, therefore, a mineraloid lacks crystalline structure. Examples of mineraloids include: a) amber (fossilized tree resin), b) opal (amorphous hydrated silica), c) jet (dense gem-quality lignite coal), d) limonite (mixture of oxides), e) mercury (liquid at normal temperatures), f) obsidian (volcanic silica glass), and g) pearl (organically produced carbonate). First usage in English c. 1910.

minette: [French *minette* = low-grade iron ore; < Middle French *minette* = small mine; < Latin *minare* = to excavate + French *-ette* = diminutive suffix.] n. A syenitic lamprophyre composed of biotite phenocrysts in a groundmass of alkali feldspar and biotite. Also, a low-grade oolitic iron ore found mainly in Luxemburg and Lorraine. Term coined in 1828 by M. Voltz, director of mines in Strasbourg, France.
Cited usage: Minette, this is a local name used by miners in the Vosges for a rock essentially composed of dark mica, orthoclase and a felspathic matrix [1888, Jethro J. Teall, British Petrography].

mini-: [Back-formation from English prefix *miniature* or *minimum* < Italian *miniatura* = small painting, illumination of manuscripts; < Latin *miniare* = to illuminate < *minire* = to color red < *minium* = red lead.] Combining prefix meaning on a small or greatly reduced scale, used in such terms as: minimicrite, miniphyric, miniskirt, and minigolf. Mini is also an integral form of many words implying small, such as: miniature, minimal, and miniator. Like the term porphyry, originally meaning purple, *mini-* is a morpheme that was originally color-based, but has come to mean something quite different (in this case size-related,

as opposed to textural). Because the red lead-based paints common in the Middle Ages were used in making small artistic portrait and landscape *miniatura*, the morpheme *mini-* evolved into a size-related form.
See also: micro-, milli-, macro-, mega-

Miocene: [Greek *meion* = less < IE *mei* = small.] adj. The fourth epoch of the Tertiary period, after the Oligocene and before the Pliocene, spanning the time from 23.0-5.33Ma. The Miocene is characterized by the development of grasses, grazing mammals, and other recognizable modern species, such as: wolves, horses, deer, whales, camels, and owls. By the early Miocene, the continents had drifted to nearly their modern positions. Term coined in 1833 by English geologist Charles Lyell (1797-1875), who devised a scheme for stratigraphically subdividing the entire Tertiary period. IE derivatives include: menu, mince, and minestrone.
Cited usage: The next antecedent tertiary epoch we shall name Miocene [1833, Charles Lyell, Principles of Geology III. 54]. The flora indicates a decidedly tropical climate in the earlier part of the Miocene [1882, Sir Archibald Geikie, Text Bk. Geol.].

miogeocline: [Back-formation from miogeosyncline; < German *miogeosynklinal* < Greek *mio* = lesser, smaller + *ge* = earth + *syn* = together, alike, same + *klinos* = to bend, lean.] n. A prograding wedge of shallow-water sediments at the continental margin. It is typical of the continental shelf, and is generally devoid of volcanic additions. Term coined in 1966 by American geologists R. S. Dietz and J. C. Holden.

miogeosyncline: [German *miogeosynklinal* < Greek *mio* = lesser, smaller + *ge* = earth + *syn* = together, alike, same + *klinos* = to bend, lean.] n. As originally proposed by Hans Stille in 1940, a narrow, downwarped basin where sedimentation is

not associated with volcanism, especially a basin situated between a larger, volcanic eugeosyncline and an area of the crust that has achieved stability (craton). With development of plate tectonic theories in the 1970's, no contemporary examples of miogeosynclines were found to exist, so the term was largely replaced by miogeocline. Term coined and published by Hans Stille in *Den Bau Amerikas* (*The Structure of America*).
Cited usage: In contrast to the Magog eugeosyncline, the Champlain belt contains dominant carbonates of shallow-water origin, unaffected by subsequent volcanism; it is a miogeosyncline [1942, Bull. Geol. Soc. Amer. LIII. 1642].
See also: syncline, eugeosyncline, eugeocline, miogeocline

mire: [Old Norse *myrr* = bog.] n. A small area of marshes, swamps, or boggy ground. First usage in English c. 1325.

miscible: [Latin *miscere* = to mix; < IE *meik* = to mix.] adj. In general, capable of being mixed into a solid solution or homogeneous liquid. Specifically, said of two or more substances in one or more phases, which, when brought together in any proportion, can be mixed to form a single phase. Copper and nickel, as well as alcohol and water are examples of single-phase miscibility, whereas carbon dioxide gas and water is an example of two-phase miscibility. First usage in English c. 1565. IE derivatives include: melange, miscellaneous, meddle, medley, mustang, and promiscuous.

mispickel: [German *mistpuckel* < (possibly) *mist* = dung + *buckel* = hill, hump.] n. An obsolete name for the mineral arsenopyrite, a silver-white to gray arsenic-bearing mineral, iron arsenic sulfide (FeAsS). The uncertain etymology, suggested by E. Taube, was made in reference to the noxious odor emitted by the mineral when heated. Mispickel is archaic, had several pronunciations and spellings, and was probably first used in its most common Germanic form *mistpuckel* by Georgius Agricola in 1546. First usage in English c. 1680.

Mississippian: [Toponymous term after the Mississippi River; < Ojibwa *missi* = great + *zibi* = big river, river of the falls.] adj. Pertaining to the fifth period of the Paleozoic era, spanning the time from 359.2-318.1Ma, corresponding to the Lower Carboniferous period of England. The term also pertains to the corresponding system of rocks. In North America, the Carboniferous period was divided into two separate periods on the basis of their relative abundance of coal, the Upper (Pennsylvanian) being richer than the Lower (Mississippian). To the Ojibwa, the source of the Mississippi River was Leech Lake in Minnesota, not Lake Itasca, as designated by the European settlers.
Cited usage: The Mountain limestone, or Lower Carboniferous mass, which I have proposed to designate the Mississippi Group, because so extensively developed in the valley of the Mississippi River [1870, James Manning Winchell, Sketches of Creation xii. 136].
See also: Paleozoic, Permian, Pennsylvanian, Carboniferous

mistral: [Old Occitan *maiestrau* = master-wind < Latin *magistralis* = of a master < *magister* = master; < IE *meg* = great.] n. A persistent, dry, cold northerly wind, especially one that blows in squalls through the Rhône Valley toward the Mediterranean coast of southern France from the Massif Central. During winter the Massif Central is commonly a center of high barometric pressure; the resulting pressure gradient between this and the warm *Golfe du Lion* to the south causes persistent currents of cold dry air from the northwest. The mistral has a variety of local names: *mangofango* in Provence; *sécaire* in Cévennes; *dramundan* in Perpignan; *cierzo* in Spain; and *cers* in the Pyrenees. First usage in English c.1600.
Similar terms: bise
See also: bise

mode: [Latin *modus* = measure, size, limit of quantity.] n. In petrology, the actual mineral composition of a rock sample, expressed in percent by weight or volume. In general statistical usage, the value or range of values in a data set that occurs most frequently. First usage in English c. 1275.

modulus: [Latin *modulus* = a unit of measure < *modus* = measured amount.] n. In general, the measure of a quantity that depends on two or more other quantities. In physics, a coefficient pertaining to a physical property, as in the modulus of elasticity, being the ratio of stress to strain (Young's modulus). Also, a quantity that expresses the degree to which a substance possesses a certain property. In mathematics, an integer that can be divided without remainder into the difference between two other integers (e.g., 1, 2, and 4 are moduli of 9 minus 5).

mofette: [French *mofette* = gaseous exhalation; < Italian *moffetta* = diminutive of *muffa* = mold, moldy smell; < Latin *mephitis* = noxious vapor < *Mephitis* = Roman goddess who averts pestilential exhalations.] n. An exhalation of carbon dioxide (CO_2) escaping from fissures, fumaroles, and the like in an area of late-stage volcanic activity. Also, the small opening in the earth from which such gases escape. The term was first used in the area around Naples, Italy, most likely due to the exhalations of Mount Vesuvius. First usage in English c. 1820.
Cited usage: Various substances have been ejected during the earthquake, as hot water, mofettes (exhalations of carbonic acid gas), mud, and black smoke [1849, E.C. Otté, translation of Alexander von Humboldt's Cosmos: A sketch of a Physical Description of the Universe I. 209].
See also: mephitic

mogote: [Spanish *mogote* = hillock, heap, haystack.] n. A residual hill in a karst region, usually honeycombed with cavities. Mogote occur

in tropical karst regions, especially Cuba and Puerto Rico. First usage in English c. 1925
Cited usage: The pepino hills of Puerto Rico are much smaller than the mogote hills of Cuba and hence more commonly rise to peaks rather than having flat summits. [1954, Wm.D. Thornbury, Principles of Geomorphology xiii 335].
Synonyms: hum, pepino, karst tower
See also: hum, pepino, karst

Moho: [Eponymous term after Croatian geophysicist Andrija Mohorovicic (1857 - 1936), who discovered the boundary.] n. Abbreviation of Mohorovicic Discontinuity, the seismic velocity discontinuity boundary between the Earth's crust and mantle. It is located at an average depth of 7km below the ocean floor and 35km below the surface of the continents. Willard Bascom coined the term mohole in 1954, after Project Mohole, which he directed. It was an attempt to penetrate the Moho. Drilling through oceanic crust began in 1961, but the project was abandoned before they could penetrate into the mantle. First usage in English c. 1935.
Cited usage: It is generally accepted that both the density and the seismic wave velocities change at the Moho [1960, New Scientist 19 May 1278/3].
Synonyms: Mohorovicic discontinuity

Mohs' scale: [Eponymous term after German mineralogist Friedrich Mohs (1773-1839), who devised the scale.] n. A scale of relative hardness (resistance to scratching) ranging from 1 to 10. The scale was designed for rating the hardness of any mineral, and was based on a standard of ten minerals. The scale is: 1) talc, 2) gypsum, 3) calcite, 4) fluorite, 5) apatite, 6) orthoclase, 7) quartz, 8) topaz, 9) corundum, 10) and diamond. (A mnemonic for remembering this list is, "The girls can flirt and other queer things can do"). Mohs devised the scale in 1822.

moiré: [French *moire* = a type of fabric, an ornamental finish (as in watery silk) < *moirer* = to water.] adj. In mineralogy, said of the appearance of feldspars that reflect a wavy, rippled pattern of light, similar to that of watered silk. In optics, pertaining to the effect produced by the appearance of a new set of wavy lines when two sets of lines are superimposed. This phenomenon also occurs with numerous other graphical shapes, such as in the dotted overlays used in the 4-color printing process. The French word is probably influenced by both the reference to shimmering water as well as the English 'mohair,' a fabric made from the hair of the angora goat. First usage in English c. 1815.

molarity: [German *Mo* = short for Molekulargewicht = molecular weight; < Latin *molecula* = diminutive of *moles* = mass.] n. The molar concentration of a solution, usually expressed as the number of moles of solute per liter of solution; a mole being the atomic or molecular weight of a substance expressed in grams. A mole is also the amount of a substance that contains Avogadro's number of molecules (6.022×10^{23}). Not to be confused with molality, which is expressed as the number of moles of solute per 1,000 grams of solvent. First usage in English c. 1835.

molasse: [French *mollasse* = flabby, flimsy; < Latin *mollis* = soft.] n. A paralic (partly marine, partly continental) or deltaic sedimentary facies consisting of a thick sequence of soft, ungraded, cross-bedded strata consisting of conglomerates, sandstones, shales, marls, and sometimes coal. Molasse is less rhythmic than the marine flysch facies, the deeper water facies that precedes deposition of molasse. They are also deposited much closer to the craton than the preceding flysch. The term molasse can represent a post-orogenic formation that contains the totality of sediments that have eroded from mountains

during and after the main uplift phase of an orogeny (diastrophism). Some people think the term should be discontinued because much so-called molasse is not post-tectonic, but syntectonic, developed from the erosion of nappes while uplift and deformation are still progressing. The term persists, however, because it describes sediments produced from the erosion of a mountain belt subsequent to the main orogenic event, not after uplift has ceased entirely.
See also: diastrophism, paralic, flysch

mold: [Middle English *molde* = shape; < Old French *modle* < Latin *modulus* < *modus* = measured amount.] n. In paleontology, an impression made in surrounding sediment by the outside or inside surface of a fossil shell, or other biologic structure. A complete mold comprises the hollow space remaining after the original fossil has dissolved. The filling of such a space produces a cast. In sedimentology, a mold is an original mark or depression made in a sedimentary surface, such as a flute mark, striation, groove, animal track, or raindrop impression. The filling of such a depression also produces a cast. First usage in English c. 1200.
Alt. spelling: mould
See also: cast

molecule: [Latin *molecula* = diminutive of *moles* = mass.] n. The smallest particle of a substance that retains the chemical and physical identity of the substance and is composed of two or more atoms. In this sense, the word first appeared in French during the 17th century, and is noted in the physical speculations and discussions initiated by philosopher Rene Descartes (1596-1650). First usage in English c. 1790.

mollisol: [Latin *mollis* = soft; < IE *mel* = soft + Latin *sol* = soil.] n. One of the twelve soil orders used in Soil Taxonomy by the U.S. Dept. of Agriculture. A Mollisol typically forms in grasslands of the steppes and prairies. They have a

dark, soft, and loamy surface horizon and a high cation exchange capacity dominated by calcium. They are extensively developed at mid-latitudes, found between the aridisols of arid climates and the alfisols and spodisols of humid climates. Mollisols are common in the Great Plains of the United States. IE derivatives include: mollusc, molasse, mollify, and schmaltz.

mollusc: [Latin *molluscus* < *mollis* = soft.] n. Any invertebrate belonging to the phylum Mollusca. This phylum is comprised of four classes: Bivalvia (clams, oysters, and mussels), Gastropoda (snails, limpets, slugs, and whelks), Cephalopoda (squid, octopus, and ammonites), and Scaphopoda (tooth-shells). The class Bivalvia is also known as Lamellibranchia or Pelecypoda.
See also: lamellibranchia, pelecypod

molybdenite: [Greek *molybdos* = lead.] n. A soft, bluish-gray to lead gray, hexagonal mineral, molybdenum sulfide, MoS_2. It the primary ore of molybdenum it is dimorphic with jordisite, and occurs in foliated to scaly masses in pegmatite dikes, quartz veins, and porphyry-type ore deposits. The name alludes to the early confusion, recorded by the Greek physician Dioscurides (40-90 CE), who gave the name *molybdos*, meaning lead, to many substances, including: galena, graphite, and molybdenite. First usage in English c. 1795.

molybdenum: [Latin *molybdaena* < Greek *molubdos* = lead.] n. A hard, silvery-white element (symbol Mo), used primarily as a minor alloy for making strong steel, as an essential trace element in agriculture, in nuclear reactors, and in various high-tech applications. The Greek physician Dioscurides gave the name *molybdos* to many substances resembling lead, such as: galena (PbS), graphite (C), and numerous lead and molybdenum ores. The element was discovered and named by Swedish chemist Carl Wilhelm Scheele in 1778.

moment: [Latin *momentum* = movement < *movere* = to move; < IE *meu* = to push away.] n. In physics, the product of a quantity and its perpendicular distance from a reference point (M monent = F force x R radius). Also, in orbital rotation, the tendency to cause rotation about a point or an axis. To make an object rotate twice as quickly one can double the force, or apply the same force from twice the distance. This is the basis for applying the mechanical advantage of levers and gears. Etymologically related terms include: motor, emotion, commotion, and promotion.

monadnock: [Toponymous term, possibly Iroquois, after Mt. Monadnock in New Hampshire.] n. A prominent rock, hill, or mountain rising above the general level of a peneplain in a temperate climate. A monadnock represents an isolated erosional remnant of a former highland. Term coined in geology in 1893 by American geomorphologist William Morris Davis. *Cited usage*: The continuity of the plateau-like uplands [of New England] is interrupted in two ways; isolated mountains rise above it, and branching valleys sink below it. Mount Monadnock is a typical example of the former, with its bold summit more than a thousand feet above the surrounding plateau [1893, W. M. Davis, Nat. Geogr. Mag. V. 70]. "His great, Monadnock hump" [1851, Herman Melville, Moby Dick].

monazite: [German *monazit* < Greek *monazein* = to be solitary, to be alone < *monos* = alone, single.] n. A yellow, brown, or reddish-brown monoclinic mineral, cerium lanthanum neodinium thorium phosphate, $(Ce, La, Nd, Th)PO_4$. Monazite is sometimes distinguished as three separate minerals; the most common cerium-rich monazite-(Ce), lanthanum-rich monazite-(La), and neodymium-rich monazite-(Nd). It is a source of the rare earth elements and thorium. Monazite is widely disseminated as an accessory mineral in granites, gneisses, and pegmatites, as well as often being naturally concentrated in detrital sand deposits. Term coined in 1829 by Johan Friedrich August Breithaupt. The etymology meaning 'to be alone' is an allusion to the typical crystal habit of primary origin for monazite, as isolated individual crystals in phosphatic pegmatites. *Cited usage*: Monazite. This name was given by Breithaupt to a mineral brought by Fielder from the Uralian mountains [1836, Thomas Thomson, Outlines of Mineralogy and Geology, I. 672 (Thomson, 1773-1852, devised the system of chemical symbols used on the periodic table. Johns Jacob Berzelius, five years younger, made a few changes to Thomsons scheme, them claimed it as his own.)]. *Synonyms*: cryptolite

mono-: [Greek *monos* = alone, solitary, single.] Combining prefix meaning solitary, single or alone, used in such terms as: monoecious, monoclinic, and mononucleosis.

monocline: [Greek *monos* = single + *klinein* = to bend, slope, lean.] n. A geologic structure in which the layers are locally steepened in an otherwise gently dipping sedimentary sequence; thus, the strata in a monocline remain parallel to one another, have the same strike, and all dip in the same direction, like sheets of plywood lifted at one end. First usage in English c. 1875. *Cited usage*: The strata are thus bent up and continue on the other side of the tilt at a higher level. Such bends are called monoclines or monoclinal folds, because they present only one fold, or one half of a fold, instead of the two which we see in an arch or trough [1879, Geikie in Encycl. Britannica X. 300/1]. *See also*: anticline, syncline, isocline

monomictic: [Greek *mono* = solitary, single + *mixis* = mixing, a mingling.] adj. In petrology, said of a sedimentary rock that is composed entirely of clasts of one mineral (e.g., a chert breccia). In limnology, said of a lake that experiences only one yearly overturn (e.g., polar lakes in the summer and tropical lakes in the winter), as opposed to dimictic lakes in temperate regions that overturn in spring and again in fall.

monsoon: [Dutch *monssoen* = monsoon; < Arabic *mausim* = season < *wasama* = to mark.] n. A seasonal wind prevailing in southern Asia, especially in the Indian Ocean, which during the period from April to October blows from the south-west, and from October to April from the north-east. The direction of the wind is dependent on periodic changes of temperature in the surrounding land-surfaces. The southwest, or summer, monsoon is commonly accompanied by heavy and continuous rainfall, and is therefore often referred to as the wet or rainy monsoon. The north-east, or winter monsoon is, in contrast, known as the dry monsoon. The Dutch, being among the earliest successful traders with Asia (especially trade-wary Japan), probably brought this Asian-sounding word from Europe to the Orient. First usage in English c.1580.
Cited usage: The proper season for Sailing on the Indian-Sea is called Mousson or Monson, by corruption of Moussem [1687, Archicald Lovell, translation of Jean de Thevenot's , The Travels of M.de Thevenot into the Levant III. I. i. 1].

montmorillonite: [Toponymous term after *Montmorillon*, a commune or district in western France.] n. A reddish, hydroxy aluminum silicate mineral group, characterized by massive texture and considerable capacity for exchanging aluminum for magnesium or a base. This group of expanding-lattice clay minerals is generally derived by alteration of ferromagnesian minerals, calcic feldspar, and volcanic glass. They are the chief constituents of bentonite and fuller's earth. Members of the montmorillonite group include: nontronite, saponite, hectorite, sauconite, beidellite, volkonskoite, griffithite, and montmorillonite.

The term is also used for a single mineral species, which is a white, grayish, or pale red, high-alumina, monoclinic end member of the group, hydrated hydroxy sodium calcium aluminum magnesium silicate, $(Na,Ca)_{0.3}(Al,Mg)_2Si_4O_{10}(OH)_2 \cdot nH_2O$.
Synonyms: beidellite
See also: saponite

monument: [Latin *monumentum* = something that reminds.] n. A natural or artificial structure used to mark the location of a survey point. In geomorphology, an isolated pinnacle, usually due to weathering, that resembles a man-made monument.

monzonite: [Toponymous term after the Tyrolean peak *Monzoni* in the Italian Alps.] n. In igneous petrology, plutonic rocks that are intermediate in composition between syenite and diorite, and have nearly equal amounts of alkali feldspar and plagioclase, little quartz, and commonly augite as the principal mafic mineral. It is the plutonic equivalent of the volcanic rock latite. Term coined in 1864 by Auguste de Lapparent, mineralogist and stratigrapher at the Catholic Institute of Paris, who helped create the first geologic map of France.
Cited usage: A special type of augite-syenite is presented by the Triassic intrusions of Monzoni in the southern Tyrol [1895, Alfred Harker, Petrology for Studentsiii 44].
Synonyms: syenodiorite
See also: latite

moor: [Old English *mor* = moor; < German *moor* = marsh; < Old Teutonic *mer* = to die.] n. A broad area of open land, often high but poorly drained, with patches of heath and peat bogs. First use in English before 900.

moraine: [French *morena* = mound of earth < *morre* = muzzle; < Italian *mora* = heap of stones.] n. An accumulation of boulders, stones, or other

debris carried and deposited by a glacier. Moraines are named in terms of their location: a) at the toe of a glacier (terminal or end moraine), b) at the flanks (lateral moraine), c) in between two glaciers (medial moraine), and d) as an extensive sheet left on glacial retreat (ground moraine). First use c. 1785.
Cited usage: If the moraine is at the extremity of the glacier it is a terminal moraine; if at the side, a lateral moraine; if parallel to the side on the central portion of the glacier, a medial moraine [1863, Sir Charles Lyell, The Geological Evidence of the Antiquity of Man].

morass: [French *marais* = marsh, swamp, bog; < German *morast* < IE *mori* = body of water.] n. In general, a wet swampy tract, bog, or marsh. In mining, a term for bog iron ore. As a technical term, morass is used primarily in the West Indies. IE derivatives include: maar, marsh, marine, meerschaum, and mermaid. First use c. 1650.
Cited usage: Only a small portion of the country was under cultivation, the rest was morass or impenetrable forest [1860, Walter Farquhar Hook, Lives of the Archbishops of Canterbury I. 355].

MORB: [English Acronym for M(id) O(cean) R(idge) B(asalt).] An acronym for basalt that erupts at divergent plate boundaries in the ocean basin. Most such lavas display pillow structure and glass selvages, characteristic of rapid cooling under water.

morpho-: [Greek *morphos* = shape, form.] Combining prefix meaning form or shape, used in such terms as: morphometry, morphostructure, morphography, morphosis, and morpheme. In Greek mythology, Morpho ("the shapely") was one of the many epithets of Aphrodite, the goddess of love; Morpheus was the god of dreams.

morphogenesis: [Greek *morphos* = shape, form + *genesis* = origin, creation < *genesthai* = to be born, come into being; < IE *gen* = to give birth,

beget.] n. In geomorphology, the origin and first stages of development of a landscape or landforms. First use c. 1880.

morphology: [Greek *morphos* = shape, form + *logos* = word, speech, discourse, reason < *legein* = to speak, to say; IE *leg* = to collect < *logia* = to speak.] n. In general, a term that deals with form and structure without much consideration of function. A term widely applied in earth and life sciences dealing with aspects of shape, form, structure or profile; such as surface land forms in geomorphology, organism shape in paleontology, soil structure in pedology, and space-time continuum in astrophysics.
Cited usage: The term morphology was introduced into science by Goethe, at least as early as the year 1817 [1879, Asa Gray, Structural Botany 6th ed.].

mortlake: [Toponymous term after the Mortlake district southwest of London, England.] n. An obsolete synonym for an oxbow lake. The Mortlake district is a drained oxbow lake of the Thames River.
Cited usage: The loop often remains as a dead river-channel or 'Mortlake.' Such loop-lakes are known in America by the special name of 'Ox~bows' [1902, Lord Avebury, The Scenery of England and the Causes to Which it is Due ix. 303].
Synonyms: cut-off lake, oxbow
See also: oxbow

mosaic: [Middle English *mosaique* < Italian *mosaico* < Latin *musaicum* = mosaic work <*museum* = mosaic work, seat of the muses; < Greek *musa* = muse.] n. In general, the process of creating pictures or decorative patterns by using small irregular pieces of material, usually variously colored, as the medium. In petrology, a texture in which individual mineral grains are roughly equant, but do not interlock. In crystallography, a term applied to the structure of crystals made up of small blocks of perfect lattices set at more or less regular grain boundaries. The possible connection between mosaic work and the ancient Greek Muses of the Arts is that an artistic mosaic was considered to be inspired by the muses. First use c. 1375.

mosor: [Toponymous term after the Mosor Mountains in Dalmatia, Yugoslavia.] n. A monadnock that has survived due to its remoteness from primary lines of drainage and erosion, often equated with a hum in a karst region.
Alt. spelling: mosore (pl.)
See also: monadnock, hum

mother lode: [English *mother* = the biggest or most significant example of its kind + *lode* = a rich source; Old English *lad* = way.] n. In mining, the richest and most productive vein of ore in a region. The Mother Lode of California, referring to a belt of gold-bearing quartz veins in Central California along the western foothills of the Sierra Nevada. The name is sometimes limited to a strip about 110km (70mi) long and from 1.5-10.5km wide, running NW from Mariposa. The discovery of alluvial gold on the South Fork of the American River led to the 1848 California Gold Rush. Mark Twain and Bret Harte helped make the Mother Lode a world famous location. First use c.1860.

mottled: [Middle English *motlei* = variegated cloth, parti-colored attire of a court jester < *mot* = speck.] adj. Said of a sediment, sedimentary rock, or a soil that is marked with multicolored spots, usually from iron compounds. Mottle is a back-formation from the word motley. First used c. 1670.

motu: [Polynesian *motu* = small coral island.] n. A small coral island, large enough to support vegetation, usually an emergent part of a larger coral reef.

moulin: [French *moulin* < Latin *molinus* = mill < *mola* = millstone < *molere* = to grind; < IE *mel* = crush, grind.] n. A cylindrical, nearly vertical shaft or cavity that is eroded into glacial ice at the downslope end of a transverse fissure. It is formed by swirling meltwater and rock debris as rushing down from the surface. Moulins can attain great size, sometimes being gradually scooped out of a deep chasm to form giant kettles. The term alludes to the sound of the gravelly cascade, resembling the sound of grinding millstones. A more figurative use of the term is in reference to living creatures that mill about in churning confusion, stemming from the herding technique used in the U.S. to swirl or mill cattle into a tight circle. IE derivatives include malleable and mylonite. First use in English c. 1855
Cited usage: These moulins occur only at those parts of the glacier which are not much rent by fissures [1860, John Tyndall, The Glaciers of the Alps ii. xxv. 363 (Tyndall was an accomplished renaissance man, having published books, being first to scale the *Weisshorn* in the Alps, discoverer of atmospheric ozone, pollutions measurement, and theories of global climate change, inventor of early fiber-optic techniques, and friend of Farraday, Pasteur, Lister, Tennyson, Bunsen, and Magnus.)].
Synonyms: glacier mill

mountain: [Latin *montanus* < *mons* = mountain; < IE *men* = to project.] n. Any part of the Earth's crust that is sufficiently elevated above the surrounding land surface that is deemed worthy of a distinctive name. A mountain is characterized by a restricted summit area. Although older usage limited the term to summit elevations at least 600m (2,000ft), modern usage reduced that minimum to 300m (1,000ft). IE derivatives include: promontory, montane, mount, mons, and mouth.
Similar terms: orogeny
See also: orogeny

muck: [Old Norse *myki* = dung < *moka* = to shovel.] n. Dark, finely divided, well-decomposed organic material, indicative of poorly drained areas. In pedology, a soil that contains at least 50% organic material. The mineral muckite has no etymological connection to muck. It is a variety of retinite, which was named after the 19th-century German mineralogist H. Muck.
See also: mud

mud: [Middle English *muddle* < German *mot* = bog, peat < *moder* = mold, decay.] n. An unconsolidated sediment consisting principally of a mixture of silt and clay and often rich in organic material. Also, used for any sticky or slippery mixture of water, silt, and clay material having a consistency ranging from semifluid to soft and plastic. First usage in English c. 1325.
See also: muck

mullion: [Latin *medianus* = that which is in the middle < medius = midst, middle, central.] n. In structural geology, a rodlike lineation creating a corrugated surface, produced during metamorphism at the interface between rocks of different ductility. The term alludes to the similarity in shape to the vertical members of stone standing between large windows, particularly those on massive structures. First usage in English c. 1565.

multi-: [Latin *multus* = much, many.] Combining prefix meaning much or many, used in such terms as: multiple, multicellular, multilingual, multiplex, and multiphase. This prefix, a corollary to the Greek *poly-*, did not become prevalent in English until the 17th century.
See also: poly-

muscovite: [Toponymous term after *Muscovy*, the Russian term for Moscow, capitol city of Russia and formerly of the entire Soviet Union.] n. A colorless to silvery to gray-brown monoclinic mineral of the mica group, hydroxy potassium aluminum silicate, $KAl_2(AlSi_3O_{10})(OH)_2$. It can be transparent, translucent, or opaque, displaying a pearly luster. Originally called Muscovy glass because residents of Moscow living near a source of large sheets of clear muscovite, used it as window panes. First usage in English c. 1550.
Synonyms: white mica

muskeg: [Cree Indian *muskeg* = Odjibway *mashkig* = Abnaki *mskakw* = bog, swamp.] n. A swamp or bog formed by an accumulation of sphagnum moss, leaves, and decayed organic matter. Muskegs are prevalent in high-latitude boreal forests, such as those found in Canada and Alaska. First usage in English c. 1770.
Cited usage: Muskegs–or level swamps–the surface of which is covered with a mossy crust five, or six inches in thickness, while a thick growth of pines and the fallen timber add to the difficulties of the road [1865, Viscount Wm. F. Milton & Walter B.Cheadle, Northwest Passage by Land 207].
See also: marsh, bog

mussel: [Old English *muscelle* = muscle; < Latin *muscula* = muscle < *mus* = mouse; < IE *mus* = mouse, muscle.] n. Any of several marine and freshwater bivalve molluscs belonging to the superfamily Unionacea, especially Mytilus edulis, known as the common mussel or blue mussel. The connection between mouse and muscle stems from the impression of the ancient Greeks that a flexing muscle resembles the movements of a mouse. First usage in English before 1000.

mylonite: [Greek *mulon* = mill < *mul* = handmill; < IE *mel* = crush, grind + Greek *-ites* < *-lite* = pertaining to rocks < *lithos* = rock, stone.] n. A fine-grained, parallel-banded microbreccia having a flow texture that occurs in narrow, planar zones of ductile deformation. Mylonites are cohesive fault rocks characterized by well-developed schistosity resulting from tectonic reduction of

grain size. They are formed by extreme granulation, grinding, and shearing of rock during overthrusting and cataclastic metamorphism. Mylonites are categorized into protomylonite, mesomylonite, and ultramylonite, based on the amount of finer-grained matrix . When known, the protolith is included in the name, such as 'granite mylonite.' The term was introduced in 1885 by Charles Lapworth. Until viewed microscopically, it is difficult to distinguish foliated fault rocks that are formed by brittle deformation (cataclasite) from those formed by plastic deformation or grain-boundary sliding (mylonite), therefore the terms are often used synonymously in the field.
Cited usage: It would appear that the mylonisation of the gneiss is probably due to the passage over it of the great mass of rock brought forward by the Moine thrust-plane [1913, B. N. Peach et al., Geol. Central Ross-shire Mem. Geol. Surv. Scot. iii.13].
See also: cataclastic, flaser

myrmekite: [Greek *myrmekia* = anthill, wart < *myrmey* = ant + *-ites* < *-lite* = pertaining to rocks < *lithos* = rock, stone.] n. An intergrowth of plagioclase with vermicular (wormlike) quartz, generally replacing potassium feldspar during late stages of consolidation. First discovered in 1874 by Michel-Levy, and named by J. J. Sederholm in 1916.
Cited usage: The intergrowth of plagioclase and 'vermicular' quartz...I have called myrmekite [1916, J. J. Sederholm, On synantectic minerals and related phenomena: Bulletin de la Commission geologique de Finlande v.9].
Alt. spelling: myrmeckite, myrmecite (obs.)

nacre: [Middle Latin *nacrum* < French *nacle* = drum; < Arabic *naqqarah* = small drum < *naqqara* = to bore, pierce; < Semitic *nqr* = bore, pierce.] n. A smooth, lustrous, variously colored deposit, chiefly calcium carbonate that makes up the shiny, iridescent substance forming the inner layer in many shells, primarily molluscs. First usage in English c. 1595.
Synonyms: mother-of-pearl

nadir: [Arabic *nadir* = opposite to, against.] n. The point directly below the observer and located on a vertical line between the observer and a point directly overhead. The point diametrically opposite the zenith. First usage in English c.1375.
Cited usage: Sun's Nadir is the axis of the cone projected by the shadow of the earth; thus called, in regard that the axis being prolonged, gives a point in the ecliptic diametrically opposite to the sun [1727-38, Chambers Cyclopedia].
See also: zenith

naif: [French *naïf* = masculine form of *naïve* = unsophisticated.] n. The natural, unpolished surface or skin of a diamond crystal; small areas of naif may be left on polished stones, especially around the girdle to indicate minimal loss in cutting and polishing. Naïf, when applied to the qualities of a person, means unsophisticated, when applied to the qualities of a gem, it means unpolished.

nakhlite: [Toponymous term after Egyptian city El-Nakhla + Greek *-lite* = pertaining to rock < *lithos* = rock, stone.] n. A basaltic achondrite type of stony meteorite containing abundant augite, and displaying a cumulate texture. The first nakhlite was found in 1911. Two other stony meteorites, one at Shergotty, India, in 1865 and another at Chassigny, France, in 1815, had similarities to the El-Nakhla stone, being young in age and having distinctive petrologic, chemical, and isotopic composition. This group of igneous achondrites acquired the abbreviated name SNC meteorites, which stands for the shergottite-nakhlite-chassignite group of meteorites, and all are considered to have originated from the planet Mars.
See also: shergottite

naled: [Russian *naled* = ice shield.] n. A Russian term for the ice layer formed in a permafrost area in winter as a result of layer-by-layer freezing of sheets of water that have seeped from the ground, a spring, or surface stream. The term naled is commonly used in Russia, because perennially frozen ground (permafrost) covers almost half the country.
Synonyms: icing, aufeis (German), taryn (Yakut)

nannoplankton: [Greek *nanos* = little old man, dwarf < *nannos* = uncle + *planktos* = to wander, roam or drift.] n. Passively floating unicellular microorganisms ranging in diameter from 5 to 60μm. They are larger than ultraplankton but smaller than microplankton. Term coined in 1902 by Lohmann.
Alt. spelling: nanoplankton

nano-: [Greek *nanos* = little old man, dwarf < *nannos* = uncle.] Combining prefix meaning one-billionth (10^{-9}), as in nanosecond or nanometer. Sometimes used more loosely to mean extremely small, as in nannoplankton.
Alt. spelling: nanno-
See also: nannoplankton

nant: [Celtic *nant* = stream.] n. A small, stream-cut valley with an active stream. In Celtic place names, the adjective comes after the noun it modifies; an example in Welsh is *Cwm nant ddu*, meaning hollow dark creek (literally, hollow creek dark).
Cited usage: There was once a time when great glacial sheets spread over the combs and glens of Snowdonia, as they spread to-day over the nants of Chamounix [1883, G. Allen, Flowers & their Pedigrees vi. 184].

naphtha: [Greek *naphtha* = liquid bitumen; < Arabic *naft* = oil for burning.] n. A name originally applied to a flammable, volatile, and liquid bituminous asphalt issuing from the ground in the Middle East; now generalized to include most of the distillates obtained from petroleum, other than gasoline. Also, a flammable liquid obtained by the distillation of certain nonpetroleum carbonaceous materials, such as: a) Boghead naphtha from Boghead coal (obtained at Boghead, Scotland), b) crude naphtha from coal tar, and c) wood naphtha from wood. First usage in English c. 1570.
Cited usage: The principal ingredient of the Greek fire was the naptha, or liquid bitumen [1788, Gibbon, Decline & Fall of the Roman Empire ii. V. 402].
Alt. spelling: naptha, neptha,
Similar terms: naphthalene, naft

nappe: [French *nappe de recouvrement* = sheet of overlaying < *nappe* = tablecloth, napkin; < Latin *mappa* = cloth.] n. A sheetlike, allochthonous rock unit that has moved on an essentially horizontal fault surface over neighboring strata as a result of thrusting; the fold limbs and axes of the thrust mass are approximately horizontal. Also, described as a sheetlike part of a broken recumbent fold that has been moved over the rock formations beneath. Nappes are typical of the Western Alps, where they occur as a complex series of imbricate thrust sheets, formed when the African foreland was pushed northward against the European foreland. First usage in English c. 1905.
Cited usage: In deciding upon the basal limit to be assigned to any particular nappe, one generally chooses some prominent thrustplane; failing this, one is entitled to select the axial plane of some recumbent anticline or syncline, according to local convenience [1922, Q. Jrnl. Geol. Soc. LXXVIII. 87].

narl: [Arabic *nar* = fire.] n. A variety of caliche that forms by alteration of permeable calcareous rocks, the type locality being in the drier parts of the Mediterranean region. The term relates to the processing of this type of caliche in limekilns.

natro-: [New Latin *natrium* = sodium; < Greek *nitron* = niter (sodium nitrate); < Egyptian *ntr* = niter.] Combining prefix meaning sodium, used in mineralogy to form the names of sodium-containing minerals, such as natrolite. Also, a prefix indicating a high proportion of sodium relative to some other metal in a mineral, as in natroalunite.

natron: [New Latin *natrium* = sodium; < Greek *nitron* = niter (sodium nitrate); < Egyptian *ntr* = niter.] n. A white, yellow, or gray monoclinic mineral, hydrated sodium carbonate, $Na_2CO_3 \cdot 10H_2O$. Natron is highly soluble in water, and occurs primarily as an efflorescence of saline residues from evaporation of alkali lakes. Borax and natron aren't entirely synonymous. The ancient mummification process depended on a salt known as natron, which came from the Natron Valley in Egypt. Natron contains borates as well as other common salts, such as sodium bicarbonate and sodium chloride. First usage in English c. 1680.

nauplius: [Latin *nauplios* = shellfish; < Greek *Nauplios* = son of Poseidon and Amymone < *naus* = ship.] n. A free-swimming larval stage of certain crustaceans, characterized by an unsegmented body having a dorsal shield, three pairs of legs,

and a single eye situated on the median. The nauplius is usually the first stage of development after leaving the egg. Formerly, a term coined by F. Mueller for a specific genus of crustacean. Nauplios was noted as 'The Wrecker of Ships,' due to his leading them astray with false lights and signals. First usage in English c. 1835.
Cited usage: The Cyclopes of our fresh waters were excluded in the Nauplius-form [1869, W. S. Dallas, translation of F. Müller's Facts for Darwin 84].
Alt. spelling: pl. nauplii

nautiloid: [Greek *nautilos* = sailor, nautilus < *nauta* = mariner < *naus* = ship; < IE *nau* = boat.] n. Any cephalopod mollusc of the genus Nautilus, having a spiral, nacre-lined shell with a series of air-filled chambers. Living examples are the chambered, pearly, and paper nautilus. IE derivatives include: navy, cosmonaut, naval, and nausea.

naze: [Toponymous term after English *Naze*, a nose-shaped promontory on the largely featureless coast of Essex, England; < Old English *naze* = nose.] n. A promontory or headland.
Alt. spelling: nase
Synonyms: ness

neap: [Middle English *neep* = back-formation from Old English *neepflood* = tide.] adj. Said of a tide that occurs when the difference between high and low tide is at a fortnightly minimum (every 14 days). Neap tides occur twice during any month, one occurring shortly after the first quarter of the Moon, and the other after the third quarter. During neap tides, high water is at its lowest level above datum, whereas low water is at its highest level. First usage in English before 900.
See also: spring, fortnight, tide

nebkah: [Arabic *nebkah* = dune formed around obstacle.] n. A small dune formed when sand is deposited on the lee side of a clump of shrubbery.

Cited usage: The spellings 'nebka' and 'nebkha' are widely used, but in this paper the author uses 'nebkah,' as suggested by Abdul Salam of Damascus University [1978, Yutaka Sakaguchi, Paleoenvironments in Palmyra District during the Late Quaternary].
Alt. spelling: nebkha, nebka
Synonyms: shrub-coppice dune

nebula: [Middle English *nebule* = cloud, mist; < Latin *nebula* = cloud; < IE *nebh* = cloud.] n. In astronomy, a cloud of gas or dust in intergalactic or interstellar space, which may condense to form stars and planets. Examples include: the Crab Nebula (M1) in the Milky Way Galaxy, the Andromeda Nebula (M31), a galaxy beyond the Milky Way, and the Whirlpool Nebula (M51). Converging at approximately 500,000km/hr, the Milky Way and the Andromeda Nebula are on a collision course, due to collide in about 2 billion years. First usage in English c.1660. IE derivatives include: nepheline, nimbus (cloud), and nebulous.
Cited usage: Modern investigations have shown that nebulae, as distinguished from ordinary star-clusters, fall into two classes having entirely different characteristics, namely, the galactic nebulae and the extra-galactic nebulae [1930, R. H. Baker, Astronomy 1st ed. xi. 465].

nehrung: [German *nehrung* = sandbar, spit.] n. A long and relatively narrow sandbar, spit, or barrier beach that forms the seaward containment for a lagoon at a river's mouth. It originates by longshore transport. The type locality is along the East German coast of the Baltic Sea, such as the *Kurishe Nehrung* (Curonian Spit), a wall of sand 100km long and 2 km wide between the sea and a long river-fed lagoon.
See also: haff, lagoon, lido

nejd: [Toponymous term after Arabic *Nejd*, a central desert plateau and province in Arabia.] n. In general, a flat desert area of exposed bedrock,

blown free of sand by the wind and typical of the Sahara Desert. Specifically, the *Nejd* is the plateau region of the central Arabian Peninsula that formed the nucleus of the modern country of Saudi Arabia (Based in Riyadh, the Kingdom of Saudi Arabia was formed in 1932 under the rule of King Ibn Saud). The Nejd is a vast plateau (762-1524 meters high) bounded on the north by the Nafud Desert, on the east by El Hasa Desert, on the south by the Dahna Desert, and on the west by the Asir Desert.
Alt. spelling: Nejd, najd, Najd
Synonyms: hammada

nek: [Afrikaans *nek* < Dutch *nek* = neck.] n. A neck or saddle connecting two peaks, frequently forming a narrow ridge that serves as a pass.

nekton: [Greek *nektos* = swimming.] n. A collective name for aquatic animals that are able to swim directionally and move about independently, such as shrimp, fish, cephalopods, seals, and whales. Nekton do not depend on water movement for transport. First usage in English c. 1890.
Cited usage: The traditional biological classification of fish and fish-like creatures into drifters or plankton, swimmers or nekton, and the bottom dwellers or benthos. The nektonic species of fish are divided into demersal and pelagic [1969, Austral. Law Jrnl. XLIII. 430].
See also: benthos, plankton

neo-: [Greek *neos* = new.] Combining prefix meaning new, current or recent, used in such terms as: neotectonics, neodymium, neoconservative, and Neolithic. Used in contrast to *paleo-*.
See also: paleo-

neoblast: [Greek *neo* = new + *blastos* = sprout, shoot, germ.] n. Grains within a metamorphic rock that are formed much more recently than other grains, frequently the result of late-stage recrystallization. Used in contrast to paleoblast.
See also: paleoblast

Neogene: [Greek *neo* = new + *genesis* = origin, creation < *genesthai* = to be born, come into being; < IE *gen* = to give birth, beget.] adj. In the geologic time scale, a subgrouping of the Tertiary period, comprising the Miocene and Pliocene epochs (23.0Ma-1.81Ma). First usage in English c. 1855.
Alt. spelling: Neocene
Synonyms: Neocene (obs.)
See also: Paleogene

Neolithic: [Greek *neo* = new + *lithos* = rock, stone.] adj. Of or belonging to the cultural period of the later stone age beginning around 10,000 BCE. The Neolithic is characterized by: a) the development of agriculture, b) the making of refined polished stone implements for use as tools and weapons, and c) artifacts identifiable as objects of art. Such cultural advances were thought by many to have originated in and radiated from a focal point in the Tigris-Euphates River Valley in the Middle East (Iraq), but there is evidence that multiple foci of advanced culture arose at nearly the same time. In organizing the extensive collection of artifacts at the National Museum of Denmark, the 19th-century Danish archaeologist Christian Thomsen proposed an innovative system based on the assumption of a progression in human technology from stone to bronze to iron. His insight that early technology had developed in chronological stages rather than concurrently at different levels of society proved essentially correct, though ultimately of limited use in describing the various progressions in other parts of the world. Once empirical study of archaeological collections began, Thomsen's Three Age system was rapidly modified into four ages by the subdivision of the Stone Age into the Old Stone Age (now Paleolithic) and New Stone Age (Neolithic). Subsequent refinement has added Mesolithic (Middle Stone Age) and Chalcolithic (Copper and Stone Ages) to the original terms, which are now known as periods rather than ages.

Even though the terms describe a cultural state more than a time frame, they can be listed chronologically; this scale of early human cultural development is—Paleolithic, Mesolithic, Neolithic, Chalcolithic, Bronze, and Iron. The term Neolithic was coined in 1865 by Sir John Lubbock (1834-1913).
Cited usage: The later or polished Stone age; a period characterized by beautiful weapons and instruments made of flint and other kinds of stone...This we may call the 'Neolithic' period [1865, Sir John Lubbock, Prehistoric Times i. 3].

neontology: [Greek *neos* = new + *ontos* = existence, being < *einai* = to be; < IE *es* = to be + Greek *logos* = word, speech, discourse, reason < *legein* = to speak, to say; IE *leg* = to collect < *logia* = to speak.] n. The investigation of extant organisms as evidence for species origin and evolution. Used in contrast to paleontology, which is the study of fossil evidence for species origin and evolution.

neosome: [Greek *neos* = new + *soma* = body.] n. The part of a migmatite that is of granitic composition and formed during the process of anatexis. Thus, the neosome is younger than the main rock mass, or paleosome.
See also: migmatite

neotectonics: [Greek *neos* = new + *tektonikos* = pertaining to building < *tekton* = builder; < IE *teks-* = to weave.] n. The study of the post-Miocene structural history of the Earth's crust, specifically aimed at investigation of active faults and other geologic structures created by crustal movement in the Pliocene, Pleistocene, or Holocene epochs.

neotype: [Greek *neos* = new + Latin *typus* = image; < Greek *tupos* = impression.] n. A single specimen of a particular species that is designated as the type specimen for that species when all holotypes, paratypes, or syntypes have been lost or destroyed. First usage in English c. 1850.

nepheline: [Greek *nephele* = cloud; < IE *nebh* = cloud, mist, dark.] n. A light-colored, hexagonal mineral, in the feldspathoid group, sodium potassium aluminum silicate, $(Na,K)AlSiO_4$. Nepheline is a major rock-forming mineral, occurring chiefly in volcanic and plutonic rocks that are deficient in silica, such as basanites and syenites. Term coined by the French mineralogist René-Just Haüy in 1800, in allusion to the cloudy appearance of the crystal fragments when placed in nitric acid. The type locality is Monte Somma, a peak of Mt. Vesuvius near Naples, Italy. IE derivatives include: nebula, nimbus, and Nibelung.
Alt. spelling: nephelite

nephrite: [Greek *nephros* = kidney.] n. A tough, compact, fine-grained, fibrous aggregate of the amphibole group mineral actinolite. It is the more common of the two forms of jade in the pyroxene group (the other form being jadeite). Most jadeite used as gemstones comes from Burma, whereas the nephrite form of jade is primarily found in the Americas. The Spanish Conquistadors of the 1500's noticed that indigenous Meso-Americans wore jade amulets to protect their kidneys. The Spanish named this green gemstone *piedra de ijada*, meaning stone of the loins, from the Meso-Americans, belief that it cured renal dysfunction. French was the official language of diplomacy and letters at the time, so *piedra de ijada* became *pierre de l'ejade*, which then became masculinized to *le jade*, and later Latinized to *nephriticus*, meaning kidney. It was finally Anglicized and shortened to the modern mineral name of nephrite. First usage in English c. 1790.
Cited usage: Nephrite or jade is in part a tough compact fine-grained tremolite [1879, Rutley, Stud. Rocks x. 131].
See also: jadeite, actinolite

nepton: [Eponymous term after Latin *Neptunus* = Neptune, Roman god of the sea.] n. The mass of sediments and sedimentary rock filling a basin, as in a geoclinal basin. Term coined in 1954 by Japanese geologist of Kyoto University, J. Makiyama.

neptunism: [Eponymous term from Latin *Neptunus* = Neptune, Roman god of the sea.] n. The obsolete theory proposed by Abraham Gottlob Werner (1749-1817) that all of Earth's crustal rocks, including basalt and granite, crystallize from mineralized water.

neritic: [Greek *Nerites* = son of *Nereus*, an uncle of Poseidon, and both gods of the sea.] adj. Said of the near-shore, usually shallow marine environment that runs parallel and adjacent to coastlines. Also defined by depth, the neritic zone extends from low-tide level down to a depth of one-hundred fathoms (200m), roughly equivalent to the maximum depth of the continental shelf. Also, said of the creatures inhabiting this depth region (e.g., the nearshore gastropod molluscs of the family Neritidae). Term coined in German as 'neritisch' in 1890 by Ernst Haeckel, while teaching at the University of Jena, in the publication *Jenaische Zeitschrift für Medizin und Naturwissenschaft* (*Jenai magazine for medicine and natural science*). *Cited usage*: The organisms living in mid-ocean in the great oceanic currents are quite different from those in the surface waters near land, and Haeckel proposes to designate the former oceanic Plankton, and the latter neritic Plankton [1891, Murray & Renard, Rep. Deep-Sea Deposits iv. 251]. *See also*: ecology

nesosilicate: [Greek *nesos* = island + Latin *silex* or *silic* = flint, hard stone.] n. The simplest of the silicate subclasses, characterized by isolated SiO_4 tetrahedra that are unbonded to other adjacent tetrahedra. The nesosilicates are characteristically hard and include several gemstones, such as: topaz, peridot, and carbuncle. *See also*: cyclosilicate, inosilicate, phyllosilicate, sorosilicate, tectosilicate

ness: [Old English *naess* = headland; < Latin *nasus* = nose; < IE *nas* = nose.] n. A cape or headland, often used in place names, such as Loch Ness. *Alt. spelling*: nose, naze *Synonyms*: naze, nose, nore, nab *See also*: naze

neuston: [Greek *neustos* = swimming < *nein* = to swim; < IE *snau* = to swim, flow, suckle.] n. Animals living in contact with the surface of any body of water, either on the air side, such as water striders, or on the water side, such as the hydra. Hydrogen bonding between water molecules is responsible for the strong surface tension at the air-water interface; this allows life forms to remain at this boundary. Neuston are primarily planktonic and microscopic, with some macroscopic motile creatures, such as the whirligig beetles. IE derivatives include: nutrient, naiad, nurse, and nymph. First usage in English c. 1825.

névé: [French *névé* = hardened snow at the head of a glacier; < Latin *niv* = snow.] n. The granular snow in high mountains, or on the upper part of a glacier. With further compaction névé is transformed into glacial ice. First usage in English c. 1850. *Cited usage*: I found grains of névé larger than a walnut [1856, Elisha Kane, Arctic Explorations I. 336]. *Synonyms*: firn *See also*: firn

nickel: [Swedish *kopparnickel* < German *kupfernickel* < *kupfer* = copper + *nickel* = demon, rascal.] n. A hard, silvery-white metallic element (symbol Ni). Nickel comprises a major portion of the Earth's core, and also occurs as the native element when alloyed with iron in meteorites. The term was coined in 1754 by A.F. von Cronstedt, after the miners' name *kupfernickel*, a copper-colored ore (niccolite-NiAs) from which he first isolated nickel in 1751. The miners named the ore mineral *kupfernickel*, meaning 'copper demon,' because it had a copper color, but yielded no copper. This deceptive connotation is also conveyed in the English cognate terms *Old Nick*, coined in 1643, and meaning the devil, and in *St. Nicholas' Clerks*, a prosaic appellation coined in 1553, and given to clever, roguish highwaymen who cheated and occasionally robbed unsuspecting travelers.

nickpoint: [French *niche* = recess, cranny, hollow, or crevice, especially in rock < Old French *nichier* = to nest; < Latin *nidus* = nest + Old French *point* = mark, place; < Latin *punctus* < *pungere* = to prick; < IE *peuk* = to prick.] n. An abrupt change in angle, such as one produced at the base of a sea cliff by wave erosion. Also, used for an inflection in a stream profile due to lithologic changes, tectonism, or erosion. IE derivatives include: puncture, pugilism, pointillism, pungent, and poignant.
Alt. spelling: nick point

Nicol prism: [Eponymous term after inventor William Nicol of Edinburgh, Scotland.] n. The Nicol prism, used in early petrographic microscopes, consists of a calcite crystal that is cut at an angle into two equal pieces and joined together again with Canada balsam. A beam of light entering the crystal undergoes double refraction. One of these parts, the so-called ordinary ray, undergoes total reflection at the Canada-balsam joint, and exits the crystal at one side. The other ray, the extraordinary ray, passes through the crystal, thus producing polarized light that vibrates in only one direction. Nicol invented the polarizing prism in 1828.

nigrine: [Latin *niger* = black.] n. A variety of the mineral rutile (TiO_2) that contains a small amount of iron that acts as a chromophore to color it black. The African Niger River has the same etymological root, being a 4,000km-long black stream colored by organic material that flows through Guinea, Mali, Niger, and Nigeria.

nilas: [Russian *nilas* = not glossy < *ni* = not + *losk* = glossy, polished, lustrous.] n. A thin, elastic crust of gray-colored ice, formed on a calm sea and displaying a dull matte surface. It bends easily under the pressure of waves and swell, thrusting into a pattern resembling interlocking fingers. Nilas is 10cm or less in thickness. It is subdivided based on appearance into: a) dark nilas (0-5cm thick), with a similar albedo to open water, and b) light nilas (5-10cm thick) with a lighter color and thus a higher albedo. In 1982, the National Climatic Data Center created a scale from 0-9 classifying sea ice by age. Shuga, the youngest ice is ranked at 0, with Nilas ranked next at 1.
Synonyms: frazil, shuga, slush, grease ice

nimbus: [Latin *nimbus* = downpour, shower, bright cloud, splendor surrounding a god; < IE *nebh* = cloud.] n. A rain cloud, especially a low, dark layer of clouds, such as a nimbostratus layer. In art, a nimbus is the halo surrounding the head of a deity or saint. IE derivatives include: nepheline, nebula, and Nibelung. First usage in English c. 1615.
Alt. spelling: nimbi (pl.)

niter: [Middle English *nitre* = sodium carbonate, natron; < Latin *nitrum* < Greek *nitron* = saltpeter; Egyptian *ntr* = niter.] n. A white, gray, or colorless orthorhombic mineral, potassium nitrate, KNO_3. In addition, the term was formerly used for various saline efflorescences, including natron ($Na_2CO_3 \cdot 10H_2O$) and soda niter ($NaNo_3$), but such use is now obsolete. First usage in English c.1400.
Alt. spelling: nitre
See also: saltpeter, natron

node: [Middle English *node* = lump in the flesh; < Latin *nodus* = knot.] n. In seismology, the location along a fault where the displacement appears to change direction. In wave analysis, the point on a standing wave or seiche where the vertical motion is at a minimum, while the horizontal velocity of the wave is greatest. First usage in English c. 1570.

nodule: [Latin *nodulus* = diminutive of *nodus* = knot.] n. In petrology, a fragment of a coarse-grained igneous rock, apparently crystallized at depth, and occurring as an inclusion in an extrusive rock, such as a peridotite nodule in a basalt flow. Also, a small, rounded mineral aggregate, often with a warty surface and having a different composition than the enclosing host rock. In sedimentary rocks such as limestone, chert nodules are common, whereas in volcanic rocks, nodules of chalcedony are common. In such sedimentary volcanic rocks, the nodules have formed later than the host rock. First usage in English c.1595.
Cited usage: It is never found crystallized, but rather, in separate irregular nodules, scattered through other strata [1794, Sullivan, View Nat. I. 439].

nomenclature: [Latin *nomenclator* = slave who accompanied his master to tell him the names of people he met < *nomen* = name; + *clator* = servant, crier < *calare* = to call; < IE *kel* = to shout.] n. The procedure of assigning names to the types, kinds, and common groupings of things, often listed in some form of taxonomic classification. Common groupings listed by taxonomic nomenclature include: organisms, soils, minerals, rocks, stars, and chemicals. First usage in English c. 1605.
Cited usage: To say that there wanteth a term or Nomenclature for it [1626, Bacon, Sylva Sylvarvm 6839].

nonconformity: [Latin *non* = not + *conformare* = to shape after < *con-* < *cum* = with, together + *formare* = to shape < *forma* = shape.] n. A type of unconformity resulting from deposition of sedimentary strata over older plutonic or metamor-phic rock, which had been exposed to erosion prior to deposition of the overlying sediments. First usage in English c. 1615.
See also: disconformity, paraconformity, unconformity

nonpareil: [French *nonpareil* < *non* = without; < Latin *non* = not + French *pareil* = equal; < Latin *par* = equal; < IE *per* = to allot, reciprocate.] n. In general, an adjective meaning without equal, unrivaled, peerless, or unique. In gemology, a large, specially cut precious gem, usually a diamond, placed in a solitary jewelry setting with no smaller supporting stones. In English, originally the name of any of various kinds of small, round, or oval clumps of gelatinized sugar containing a fruit or seed, and used to decorate larger confections; a comfit. First usage in English c. 1425.
Synonyms: solitaire

nontronite: [Toponymous term after French *Nontron*, an arrondissement (county) in France.] n. A pale-yellow to green, iron-rich, dioctahedral phyllosilicate of the clay mineral group, hydrated hydroxy iron aluminum silicate, $Na_{0.3}Fe_2(Al,Si)_4O_{10}(OH)_2 \cdot n(H_2O)$. It is in the smecktite family of expandable clays, and is one of the dominant minerals making up sediments along the mid-ocean ridges. Term proposed by Berthier in 1827.
Cited usage: A large number of names have been proposed for the hydrous silicates of ferric iron; and of these nontronite, chloropal, pinguite, fettbol, and graminite are recognized by most mineralogists as belonging to the species nontronite [1928, Amer. Jrnl. Sci. CCXV. 10].
See also: montmorillonite, beidellite, saponite

norite: [Toponymous term after Norway.] n. An igneous plutonic rock, similar to gabbro in composition, but different from it in that orthopyr-oxene rather than clinopyroxene is the dominant mafic mineral. Term coined c. 1875 by petrologist Harry Rosenbusch.

Cited usage: Norite, under this name Rosenbusch has proposed to group all the older gabbro-like rocks in which any rhombic pyroxene is conjoined with a plagioclase felspar [1885, Geikie, Textbook Geol. 2nd ed. 154].

normal: [Latin *normalis* = made according to the square < *norma* = carpenter's square; IE *gno* = to know.] adj. In structural geology, said of a fault in which the hanging wall has moved downward relative to the footwall, generally as a result of application of tensional force. Such faults are particularly characteristic of divergent plate boundaries and back arc basins. In geometry, a term meaning perpendicular or at right angles. In statistics, referring to a bell-shaped curve that graphically represents a normal distribution. In general English usage, conforming with, or adhering to, a standard pattern, level, or type. IE derivatives include: annotate, narrate, prognosis, notorious, connoisseur, and diagnosis. First usage in English c. 1525.
Cited usage: Normal, the same with Perpendicular, or at Right Angles, and 'tis usually spoken of a Line or a Plane that Intersects another Perpendicularly [1704, J. Harris, Lex. Techn. I].

noselite: [Eponymous term after K. W. Nose, a German geologist.] n. A rare, gray, blue, or brown isometric mineral of the feldspathoid group, sodium aluminium silicate sulphate, $Na_8(AlSiO_4)(SO_4)$. Term coined by Martin H. Klaproth in 1815.
Cited usage: Sodalite, hauyne, and nosean are all silicates of alumina and soda [1879, Rutley, Stud. Rocks].
Alt. spelling: nosean
Synonyms: nosin, nosite

notch: [English *a notch* = alteration by metanalysis of the term plus its article *'an otch'* < French *oche* < Old French *ochier* = to notch.] n. In geomorphology, used in the U.S. Northeast for a narrow pass between mountains or through a ridge. In coastal geology, a deep, narrow cut or hollow along the base of a sea cliff near the high-water mark. In general English usage, any angular or V-shaped indentation or incision in an object. Metanalysis is the transposition of a letter between a noun and its indefinite article. It occurs in many common English words, such as 'a newt' < 'an ewt,' or the reciprocal 'an umpire' < 'a numpire.' First usage in English c. 1575.

notochord: [Greek *noton* = back + *khorde* = gut, entrail.] n. A flexible, rodlike structure that is the chief axial support for the body of the lowest chordates, such as the lancelet and amphioxus. There is an analogous structure in embryos of the higher vertebrates, from which their spinal columns develop. First usage in English c. 1845.

novaculite: [Latin *novacula* = a sharp knife, a razor + *ite* < *lithos* = rock.] n. A hard, dense, evenly textured, siliceous sedimentary rock similar to chert, but characterized by dominance of microcrystalline quartz over cryptocrystalline quartz (chalcedony). There is some disagreement as to whether chalcedony in the chert protolith became coarser-grained during diagenesis or during low-grade thermal metamorphism. The type occurrence is found in Paleozoic beds of the Ouachita Mountains of Arkansas and Oklahoma. Novaculite is used for making sharpening stones (hones and whetstones). First usage in English c. 1795.
Cited usage: In Arkansas, the novaculite used extensively for hones occur[s] in beds referred to this (Millstone soil) epoch [1863, James D. Dana, Manual of Geology 322].

nucleus: [Latin *nucleus* = kernel < *nucula* = little nut < *mux* = nut.] n. In physics, the positively charged center of an atom, composed of protons and neutrons, and possessing most of the mass, but only a small fraction of the volume of an atom.

In biology and paleontology, a large, membrane-bound, nearly spherical protoplasmic organelle within a living cell. The nucleus usually contains the cell's hereditary material, in addition to controlling metabolism, growth, and reproduction. The earliest and most primitive single-celled life forms to appear on Earth, the prokaryotes, lacked a nucleus, whereas the later eukaryotes possessed a nucleus. First usage in English c. 1700.
See also: prokaryote, eukaryote, proton, neutron

nudibranch: [Latin *nudi-* < *nudus* = naked; < IE *nogw* = naked + Greek *brankhia* = gills.] n. Any of various, highly colorful, marine gastropods of the suborder Nudibranchia. They lack a shell and gills, but have fringelike projections that serve as respiratory organs. First usage in English c.1840. IE derivatives include gymnosperm and gymnast.

nuée ardente: [French *nuée ardente* = burning or glowing cloud.] n. A term that describes the complete phenomenon of a basal pyroclastic flow and its overriding ash cloud, typically emitted horizontally from the volcano and travelling at great speed down its flanks. The hot, dense cloud of ash and fragmented lava, suspended in a mass of gas, flows downhill like an avalanche. Term attributed to Alfred Lacroix in 1903, in reference to Mt. Pelée, but Lacroix said that he realized the expression had earlier been used by the inhabitants of San Jorge in the Azores. He also said that by *ardent* he implied a cloud of 'burning' rather than incandescent 'glowing' gases. However the expression *nuee ardente* is usually rendered into English as 'glowing cloud' rather than 'burning cloud.' The eruption of Mt. Saint Helens in Oregon in 1980 produced a nuée ardente. The most famous example of a nuée ardente in historical times is that of Mount Pelée, Martinique in the Greater Antilles, where, on August 30, 1902, a cloud of incandescent ash accompanying a glowing avalanche overwhelmed the town of Saint Pierre, killing over 30,000 people.

Cited usage: The highly viscous lavas of andesitic and trachytic nature might explode subaërially into gas and divided solid material, causing such effects as the 'Nuées Ardentes' of Mt. Pelée [1912, Amer. Jrnl. Sci. CLXXXIV. 413].

nullah: [Hindi *nala* = brook, rivulet, ravine.] n. A sandy-bottomed river channel occupied by intermittent streams, as found in the deserts of India and Pakistan. The term is similar to arroyo, which is a Mexican-Spanish term popular in the Americas.
Cited usage: Here and there are nullahs, with high stiff earthbanks for the passage of rain torrents [1859, Sir R. F. Burton, The Lake Regions of Central Equatorial Africa, with Notices of the Lunar Mountains and the Sources of the White Nile; Being the Results of an Expedition Undertaken under the Patronage of Her Majesty's Government and the Royal Geographical Society of London in the Years 1857-1959 XXIX. 206].
Alt. spelling: nulla, nallah, nalla
See also: arroyo, wadi

nummulite: [Latin *nummus* = coin.] n. Any foraminifer belonging to the family Nummulitidae, characterized by a relatively large, planispiral, and chambered test. Nummulites range from Upper Cretaceous to present and are especially abundant in Eocene limestones, attaining great thicknesses around the Mediterranean basin. The pyramids of Egypt are built of nummulitic limestone. Herodotus (c.300 BCE) mistook the fossils for lentils left over from the meals of the pyramid builders. Piled en masse, the disk-shaped nummulite forams resemble a pile of coins or legumes. Term coined by A. d'Orbigny.
Cited usage: Secondary rocks, composed in great part of grey and greenish sandstone and conglomerate, with some thick beds of nummulitic lime~stone [1833, Lyell, Principles of Geology III. 185].

nunatak: [Inuit *nunatak* = lonely peak.] n. An isolated ridge, peak, hill, or knob of bedrock that projects above the surface of a glacier, and is completely surrounded by ice. Nunataks are generally angular and jagged, due to repeated freeze-thaw cycles. After the ice has retreated, nunatakker contrasts with the rounded contours of the glaciated landscape below. A 'trim line' marks the boundary between the smooth glaciated rock originally below the ice, in contrast to the jagged outcrops above the ice. The Inuit language is spoken by many Eskimo people of Athabascan heritage. The term *Inuit*, meaning people, derives from *inuk*, meaning man.
Cited usage: At Kangerdlugssuak (Greenland), where the mountains are lofty and the sides of the fjords steep, there are three Nunataks equalling the height of the neighbouring land; but above the Inland Ice of Disko the Nunataks are lower Knolls [1877, Q. Jrnl. Geol. Soc. XXXIII. 145].
Alt. spelling: nunatakker (pl), nunatag
Similar terms: horn
See also: pingo

nutation: [Latin *nutare* = frequentative indicating repetition of action < *nuere* = to nod.] n. A slight oscillation of the Earth's axis of rotation, having an 18.6-year period. It is caused by the gravitational effect of the Moon's precessional orbital axis (which is in turn caused by the Sun's gravitational pull). This oscillation is superimposed upon the steady 26,000-year precessional wobble of the Earth's rotational axis. Both of these irregularities are components of the Milankovich cycles. First usage in English c. 1610.
Cited usage: Another nutation arising from another cause may produce all this diversity in the distance of the Pole-Star from the Pole [1702, David Gregory, Astronomica Physicae et geomatrias elementa I. 502].
See also: frequentative

oasis: [Toponymous term from *Oasis*, a popular fertile site located in the Libyan desert; < Greek *Oasis* < Coptic *ouahe* < Egyptian Arabic *wah* = dwelling-place, oasis < *ouih* = to dwell.] n. A fertile, vegetated area surrounded by desert where the water table has come close enough to the surface for either natural springs to exist or wells to be dug, thus making it suitable for a dwelling place. Due to its adoption into the English language, the term oasis has largely changed in pronunciation from the traditional short *a* to the now familiar long *a* sound. First usage in English c. 1610.
Cited usage: Near it is a model of the pyramids, accompanied by an oasis with its grove of palms, and a caravan of camels [1816, J. Scott, Visiting Paris (ed. 5) 239].

ob-: [Latin prefix *ob-* = in the direction of, towards, against, in the way of, in front of, in view of, on account of.] Combining prefix meaning a) in the direction of or towards, as in obduction; b) meaning against, as in obsequent and obstinate; or c) meaning in the way of or in front of, as in obstruction, obvious and obnoxious.

obduction: [Latin *obducere* = to draw over, cover over < *ob-* = in the direction of, towards, against + *ducere* = to lead, draw.] n. The process of overthrusting of oceanic crust onto the leading edge of a continental plate. Term coined in 1971 by American geologist Robert G. Coleman.

oblate: [Latin *oblatus* < *ob-* = toward + (pro)latus = to stretch out < *pro* = forth + *latus* = brought.] adj. In geometry, referring to the shape of a spheroid generated by rotating an ellipse about its shorter axis, such that the equitorial diameter is greater than the distance between the poles. Earth is an example of an oblate spheroid, being slightly compressed or flattened at the poles. First usage in English c. 1700.

oblique: [Latin *ob-* = in the direction of, towards, against + *licinus* = bent upward.] adj. Having a slanting, inclined, or sloping direction, course, or position. In structural geology, fault movement that is intermediate between dip-slip and strike-slip is called oblique-slip. In mathematics, designating geometric lines or planes that are neither parallel nor perpendicular. First usage in English c. 1425.

obsequent: [Latin *ob-* = against + *sequi* = to follow.] adj. In geomorphology, primarily said of a stream or drainage system whose course or direction is opposite to that of the original consequent direction, and thus opposite to the dip of the local strata. Also, said of a geologic or topographic feature that does not resemble the original consequent feature from which it developed. An example being a ridge of basalt located where a lava flow originally filled a stream valley, but became a ridge or inverted valley by more rapid differential erosion of the stream banks. Term coined in 1895 by American geomorphologist William Morris Davis.
See also: post-, consequent, anteconsequent, antecedent

obsidian: [Eponymous term after Obsius, a Roman who supposedly discovered this rock in Ethiopia.] n. A black or brownish-red, silica-rich volcanic glass, usually of rhyolitic composition displaying a characteristic conchoidal fracture. The early printed editions of *Historis Naturalis* (*Natural History*) by Pliny the Elder erroneously named the discoverer Obsidius, and therefore the rock *obsidianus lapis*, instead of *obsianus lapis*. The erroneously named term was used by medieval natural history writers, and passed on into French as *obsidiane*. First usage in English c. 1375.

Cited usage: There may be ranged among the kinds of glasses, those which they call Obsidiana, for that they carry some resemblance of that stone, which one Obsidius found in Aethyopia [1601, Holland, Pliny II. 598].

See also: holohyaline, pumice

occlude: [Latin *occludere* = to shut up < *oc-* = variation of *ob-* = against + *claudere* = to close.] v. In general usage, to block or close a passage, or to prevent the passage of, often with regard to light and vision. In meteorology, to force air upward from the Earth's surface, as when a cold front overtakes and flows under a warm front. In chemistry, to absorb and retain a substance within the crystal lattice of another substance, as in the retention of liquid in a zeolite. First usage in English c. 1595.

ocean: [Eponymous term after *Okeanos*, the god of the oceans in Greek mythology, personified by Homer as 'the god of great primeval water', and manifested as a great river encircling the Earth.] n. The entire body of saltwater that surrounds the continents and covers more than 70% of the Earth's surface. Okeanus was the son of Uranus and Gaia. He was the brother and husband of Tethys, titaness of the sea, hence the name Tethys Sea, given by the Greeks to the great outer sea (the modern world ocean), being all the seas of the Earth lying beyond the inner, or local, Mediterranean Sea. First usage in English c. 1275.

Cited usage: The ocean, is that generall collection of all waters, which environeth the world on every side [1635, Swan, Spec. M. vi. 2 187].

See also: Gaia hypothesis

oceanography: [Greek *okeanos* = ocean + Latin *graphicus* < Greek *graphikos* < *graphe* = writing <

graphein = to write; < IE *gerbh* = to scratch.] n. The study of the world ocean, including the physical, biological, chemical, and geological properties of the ocean, and the resultant phenomena which occur.

See also: ocean, Mediterranean

oceanology: [Greek *okeanos* = ocean + *logos* = word, speech, discourse, reason < *legein* = to speak, to say; IE *leg* = to collect.] n. A seldom used synonym for oceanography, coined in the mid-nineteenth century and previously intended to have an even broader meaning.

Cited usage: The use of the words Oceanology and Oceanologist was well and truly pondered in days past. Today they are little used save in Russia [1955, Deep-Sea Research II. 247].

ocellar: [Latin diminutive of *oculus* = eye; < IE *ok* = to see.] adj. In petrology, said of the texture in igneous rocks in which radiating crystals of one mineral are arranged in aggregates around larger, euhedral crystals of another mineral, resulting in an eyelike display. A common example is the radial arrangement of minute crystals of biotite or acmite around larger crystals of nepheline or leucite. In paleontology, pertaining to ocelli, the minute simple eyes of an insect or other arthropod. Term coined by petrologist Harry Rosenbusch in 1887. Some authors regard the ocellar structure as simply a variety of the centric structure described by Friedrich Johann Karl Becke in 1878. IE derivatives include: augen-schist, phlogopite, ocular, diopter, triceratops, and pinochle.

Cited usage: The structures which specially distinguish these granophyric rocks are the centric or ocellar structure and the drusy or miarolitic structures [1889, Judd, in Q. Jrnl. Geol. Soc. May 176].

ochre: [Latin *ochra* < Greek *okhra* = yellow earth < *okhros* = pale yellow.] n. Any of several earthy oxides of iron, and sometimes other metallic oxides, usually hydrated and mixed with varying amounts of clay. They occur in various colors, and are widely used as pigments. Examples named by color include: yellow and brown ochre (limonite), red ochre (hematite), black ochre (manganese oxide or pyrolusite), and white ochre (china clay); whereas examples named by element are: antimony ochre (stibiconite), lead ochre (massicot and litharge), and tungsten ochre (tungstite). First usage in English c. 1375.

Cited usage: Yellow Ochre is sometimes called Oxford Ochre, being abundant in that neighbourhood [1854, T. H. Fielding, Painting in Oil Water Color (ed. 5) 179].

Alt. spelling: ocher

octa-: [Greek *octo-* = eight.] Combining prefix meaning eight, e.g., octahedral, octogenarian, octahedrite, and octopus.

odometer: [Greek *hodometron* = odometer < *hodos* = journey, way + *metron* = measure.] n. An instrument for measuring distances traversed, e.g. a pedometer can be considered a type of odometer. First usage in English c. 1790

oghurd: [Uncertain origin (possibly) Arabic *oghurd* = star.] adj. Referring to star-shaped sand dunes formed by radially inblowing wind converging as vertical drafts, possibly created by rising air that creates ascending vortices with vertical axes. This ascending spiral motion resembles the Langmuir circulation cells found in the surface layer of the ocean affected by wind. The term is also used in the Sahara Desert for a mountainous dune built upon an underlying, elevated bedrock feature, and towers above the level of surrounding dunes. Some oghurd dunes are stationary for periods of time long enough for them to have long-lived geographic names, such as the star dunes of the Grand Erg Oriental at the junction of Algeria, Tunisia, and Libya.

ogive: [Middle English *oggif* = stone comprising an arch; < French *augive* = diagonal rib of a vault, to resist; (possibly) < Latin *augere* = to augment, increase.] n. A dark, arcuate band of accumulated material stretching from laterally across the surface of a glacier, being arched in the direction of flow due to greater flow velocity in the middle. They are often found in multiple, parallel series of similar bands. The etymological connection stems from the glacial bands that, in plan view, appear like the diagonal ribs stretching across the vaulted ceiling of a Gothic cathedral. These diagonal ribs are termed arc ogive; the other principal ribs involved in forming this distinct gothic architectural pattern are the transverse ribs (arc doubleau) and the wall ribs (foe'meret).

Cited usage: Vareschi's latest and most important researches were made upon various systems of banding which appear in the tongue of the Great Aletsch Glacier, in particular what are locally called Ogiven. These are curved bands visible on the glacier surface, often in regular longitudinal series. The ogive itself is generally darker block ice, whilst between one ogive and the next is paler and higher Buckel ice [1947, Jrnl. Glaciology I.327].

Alt. spelling: ogiven (pl.)
Synonyms: Forbes's Bands

oikocryst: [Greek *oikos* = house + *krustallos* = crystal; <IE *kreus* = begin to freeze, form a crust.] n. In a poikilitic rock texture, an enclosing crystal that creates a confining houselike structure around an inner crystal. IE derivatives include: crustacean, crust, and crouton.

See also: ubac

ojo: [Spanish *ojo* = eye.] n. In geomorphology, used in the U.S. Southwest for a small lake or pond. In ore deposits, often used in names of various clumped ore deposits, such as the El Ojo Deposit in Argentina, and the Ojo Caliente in Zacatecas, Mexico.

olenellid: [Eponymous term after *Olenus*, the Roman man who was turned to stone, as told in Ovid's *Metamorphoses*.] n. A trilobite belonging to the primitive Lower Cambrian family Olenellidae. Term coined by E. Billings in 1861, based on the Greek myth of Olenus, who lived with his wife on Mt. Ida. The wife claimed she surpassed any goddess in beauty. For this assumption they were both turned to stone. The trilobite faunas of the early Cambrian can be divided into two main regional groups: the olenellid fauna found in northwestern Europe and North America and the redlichiids in Asia, Australia, and North Africa.

oligo-: [Greek *oligos* = little, few.] Combining prefix meaning few or little, as in Oligocene (little recent), oligotrophic (few nutrients), oligomictic (little mixing), and oligoclase (little fracture).
Alt. spelling: olig-
See also: oligoclase

Oligocene: [Greek *oligon* = very little + *kainos* = recent.] adj. The third epoch of the Tertiary Period (33.9-23.0Ma). Term coined in 1833 by Charles Lyell, who devised a scheme for stratigraphically subdividing the entire Tertiary Period. His divisions were based on the percentage of fossil molluscs found in ancient strata that are represented by creatures still alive today. The oldest Tertiary rocks contain the fewest species living today, whereas the youngest would have proportionally more. In his division of the Tertiary, Lyell created the Eocene (dawn of the recent), Oligocene (little recent), Miocene (middle ages of the recent), Pliocene (nearly recent), and Newer Pliocene, which he later called the Pleistocene (most recent). The Paleocene was added later, first as the lower part of the Eocene, and finally as a distinct epoch.

oligomictic: [Greek *oligos* = little, few + *miktos* = mixed < *mig* = to mix.] adj. In petrology, said of

rocks consisting of one to two dominant minerals (those composed of several minerals are polymictic). In limnology, said of a lake that exhibits stable thermal stratification, only rarely undergoing mixing or overturning.
Cited usage: Oligomictic rocks are the characteristic deposits of epicontinental seas and are found rarely in geosynclinal depressions, whereas polymictic rocks are characteristic of geosynclinal regions [1949, F. J. Pettijohn, Sedimentary Rocks xiv. 438 Schwetzoff].

oligotrophic: [Greek *oligos* = little, few + *trophos* = nourishment, feeder.] adj. Said of a body of water that is deficient in basic nutrients needed by autotrophs to perform photosynthesis. In the case of a lake, the term also implies an abundance of oxygen in the deeper hypolimnion. First usage in English c. 1925.
Cited usage: The typical oligotrophic lakes are deep, with submerged beaches narrow or absent, inconsiderable or no littoral vegetation, and an indistinct littoral zonation [1928, Proceedings Linnaean Soc. CXL 100].

olistostrome: [Greek *olistomai* = to slide + *stroma* = bed.] n. A sedimentary deposit that consists of a chaotic mass of rock and also contains large clasts composed of material older than the enclosing sedimentary sequence. The clasts may be gigantic and are then called olistoliths. Such deposits are generally formed by submarine gravity sliding or slumping of unconsolidated sediments. Term coined by G. Flores in 1955.
Cited usage: By Olistostromes we define those sedimentary deposits occurring within normal geologic sequences that are sufficiently continuous to be mappable, and that are characterized by lithologically and/or petrographically heterogeneous materials [1956, G. Flores, Proc. 4th World Petroleum Congress].
Synonyms: sedimentary melange

olivine: [English *olive* (from its color) < Latin *oliva* = olive green.] n. An olive-green, grayish-green or brownish, orthorhombic mineral, magnesium iron silicate, $(Mg,Fe)_2SiO_4$. Endmembers of the isomorphous series in the olivine family are forsterite and fayalite. Olivine occurs in igneous, mafic, and ultramafic rocks, as well as in meteorites. Gem quality crystals of olivine are called peridot. The island where olivine was extracted in ancient times was well guarded by the Egyptians, therefore the stone was named by the Greeks as *laiggourios*, meaning guarded stone. But the term allowed for a play on words with *oureo*, meaning to urinate, and *lunx*, meaning the lynx cat, hence the legend that this stone was the solidified urine of the lynx. First usage in English c. 1790.
Synonyms: chrysolite
See also: chrysolite, forsterite, fayalite, peridot

ombro-: [Greek *ombros* = rain, wet, shade.] Combining prefix meaning rain or wet; as in ombrogenous, ombrophilic, or ombrophobic. The corresponding prefix in Latin is *pluvi*, meaning to rain, as in pluvial lake.
See also: pluvial

ontogeny: [Greek *ont* = being < *einai* = to be + *genea* = generation, race < *genos* = race, kind < *gonos* = birth, offspring; < IE *gen* = to give birth, beget.] n. The origin and development of an individual organism from embryo to adult. Term coined in 1866 by Ernst Haeckel as part of his biogenetic theory, in which he inspired wide usage of the term with the euphonic phrase "ontogeny recapitulates phylogeny." Despite Haeckel's tarnished reputation, his "politics is applied biology" maxim later being embraced by the Aryan Nazis, Stephen Jay Gould resurrected the atheist Haeckel's ladder theory of evolution in 1977 with his book *Ontology and Phylogeny*.
Cited usage: The ontogeny of every organism repeats in brief its 'phylogeny', i.e. the individual development of every organism repeats approximately the development of its race [1872, Microscopy Jrnl. July 185].
See also: recapitulation, phylogeny, ecology

onyx: [French *onique* = onyx-stone; < Greek *onux* = (finger)nail, onyx; < IE *nogh* = nail, claw.] n. A variety of chalcedony (cryptocrystalline quartz), similar to agate, consisting of banded planar layers of different colors. Certain types of onyx are pink with white streaks, resembling the color and layering of fingernails. The term now applies mainly to the banded, deeply colored black and light variety of chalcedony. In Greek mythology, while the goddess Aphrodite (Roman goddess Venus) was asleep, the god Eros (Roman god Cupid) playfully cut her fingernails with his arrow and let the clippings fall into the waters of the Indus river. Because even the smallest part of the body of an immortal cannot die, Aphrodites' nail clippings lived on, transformed into onyx by the three Fates (Clothos, Lachesis, and Atropos). First usage in English c. 1275.

oolith: [Old English *oeg* < Greek *oion* = egg + *lithos* = rock, stone.] n. A subspherical, sand-sized, accretionary sedimentary particle resembling fish roe. Ooliths display concentric layers of calcium carbonate that have deposited in shallow, wave-agitated water around tiny nucleii of mineral grains, fecal pellets, or other substances. Deposits of ooliths aggregate together and become lithified to form concretionary limestone called oolite. A well-known oolitic limestone series of Jurassic age occurs in England between the Chalk and the Lias members. It is subdivided into the Portland Oolite (Upper), the Oxford Oolite (Middle), and the Bath Oolite (Lower). First usage in English c. 1785.
Cited usage: The grains are called ooliths and the rock containing them oolite or oolitic limestone. Ooliths generally show a series of

concentric coats of calcareous material in which a radiating crystalline structure can often be made out [1926, G. W. Tyrrell, Princ. Petrol. xiii. 227].
Synonyms: ooid
Similar terms: oolite, roe stone
See also: pisolite, oolite

ooze: [Middle English *wose*: < Old English *wase* = mud.] n. A layer of mudlike sediment on the floor of oceans and lakes, composed chiefly of the remains of microscopic zooplankton, often lithifying into calcareous or siliceous sedimentary rock. First usage in English before 1000.

opacite: [French *opacite* < Latin *opacitas* = darkness < *opacus* = dark + Greek *-ites* < *-lite* = pertaining to rocks < *lithos* = rock, stone.] n. A term applied to myriad microscopic, opaque mineral grains and flakes that develop as rims on biotite and hornblende phenocrysts in volcanic rocks, as a result of post-eruption oxidation and dehydration. They are composed chiefly of magnetite dust. Term coined in 1872 by Vogelsang.
Cited usage: Opacite, a name proposed by Vogelsang for the black opaque scales or grains which cannot be identified with magnetite, menaccanite, or any other mineral [1880, Dana's Min. Appendix ii. 42].

opal: [Greek *opallios* = opal; < Sanskrit *upala* = gem, precious stone; < IE *upo* = under, up from under.] n. A variously-colored, amorphous mineraloid, hydrated silicon dioxide, $SiO_2 \cdot nH_2O$. It differs from quartz in that it is isotropic, has a lower refractive index, is softer and less dense. Scanning electron microscope images show that certain precious varieties of opal are composed of layers of tightly spaced submicroscopic spheres of cristobalite. These varieties show a complex play of color called 'fire', and are highly regarded as gems despite their softness and instability upon dehydration. The IE *upo*, meaning up from

under, relates to the way light seems to emanate from deep within the stone, as in fire opal. Opal was first recognized as a precious gem in India before being brought back to Europe. One of the more fanciful powers attributed to opal in ancient times was the belief that, when carried on the person wrapped in a bay-leaf, it conferred invisibility. An IE derivative is the prefix hypo- meaning under.
Cited usage: In the Opal you shal see the burning fire of the Carbuncle or Ruby, the glorious purple of the Amethyst, the greene sea of the Emeraud, and all glittering together [1601, Holland, Pliny II. 614].

operculum: [Latin *operire* = to cover < *ob-* = towards, against, in the way of, in front of + *aperire* = to open.] n. The usually calcareous plate secreted by some gastropods and other molluscs, which serves to close the aperture of their shell when the animal retracts.
Cited usage: Most spiral shells have an operculum, or lid, with which to close the aperture when they withdraw for shelter [1856, Woodward, Mollusca 47].

ophicalcite: [Greek *ophis* = serpent + Latin *calcem, calx* = lime, limestone, pebble; < Greek *khalix* = pebble.] n. A recrystallized metamorphic rock composed of a mixture of serpentine and crystalline limestone.
Cited usage: A beautiful variety of ophicalcite or serpentine-marble [1875, Dawson, Dawn of Life vi. 147].
Synonyms: calcitic ophiolite, serpentine marble

ophiolite: [Greek *ophis* = snake, serpent + *lithos* = rock, stone.] n. A group of mafic and ultramafic igneous and metamorphic rocks, which are enriched in ferric and magnesium minerals, such as serpentine, chlorite and epidote. Ophiolites contain mantle material brought to the sea floor crust in the forms of basalt, gabbro, and peridotite.

The origin of ophiolite was originally associated with early phase development of a geosyncline. Term coined by G. Steinman in 1905.
Cited usage: The whole assemblage of sea-floor basalts and associated upper-mantle peridotites is known as the ophiolite association [1977, A. Hallam, Planet Earth 166].

ophitic: [Greek *ophis* = snake, serpent + *lithos* = rock, stone.] adj. Said of various eruptive or metamorphic rocks, such as serpentinite and serpentine marble, which are usually green, and have spots or markings resembling a serpent.
Cited usage: The names Serpentine, Ophite, Lapis colubrinus, allude to the green serpent-like cloudings of the serpentine marble [1868, James D. Dana, A System of Mineralogy 5th ed. 468].

opisthobranch: [Greek *opistho-* = behind + *brankhia* = gills.] n. Any of various marine gastropod molluscs of the subclass Opisthobranchia. They are characterized by a set of gills located behind the heart, hence the name. Also, they generally have a reduced shell, or in some cases, the shell being entirely absent. Another distinguishing feature is their two pairs of tentacles.

order: [Latin *ordo* = order, row, series; < IE *ar* = to fit together.] n. In the taxonomic classification of living creatures, the grouping smaller than class and larger than family. In the post-Linnaean natural order of plants, an order is a group of genera or families showing a common basic structure; this is a refinement of Linnaeus's system that only needed a single common trait, usually related to reproductive apparatus. IE derivatives include: army, harmony, aristocracy, arithmetic, and rhyme.

Ordovician: [Celtic *Ordovices* = those who fight with hammers; < *ordo* = hammer + *wik* = to fight, conquer.] adj. Referring to the second period of the Paleozoic era, from 488.3-443.7Ma, after the Cambrian and before the Silurian. The term is derived from the ancient tribe inhabiting Wales, the Ordovices, who were the northern neighbors of the Silures and the southern neighbors of the Degeangli. Their resistance to Roman supremacy was finally vanquished by Agricola in 78 CE, their territory then becoming a *civitas* of the Roman Empire. Term coined by C. Lapworth in 1879.
Cited usage: The whole of the great Bala district where Sedgwick first worked out the physical succession among the rocks of the intermediates or so-called Upper Cambrian or Lower Silurian system lay within the territory of the Ordovices; a tribe as undaunted in its resistance to the Romans as the Silures. Here, then, we have the hint for the appropriate title for the central system of the Lower Palaeozoics. It should be called the Ordovician System. [1879, C. Lapworth, Geol. Mag. Decade II. VI. 14].

ore: [Old English *ar* = copper, brass, bronze, unwrought metal; Latin *aera* = counters < *aes* = piece of metal, brass, money.] n. A native mineral containing a precious or useful metal or non-metallic compound in such quantity and in such chemical combination as to make its extraction profitable.

orient: [Latin *oriens* = rising sun, east < *oriri* = to arise, be born.] n. That part of the Earth's surface situated to the east of some recognized point of reference. Also used to describe the eastern part of a country, eastern countries in general, and the countries of Asia in particular. The term has been associated with racist terminology by some, especially when referring to people as "Orientals."

oro-: [Greek *oros* = mountain.] Combining prefix meaning mountain, used in such terms as: orogeny and orocline.

orocline: [Greek *oros* = mountain + *klinein* = to incline, bend.] n. A linear orogenic belt that has been sharply bent in plan view due to crustal deformation. Term coined by S.W. Carey in 1955. *Cited usage*: For an orogenic system which has been flexed in plan to a horse-shoe or elbow shape, the name orocline is proposed [1955, S. W. Carey, Papers & Proc. R. Soc. Tasmania LXXXIX. 257].

orogeny: [Greek *oros* = mountain + *genesis* = origin, creation < *genesthai* = to be born, come into being; < IE *gen* = to give birth, beget.] n. The process of mountain building and formation, especially by folding and faulting of the Earth's crust. IE derivatives include: genus, gender, genius, germ, nature, connate, native, pregnant, and naïve.

orpiment: [Latin *auripigmentum* = orpiment < *aurum* = gold + *pigmentum* = pigment.] n. A bright yellow mineral, arsenic trisulfide, As_2S_3. Orpiment is found native in soft masses and resembles the element gold in color. Also called yellow orpiment to distinguish it from red orpiment (realgar, As_2S_2). Both minerals are used as pigments, especially in the fine arts.
Synonyms: yellow orpiment, King's Yellow, yellow arsenic
See also: realgar

orrery: [Eponymous term after Charles Boyle, Earl of Orrery.] n. A mechanized model made to represent the relative motions of the planets and their moons about the sun by means of mechanical clockwork. In the hierarchical times of 17th-century England, inventors and artisans were often obliged to name their creations after rich, noble, and eminent countrymen. So it was with the model invented by George Graham and crafted by J. Rowley around 1700, that they felt compelled to name it after the Earl of Orrery.

Cited usage: Mr. John Rowley calls his Machine the Orrery, in Gratitude to the Nobleman of that Title [1713, Steele, Englishm. No. 11].

ortho-: [Greek *ortho* = straight, right.] Combining prefix meaning straight, right, correct, or at a right angle. This prefix has stirred up confusion because *ortho-* has both a physical and an ethical sense. In ethics, straight can be synonymous with right and correct, as in 'straight shooter' or 'straight and narrow.' In a physical sense, straight and right have completely different meanings. The evolution of this duality of meaning allows for the term orthogonal to mean either at right angles, or in a straight line. For example, orthoclase has cleavage at right angles, but orthokinesis means movement in a straight line. This dichotomy of meaning is expressed in the term orthogeosyncline, which is a linear feature between a kraton and oceanic kratogen, at the same time having repeated folds at right angles to the synclinal plane.

orthoclase: [Greek *ortho-* = straight, right + *klasis* = a breaking < *klan* = to break.] n. A cream, pinkish, or gray monoclinic mineral of the feldspar group, essentially potassium aluminum silicate, $KAlSi_3O_8$. Orthoclase is polymorphous with sanidine and microcline, and is common in granites, granodiorites, and syenites. The name alludes to the two planes of cleavage oriented at right angles to one another. First usage in English c. 1845.
Similar terms: potassium feldspar
See also: sanidine, microcline

orthogenesis: [Greek *ortho* = straight, right + *genesis* = origin, creation < *genesthai* = to be born, come into being; < IE *gen* = to give birth, beget.] n. In evolution, a specific trend within a group of organisms that continues in the same way for multiple generations, unaltered by outward influence. Orthogenesis appears to be independent of external factors, such as natural selection. In anthropology, the theory that all cultures pass through sequential periods in the same order. The orthogenetics theory mirrors the Lamarckian view that the evolution of a species is influenced most strongly by internal factors not subject to natural selection. In this term, the prefix *ortho-* means straight as it refers to the straight-line sequential lineage of evolution. First usage in English c. 1890.

orthopyroxene: [Greek *orthos* = straight, right + *pyros* = fire + *xenos* = stranger.] n. Any mineral in the pyroxene group that crystallizes in the orthorhombic system. They are usually void in calcium and have little aluminum, such as enstatite and hypersthene. As the term is restricted to minerals in the orthorhombic system, the use of the prefix *ortho-* refers to the crystallographic axes oriented at right angles to each other.
See also: pyroxene, clinopyroxene

orthorhombic: [Greek *orthos* = straight, right + *rhombus* = flatfish, magician's circle.] adj. Of the six crystal systems, said of that system in which the three crystallographic axes are at right angles to each other, but are of different lengths. An orthorhombus is a special case of a rhombus, being an equilateral parallelogram in which all the sides are at right angles to each other. First usage in English c. 1865
Similar terms: rectangular, prismatic, trimetric, orthosymmetric

orycto-: [Greek *oryktos* = dug up.] Combining prefix referring to fossils. The term had a broad meaning in the 18th century, referring to anything dug up or excavated, including minerals and artifacts as well as fossils. Used in such terms as: oryctology, oryctofacies, oryctophile, and oryctozoology.
Cited usage: Oryctology is the part of physics which treats of fossils [1753, Chambers Cyclopedia Suppl.].

os: [Swedish *as* = esker.] n. An elongated narrow ridge of gravel. Os are common in Sweden, there

sometimes extending over 160km long. They are usually formed by a stream flowing through a tunnel in a continental glacier.
Alt. spelling: osar, as
Synonyms: as (Swedish), esker (Irish), kame (Scottish)
Similar terms: os, esker, kame

osmosis: [Greek *osmos* = thrust, push < *othein* = to push.] n. Diffusion of fluid through a semipermeable membrane from a relatively dilute solution with a low solute concentration to a solution with a higher solute concentration until there is an equal concentration of fluid on both sides of the membrane. First usage in English c. 1865.

ostracod: [Greek *ostrakodos* = testaceous, hard-shelled < *ostrakon* = shell, earthen vessel, potsherd; < IE *ost* = bone.] n. Any of various minute, chiefly freshwater crustaceans of the subclass Ostracoda, shaped like a shrimp, and having a body enclosed in a bivalve carapace. Etymologically related terms include: oyster, teleost, osteo-, and ostracize. First usage in English c. 1860.

otolith: [Greek *ot* = ear + *lithos* = rock, stone.] n. The calcareous precipitate bone found in the inner ear of some vertebrate animals, and occasionally found in invertebrates. They are often in the shape of rhombic crystals and can become quite large in certain species of fish and cetaceans. First usage in English c. 1830.
Cited usage: The otoliths are two small rounded bodies, consisting of a mass of minute crystalline grains of carbonate of lime, held together in a mesh of delicate fibrous tissue [1883, H. Gray, Gray's Anatomy 10th ed. 618].
Synonyms: statolith

oule: [Spanish *olla* = pot, kettle.] n. A glacial term originating in the Pyrenees of Spain for an ice-cut, recessed, half-bowl-shaped hollow, often containing a lake.
Synonyms: cirque

outlier: [English *out* < Old English *ut* = out + English *lie* < Middle English *lien* < Old English *licgan* = to lie, lay.] n. A rock mass in the position where it originally lithified as part of a laterally continuous sedimentary formation, but is now situated at a distance from the main body due to localized erosion and denudation of intervening rock. An outlier is an isolated mass surrounded by older rocks. In statistics, an outlier is a data point lying well outside the expected range. First usage in English c. 1605.
Cited usage: When a portion of a stratum occurs at some distance detached from the general mass some practical mineral surveyors call it an outlier, and the term is adopted in geological language [1833, Lyell, Principles of Geology III. Gloss. 76].
See also: inlier

oxbow: [English *oxbow* = The bow-shaped piece of wood that forms a collar for a yoked ox and has its upper ends fastened to the yoke.] n. Referring to a curved lake that was originally formed as a meander of an adjacent river. An oxbow lake is created after a river changes its course and cuts through the narrow neck of a meander. Eventually, siltation causes the meander to be isolated as a lake. First usage in English c. 1750.
Cited usage: If a flood occurs when only a narrow neck of land is left between adjoining loops, the momentum of the increased flow is likely to carry the stream across the neck. A deserted channel is left, forming an ox-bow lake which soon degenerates into a swamp [1944, Arthur Holmes, Princ. Physical Geol. x. 165].
Synonyms: mortlake
See also: mortlake

oxide: [Greek *oxus* = sharp, acid; < IE *ak* = sharp.] n. A compound of oxygen with another element or radical. Also used as a combining form for such terms as: hydroxide, peroxide, dioxide, and sesquioxide. The historical reference to sharp

refers to oxygen being named by Antoine Laurent Lavoisier as the principal element forming acids, characterized by a sharp taste. IE derivatives include: acid, vinegar, egg, acupuncture, acrid, acerbic, and ester. First usage in English c. 1785.
See also: sesquioxide, hydroxide, oxygen

oxisol: [Greek *oxygen* = acidifying < *oxus* = sharp, keen, acute, acid + *sol* = soil.] n. A type of stable, highly weathered mineral soil found in tropical regions, having an oxic soil horizon within two meters of the surface. First usage in English c. 1955.
Cited usage: The Oxisols include the soils that, in recent years, have been called Latosols, and many, if not most, of those that have been called Ground-Water Laterite soils. All soils that have oxic horizons are included in the order [1960, Soil Classification: 7th Approximation (U.S.D.A.) xvi. 238/1].

oxy-: [Greek *oxus* = sharp, keen, acute, acid.] Combining base or prefix meaning the presence of oxygen, sharp in taste, or simply sharp. *Oxy-* is the combining form used chiefly in scientific terms, and usually refers to oxygen, such as in: oxysphere, deoxyribonucleic acid, and oxisol. The root is also used in non-oxygen-related terms meaning sharp, such as: *oxyrhinous* meaning sharp-nosed, *Oxystomata*, meaning division of crabs with sharp, projecting mouths, and *oxymoron*, meaning pointedly foolish.

oxygen: [Greek *oxus* = sharp, keen, acute, acid + *genesis* = origin, creation < *genesthai* = to be born, come into being; < IE *gen* = to give birth, beget.] n. A nonmetallic gaseous element constituting 21% of the Earth's atmosphere by volume, and naturally occurring as the diatomic gas O_2 (symbol O). It combines with most elements, is essential for plant and animal respiration, and is required for nearly all combustion processes. Oxygen was named in 1785 by Antoine Laurent Lavoisier,

because he considered it to be the principal
element in the formation of acids, which have a
sharp taste.
Cited usage: Lavoisier endeavoured to show
that vegetable and other matters consist of air,
charcoal, and inflammable gas, or, in his language,
oxygene, carbone, and hydrogene [1789, J. Keir,
Dict. of Chemistry].

ozocerite: [Greek *ozokeros* = odiferous wax <
ozein = to smell + *keros* = bees-wax.] n. In
sedimentology and petroleum geology, a brown to
jet-black paraffin wax consisting of a mixture of
natural hydrocarbons occurring in some bituminous shales and sandstones. Ozocerite usually
occurs as thin stringers and veins that fill rock
fractures in areas of mountain building. In paleontology, a non-petroleum-based, waxlike fossil
resin having a brown to yellowish color and
distinctive aromatic odor. Term coined by Meyer c.
1835, after discovering it in Moldavia.
Cited usage: Ozokerite, a variety of black
bitumen lately discovered by Meyer [1837, James
D. Dana, A System of Mineralolgy 441].
Alt. spelling: ozokerite
Similar terms: native paraffin, mineral tallow,
mineral wax, earth wax
See also: paraffin

padang: [Malay *padang* = big field.] n. An open grassy space; a field.

padparadscha: [Sinhalese *padmaragaya* = lotus blossom color.] n. A pinkish-orange to reddish-yellow variety of sapphire. Sinhalese is the native language of Sri Lanka (formerly Ceylon), a major producer of precious gems. The word tourmaline is also derived from Sinhalese.
Cited usage: Ever notice how gemologists and dealers toss the term padparadscha around like some kinda überrock. If the gem isn't buff enough to make the centerfold of Gem & Gemology, it is sniffed at as a lesser being unworthy of the name [1984, Peter Bancroft, Gem & Crystal Treasures, Western Enterprises and The Mineralogical Record].
Synonyms: padmaragaya
See also: tourmaline

paha: [Dakota *paha* = turtle hill.] n. A small dome-shaped mound left after late glacial melting, composed chiefly of drift, but capped by loess. Also, a Malaysian term as a unit of weight for measuring gold, equal to about 9 grams (0.33 ounces).
Similar terms: drumlin

pahoehoe: [Hawaiian *pa* = nominative prefix + *hoehoe* = reduplication of *hoe* = to paddle.] n. Hawaiian term for a type of basaltic lava that is characterized by a smooth, ropy, billowy surface. The name probably alludes to the wavy swirls and marbling seen on the lava's surface. There are several varieties of pahoehoe lava whose names conjure up useful visual images, such as: corded, elephant-hide, entrail, festooned, filamented, sharkskin, shelly, and slab pahoehoe pahoehoe. The widespread use of the term pahoehoe has rendered the synonym dermolith obsolete. Pahoehoe is contrasted to clinkery to blocky-surfaced aa lava. First usage in English c. 1855.
Cited usage: From this a stream of the smooth satin-like lava called 'pahoehoe' in Hawaii flowed for a few hours [1869, Q. Jrnl. Geol. Soc. XXV]. One is pahoehoe–a taffy-like lava that has hardened into folds and creases that give it a smooth, ropy look, like frosting that has spilled over the top of a cake [1972, Islander (Victoria, B.C.) 24 Sept. 4/2].
Synonyms: ropy lava, dermolith (obs.)
See also: aa

pakihi: [Maori *pakihi* = barren soil < *paki* = open land.] n. The type of waterlogged, barren soil associated with open, swampy lands. Pakihi is characteristic of the northwestern parts of South Island, New Zealand. The mild, humid and wet climate encourages growth, but the high moisture content of the soils limits soil organisms and retards decomposition. Thus, organic matter tends to accumulate in a peaty layer above the compact, older mineral subsoils, leading to waterlogging and acid leaching of essential nutrients, such as calcium and nitrogen. Along the New Zealand coast, pakihi soils are an important agricultural asset, yet to the naïve visitor the often unkempt appearance of pakihi-soil farmland suggests impoverished or poorly managed land.
Cited usage: The 33 million acres of sour and barren 'pakihi' soil on the West Coast of New Zealand [1970, N.Z. Listener, 7 Dec. 6/3].

palagonite: [Toponymous term after *Palagonia*, a volcanic area in Sicily.] n. Hydrothermally altered tachylyte (basaltic glass). It is often formed as a result of subaqeous eruptions, and is common in the subglacial thuyas of iceland as palagonite tuff. It is also common in the interstices of pillow basalt. Palagonite is usually green to black, and resembles obsidian. Term coined by Sartorius Von Waltershausen in 1846.
Cited usage: Under the microscope palagonite appears as a perfectly amorphous substance [1879, Rutley, Stud. Rocks xiii. 272].

paleo-: [Greek *palaios* = ancient < *palai* = long ago.] Combining prefix meaning ancient, used in words such as: paleontology, paleolithic, and paleoarcheology. *Paleo-* is used in contrast to the prefix *neo-*, meaning new.
Alt. spelling: palaeo-
See also: neo-

paleoblast: [Greek *palaios* = ancient < *palai* = long ago + *blastos* = sprout, shoot, germ.] n. A crystal or crystal remnant, especially in a metamorphic rock, which is older than the other crystal grains in the rock, usually representing former equilibration conditions.

Paleocene: [Greek *palaios* = old + *kainos* = recent.] adj. The first epoch of the Tertiary period (65.5-55.8Ma). In 1833, Charles Lyell subdivided the Tertiary period, calling the oldest epoch Eocene. This was later also subdivided, and the earliest part of the Eocene was named Paleocene, which was later made a distinct epoch. All the Cenozoic epochs end in '-cene', the same root that is used in the prefix of the term Cenozoic. The Paleocene epoch is an oxymoron, like neoclassic or soft rock. First usage in English c. 1875.
See also: paleo-

Paleogene: [Greek *palaios* = ancient < *palai* = long ago + *genesis* = origin, creation < *genesthai* = to be born, come into being; < IE *gen* = to give birth, beget.] adj. In the geologic time scale, a subclass of the Tertiary period encompassing the Paleocene, Eocene and Oligocene epochs; a span of time from 65.5-23.0Ma. Giovanni Arduino first used the term Tertiary in 1759 to denote the third of the four geologic eras: Primary, Secondary, Tertiary, and Quaternary. Later, Charles Lyell reclassified Tertiary and Quaternary as geologic periods in his detailed classification of time. This left the the former Paleogene and Neogene Periods to be demoted to an intermediate rank.
See also: Neogene

Paleolithic: [Greek *palaios* = ancient < *palai* = long ago + *lithos* = rock, stone.] adj. Of or relating to the cultural period of the Stone Age beginning with the earliest chipped stone tools, about 750Ka, until the beginning of the Mesolithic Age, about 15Ka. Human cultural development has been classified into distinct archeological groupings: the Paleolithic (Old Stone Age), Mesolithic (Middle Stone Age), Neolithic (New Stone Age), Chalcolithic (Copper/Stone Age), Bronze, and Iron Ages. First usage in English c.1860.
Synonyms: Stone Age, Old Stone Age
See also: neolithic

paleontology: [Greek *palaios* = ancient < *palai* = long ago + *onto* = existence, being < *einai* = to be + *logos* = word, speech, discourse, reason < *legein* = to speak, to say; IE *leg* = to collect < *logia* = to speak.] n. The study of life in the geologic past, as represented by the fossils of plants, animals, and other organisms. The term was first coined by Ducrotay de Blainville in 1832. Two years later, Fischer von Waldheim apparently devised the same term independently. Prior to 1832, fossil study was called oryctology, from Greek *oryctos*, meaning dug, formed from digging.
Cited usage: Palaeontology is the science which treats of fossil remains, both animal and vegetable [1838, Lyell, Elements of Geology II. xiii. 281].

Alt. spelling: palaeontology
See also: neontology

paleosol: [Greek *palaios* = ancient < *palai* = long ago + Latin *solum* = soil < *solium* = seat; < IE *sed* = seat.] n. A soil formed during an earlier period of pedogenesis, most commonly by burial. A paleosol may have been subsequently exhumed by erosion, or it may have been continuously present on the landscape until the current state of pedogenesis. IE derivatives include: sediment, sessile, and sedentary.
Cited usage: An ancient soil, hereafter referred to as *paleosol, has been dated..as pre-Wisconsin in the Lake Bonneville and Denver basins [1950, Hunt & Sokoloff, Prof. Papers U.S. Geol. Survey No. 221. 109/1]. The expressions "buried soil" and "paleosol" are synonyms, and can replace "geosol" in every case without loss of understanding [1999, Donald L. Johnson, 07 Oct.].
Synonyms: fossil soil, buried soil

paleosome: [Greek *palaios* = ancient < *palai* = long ago + *soma* = body.] n. The unaltered and relatively immobile original part of a migmatite, being the dark-colored metamorphic component (paleosome), as opposed to the light-colored granitic component (neosome).
See also: anatexis, migmatite

Paleozoic: [Greek *palaios* = ancient; < IE *kel* = far away (in time or place) + Greek *zoikos* = pertaining to animals < *zoa* = plural of *zoon* = animal, living being.] adj. The first era of the Phanerozoic eon from the Cambrian through the Permian periods (542-251Ma). The Greek prefixes *tele-* and *paleo-* both mean far away and derive from the same IE root word *kel*, but *tele-* usually refers to distance, whereas *paleo-* is used in reference to time. First usage in English c. 1835.
See also: paleontology

palimpsest: [Greek *palimpsestos* = scraped again < *palin* = again; < IE *kel* = revolve, move around + Greek *psen* = to scrape.] adj. In general, said of an object, place, or area that reflects previous history and leaves a record of change. In geology, said of relict structures, formations, morphology and lithology upon which subsequent processes have superimposed their imprint. Reworked shelf sediments, multiple-episode metamorphism, superimposed drainage patterns, and prograde/retrograde shoreface sequences are examples of palimpsest features. As a noun, a group term for objects, places, or areas that reflect their history. The original palimpsest was a papyrus or parchment manuscript that had been written on more than once, with earlier writing incompletely erased and often legible. They were also sometimes wax tablets that could be scraped clean and written on again. IE derivatives include cyclone and palingenesis. First usage in English c. 1660.
Cited usage: Spaniards in the sixteenth century saw an ocean moving south through a palimpsest of bayous and distributary streams in forested paludal basins [John McPhee]. The name "pseudotachylyte" is a palimpsest word. It was coined early in this century to name a peculiar glassy rock that looks sort of like tachylyte but isn't [1997, John Spray, Geology Magazine, July issue].

palingenesis: [Greek *palin* = again + *genesis* = origin, creation < *genesthai* = to be born, come into being; < IE *gen* = to give birth, beget.] n. In general use, a term meaning rebirth and regeneration. In geology, formation of a new magma by the melting of pre-existing igneous rock in situ. Not to be confused with anatexis, which signifies partial melting of any parent material. First usage in English c. 1615.
See also: anatexis

palinspastic: [Greek *palin* = again + *spastikos* = drawing in < *span* = to draw, tug.] adj. In cartography, said of a paleogeographic or paleotectonic map that shows rock units returned to their original positions before deformation. Term coined in 1937 by geologist G. Marshall Kay.
Cited usage: Palinspastics concern the placement of rocks in their relative original positions. Palinspastic maps have been used to illustrate concepts of continental drift and of intracontinental movements in the development of mountain systems [1945, Bull. Amer. Assoc. Petroleum Geologists XXIX. 426, 435].

palisade: [French *palissade* < *palissa* = stake; < Latin *palus* = stake fixed in the ground; < IE *pak* = to fasten.] n. In geomorphology, a line of lofty steep cliffs, usually along a river or lake margin, especially where the cliffs are composed of vertical basalt columns. The lofty New Jersey Palisades are world-class examples; these cliffs extend for about 15 miles along the western bank of the Hudson River, overlooking New York on the opposite shore. In general usage, a fencelike fortification made of close-spaced poles fixed in the ground and spaced tightly together forming a barrier. IE derivatives include pegmatite and peace (a peaceful 'binding together' by treaty or agreement). First usage in English c. 1595.
Cited usage: High cliffs of basaltic columns, like those exposed on the Hudson and Columbia rivers, are often called palisades [1886, A. Winchell, Walks Geol. Field 96].

pallasite: [Eponymous term after English *Pallas*, a large asteroid in the asteroid belt; < Greek *Pallas* = an alternative name for the goddess Athena.] n. A stony-iron meteorite composed of large single crystals of olivine embedded in an iron-nickel matrix. Pallasites are believed to have formed in a layered planetoid at the boundary of the silicate mantle and the iron-nickel core. Pallasite was named after the asteroid Pallas (diameter = 540km)

because it is composed of stone and iron with embedded olivine crystals, just like the few pallasite meteorite examples found on earth.

palsa: [Swedish *pals* = elliptical, lens.] n. An elliptical, domelike frost mound containing lenses of ice, and formed in areas of peat. Palsa typically occur in subarctic bogs and range in size from 3-6m high, and 2-25m long. In Canada, palsa are defined as permafrost mounds composed of alternating layers of segregation ice and peat or mineral soil. Term coined by Fries & Bergström in 1910.
Cited usage: Peat knobs in swamps and bogs and hillocks in tundra, commonly called Palsen, are described from northern Europe and Siberia in an extensive foreign literature [1942, Geogr. Rev. XXXII. 420].
Alt. spelling: palsen (pl.)

paludal: [Latin *palud* = marsh; < IE *pel* = to fill.] adj. Of or relating to a swamp or marsh. IE derivatives of *pel* include: Pliocene, Pleistocene, plural, and pleiotropism. First usage in English c. 1815.
Cited usage: The Hydrophytes or Water-loving plants include first the Paludal or Marsh and Bog plants [1932, G. C. Druce, Comital Flora of the British Isles p. xxv].

palynology: [Greek *palunein* = to sprinkle + *logos* = word, speech, discourse, reason < *legein* = to speak, to say; IE *leg* = to collect < *legein* = to speak.] n. The scientific study of the structure and dispersal of spores and pollen. Because they can be good indicators of plant geography and taxonomy, palynology applies to stratigraphy, paleoecology, and archeology. Term coined by Hyde and Williams in 1944.
Cited usage: We would therefore suggest palynology : the study of pollen and other spores and their dispersal. We venture to hope that the sequence of consonants p-l-n, (suggesting pollen) and the general euphony of the new word may commend it to our fellow workers

in this field [1944, Hyde & Williams, Let. 15 July in Pollen Analysis Circular 28 Oct. 6].

pampas: [Spanish < Quechua *pampa* = flat field.] n. A term used in Argentina and Uruguay for an extensive treeless grassy plain in a temperate region. Comparable regions to the pampas are: the prairies of North America, the steppes of Russia, the *llanos* of Central America, the *bambas* of Peru (e.g., *Moyo-bamba*), and the subtropical savannahs of Africa.
Synonyms: llano, bamba, savannah
See also: savannah

Pangea: [Greek *pan* = all, entire + *gaea* = land, earth.] n. A supercontinent that existed during late Paleozoic and early Mesozoic time, and formed by the combination of two earlier protocontinents, Laurasia in the north, and Gondwanaland in the south. Pangea represents the most recent coalescing of all the Earth's land masses into a single supercontinent. The name Pangea is frequently stated to have been coined by Alfred Wegener in 1914 in *Die Entstehung der Kontinente und Ozeane* (*The Origin of the Continents and Oceans*), but the term does not appear in the 1st edition. The term *Pangäa* does appear, however, in the 2nd edition of 1920, but with no indication that Wegener has coined it.
Cited usage: The rifting of Pangaea and the floating away of Australasia, Antarctica, and the Americas are said to have begun east of Africa in Jurassic time and west of Euro-Africa in early Cretaceous time. [1928, C. Schuchert, Theory Continental Drift (Amer. Assoc. Petroleum Geol.) 106]
Alt. spelling: Pangaea
See also: Gondwanaland, Laurasia

panmixia: [Greek *pan* = all + Latin *mixtus* = past participle of *miscere* = to mix; < IE *meig* = to mix.] n. The random mating of individuals within a population, with no apparent tendency to chose

partners with particular traits. This implies a lack of natural selection in the reproductive aspect of evolution. Term coined by Weisman in 1889 in his work *Essentials of Heredity*. IE derivatives include melange and promiscuous.

Panthalassa: [Greek *pan* = all, entire + *thalassa* = sea.] n. The ocean surrounding Pangea, being a single, contiguous, largely equatorial ocean. Eduard Suess may have coined the term in 1893. *Cited usage*: The Pacific Ocean may be regarded as the remnant of Panthalassa, with the Atlantic, Arctic, and Indian Oceans being mostly rift oceans. [1970, R. S. Dietz, in Johnson & Smith Megatectonics Continents & Oceans iii. 36]

para-: [Greek *para-* = beside, past, beyond, adjacent, parallel.] Combining prefix meaning beside or side by side, used in such terms as parallel, paraconformity, parasite, and paragraph. Also, as a prefix meaning beyond, used in such terms as paradox, parable, parapsychology, and paradigm. The more abstract applications of this prefix imply proximity or similarity to the root being modified, with the added meaning of being separate from or going beyond that which is denoted by the root word.
Alt. spelling: par-

paracme: [Greek *para* = beside, past, beyond + *akme* = summit; < IE *ak* = sharp.] n. In evolution, a point or period at which the peak vitality and vigor of an individual, species, or other grouping has recently passed.
Cited usage: Paracme, that part of life, in which a person is said to grow old, and which, according to Galen, is from 35 to 49 [1730-6, Bailey's Folio].

paraconformity: [Greek *para-* = beside, beyond, parallel + Latin *conformare* = to shape after < *com-* < *cum* = together, in union, altogether + *formare* = to shape < *forma* = shape.] n. An obscure unconformity where no erosional surface

is discernible; the break is a bedding plane, and the beds above and below the contact are parallel. Term coined in 1957 by Dunbar and Rodgers, formerly classified by them as a type of disconformity.
Cited usage: We propose to restrict the term disconformity to the third type, in which two units of stratified rocks are parallel but the surface of unconformity is an old erosion surface of appreciable relief, and to introduce a new term *paraconformity for the fourth type, in which the beds are parallel and the contact is a simple bedding plane [1975 Nature 3 Jan. 15/1]. Here we use the term unconformity to refer to a significant gap (demonstrated or inferred) in the stratigraphic record (disconformity or paraconformity) [1957, Dunbar & Rodgers, Princ. Stratigr. VI. 119/2].
See also: disconformity, nonconformity, unconformity

paraffin: [Latin *parum* = little, not very; < IE *pau* = little, few + Latin *affinis* = associated with.] n. As a naturally occuring substance in its native state, a mixture of hydrocarbons that are colorless to white, tasteless, waxy, and odorless. Parafin is solid at room temperature, and may be obtained by dry distillation from a variety of organic materials, including petroleum. Term coined in 1830 by its discoverer Reichenbach, in reference to its neutral quality and the small affinity it possesses for other bodies.
Cited usage: Paraffin was discovered about the same time (1830) by Dr. Christison and Dr. Reichenbach; the former called it petrolin [1838, Penny Cyclopedia XII. 396].
See also: ozocerite

paragonite: [Greek *paragon* = leading astray, mislead.] n. A yellowish to greenish monoclinic mineral of the mica group, hydroxy sodium aluminum silicate, $NaAl_2(AlSi_3)O_{10}(OH)_2$.

Paragonite is the sodium analogue to muscovite, and usually occurs in metamorphic rocks. The name alludes to it being mistaken for muscovite or talc. First usage in English c. 1845.
Cited usage: Paragonite constitutes the mass of the rock at Monte Campione [1868, James D. Dana, A System of Mineralogy 5th ed. 488].

paralic: [Greek *paralios* = by the sea, maritime < *para* = beside + *als* = the sea.] adj. Said of deposits formed in shallow water near the sea. The etymology of paralic implies nonmarine, but the term is often used in reference to inter-tongued marine and continental deposits, such as those formed in coastal lagoons, marshes, and basins.
Cited usage: A distinction is, however, sometimes made between limnetic coals and paralic coals, or those derived from plant remains which collected in marshes near the sea border [1914, H. Ries, Econonmic Geology 4th ed. i. 13].
See also: molasse

paramorph: [Greek *para* = beside + *morphos* = shape, form.] n. A pseudomorph with the same chemical composition as the original crystal, being formed by a change of internal characteristics without a change in external features, as in calcite after aragonite. First usage in English c.1875.
See also: pseudomorph

paramos: [Spanish *paramo* = wasteland.] n. A treeless alpine plateau of the Andes and tropical South America. It is exposed to considerable wind and thick cold fogs. Paramos was probably assimilated into Spanish from a native language in the northern highlands of South America.
Cited usage: The Indian of the Andes through whose rude straw hut the piercing wind of the paramos sweeps, and chills the white man to the very bone [1875, Encycl. Britannica I. 89/2]. The Venezuelan and Columbian Paramo–a narrow zone of cold bleak terraces [1901, A. H. Keane, Central and South America. I. 193].

parasite: [Greek *parasitos* = one who eats at another's table < *para* = beside + *sitos* = food.] n. In biology, a relationship where one organism derives sustenance at the expense of another, without providing anything useful in return. In volcanology, a cone or crater that develops on the side of a larger volcano. In structural geology, a fold that develops on the limb or hinge of a larger fold. The original use of the term in ancient Greece refers to one who eats at the table of another, living at their expense and repaying them with flattery or amusing conversation. This meaning was popularized in European literature by Rabelais in *Gargantua et Pantagruel* (*Gargantua and Pantagruel*) in 1535.
Cited usage: So we find in Plautus a certain parasite making a heavy do, and sadly railing at the inventors of hour–glasses and dials as being unnecessary things, there being no clock more regular than the belly [1535, Francois Rabelais (1495-1553), Five books of the lives, heroic deeds and sayings of Gargantua and his son Pantagruel]. It is the fashion of a flatterer and parasyte to lyue of an other man's trencher [1539, Richard Taverner tranlsation, Erasmus Proverbs or Adagies 71].

parazoan: [Greek *para* = beside + *zoa* plural of *zoon* = animal, living being.] n. In paleontology and biology, any member of a group of invertebrates, similar to metazoans by having differentiated cells, but differing from them by not having differentiated tissues. They differ from protozoans by being multicellular and macroscopic. The only surviving parazoans are the sponges in the phylum Porifera.
See also: metazoan

parma: [Russian *parma* = rolling hills.] n. Broad, low-dipping domes or plateaus formed by structural deformation and subsequent erosion of secondary, or parasitic, folds, as is the case in the Urals and the Appalachians. The type feature is found in the Ural Mountains. Also referred to as a quaquaversal.

Cited usage: Many of the low-dipping domes are perceptible as such only by the erosion which has removed their central portion, often leaving a topographic depression. Such low domes have also been called parmas [1913, A. W. Grabau, Princ. Stratigr. xx. 808].
Similar terms: quaquaversal
See also: quaquaversal

parvafacies: [Latin *parva* = small + *facia* < *facies* = form, figure, appearance < (perhaps) *facere* = to make, or < *fa-* = to appear, shine.] n. A laterally limited chronostratigraphic unit that lies between designated chronostratigraphic planes of a magnafacies. The terms magnafacies and parvafacies were coined by Caster in 1934. They emphasize the distinction between lithostratigraphic and chronostratigraphic units in sequences displaying marked facies variation, but have remained informal despite their impact on clarifying the concepts involved.
Synonyms: heteropic facies (Europe)
See also: magnafacies

paternoster lake: [Latin *pater* = father + *nostre* = our.] n. Each lake in a line of connected lakes in a glaciated valley. The "Lord's Prayer" with pater noster as the first two words in Latin is part of the Catholic ritual, in which a series of prayers is counted on a string of beads called a rosary. The name alludes to the similarity between the sequence of lakes and a string of beads. The term is a slight misnomer; it should really be 'Hail Mary lake' because ten "Hail Marys" are recited in continuous rows separated by just one Our Father.
Cited usage: If the step-tread basins are occupied by lakes these may follow one another like beads upon a string 'paternoster lakes' [1942, C. A. Cotton, Climatic Accidents in Landscape-Making xix. 256].

patina: [Latin *patina* = plate, paten; < Greek *patan* = platter; < IE *pet* = to spread.] n. A surface film or encrustation formed on rock long exposed to in situ weathering, such as desert varnish. Originally, a patina was strictly the greenish film produced by oxidation on the surface of copper and old bronze by the formation of a carbonate when exposed to a moist environment. The term is derived from the thin bronze platters held under the chin when receiving the eucharist during Christian religious services. IE derivatives include: compass, fathom, pass, passport, and patent. First usage in English c.1745.

pearl: [French *perle* = pearl; < Latin *perla* = pearl, seashell < *perna* = ham, leg of mutton, leg-of-mutton-shaped marine bivalve mentioned by Pliny.] n. A dense, spheroidal, nacreous concretion that forms around an irritating foreign body, such as a grain of sand, within or beneath the mantle of various marine and freshwater molluscs, the pearl-oyster (Meleagrina margaritifera) of the Indian Ocean being the most famous. Pearls are composed of filmy layers of aragonite interspersed with lamellae of the animal shell membrane material called conchiolin. They are smooth and lustrous, usually white to bluish-gray, but they can also be variously colored to black. First usage in English c. 1325. *See also*: conchiolin

peat: [Middle English *pete* < Medieval Latin *petia* or *pecia* = piece; (perhaps) < Welsh *peth* = portion.] n. An early stage in the development of coal, peat consists of partially carbonized plant matter that accumulates in anoxic conditions beneath a persistently moist mossy surface layer growing in bogs and fens. The etymology alludes to the practice of cutting up peat into pieces used for fuel, thus the saying "to make peats." Many peat bogs, such as those composed of the mosses of the genus Sphagnum, consist largely of a single type of moss. The two most common forms of peat are: a) "Hill peat" (the mountain or brown bogs of Ireland), found in mountainous districts, and consisting mainly of Sphagnum and Andromeda,

and b) "Bottom peat" (the lowland or red bogs of Ireland), found in lakes, rivers, and brooks. First usage in English c.1325.

pebble: [Middle English *pobble* < Old English *papol* < Latin *papula* = pimple, mote.] n. A rock fragment between 4 and 64 millimeters (0.16 and 2.51 inches) in diameter, especially one that has been naturally rounded. Different size ranges for pebble are used in Great Britain (10-50mm), and by the U.S. Bureau of Soils (2-64mm). In the heirarchy of sediment sizes, a pebble is larger than a granule but smaller than a cobble. First usage in English c. 1275.

ped: [Greek *pedon* = soil, earth.] n. A unit of soil structure such as a prism, block, aggregate, or granule, which is formed by natural processes. *Cited usage*: An individual natural soil aggregate is called a ped, in contrast to (1) a clod, caused by disturbance, (2) a fragment caused by rupture, or (3) a concretion caused by local concentrations of compounds that irreversibly cement the soil grains together [1951, Soil Survey Manual (U.S.D.A. Handbook. No. 18) 225].

ped-: [Combining prefix having various distinct meanings and roots: a) Greek *pedon* = ground, soil, earth; b) Greek *paes* = boy, child; c) Latin *pedem* = foot.] Combining prefix meaning: a) soil, as in pedalfer, pedology, and pedon; b) young, as in paedomorphism and pediatrics; and c) foot, as in pediment and pedometer.

pedalfer: [Composite term from *ped* + *al* + *fer* < Latin *pedon* = soil, earth + *al* = abbr. of aluminum + *fer* = abbr. of *ferrum* = iron.] n. Prior to the modern U.S. Soil Survey classification system, a general term for a leached soil characteristic of humid climates, in which the sesquioxides of iron and aluminum accumulate in the B horizon, and, due to acidic conditions, a calcium carbonate layer does not accumulate. First usage in English c.1925.

Cited usage: The significant feature of pedalfers is that moisture is sufficient to wet the soil to its capacity and to allow excess water to remove the soluble constituents [1972, C. B. Hunt, Geol. Soils viii. 167].
See also: pedocal

pediment: [Latin *pediment* = alteration of earlier *perement* < *pyramis* = pyramid (influenced by Latin *pedem* = foot).] n. A large-scale, broad, gently sloping rock surface at the base of a steeper mountain slope, underlain by bedrock, and often covered with a thin veneer of alluvium. Pediments are formed primarily by erosion at the base of an abrupt and receding mountain front or plateau escarpment. The term pediment is a corruption or alteration of *periment*, originally an architectural word applied to a 17th-century, Grecian-style, low, triangular gable over a gateway portico. The etymological connection between the erosional feature and the architectural structure is that they are both relatively low, broadly crowned, and form an obtuse triangular shape. First usage in English c.1660.

pedion: [Greek *pedion* = a plain, a flat surface.] n. An open crystal form, unique to the triclinic system and having only one crystal face, with no symmetrically equivalent face. Thus, there are no necessarily parallel faces. Term coined in 1899 by W.J. Lewis .
Cited usage: Each form consists of a single face, and will be called a pedion [1899, W. J. Lewis, A Treatise on Crystallography xi. 148].
Alt. spelling: pedia (pl.)

pediplain: [Greek *ped* = foot + Latin *planum* = plain.] n. An extensive plain formed in the mature stage of an arid erosional cycle by the coalescence of neighboring pediments. A pediplain is to pediment what a bajada is to an alluvial fan. In 1942, Arthur D. Howard suggested the term *pediplane* in a

semantic argument over Maxson and Anderson's 1935 coinage of *pediplain*, largely because it is not a true plain in the geomorphic sense; however, Howard's poposal can be discounted using the same logic, because it is not a true plane in the geometric sense.
Cited usage: Widely extending rock-cut and alluviated surfaces of this type formed by the coalescence of a number of pediments and occasional desert domes may be called 'pediplains' [1935, Maxson & Anderson, Jrnl. Geol. XLIII. 94]. The writer proposes the term pediplane as a general term for all degradational piedmont surfaces produced in arid climates which are either exposed or covered by a veneer of contemporary alluvium no thicker than that which can be moved during floods [1942, A. D. Howard, Jrnl. Geomorphol. V. 11].
Alt. spelling: pediplane
Synonyms: panfan, desert peneplain

pedocal: [Greek *pedon* = soil, earth + *cal* = abbr. of calcium.] n. In pre-U.S. Soil Survey taxonomy, a soil that contains an accumulated layer of calcium carbonate, $CaCO_3$. It is the characteristic soil of dry climates, and is the arid analogue of pedalfer. First usage in English c. 1925.
See also: pedalfer

pedology: [Greek *pedon* = soil, earth + *logos* = word, speech, discourse, reason < *legein* = to speak, to say; IE *leg* = to collect.] n. The scientific study of soil, especially its morphology, genesis, and classification. Term coined by G.W. Robinson in 1924. The geologist Frederick Augustus Fallon suggested an alternative term in German, *bodenkunde*, meaning soil lore, while the Russians use the term *pochvovedenie*, meaning soil science.
Cited usage: The writer ventures to hope that this convenient term (pedology as a branch of geology) will be more generally used to describe the scientific study of soils [1924, G. W. Robinson, Geol. Mag. LXI. 444].

pedometer: [Greek *ped-* = foot + *metron* = measure.] n. An instrument for recording the number of steps taken, and thus calculating the distance travelled by multiplying the steps taken by the length of one's pace. First usage in English c. 1725.

pedomorphism: [Greek *paedo-* = combining form of *pais* = boy, child + *morphosis* = formation < *morphoun* = to form < *morph* = form.] n. Phylogenetic change that involves retention of juvenile characteristics by the adult.
Alt. spelling: paedomorphism
Synonyms: neoteny

pegmatite: [Greek *pegma* = something fastened together < *pegnunai* = to fasten; < IE *pak* = to fasten.] n. An exceptionally coarse-grained igneous rock characterized by large interlocking crystals which may include rare minerals rich in elements, such as lithium, boron, beryllium, tantalum, and the rare earths (such as uranium). Pegmatites usually form near the margins of large igneous plutons as dikes, pods, or lenses, and are the product of late-stage crystallization of a magma, generally in the presence of an aqueous fluid phase. Pegmatites are evidently formed under fairly high confining pressure, as there is no analogous feature in extrusive systems. Although there are pegmatites similar in composition to a wide range of igneous rocks, including gabbros and syenites, most pegmatites have a composition close to that of granite. The term was first suggested by René-Just Haüy in 1822 as a synonym for graphic granite, but in 1845 Haidinger initiated its use with the present meaning. IE derivatives include: palisade, impact, peace, and peasant.

pelagic: [Greek *pelagikos* = the sea.] adj. Of or pertaining to the open ocean or 'high seas,' as distinguished from the environment between low tide and the continental shelf (neritic zone), the intertidal coastal environment (littoral zone), and the bottom environment (benthic zone). Various prefixes are added to the base term pelagic with increasing depth from the sea surface to distinguish the different open ocean depth zones; these prefixes are: epi-, meso-, bathyo-, abysso-, and hado-. First usage in English c. 1655.

peléan: [Toponymous term after Mount Pelée on the Caribbean island of Martinique; < French *pelé* = bald, peeled.] adj. Pertaining to a type of volcanic eruption characterized by the lateral emission of pyroclastic flows from the summit area (nuées ardentes) and the vertical extrusion of very viscous lava from the central vent which tends to become consolidated as a solid plug. The rhyolite lava dome Mount Pelée, on the island of Martinique, erupted catastrophically on May 8, 1902, with pyroclastic flows killing over 30,000 people. Subsequently a spine of viscous rhyolite projected upward to a height of 400 meters. In allusion to this eruption, T. Anderson and J. S. Flett coined the term peléan in 1903. In 1932, Frank Perret wrote, "It is still an unsettled question whether the name 'Pelée' is derived from the French *pelé*, meaning bald (literally peeled) mountain, or whether the Carib inhabitants, who called it the 'mountain of fire,' named it for 'Pele' the Hawaiian goddess of fire." There are other volcanoes named Pele: *Mont Pele* (also called *Le Djungo*) in West Africa; *Pu'u o Pele*, a cinder cone in Haleakala crater on Maui; *Pele* (also called *Fatouleo-Kakoulo*), in Vanuatu, a group of 83 islands in the South Pacific; and *Pele's vents*, the unofficial name for the summit of Lo'ihi, Hawaii's youngest volcano which has yet to break the sea surface.
Cited usage: We propose to adopt the term 'A Peléan Eruption' to designate this group of phenomena [1903, Anderson & Flett, in Phil. Trans. R. Soc. A. CC. 499]. The Peléean phase is the most violently explosive of all. The extraordinary features of a Peléean cloud are that it is emitted as a horizontal blast from beneath the lava

plug in the summit of the volcano; that it carries with it an enormous amount of rock fragments [1934, C. R. Longwell et al., Outl. Physical Geol. x. 189-191].

pelecypod: [Greek *pelekus* = hatchet + Latin *poda* = foot.] n. Any animal in the class Pelecypoda of the phylum Mollusca. Pelecypods have a hatchet-shaped foot. First usage in English c. 1855.
Synonyms: lamellibranch
See also: lamellibranch

Pele's hair: [Hawaiian *ranoho o Pele* = hair of Pelé, goddess of the volcano Kilauea.] n. Volcanic glass formed into fine strands and threads that resemble hair, formed by strong volcanic winds. Pele's hair is naturally spun volcanic glass that is blown away from lava fountains, cascading lava falls, or turbulent lava flows and takes its shape due to solidification in an airborne molten state, its low viscosity, and light weight. It is usually gold in color and has a diameter of less than half a millimeter. Pele (Pele-kumu-ka-lani) has a life history encompassing time spent on each island of the Hawaiian chain. Her story, along with the life of her older sister Na-maka-o-ka-hai, goddess of the ocean, with whom she had an ongoing sibling rivalry, reflects the Hawaiian settlers, understanding of the relative ages of the islands.
Cited usage: Pele's Hair, lava blown by the wind into hair-like fibers [1861, Bristow, Gloss. Min. 276].
See also: Pele's Tears, lapilli

Pele's tears: [Hawaiian *Pele* = Hawaiian goddess of earth's raging fire whose abode is the volcano Kilauea.] n. Pele's tears are solidified drops of volcanic glass. They range in size from a few millimeters to several centimeters and can be trailed by Pele's hair.
See also: Pele's hair, lapilli

pelite: [Greek *pelos* = clay, earth, mud.] n. A fine-grained, sedimentary rock composed of clay- or mud-sized detrital particles. Psephite (conglomerate), psammite (sandstone), and pelite (claystone) are derived from Greek, whereas the Latin equivalent terms are: rudite, arenite, and lutite. In 1921 Tyrell proposed that the Greek terms also be used to describe metamorphosed sedimentary rocks. First usage in English c. 1875.
See also: arenite, lutite, psephite, psammite, rudite

pellucid: [Latin *pellucere* = to shine through < *per* = through + *lucere* = to shine; < IE *leuk* = light, brightness.] adj. Able to transmit light, either wholly - as in transparent, or partially - as in translucent. Etymologically related terms include: luster, lumen, illustrate, lunar and lunacy. First usage in English c.1615.
Cited usage: The comet's taile is nothing else but an irradiation of the sunne through the pellucide head of the comet [1619, Bainbridge, An Astronomical Description fo the Late Comet].

pelmato-: [Greek *pelmato* = sole of the foot.] Combining prefix relating to marine organisms that have an attachment column or foot to serve as a holdfast (e.g., the Pelmatozoan branch of echinoderms, such as crinoids or the extinct Blastoidea and Cystoidea).
Alt. spelling: pelma-

peloid: [Greek *pelos* = mud, clay.] n. An ovoid-shaped particle of microcrystalline or cryptocrystalline carbonate material.
Cited usage: Petrography of carbonate grains: oöids, pisolites, peloids and other micritic fabrics [1972, Nature 25 Feb. p. ix/1].

peñasco: [Spanish *peñasco* = large rock < *peña* = rock.] n. A projecting rock, especially one that has been isolated due to recession of the surrounding topography.

pendant: [French *pendre* = to hang; < Latin *pendere* = to hang; < IE *pen* = to draw, stretch, spin.] n. In petrology, a downward projection of country rock into a batholith, or igneous intrusion. The impaction of roof rocks in the melt leaves remnants of roof pendants which resemble xenoliths in the intruding rock. In speleology, one of many closely spaced solution remnants hanging within a cave. The geologic usage fits closely with the term's use in architecture as a hanging ornament on roofs and ceilings, especially in the later styles of Gothic architecture, where it is of stone, and integral to the construction. Etymologically related terms include: pendulum, perpendicular, appendix, and pension. Not to be confused with the geomorphologic term pendent. First usage in English c. 1325.
Synonyms: roof pendant
See also: pendent, speleology, xenolilth, batholith

pendent: [French *pente* = slope, declivity, inclination (of a hill, etc.); < Latin *pendere* = to hang.] n. In geomorphology, a landform that slopes steeply downward, as in a hillside, or overhangs, as in a cliff. Pendent and pendant are easily confused because they have the same Latin root, have a similar sense, and share spelling variations. Their meanings began to diverge in French and became even more distinct in geologic usage. First usage in English c. 1300.
Alt. spelling: pendant
See also: pendant

pene-: [Latin *paene* = nearly, almost, all but.] Combining prefix meaning almost, used in such terms as: penecontemporaneous, peneplain, and peninsula.
Alt. spelling: paen-, pen- (before a vowel)

penecontemporaneous: [Latin *paene* = nearly, almost, all but + *con-* = with + *tempus* = time + *-aneus* = abounding in, full of, characterized by, of the nature of.] adj. Said of a geologic process, or resultant structure or mineral, occurring immediately after deposition but before consolidation of the enclosing rock. This lithostratigraphic term was coined to provide a term that encompasses the temporal depositional characteristics of both syngenetic and epigenetic sedimentary structures. First usage in English c. 1900.
Cited usage: The present trend of opinion considers that much chert and flint were deposited contemporaneously with the enclosing rocks. The views of origin may be placed in three classes: contemporaneous (syngenetic), penecontemporaneous, and subsequent (epigenetic) [1939, W. H. Twenhofel, Princ. Sedimentation x. 375].

peneplain: [Latin *paene* = almost + *planus* = plain, flat; < IE *pel* = flat, to spread.] n. A low, nearly featureless, slightly undulating land tract of great area, especially one held to be the product of long-continued subaerial erosion of land undisturbed by crustal movement. A peneplain represents the penultimate stage in the cycle of erosion in a humid climate. The term also applies to a former surface of this kind as it exists today, and after undergoing subsequent change, such as tectonic uplift, glacial erosion, or burial as an unconformity. Term coined in 1889 by Davis. IE derivatives include: planet, field, palmate, esplanade, explain, and polka.
Cited usage: Given time enough, and the faulted ridges of Connecticut must be reduced to a low base-level plain. I believe that time enough has already been allowed, and that the strong Jurassic topography was really worn out somewhere in Cretaceous time, when all this part of the country was reduced to a nearly featureless plain, a 'peneplain', as I would call it, at a low level [1889, W. M. Davis, Amer. Jrnl. Sci. XXXVII. 430].
Synonyms: rumpfflache, endrumpf
See also: rumpfflache, endrumpf

peninsula: [Latin *paene* = almost, nearly + *insula* = island.] n. Any piece of land projecting into the sea so that the greater part of its boundary is coastline, e.g., southern India, the Balkan Peninsula, Florida, and San Francisco. The visual effect from above is like a finger projecting into the sea. First usage traced to 1538. 'Peninsula', as used by Pliny, was also called *demie island* or *ile demi* in French. First usage in English c. 1535.
Cited usage: This Peninsula to cumpace it by the Rote lakkith litle of a Mile [1538, Leland, Itin. III. 21].

Pennsylvanian: [Toponymous term after the state of Pennsylvania; < English *Penn*, after William Penn + Latin *sylvan* = of the forest.] adj. In North America, referring to a period in the Paleozoic Era from 318.1-299Ma, before the Mississippian and after the Permian. It corresponds more or less to the Upper Carboniferous in Europe, started with a marine transgression, and is characterized by extensive coal measures. William Penn (1644-1718) was an English Quaker who founded Pennsylvania largely on the success of his 'peace treaties' with the Native Americans of that area. The Latin *Silvanus* was used as the proper name of a divinity of the fields and forests identified with Pan; it was anglicized to Silvan by John Milton in the 17th century.
Cited usage: The need of a name to distinguish this system of rocks from those which have been described under the name Mississippian has long been felt, and the name Pennsylvanian, which has recently come into wide use in this country, was adopted because the system is well developed and well known in Pennsylvania. [1906, Chamberlin & Salisbury, Geology: Earth History II].
Synonyms: Age of Ferns
See also: Mississippian, Paleozoic, Permian, Carboniferous

peñon: [Spanish *peñon* = large rock, rocky mountain.] n. Used in the U.S. Southwest and Mexico to describe a high, rocky point.

penumbra: [Latin *paene* = almost, nearly + *umbra* = shadow.] n. A partial shadow, as in an eclipse, between regions of complete shadow and complete illumination.

peperino: [Italian *piperno* = peperino < *pepere* = pepper.] n. An unconsolidated, light and porous volcanic tuff, usually brown in color, composed of sand, cinders, and crystal fragments, often of leucite. The name was first given to the volcanic tuffs of Monte Albano near Rome, named after its texture consisting of small, peppery grains.
Cited usage: The stone in general use for building here, is a hard volcanic tuffa of the sort called Piperno in Italy [1777, Hamilton, Phil. Trans. LXVIII].
Alt. spelling: piperno (Italian dialect)

pepino: [Spanish *pepino* = cucumber; < Latin *pepo* = large melon, gourd; < Greek *pepon sikyos* = ripe gourd.] n. In geomorphology, a small, rounded, conical, residual hill, characteristic of tropical and subtropical karst regions, especially one found in Puerto Rico. The word pepino is still used in numerous Latin American countries, especially Mexico, where it refers to the common cucumber.
Cited usage: Also known as 'haystack hills',... the Pepino hills are a characteristic feature of a mature karst landscape in tropical to subtropical latitudes [1968, R. W. Fairbridge, Encycl. of Geomorphology].
Synonyms: hum, mogote, pepino hill
See also: hum, mogote, karst

per-: [French *par* = by means of; < Latin *per* = thoughout, thoroughly, to completion, through, by means of.] Combining prefix meaning completely, the most, or the maximum amount, as in perchlorate, percrystalline, perfect, and perdition; meaning

through or by means of, as in percolate, persist, or percussion. In chemical nomenclature, the prefix forms nouns and adjectives denoting the maximum (or supposed maximum) of some element in a chemical combination.
Alt. spelling: par (used in borrowed French phrases)

peralkaline: [French *par* = by means of; < Latin *per* = throughout, thoroughly, to completion, through, by means of + Arabic *alqily* = ashes, lye, potash < *al* = the + *qily* = ashes < *qalay* = to fry, roast.] adj. Said of a rock containing a high proportion of the alkalies soda and potash, where total soda ($Na_2O + K_2O$) exceeds alumina (Al_2O3).
Cited usage: The following groups of rocks stand out as chemically distinct: A peraluminous group, characterized by primary muscovite, biotite, corundum, tourmaline, topaz, almandine, or spessartite. A peralkaline group, characterised by soda-pyroxenes or soda-amphiboles, and by the virtual absence of anorthite [1927, S. J. Shand, Eruptive Rocks vii. 128].

perch: [Latin *pertica* = pole, long staff, measuring-rod.] n. A rod of a definite length used extensively in early land surveys, equal to 16.5 ft (5m).
Cited usage: An acre of grounde by the statute, that is to say xvi (sixteen) fote and a halfe, to the perch or pole, foure perches to an acre in bredth, and fortye perches to an acre in lengthe [1523. Fitzherb, Husb. 12].
Synonyms: rod, pole

percolation: [French *par* = by means of; < Latin *per* = throughout, thoroughly, to completion, through, by means of + *colare* = to filter < *colum* = sieve.] n. Slow laminar movement of liquid through the interstitial space between closely packed particles. Also, a legal term for water that "oozes, seeps, or filters" through soils in an unchanneled manner. First usage in English c. 1610.
Synonyms: infiltration

pereletok: [Russian *pereletok* = survives over the summer < *perel* = ice + *etok* = age, old.] n. A layer of frozen ground between the top layer of seasonally frozen ground (active layer) and the perenially frozen ground below (permafrost). Pereletok remains frozen for periods up to several years before thawing. Because there is no perceptible difference between a long-duration pereletok and a recently formed permafrost, some geologists discount the prevailing use of pereletok, or the synonym climafrost. Instead they recommend that all ground that stays frozen over summer be called permafrost and all ground thawing at some point in the year as seasonally frozen ground.
Synonyms: climafrost, intergelisol

pergelisol: [Latin *per* = throughout, thoroughly, to completion, through, by means of + *gel* = solid (back-formation of gelatin); < Latin *gelare* = to freeze; < IE *gel* = cold, to freeze + Latin *solum* = floor, ground, soil.] n. One of the twelve orders in soil taxonomy indicating permanently frozen ground. Term coined by K. Bryan in 1946. As used in science, the terms *gel*, *sol*, and *colloid* were coined by Thomas Graham c. 1860. The suffix *-sol* is used to form the names of different kinds and states of soil. Etymologically related terms include: glacier, jelly, chill, and congeal.
Cited usage: It is impossible to make a verb or a verbal noun from 'permafrost' as 'permafrosting' and 'permafrosted' imply that a permanent surface or coating has been applied. Further, the term cannot be easily converted into other European languages. These various objections can be met by a new term. Such a word is 'pergelisol' [1946, K. Bryan, Amer. Jrnl. Sci. CCXLIV. 635].

peri-: [Greek *peri* = around, near.] Combining prefix meaning around, about, or enclosing, used in such terms as: pericline, peridot, periglacial, peripheral, perisome, periscope, and peritidal.

pericline: [Greek *peri* = around + *clinos* = sloping.] n. In mineralogy, a variety of albite, elongated and twinned along the b-axis. In structural geology, a term used mainly in Britain for a fold that has beds either a) dipping outwards from a center, as in a dome, or, b) dipping inwards toward a central point, as in a basin.
Cited usage: Pericline is in large, opaque, white crystals [1868, James D. Dana, A System of Mineralogy 5th ed. 350].

peridot: [French *peritot* = peridot; < (possibly) Arabic *faridat* = gem, pearl, precious stone.] n. A pale-green, translucent to transparent, gem-variety of the mineral olivine. Peridot is associated with mafic igneous rocks, and some of the best localities are Saint John's Island in the Red Sea, the state of Arizona in the U.S. Southwest, and in Myanmar (Burma). Though mined on St. John's Island for thousands of years, peridot became popular in Western Europe during the Crusades, and is sometimes called "Emerald of the Crusaders." When the Spaniards brought emeralds from the Americas, interest in peridot waned. The name chrysolite was formerly applied to peridot and many other green-colored gemstones. First usage in English c. 1325.
Alt. spelling: peridote
See also: chrysolite

period: [Greek *periodos* = going round, way round, circuit, revolution, cycle of years < *peri* = around + *odos* = way.] n. A geochronologic time unit, being shorter than an era, but longer than an epoch. In ancient Greece, a *periodos*, or period, was the term reserved strictly in reference to the time span between each of the four pan-Hellenic sport festivals. The span of one Olympiad then, was equal to four periods, comprising each of the pan-Hellenic games: the Olympics (dedicated to Zeus), the Pythian Games (to Apollo), the Isthmean Games (to Aphrodite), and the Nemean Games (also dedicated to Zeus). First usage in English c. 1400.

Cited usage: In the scheme of nomenclature proposed by the International Geological Congress, period is the chronological term of the second order, to which system is the corresponding stratigraphic term; as, Silurian period or system [1895, Funk's Stand. Dict. s.v.].
See also: era, epoch

peritectic: [Greek *peri-* = about, around + *tektos* = melted < *tekein* = to melt.] adj. In petrology, said of a temperature at which a solid will melt incongruently to form both a liquid and a solid havinf different compositions. The peritectic is also called the reaction temperature, or incongruent melting temperature. The etymology may allude to the observation that, if equilibrium is not maintained, on cooling to the peritectic temperature, crystals of the low-temperature phase may grow around crystals of the high-temperature phase. An example being augite crystals that envelop crytals of olivine. Conversely, upon heating, the low-temperature crystals will melt, as sanidine crystals will melt to release previously exposed crystals of leucite. First usage in English c. 1920.
Cited usage: The reverse type of change to a eutectic reaction (is) called a peritectic reaction [1973, J. G. Tweeddale, Materials Technol. I. vi. 163].
See also: eutectic

perlite: [French *perle* = pearl; < Latin *perla* = pearl, seashell < *perna* = ham, leg of mutton, leg-of-mutton-shaped marine bivalve mentioned by Pliny + Greek *lithos* = rock.] n. A natural volcanic glass similar to obsidian but having a higher water content. Perlite has a characteristic texture with numerous curving cracks roughly concentric around closely spaced centers. It is often commercially prepared for use as a lightweight soil aggregate. The name may allude to the similarity of the perlitic texture to the spheroidal appearance of a natural pearl. First usage in English c. 1830.

Cited usage: Perlite must be regarded as the vitreous condition of the felsitic rhyolites [1879, Rutley, Study Rocks xi. 193].

permeability: [Latin *permeare* = to penetrate < *per* = through + *meare* = to pass; < IE *mei* = to change, go, move.] n. The ability of a porous rock, sediment, or soil to transmit a fluid. Permeability is the measure of the relative ease of fluid flow. The standard unit of measurement for permeability in the c.g.s. system is the darcy, which is the permeability of a medium that allows a flow of 1cc/sec of a liquid of 1 centipoise viscosity under a pressure gradient of 1atmosphere/cm. IE derivatives include: zenith, azimuth, migrate, mistake, mutate, and communism. First usage in English c. 1755.

Permian: [Toponymous term after *Perm Oblast*, a region in west-central Russia.] adj. Final period of the Paleozoic era from 299-251Ma. Term coined in 1841 by Sir Roderick Murchison, due to prevalence of Late Paleozoic sedimentary strata in the province of Perm, Russia. The Permian Period is characterized by the formation of the supercontinent Pangaea, the rise of coniferous plants, and the diversification of reptiles. The end of the Permian marks the largest known mass extinction in the history of life, which also marks the end of the Paleozoic era. First usage in English c. 1840. *Cited usage*: The carboniferous system is surmounted, to the east of the Volga, by a vast series of beds of marls, schists, limestones, sandstones and conglomerates, to which I propose to give the name of 'Permian system' [1841, Murchison, Lond. & Edin. Phil. Mag. XIX. 419].

perthite: [Toponymous term after English *Perth*, located in Ontario, Canada.] n. A variety of alkali feldspar in which lamellae of sodium-rich feldspar (usually albite) has exsolved from a potassium-rich host (usually microcline). The layered intergrowths of perthite form at high temperatures, then differentiate as they slowly cool. Term coined

in 1832 by T. Thomson, after Perth, Ontario, the type locality. Perthitic intergrowths have textures ranging from macroperthite (visible to the naked eye), to microperthite (visible with a light microscope), to cryptoperthite (visible with a scanning-electron microsope or by X-ray diffraction). *Cited usage*: Perthite, a flesh-red aventurine feldspar, consisting of interlaminated albite and orthoclase [1868, James D. Dana, A System of Mineralogy (ed. 5) 356].

pervious: [Latin *per* = through + *via* = way.] adj. A synonym for permeable. The word impervious is common in modern usage, but pervious is a rare case of preservation of a formerly common Old English positive. Examples of other nearly obsolete Old English positives are: ept, couth, hap, and kempt. *Cited usage*: A coarse argillaceous gravel pervious to water [1807, Vancouver Agric. Devon 22]. *Synonyms*: permeable *See also*: permeability

petra: [Greek *petrai* = small mass of naturally occurring rock < *petra* = rock.] n. The rock materials produced in specific sedimentary organic environments. Term coined by Swain in 1958.

petro-: [Greek *petros* = stone or *petra* = rock.] Combining prefix meaning rock, used in such terms as: petrology, petrography, petroglyph, and petrogenesis. In petroleum geology, a prefix meaning petroleum hydrocarbons, used in terms such as: petrochemical, petropolitics, petrodollar, and petro-wars.

petroglyph: [Greek *petros* = rock, stone + *gluph* = carving < *gluphein* = to carve; < IE *gleubh* = tear apart, cleave.] n. A picture, symbol, grapheme, graffiti, or intentional mark carved into rocks and rock walls by humans. Petroglyphs are recessed beneath the surface, and do not include writing,

drawing, or painting on top of rock surfaces. IE derivatives include: clever, cleft, fissure and hieroglyphic.
See also: pictograph

petrography: [Greek *petra* = rock + Latin *graphicus* < Greek *graphikos* < *graphe* = writing < *graphein* = to write; < IE *gerbh* = to scratch.] n. The scientific description of the composition and formation of rocks, especially by means of microscopic examination of thin sections.
See also: petrology

petroleum: [Greek *petra* = rock + *oleum* = oil.] n. A naturally occurring complex hydrocarbon, usually liquid but also occurring in more viscous tarry or waxlike forms. Petroleum occurs as trapped, anaerobically decomposed organic residue. It occurs worldwide, but is found in large quantities in sedimentary rocks where specific paleo-environmental conditions conducive to its formation and entrapment existed. Typical sites include continental shelves and areas with diapiric structure composed of numerous salt domes. Petroleum is an extremely versatile, but non-renewable resource.

petrology: [Greek *petra* = rock; < Sanskrit *parvata* = mountain + *logos* = word, speech, discourse, reason < *legein* = to speak, to say; IE *leg* = to collect.] n. That branch of geology that deals with the origin, structure, composition, occurrence, and evolutionary history of rocks.

phacolith: [Greek *phakos* = lentil + *lithos* = rock, stone.] n. An intrusive mass of igneous rocks situated between consecutive strata, either at the top of an anticline or at the bottom of a syncline. Term originally coined as phacolite in 1909 by Harker; later changed to phacolith.
Cited usage: These long narrow intrusions are called by Mr. Harker phacolites (more correctly phacoliths) [1910, Lake & Rastall, Textbook Geol.

xiii. 225]. The phacolith might be described as a saddle-shaped laccolith; but there is this important difference between a laccolith and a phacolith that the former is the cause of the folding of its country rocks while the latter is a consequence of folding [1947, S. J. Shand, Study of Rocks (ed. 2) ii. 16].
Similar terms: harpolith

phan-: [Greek *phaneros* = visible, evident < *phainein* = to cause to appear < *phainesthai* = to be brought to light; < IE *bha* = to shine.] Combining prefix meaning visible, used literally, as in phaneritic, meaning rocks with visible crystals; or used figuratively, as in Phanerozoic, meaning the eon of visible life forms. IE derivatives include: banner, phantasmagoria, fantasy, epiphany, phanerogam (flowering plant), and hierophant (an interpreter of sacred mysteries and esoteric principles).
See also: crypto-, micro-

phaneritic: [Greek *phaneros* = visible, evident < *phainein* = to cause to appear < *phainesthai* = to be brought to light; < IE *bha* = to shine.] adj. Said of the texture of an igneous rock in which the individual grains are large enough to be identified with the naked eye.

Phanerozoic: [Greek *phaneros* = visible, evident < *phainein* = to cause to appear < *phainesthai* = to be brought to light; < IE *bha* = to shine.] adj. The most recent of the highest-ranking geochronologic time units, the Phanerozoic eon comprises the Paleozoic, Mesozoic, and Cenozoic eras. The term alludes to the abundance of macroscopic fosils found in rocks of this age.
Cited usage: Cryptozoic fauna imperceptibly blends with what by way of contrast may be called the phanerozoic [1896, A. Dendy, Natural Sci. July 8].

phantom: [Latin *phantasma* < Greek *phantazein* = to make visible < *phantos* = visible < *phainein* = to show.] n. A bed that is missing from a

sedimentary sequence, but that appears elsewhere in a nearby stratigraphic sequence of similar age. Phantom beds usually have no clear reason for their disappearance.
Similar terms: ghost member

pheno-: [Greek *phainein* = to cause to appear < *phainesthai* = to be brought to light; < IE *bha* = to shine.] abbr. Combining prefix meaning to appear, but often referring exclusively to substances derived from the hydrocarbon phenol. The term was first used in 1841 by the French chemist Laurent, alluding to phenolic substances that he called 'hydrate de phényle,' 'acide phénique,' or simply 'phéne.' The etymology is in allusion to the fact that the substance was a coal-tar product, arising from the manufacture of illuminating gas.

phenocryst: [Greek *phaino* = showing < *phanein* = to show, appear, to shine + *krystallos* = crystal.] n. Each of the relatively large, conspicuous crystals in a fine-grained ground mass that are found in either a phaneritic or aphanitic porphyritic igneous rock. Term coined in 1889 by Joseph Paxson Iddings.
Cited usage: Two phases of consolidation to be observed, the first (porphyritic) marked by the formation of large crystals (phenocrysts) which were often broken and corroded by mechanical and chemical action [1893, Geikie, Textbook. Geol. ii. 3rd ed. 155].

phlogopite: [German *phlogopit* < Greek *phlogopos* = fiery-looking < *phlog* = fire + *ops* = eye, face.] n. A usually brownish mica mineral with a pearly to copper-like submetallic luster, hydroxy potassium magnesium aluminum silicate $K(Mg,Fe)_3AlSi_3O_{10}(OH)_2$. It is commonly found in kimberlites, metamorphosed dolostone, and serpentine.
Cited usage: Phlogopite crystallizes in the same system, and has the same cleavage as muscovite [1879, Rutley, Stud. Rocks x. 135].

phonolite: [Greek *phon* = voice, sound + *lithos* = rock, stone.] n. A fine-grained, porphyritic, volcanic rock consisting of alkali feldspar (usually sanidine or anorthoclase) and nepheline as the main feldspatoid. Phonolites are the extrusive equivalent of nepheline syenites and are found in continental regions subjected to anorogenic upwarping and rifting, and on off-axis islands in ocean basins that are separated from ridges or hot spots. Originally named *klingstein* in German from *klingen*, meaning clink, and *stein*, meaning stone. The name refers to the fact that thin slabs of phonolite reverberate with a bell-like ring when struck with a hammer.
Cited usage: Phonolite, sounding stone; a name proposed as a substitute for klingstein (jingling stone) [1828-32, Webster's Dictionary]. Phonolyte (or clinkstone), a compact grayish rock, often containing crystals of glassy feldspar, and having a zeolite in the base along with orthoclase [1868-80, James D. Dana, A System of Mineralogy 359].
Synonyms: clinkstone

phosphate: [Greek *phosphorus* = morning star, light bringer < *phos* = light + *phoros* = bringing.] n. A salt, ester, or other organic derivative of phosphoric acid, being a mineral compound containing the tetrahedral phosphate group (PO_4^{3-}). In geology, phosphates are usually found as metallic salts, such as those found in the mineral apatite, $Ca_5(PO_4,CO_3)_3(F,OH,Cl)$.
See also: phosphorous

phosphorus: [Greek *phosphorus* = morning star, light bringer < *phos* = light; < IE *bha* = tp shine + Greek *phoros* = bringing.] n. A highly reactive, poisonous to humans, nonmetallic element (symbol = P), that occurs in three allotropic forms: a) a yellow, poisonous, flammable, and luminous in the dark form, b) a red form that is less flammable, and c) a black form that is relatively insoluble and is least flammable. It is widely distributed in

nature, and occurs combined with other elements, as in the mineral apatite. It is an essential constituent of protoplasm. The name alludes to the luminous nature of yellow phosphorus, resulting from slow oxidation on exposure to light. The prefix *phos-*, as in phosphorescence, is used as a combining form in many phosphorus-containing compounds. IE derivatives include diaphanous, epiphany, and Phanerozoic.
Cited usage: As a result of exposure to heat or light, phosphorus sometimes acquires a red colour, and this red substance is allotropic or amorphous phosphorus [1866, Brande & Cox, A Dictionary of Science, Literature, and Art].
See also: phosphate, Phanerozoic

phreatic: [Greek *phreat* = a well, spring; < IE *bhreu* = boil, bubble, effervesce, burn.] adj. In hydrology, said of water occurring in the zone of saturation below the water table, especially that water capable of moving through the zone. In volcanology, said of eruptions caused by groundwater flashing to steam as it is heated by magma, often with little or no lava ejection. Phreatic eruptions are a common precursor of volcanic activity. Term coined by R. Hay c.1890 while in the service of the American Artesian and Underflow Office.
Cited usage: The new word phreatic, which is a very convenient term for underground waters [1892, R. Hay, Final Geol. Rep. U.S. 52nd Congress, #41, iii. 8].

phyllite: [Greek *phyllos* < *phullon* = leaf; < IE *bhel* = to thrive, bloom + Greek *lithos* = rock.] n. In petrology, the name of a metamorphic rock, intermediate in grade between slate and schist, in which microscopic crystals of mica, chlorite, or graphite impart a silky sheen to the cleavage surfaces. In mineralogy, a now obsolete term for certain magnesia-mica minerals, such as: micas, clays, chlorites, and vermiculites, often occurring as small scales in argillaceous schist or slate. IE derivatives include: flower, exfoliate,

blade, folio, and podophyllin.
Cited usage: By increase of its mica-flakes a clay-slate passes into a phyllite [1886, Geikie, Geology 223].

phyllosilicate: [Greek *phyllos* < *phullon* = leaf + Latin *silex* or *silic* = flint, hard stone.] n. The subclass of silicates characterized by SiO_4 tetrahedra linked in thin, broad, flat sheets of indefinite extent in which the ratio of silicon and aluminium to oxygen is 2:5. Common phyllosilicate minerals include those from the following groups: mica, clay, serpentine, and chlorite. Examples include biotite, muscovite, lepidolite, phlogopite, and sericite. The leafy micaceous habit of phyllosilicates might be better visualized if associated with the Greek foods baklava and spanakopita, which are made using the distinctively thin and leafy pastry dough called phyllo (filo).
Cited usage: The phyllosilicates as a group typically have a platy crystal habit, with a cleavage parallel to the plane of layering of the structure, and are optically negative with rather high birefringence [1966, McGraw-Hill, Encycl. Sci. & Technology XII. 312/1].
See also: cyclosilicate, inosilicate, nesosilicate, sorosilicate, tectosilicate, silicate

phylogeny: [Greek *phylon* = race, class + *genea* = generation, race < *genos* = kind, race < *gonos* = birth, offspring; < IE *gen* = to give birth, beget.] n. The genesis and evolution of the characteristic features of a group of organisms. Term used in contrast to ontogeny, which refers to the development of an individual organism. Theory and term coined in 1866 by Ernst Haeckel .
Cited usage: Professor Häckel in his Generelle Morphologie has recently brought his great knowledge and abilities to bear on what he calls Phylogeny, or the lines of descent of all organic beings [1872, Darwin, Origin of Species 5th ed. xiv].
See also: ontogeny, recapitulation, ecology

phylum: [Greek *phylon* = class, race or stock; < IE *bheu* = to be, grow, exist.] n. A division of taxonomic classification, smaller than a kingdom and larger than a class. A phylum is often defined as a group of animals or plants genetically related and sharing a common ancestral form. Term first used in taxonomy by Carl Linnaeus (1707-1778). IE derivatives include: be, physics, future, and neighbor.
Alt. spelling: phyla (pl)

physical: [Middle English *phisik* = natural science < Greek *phusis* = nature < IE *bheu* = to be, exist, grow.] adj. Pertaining to the phenomenal universe perceived by the senses; said of that which is connected with matter and energy, or the flux between them. In ancient Greece, Aristotle defined physical to include the entire study of nature, things both organic and inorganic. By the 17th century, the philosopher John Locke even included spirits (gods and angels) as residing in the physical realm. In a geologic sense however, beginning in the 18th century, the term became more limited to inorganic matter and the processes affecting it.
Cited usage: Everything physical is measurable by weight, motion, and resistance. [1832-4, De Quincey, Caesars Wks. 1859 X. 14] There may be a physical break–unconformity–and also a palaeontological break, between two successive groups of strata. [1885, Lyell's Elements of Geology 100]

phytoplankton: [Greek *phytos* = plant, that which is grown + *planktos* = to wander, roam or drift.] n. That part of the marine planktonic mass composed of organisms capable of photosynthesis (plants and algae). Phytoplankton are the photo-autotrophs, such as: diatoms, coccoliths, and dinoflagellates.
Cited usage: In the spring months there is a great development of bacteria and other phytoplankton, which render the water less transparent than at other times of the year [1900, Geogr. Jrnl. XV. 336].

picacho: [Toponymous term after English *Picacho Peak* < Spanish *picacho* = peak.] n. Term used in the U.S. Southwest desert region for a sharply pointed, isolated peak. Named after the 500m (1,500ft) high Picacho Peak, located in the light-colored Picacho Mountains of Arizona. The name Picacho Peak is, therefore, redundant; translated literally into English it is "Peak Peak."

picrite: [Greek *pikros* = bitter.] n. A blackish-green volcanic rock, usually porphyritic, and consisting chiefly of olivine and augite, with possibly lesser amounts of biotite, amphibole, and plagioclase. Typical picritic texture consists of a dark, compact groundmass containing porphyritic phenocrysts.
Synonyms: bitterspath (Ger.)

pictograph: [Latin *pictus* = past participle < *pingere* = to paint; < IE *peik* = to cut, mark + Latin *graphicus* < Greek *graphikos* < *graphe* = writing < *graphein* = to write; < IE *gerbh* = to scratch.] n. A picture, symbol, grapheme, or intentional mark made on the surface of a rock or rock wall by human beings. Pictographs are not carved into rock, but are surface markings made by paints, dyes, chalk, or other substances capable of leaving a lasting film on the surface. IE derivatives include: poikilitic, picrite, and pigment.
See also: petroglyph

piedmont: [Toponymous term after *Piemonte*, a region in NW Italy at the foot of the Alps; < French *piedmont* = foot of the mountain < *pied* < Latin *ped* = foot + French *mont* < Latin *monte* = mountain.] n. An area of land formed or lying at the foot of a mountain or mountain range. The historic Piemonte region in Northwest Italy consists of eight provinces bordering France and Switzerland.

piezo-: [Greek *piezein* = to press tightly, squeeze.] Combining prefix meaning mechanical pressure, used in such terms as: piezoelectric, piezometric, and piezometer.

pike: [French *pique* < *piquer* = to prick.] n. General term used in the United Kingdom for a sharp-pointed peak, mountain, or hill.

pilotaxitic: [Greek *pilos* = felt, or hair made into felt + *taxis* = arrangement < *tassein* = to arrange.] adj. Said of a texture of holocrystalline volcanic rocks in which lath-shaped microlites (typically plagioclase) occur in the groundmass with a felt-like appearance, generally interwoven and in a random distribution. Term coined in 1888 by petrologist Harry Rosenbusch (1836-1914).
Cited usage: Pilotaxitic, the name given by Rosenbusch...to a holocrystalline structure especially characteristic of certain porphyrites and basalts [1888, F. H. Hatch, Teall British Petrographic Glossary].
Synonyms: felty

pinacoid: [Greek *pinak* = slab + *oidos* = having the likeness of.] adj. Term for an open crystal form, having a face (plane) that intersects the *c* axis, but is parallel to the *a* and *b* axes. (The *a*, *b*, and *c* axes in crystallography are synonymous with the Cartesian *x*, *y*, and *z* axes, respectively.) This crystal form results in platy or tabular crystals. The open crystal form, as in the pedion, pinacoid, dome, sphenoid, prism, and pyramid do not form an entire crystal on their own, but need another form to complete the crystal. This is in contrast to closed forms, such as the cube, tetrahedron, and octahedron.

pingo: [Eskimo *pingok* = conical hill.] n. In Arctic regions, a perennial, conical, soil-covered, dome-shaped ice mound, formed in part by increased hydrostatic pressure within or below the permafrost layer. Term coined in 1928 by L. Koch.
Cited usage: I have, however, introduced the name Pingo to designate a mountain entirely submerged by the Inland Ice, but setting its mark upon the surface of the ice. The best example of a Pingo is Mt. Haffner, Pingorsoak [1928, L. Koch, Meddelelser om Grønland LXV. 196-7].
Synonyms: boolgoonyakh
See also: nunatak

pinnacle: [Latin *pinna* = wing; < Greek *pterux* = wing.] n. In geomorphology, a tall, slender, tapering, and spire-shaped pillar of rock. At the top or summit of a pinnacle, there is often a resistant caprock, which serves to preserve the material beneath. In oceanography, a small, isolated reef patch or spire made of rock or coral that is awash or near the surface. In glaciology, a weathered and irregular iceberg topped with spires.

pinnate: [Latin *pinna* = wing; < Greek *pterux* = wing.] adj. In general usage, referring to a shape having bilateral symmetry of flat planes of grouped strands on each side of a common axis; like a feather or palm frond. In geomorphology, said of a symmetrically aligned dendritic drainage pattern. In biology and paleontology, said of lateral branches that grow in the same plane on either side of a main supporting axis.
Cited usage: All the other recognized types of drainage pattern: rectangular, trellis, annular, pinnate, contorted, are responses to structure [1942, O. D. von Engeln, Geomorphol. xi. 215].

piperno: [Italian *piperno* = peperino < *pepere* = pepper.] n. An unconsolidated, light, and porous volcanic tuff, usually brown in color, composed of sand, cinders, and crystal fragments, often of leucite. The name was first given to the volcanic tuffs of Monte Albano near Rome, alluding to its texture consisting of small, peppery grains.

Cited usage: The stone in general use for building here, is a hard volcanic tuffa of the sort called Piperno in Italy [1777, Hamilton, Phil. Trans. LXVIII. 3].
Alt. spelling: peperino. Piperino
See also: peperino

pipkrake: [Swedish *pipkrake* = needle ice < *pip* = pipe, musical tube + *klake* = hard as frozen ground.] n. One of the ice needles in needle ice. Pipkrakes are small crystals from 2.5-6cm long, and form perpendicular to the ground surface where ambient temperatures fluctuate above and below 0°C.
Cited usage: Water expands about 10% upon freezing (ice being characterized by a high expansion coefficient, as seen in the growth of ice needles or pipkrakes) [1968, R. W. Fairbridge, Encycl. Geomorphology 370/1].

pisolite: [Greek *pisos* = pea + *lithos* = rock, stone.] n. A rock, usually limestone, composed primarily of small, rounded accretionary masses called pisoliths.
Cited usage: Applied to an individual grain of this (pisolite limestone) similar to an oolith but larger (in mod. use applied to grains of diameter 2 mm. or more) [1884, Lyell, Elements of Geology (ed. 4) 12]. Oolite and pisolite are limestones built up of little spheroidal bodies resembling the roe of fishes or heaps of peas [1931, S. J. Shand, Study of Rocks xi. 153].
See also: pisolith, oolith

pisolitic: [Greek *pisos* = peas + *lithos* = rock, stone.] n. In sedimentology, referring to the texture of a sedimentary rock composed of pisoliths. Pisoliths are small, rounded, accretionary grains usually of calcium carbonate. Thay are larger than ooids, being greater than 2mm in diameter. In pedology, referring to the texture of a laterite soil containing pisoliths of bauxite, possibly formed by collooidal deposition during

soil genesis. In volcanology, referring to the texture of a tuff containing pisoliths (accretionary lapilli or mud balls).
Synonyms: pisolite
See also: pisolite, oolith

piton: [French *piton* = eyebolt, nail.] n. In general usage, a metal spike with an eye at one end that is driven into ice or a rock crevice, and used by mountain climbers as a temporary attachment point. In volcanology, a term for volcanic peaks, especially those in French-speaking islands, such as the West Indies. The etymology is uncertain whether the name for a volcanic peak is derived from mountain climbers using pitons to ascend peaks, or vice versa.

placer: [American Spanish *placer* = deposit, shoal < *placel* = sandbank < *plaza* = place.] n. A surface mineral deposit formed by the natural mechanical concentration of dense mineral particles from weathered debris. Examples of minerals found in placer deposits include: gold, platinum, diamonds, rutile, and cassiterte. In U.S. mining law, the term placer includes all forms of metalliferous mineral deposits except for veins in place.
Cited usage: They have at last discovered gold [in California]. Those who are acquainted with these 'placeres', as they call them, (for it is not a mine), say it will grow richer, and may lead to a mine [1842, Niles' Register 8 Oct. 96].

plagio-: [Greek *plagios* = oblique, slanting, inclining < *plagos* side; < IE *plak* = to be flat.] Combining prefix meaning slanting or inclining, used in such terms as plagioclase and plagiotropism.

plagioclase: [Greek *plagios* = oblique, slanting, inclining < *plagos* = side; < IE *plak* = to be flat + Greek *klasis* = fracture, cleavage < *klan* = to break.] n. Any mineral of the common rock-

forming series of triclinic feldspars having the general formula, $(Na_x, Ca_{x-1})(Si_{x+2}, Al_{x-2})O_8$. Term coined by Breithaupt in 1847.
Cited usage: Plagioclase, Breithaupt's name for the group of triclinic feldspars, the two prominent cleavage directions in which are oblique to one another [1868 James D. Dana, A System of Mineralogy (ed. 5) Suppl. 802].

plain: [Latin *planum* = plane.] n. An area of land that is comparatively flat, featureless, and of few trees relative to the surrounding region, such as a meadow or prairie.

planation: [Latin *planum* = plane.] n. The levelling of a landscape by erosion.
Cited usage: The process of carving away the rock so as to produce an even surface, and at the same time covering it with an alluvial deposit, is the process of planation [1877, G. K. Gilbert, Rep. Geol. Henry Mts. 127].

plankton: [Greek *planktos* = to wander, roam or drift.] A generic name for all species and developmental forms of free-floating organic life found in all depth zones in the ocean, or in bodies of freshwater. Categories of plankton include: phytoplankton, zooplankton, holoplankton, and meroplankton. Term coined in 1887 by Victor Hensen.

planosol: [Latin *planum* = plane + French *sol* < Latin *solum* = ground.] n. An obsolete term for an intrazonal soil characterized by a thin, strongly- leached surface horizon overlying a compacted hardpan or claypan, and occurring on flat uplands with poor drainage. Term coined in 1938 by M. Baldwin.
Cited usage: The term 'Planosol' is being proposed to cover those soils with claypans and cemented hardpans not included with the Solonetz, Ground-Water Podzol, and Ground-Water Laterite [1938 M. Baldwin et al. in U.S.D.A. Yearbook. 991].

plastic: [Latin *plasticus* = moulder, sculptor; < Greek *plastos* = formed < *plasis* = formation.] adj. In structural geology, said of a body that undergoes permanent deformation under strain without rupturing.

plateau: [French *plattel* = platter < *plat* = flat; < Greek *platus* = flat; < IE *plat* = to spread.] n. An elevated tract of comparatively flat or level land, usually being an extensive flat upland area of at least 150 meters elevation above the adjacent land. Plateaus can be formed: a) structurally, from regional uplift, as an elevated plateau, b) residually, from horizontal strata resistant to erosion, as a remnant plateau, or c) volcanically, from the outpouring of lavas, as a Deccan-type plateau. A plateau is larger than a mesa, has much of its total surface at summit level, has at least one precipitous side, and is often dissected by deep stream cuts. IE derivatives include: platy, platinum, plantigrade, flounder and pancake.
Cited usage: On the chalk of Berkshire, extensive plateaus, six or seven miles wide, would again be formed [1830, Lyell, Principles of Geology I. 375].
Alt. spelling: plateaux (pl)
Synonyms: tableland

platinum: [Spanish *platina* = diminutive form of *plata* = silver; < Greek *platus* = flat, broad; < IE *plat* = to spread.] n. A dense, silver-white to gray, metallic element (symbol Pt). It occurs widely in nature as the cubic mineral and is usually mixed with other metals such as palladium, iridium, osmium, and nickel. It is ductile and malleable, does not oxidize in air, and is a good catalyst. Platinum occurs in alluvial deposits as the native mineral. Disseminated in mafic and ultramafic igneous rocks platinum combines with other elements in such minerals as sperrylite $(PtAs_2)$ and cooperite $(Pt, Pd, Ni)S$. IE derivatives include: plate, plateau, platy, platitude, and plaza.

playa: [Spanish *playa* = shore, beach; < Latin *plagia* = hillside, shoreline; < (probably) Greek *plagios* = oblique, sideways.] n. A term used in the U.S. Southwest for a dry lake bed which is characteristically flat and barren, being composed of layers of sand, silt, clay, and soluble salts. Playas are located at the lowest part of an undrained desert basin. After a rainfall, a playa becomes a temporary, or ephemeral, lake.
Synonyms: sabkah
See also: plagio-

plaza: [Spanish *plaza* = marketplace; < Latin *platea* = broad street.] n. A term used in desert regions, especially the U.S. Southwest, for a flat, wide floor of an open valley.

Pleistocene: [Greek *pleistos* = most < IE *pel* = to fill + *kainos* = recent, new.] adj. First epoch of the Quaternary period (1.81Ma-10Ka), which, compared to other epochs of the Cenozoic Era, has the greatest number of fossils of extant species. It is characterized by cyclic high-latitude glaciation, worldwide spread of hominids, and the post-Wisconsin extinction of numerous land mammals, such as mastodons and saber-toothed cats. The term Pleistocene was coined in 1833 by Charles Lyell, in allusion to a section of strata in eastern Sicily, which contains a high percentage of living species of mollusc shells. Strata with greater than 70% fossils of extant species were originally designated Pleistocene in age. This was discovered to be too arbitrary a criterion, because many rock units contain no fossils. Therefore, the Pleistocene is now subdivided on the basis of paleomagnetism and oxygen-isotope levels. The Latin equivalent to the Pleistocene epoch is the Diluvium epoch, a term still used in parts of Europe.
Synonyms: Diluvium
See also: Diluvium, Holocene

pleochroism: [Greek *pleion* = more; < IE *pele* + *khros* = color.] n. The property possessed by some anisotropic crystals to differentially absorb certain wavelengths of transmitted light in different crystallographic directions. The resultant effect is that a mineral will exhibit different colors when viewed from different directions. The effect is most easily seen under polarized light. Etymologically related terms include: Pliocene, Pleistocene, pleiotropism, and paludal.

pleura: [Greek *pleura* = side of the body.] n. One of two laterally symmetrical parts of an invertebrate; such as the attachment point in arthropods, the lateral thoracic or pygidial lobes of a trilobite, or the toothed part of a mollusc.

plica: [Latin *plica* = fold < *plicare* = to fold.] n. One of many dominant parallel ridges on a bivalve or brachiopod shell extending radially outward from the hinge beak to the outer edge.

plinian: [Eponymous term named after the Roman, Pliny the Elder (23 - 79AD).] adj. A term describing the most vigorous stage of a volcanic eruption in which a narrow jet of gas is ejected with great violence from a central vent to a height of several kilometers before it expands sideways, looking much like an atomic bomb. So called because Mount Vesuvius in Italy experienced this kind of eruption in 79 CE, killing the Roman explorer, soldier, statesman, and encyclopedist Pliny the Elder. The event was later described by his nephew, Pliny the younger.
Cited usage: Four days after the paroxysm [of Vesuvius in 1906]–the Vesuvian phase–began, it culminated in a mighty uprush of gases–the Plinian phase –which continued for the greater part of a day, tearing away the upper portions of the cone, and reaching a height of 8 miles [1944, A. Holmes, Princ. Physical Geol. xx. 466]. Probably the best modern example of a Plinian eruption was that of the Bezymianny volcano in Kamchatka [1976, P. Francis, Volcanoes iii. 114].

plinthite: [Greek *plinthos* = tile, brick, stone squared for construction.] n. In a highly weathered lateritic soil, material rich in iron and aluminium oxides, clay, and quartz. Plinthite occurs as red mottles in a globular, reticulate, platy, or polygonal pattern. Repeated wetting and drying of this laterite changes plinthite into in irreversible, impermeable ironstone hardpan. Term coined in 1836 by T. Thomson, for the lateritic, brick-red clays occurring among the trap rocks of Antrim and the Hebrides. The platy shapes of this plinthite resemble the slablike shape of an architectural plinth.
Cited usage: Plinthite. I give this name to a mineral which occurs in the County of Antrim, from its brick red colour [1836, T. Thomson, Min. I. 323].

Pliocene: [Greek *pleion* = more; < IE *pel* = to fill.] adj. The final epoch of the Tertiary Period (5.33-1.81Ma), distinguished from other epochs of the Tertiary period by having the greatest number of fossil species still alive today. Term coined in 1833 by Charles Lyell, who devised a scheme for stratigraphically subdividing the Tertiary period based upon the percentage of fossil molluscs found in ancient strata that remain extant species today. The oldest Tertiary rocks contain the fewest extant species, whereas the youngest have proportionally more species alive today. In his division of the Tertiary, Lyell created the Eocene (dawn of the recent), Oligocene (little recent), Miocene (middle recent), Pliocene (nearly recent) and Newer Pliocene (which he later termed the Pleistocene - most Recent). Paleocene was later added, first as the lower part of the Eocene, and finally as a distinct epoch.
Cited usage: The Pliocene rocks of England include the red crag and coralline crag of the eastern counties. [1866, Brande & Cox, Dict. Sci.].
Synonyms: Upper-Tertiary (Great Britain)

plumbago: [Latin *plumbago* = lead ore < *plumbum* = lead.] n . An obsolete term for graphite, and for minerals that resemble graphite, such as: molybdenite, litharge (a yellow oxide of lead), or galena (the sulfide of lead).

plume: [Latin *pluma* = a small soft feather, down.] n. In tectonic and geodynamic theory, an upwelling column of hot, partially molten material rising from the Earth's lower mantle, then spreading laterally upon meeting the lithosphere. In oceanography, hydrology, and limnology, a long streamer of fluid issuing from a localized source and spreading out into another fluid with different characteristics (e.g., upwelling plume, pollution plume, thermal plume, or a halo-plume). In mineralogy, a featherlike inclusion in a gem.
Cited usage: The plume effectively burns a hole through the overlying crustal plate and a volcano results [1976, P. Francis, Volcanoes i. 49].

pluton: [Eponymous term after the Roman god *Pluto* < Greek *Plouton* = god of the underworld < *ploutos* = wealth; < IE *pleu* = to flow.] n. A body of intrusive igneous rock formed by cooling and crystallization of magma. The etymological connection stems from the ancient belief that the underworld is the source of wealth that permeates to the surface through the ground. IE derivatives include: flow, pluvial, and pneumatic.
Cited usage: The unstratified crystalline rocks have been very commonly called Plutonic, from the opinion that they were formed by igneous action at great depths [1833, Lyell, Principles of Geology. III. 353].

pluvial: [Latin *pluvia* = rain < *pluere* = to rain; < IE*pleu* = to flow.] adj. In meteorology, pertaining to precipitation of water in the atmosphere. In climatology, said of a climate characterized by high rainfall. In geomorphology, said of a geologic process, feature, or time period marked by abundant rainfall, such as the pluvial lakes that occupied the Great Basin of Nevada during the wetter times of the last glacial period. As a noun, a

period of relatively high average rainfall at low and intermediate latitudes during the geologic past, especially during the Pleistocene epoch. Pluvials alternated with interpluvial periods in a cycle analogous to the cycle of glacial and interglacial periods at higher latitudes. IE derivatives include Pluto, pluton, and flood.
Cited usage: Prolonged periods of high rainfall are called pluvials, and are marked by changes in lake levels and in flora and fauna [1970, Bray & Trump, Dict. Archaeol. 184/1]. The cold phases are called 'glacial periods' in northern latitudes and 'pluvial periods' in the latitude of Palestine, where there was no glaciation, but instead a greatly increased rainfall [1949, Wm. Foxwell Albright, Archaeology of Palestine 50].

pneumatolytic: [Greek *pneuma* = wind, spirit + *lutikos* = able to loosen.] adj. In petrology, applied to the stage of magmatic differentiation between pegmatitic and hydrothermal. Also, said of contact metasomatic effects that are adjacent to deep-seated intrusions. In ore deposits, said of a rock or mineral deposit formed from gaseous causes, fumerolic deposits near volcanoes for example. The term alludes to the gaseous nature of the fluid involved in pneumatolytic processes.
Alt. spelling: pneumatolitic

pneumo-: [Greek *pneuma* = wind, spirit.] Combining prefix for various terms relating to air, and, by extension, any gaseous substance; used in such terms as: pneumatolysis, pneumotectic, pneumonia, and pneumatocyst.

pocosin: [Algonquin *pocosin* = swamp on a hill, near an opening or widening *poquo* = to widen, open out.] n. The elevated, perennially wet, and peat-rich areas covered with evergreen trees and shrubs that grow on sandy soils along the U.S. Southeastern seaboard in coastal Virginia, Maryland, Delaware, and the Carolinas. The etymological connection relates to a locality where

water backs up in spring freshets, and becomes more or less marshy or boggy. As the name of a river in Virginia, the term was used as early as 1635.
Cited usage: There is a large aggregate of territory bordering on the seas and the sounds, known as Swamp Lands. They are locally designated as 'dismals' or 'pocosins', of which the great Dismal Swamp on the borders of North Carolina and Virginia is a good type [1875, W. C. Kerr, Rep. of the Geol. Survey of N. Carolina I. 15].
Alt. spelling: poquosin
See also: marsh, swamp

pod: [Origin unknown, possibly from Russian *pod* = under; < IE *ped* = foot.] n. In geology, a body of ore or rock whose length greatly exceeds its other dimensions. In geomorphology, a term used in the steppes of southern Russia for a broad, shallow, depression that is host to numerous intermittent lakes. First usage in English c. 17th century.
Cited usage: Chromite deposits of the sack-form variety are notable for their variation in size and shape. The majority might be termed lenses or pods, as their length greatly exceeds their width [1942, T. P. Thayer, Bull. U.S.G.S. No. 935. 23].

podzol: [Russian *pod* = under; < IE *ped* = foot + *zol* = ashes; < IE *ghel* = to shine.] n. In soil taxonomy, a leached, infertile, acidic soil formed mainly in cool, humid, temperate climates that are characterized by coniferous forests or heath vegetation. Podzols have a distinctive white or gray ashlike subsurface layer composed of residual quartz grains left after intense leaching of other soil consatituents. The type locality is in Northern Russia. The term was originally applied only to the ashlike layer itself, instead of to the entire soil.
Cited usage: If the American view is accepted, a lateritic soil can also be podzolic [1952, P. W. Richards, Tropical Rain Forest ix. 209].
Alt. spelling: podsol

poikilitic: [Greek *poikilos* = mottled, spotted, variegated.] adj. Said of the texture of an igneous rock, in which small crystals of one mineral are chaotically scattered within a much larger anhedral crystal of another mineral, such as grains of plagioclase feldspar within a pyroxene crystal. A now obsolete use of the term applied to the Triassic and Permian periods because those rock systems are composed mainly of variegated and mottled rocks. Term coined by G. H. Williams in 1886.
Cited usage: This structure is so common in many massive rocks, especially in the more basic kinds, that I would venture to suggest the use of the term 'poicilitic' [1886, G. H. Williams, Amer. Jrnl. Sci. CXXXI]. (In reference to the obsolete meaning) The term 'Poikilitic' was formerly proposed for them, on account of their characteristic mottled appearance [1885, Geikie, Textbook of Geol. 2nd ed. 748].
Alt. spelling: poecilitic

poikilo-: [Greek *poikilos* = various, mottled, variegated, spotted.] Combining prefix meaning variety or change. The term can relate to visible textures, such as poikilitic, poikiloblastic, poikilocrystallic, and poikilophitic. Also used to mean varying values, such as the temperature fluctuations of poikilothermic (cold-blooded) organisms.

poikiloblastic: [Greek *poikilos* = mottled, spotted, variegated.] adj. Said of the texture of a metamorphic rock, or to the rock itself, in which small crystals of an original mineral occur within crystals of its metamorphic product.
Alt. spelling: poeciloblastic
Synonyms: sieve texture

poikilothermic: [Greek *poikilos* = various, spotted + *therme* = heat < *thermos* = hot.] adj. Said of an animal that exhibits passive variation in its internal body temperature, changing with the temperature of its environment. With the possible exception of some species of dinosaurs, all animals except birds and mammals are poikilotherms.
Synonyms: ectothermic
See also: endothermic, ectothermic, homoiothermic

pokelogan: [Ojibwa *pokenogun* = pokelogan.] n. A relatively stagnant bay, marsh, or pocket of water connected to a river. Loggers in the U.S. Northeast and Midwest used this term to describe a place into which logs may float during a lumber drive on a river or lake. This prompted Henry David Thoreau, in his essay '*Ktaadn*,' to coin the folk etymology "poke-logs-in." Ojibwa is in the Algonquin family of languages.
Cited usage: The term pokeloken, an Indian term, signifying 'marsh', is still largely used by the lumbermen in Maine, and in the Northwest [1872, Schele de Vere, Americanisms 20].
Alt. spelling: pokeloken
Synonyms: logan, set-back, bogan, peneloken

polder: [Dutch *polder* = a reclaimed tract of land from a body of water.] n. Used especially in the Netherlands, an area of low-lying land that has been isolated from the sea, a lake, or a river, drained of standing water, and then protected from flooding by dikes, levees, and dams. The world-renowned linguist Otto Jespersen called the Low Dutch dialect "Polder Dutch," due to the low geographical elevations where it is commonly spoken.
Cited usage: The soil is moorish, boggy and fenny, such as our Ancestors have usually called Polder: i.e. a marsh fenn, a meadow by the shore side, a field drain'd or gain'd from a river or the sea, and inclosed with banks [1669, William Somner, A Treatise of the Roman Ports & Forts in Kent].

polje: [Serbo-Croation *polje* = field.] n. In karst topography, a usually flat-floored, enclosed

depression that drains underground, commonly with a layer of alluvium across which a stream may course, and occasionally damming to form a lake.
Cited usage: The largest depressions of Yugoslavia, the poljes, are probably not solution forms at all but tectonic depressions modified by solution of the limestone preserved in them [1960, B. W. Sparks, Geomorphol. vii. 155].
Alt. spelling: polye; polje
See also: karst, uvala

poly-: [Greek *polu* = poly- < *polus* = much, many; < IE *pele* = to fill.] Combining prefix meaning many, used in such terms as: polymorph, polymictic, polymodal, and polyplacophoran.
See also: multi-

polymorphism: [Greek *polus* = many + *morphos* = shape, form.] n. In mineralogy, denoting a substance that exists in more than one form or crystal structure (e.g., diamond and graphite, or calcite and aragonite). In paleontology and biology, multiple alleles of a gene within a population, usually expressing different phenotypes (e.g., human hair and skin color). Polymorphism pertains to the existence of two or more distinctly different groups of a species that still belong to the same species (e.g., dog or horse breeds and human ethnicities). Frequently, each of the different forms of a polymorphic substance is called a morph; when only two morphs exist for a single species, they are dimorphic.
Cited usage: The perplexing subject of polymorphism [1857, Charles Darwin, Life and Letters].
See also: isomorphism, heteromorphism

polyp: [Middle English *polip* = nasal tumor; < Old French *polipe* < Latin *polypus* = cuttlefish, nasal tumor; < Greek *poly* = many + *pous* = foot.] n. In marine biology, the sedentary stage of a marine invertebrate of the phylum Cnidaria, characterized by a cylindrical or saclike body and numerous tentacles attached near the ventral cavity. The polyp stage usually alternates with the free-swimming medusa stage. The etymology alludes to the original meaning of a cephalopod having many tentacles, such as a squid (cuttlefish) or octopus.

polyploidy: [Greek *polus* = many + *ploidos* < *eidios* = form.] adj. A cell or organism containing multiple copies of its homologous chromosomes, that is to say, more than twice the haploid number. Polyploidy occurs in animals, but is more common among flowering plants. Wheat, for example, after hybridization by humans, has strains that are diploid, tetraploid (durum or macaroni wheat), and hexaploid (bread wheat). Term coined by H. Winkler in 1916.

ponor: [Serbo-Croatian *ponor* = swallet, swallow hole.] n. A hole in the bottom or side of a natural depression in karst terrain through which water passes to and from underground channels.
Cited usage: The funnel-shaped hollows which are so frequently met with on the surface of the karst are termed ponors [1922, Geol. Mag. XIX. 406].
Synonyms: swallow hole, stream sink, aven, gouffre
See also: karst, polje, dolina, uvala

pontic: [Toponymous term after Pontus, the Black Sea; < Greek *pontos* = sea (specifically the Black Sea).] adj. Said of deep, low-energy, sedimentary environments and the sedimentary facies that accumulate in them, such as carbon-rich shales and limestones. The ancient country of Pontus was located on the southern coast of the Black Sea in Asia Minor (now Turkey). It initially developed in the 4th century BCE, then expanded under the leadership of Mithridates VI to include most of Asia Minor and Greece. Pontus ultimately succumbed to the Roman army, first led by Lucullus, and finally by Pompey of Rome in 66 BCE.

ponzite: [Toponymous term after *Ponza*, a group of Italian islands in the Mediterranean Sea + Greek *-ites* < *-lite* = pertaining to rocks < *lithos* = rock, stone.] n. An alkali-rich, low-quartz igneous extrusive rock; a feldspathoid-free trachyte, Term coined by Washington in 1913 after the type locality in the Ponza Islands.

porcelain: [French *porcelaine* = cowry or Venus shell, porcelain; < Italian *porcellano* = of a young sow < *porca* = sow; < Latin *porcus* = pig; < IE *porko* = young pig.] n. A hard, white, translucent, earthenware ceramic made by firing pure clay and then glazing it with variously colored fusible materials; specifically, a hard, plastic paste composed of kaolin and petuntze, or a related siliceous material, which, when high-fired at about 1300°C, becomes a hard, vitrified, strong, and translucent ceramic. Porcelain is considered intermediate between glass and earthenware. Used as an adjective to describe a hard, white, translucent lustre, usually of a siliceous chert, and sometimes naturally baked (as in porcelain jasper) making it appear like pure clay-fired ceramic. The term porcelain was supposedly coined by Marco Polo in the 13th century. The term comes from the cowry shell's resemblance to a pig's back. IE derivatives include: porcupine, porpoise, pork, and aardvark.

pore: [Latin *porus* < Greek *poros* = passage; < IE *per* = to lead, pass over.] n. In general usage, a small opening or passageway; used in geology with various meanings, such as: an interstice in rock or soil, one of many small openings in the surface of an organism linking it to the environment, or a thinning in a pollen grain. As an adjective in compound terms, such as: pore canal, pore diameter, pore ice, pore space, and pore pressure. IE derivatives include: petrify, petroleum, diapir, and fjord.
Synonyms: interstice
See also: Porifera

poriferan: [Latin *porus* = pore + *-fer* = bearing.] n. Any of various members of the phylum Porifera constituting the sponges. Poriferans are sessile, mostly marine, water-dwelling filter feeders. There are over 15,000 modern species of sponges, with the fossil record dating back to Precambrian time.
Synonyms: sponge

poro-: [Latin *porus* < Greek *poros* = passage; < IE *per* = to lead, pass over.] abbr. As a combining prefix meaning opening or hole, as in: porous, porosity, porolith, and poriferan.

porphyro-: [Greek *porphyrites* = purple, hard, crystalline rock quarried in Egypt < *porphyros* = purple < *porphyra* = the purple-whelk.] Combining prefix meaning porphyritic, an igneous texture characterized by relatively large phenocrysts set in a fine-grained or glassy groundmass; used in such terms as porphyroblast, porphyroclast, porphyroid, and porphyro-aphanitic.
See also: porphyry

porphyry: [Greek *porphyrites* = purple, hard, crystalline rock quarried in Egypt < *porphyros* = purple < *porphyra* = the purple-whelk.] n. An igneous rock containing relatively large conspicuous phenocrysts of any composition embedded in a fine-grained or glassy groundmass. This term, derived from the name of a color, has transformed into a strict textural term in geology. The color name was adapted and given to a highly prized hard rock quarried in ancient Egypt. It was composed of white to red phenocrysts of plagioclase feldspar contained in a microcrystalline purple ground-mass consisting of hornblende, plagioclase, apatite, thulite, and withamite, the last two imparting the distinctive purple-red coloration. Eminent Roman leaders employed sculptors to make lasting busts of themselves from this special rock, especially because its color was that of the imperial robes (*porphyrogenitoss*, meaning born of the purple). These purple busts, composed of two very differently sized crystal masses, became known as porphyries. The purple-whelk shellfish, or common whelk, is so named because of the distinctive purple dye it yields when boiled. The site of the lost ancient quarries was discovered by Burton and Wilkinson in the early 1800's at Gebel Dokhan, near the Red Sea (about 27°N 33°E). *Cited usage*: Porphyry is hence applied to every species of unstratified rock, in which detached crystals of feldspar are diffused through a base of other mineral composition [1833, Lyell, Principles of Geology]. *See also*: porphyro-

portal: [Latin *porta* = gate.] n. In mining, the opening of an adit or tunnel.

portland cement: [English *Portland stone* = naturally occurring building stone quarried on the English Isle of Portland.] n. A high-quality cement made by mixing proportioned amounts of finely ground limestone and shale, then firing them to incipient fusion to drive off water. Invented in England by Joseph Aspdin in 1824, the resultant product looks like Portland stone. *See also*: pozzolan

post-: [Latin *post* = after, behind.] Combining prefix meaning after or behind, used in such terms as: posterior, postglacial, postorogenic, posthumous, and postobsequent. *See also*: obsequent, orogeny, glacier, humus

potable: [Latin *potabilis* = drinkable < *potare* = to drink.] adj. Said of water, and occasionally other liquids, that are suitable for drinking. *Cited usage*: They bore the tree with an awger, and there issueth out sweet potable liquor [1645, Howell, Lett. I. 369].

potam-: [Greek *potamos* = river.] Combining prefix relating to rivers, used in such terms as: potamic, relating to rivers; potamology, the

scientific study of rivers; and potamoclastic, sedimentary clasts produced by fluvial processes. *Cited usage*: In the school of Carl Ritter, much has been said of three stages of civilization determined by geographical conditions, the potamic which clings to rivers, the thalassic which grows up around inland seas, and lastly the oceanic [1883, Sir J.R. Seeley,The Expansion of England 87]. *Synonyms*: fluvio- *See also*: fluvial

potash: [Dutch *pot-asschen* = crude wood ashes leached in a pot.] n. Potash is a catchall term for various compound of potassium: potassium carbonate (K_2CO_3), potassium hydroxide (KOH), anhydrous potassium oxide (K_2O), and potassium chloride (KCl). It is occasionally used for the element potassium in its free state. In general industrial usage, potash generally refers to the carbonate, whereas chemists usually refer to it as the hydroxide or anhydrous oxide. In 1756 *pot-asschen* was proven by Dr. Joseph Black of Edinburgh to be a compound substance, which, in 1807, Sir Henry Davy found to contain a new element, the metal potassium.

potassium: [English *potash* < *pot ashes* the chemical symbol *K* < Latin *kalium* < Arabic *qali* = alkali < *al-qali* = the substance that has been roasted, ashes of saltwort.] n. A silver-white, explosively reactive, metallic element of the alkali series (symbol K). Potassium lies directly below sodium on the periodic table, making it more reactive. Although widely distributed, being the seventh most abundant metal in the Earth's crust, it is not found as a free element in nature, but is always bound in compounds. Potassium was discovered and named in 1807 by Sir Humphry Davy, who isolated it from dried and leached plant ashes. The name *kalium* was suggested for this element by Klaproth and Gilbert in 1813; it is the preferred term in most languages other than

the Germanic and Romantic languages, which adopted only the first letter for use as the elemental symbol.
Cited usage: Potassium and Sodium are the names by which I have ventured to call the two new substances [1807, Sir H. Davy, in Phil. Trans. XCVIII. 32].
Synonyms: ashlagan
Similar terms: potash

potrero: [Spanish *potrero* = corral.] n. In beach processes, a long, extended beach ridge, resembling an island and separated from the coast by a lagoon and barrier island. Usually surrounded by mud flats, a potrero consists of an accumulation of continuously growing dune ridges. In geomorphology, a type of oval-shaped valley that is entirely encircled by steep mountains except for a narrow and deep canyon that drains from within, called *boca de la potrero*, meaning mouth of the potrero. The etymological connection refers to the corral-shaped valley defined by steep mountains. Because steep walls abound along the periphery of a potrero, social development, such as: trails, roads, and settlements occurs close to the mouth of the canyon (*boca de la potrero*).

pozzolan: [Toponymous term after *Pozzuoli*, a town and bay near Naples, Italy; < Latin *puteoli* = little springs.] n. A siliceous, volcanic, leucite tuff that, when mixed with portland cement, reacts with $Ca(OH)_2$ to enhance the concrete's adhesive and cohesive properties. The town of Pozzuoli, Italy was chosen for the rock name due to its proximity to the stratovolcano Mt. Vesuvius, where this type of volcanic rock occurs. The eruption of Vesuvius in 79 CE. buried the towns of Pompeii and Herculaneum; among the many killed was Pliny the Elder. The Romans used pozzolana cement to build many famous structures including the Appian Way, the Roman Baths of Caracalla, the Basilica of Maxentius, the Colosseum and the Pantheon in

Rome, and the Pont du Gard aqueduct in southern France.
Cited usage: Rome is built, one may say, of pozzolana [1900, The Quarterly Review Jan. 33]. Pozzolans are natural and artificial siliceous and aluminous substances which are not cementitious themselves but which react with lime in the presence of water at atmospheric temperatures to produce cementitious compounds [1951, Economic Geology XLVI. 311].
Alt. spelling: puzzolan, pozzolana (It.), puzzuolana
See also: portland cement, travertine

prase: [Greek *prasios* = leek-green < *prason* = leek.] n. A dull, light-green variety of translucent chalcedony (cryptocrystalline quartz). The dull green color is due to the presence of masses of hairlike inclusions of actinolite.

pre-: [Latin *prae* = before, in front, in advance.] Combining prefix meaning: a) prior occurrence in time or development, as in Precambrian, precognition, and preview; or b) the state of being in front, as in preeminent, preamble, and precipitation.

Precambrian: [Latin *prae* = before, in front, in advance + Welsh *Cambria* (Latinized form of Cymry) = Wales < Old Celtic *combroges* = compatriots.] adj. All geologic time prior to the Phanerozoic eon, from the time Earth originated c. 4.6Ga to the start of the Cambrian Period in the Paleozoic era 542Ma. The term Precambrian is an informal time unit encompassing the Hadean, Archean, and Proterozoic eons.
Cited usage: In Precambrian nomenclature almost every geochronologic and chronostratigraphic subdivision should probably be regarded as informal, except for Archean and Proterozoic and their subdivisions into Early, Middle, and Late, which have been formalized as geochronometric units [Harrison and Peterman 1980, 1982, Geol. Soc. of Amer. Bull. V91 No 6 P. 377-380].

Synonyms: Cryptozoic (obsolete)
See also: Archean, Hadean, Proterozoic, Cryptozoic, Phanerozoic

precipice: [French *précipice* < Latin *praecipit-* = headlong < *prae* = before, in front + *caput* = head.] n. An overhanging, vertical, or very steep face of rock, such as a cliff, crag, or steep face of a mountain.

precipitation: [Latin *praecipitationem* = noun of action < *praecipitare* = to go headlong, to fall < *prae* = before + *caput* = head.] n. In meteorology, water that descends from the atmosphere to the surface of the earth in any form (e.g., rain, snow, hail, sleet, or fog drip). In chemistry, the creation of a solid substance directly out of liquid solution on addition of some causative agent or reagent.

precise: [Latin *praecdere* = to shorten < *pre* = before + *caedere* = to cut; < IE *kaid* = to strike.] adj. Strict in execution or performance; scrupulous and exacting. Not to be confused with accurate. IE derivatives include: scissors, circumcise, incise, and chisel.
See also: accurate

presque isle: [French *presque* = almost + *isle* = island.] n. A peninsular point extending into a lake, feebly connected to the mainland so that it nearly forms an island. The type example is Presque Isle, Michigan, in the U.S., which protrudes into Lake Huron.

primarrumpf: [Latin *primarius* = of the first rank, chief, principal < *primus* = first + German *rumpf* = body, torso.] An obsolete term for a flat and steplike remnant structure that has not been destroyed by a prevailing cycle of erosion in the area. Term coined in 1924 by W. Penck, in an attempt to name the universal initial geomorphic unit for all the topographic sequences that follow.

pro-: [Latin *pro-* < Greek *pro* = before (of time, position, preference, priority), in front, rudimentary.] Combining prefix meaning before, prior, or in front, with reference to such ideas as: a) time, as in Proterozoic and prologue, b) position, as in prosome and prostrate, c) priority, as in prorate, and d) preference, as in prospector and pro-choice.

prodelta: [Latin *pro-* = forward + Greek *delta* = 4th letter of Greek alphabet in the shape of a triangle.] n. The part of a river delta that lies beneath and extends beyond the delta front. The prodelta lies deeper than the effects of wave erosion and extends to the basin floor where clastic river sediment is negligent.
See also: delta

progradation: [Latin *pro-* = forward + *gradus* = step.] n. The seaward advance of a coastline, delta, or fan due to: a) continued sediment supply from terrigenous sources, b) a lowering of sea level, or c) regional change in the environment.
Cited usage: The virtual elimination of shelf seas, during a prolonged phase of tectonic stability and peneplanation following rapid build-up of evaporites and progradation of coastal plain sediments [1971, Nature 10 Sept. 91/2].

proluvium: [Latin *pro* = forwards, downward, in front of + *luere* = to wash.] abbr. A deltaic sediment of varied composition that accumulates at the toe of a slope due to flooding or intense washing of the sediments. Also, in arid climates, proluvium refers to mudflow and flash-flood deposits of a bajada-type alluvial fan.
Similar terms: alluvium, diluvium, eluvium, fluvial
See also: alluvium, diluvium, eluvium, illuviation

Proterozoic: [Greek *proteros* = former, anterior, before in place, time, order or rank.] adj. The most recent of the three eons in Precambrian time; after

the Archaean eon and prior to the Phanerozoic eon. The Proterozoic eon lasted from 2.5Ga-542Ma and is divided into the Paleopreoterozoic, Mesoproterozoic, and Neoproterozoic eras. There is possible confusion regarding time units of the Precambrian, due to its sometimes erroneously being called an eon. When the Precambrian is considered an eon, then the Hadean, Archaean, and Proterozoic would logically all be eras, and, following this logic, the Paleo-, Meso-, and Neoproterozoic would all be periods.
Cited usage: To the Proterozoic Era is assigned the time that elapsed between the close of the formation of the igneous complex and the beginning of the lowest system which is now known to contain abundant well-preserved fossils [1905, Chamberlin & Salisbury, Geol. I. i. 162].
Synonyms: Algonkian
See also: Precambrian, Archaean, Hadean, Cryptozoic

protist: [Greek *protistos* = the very first, superlative of *protos* = first.] n. Any organism in the kingdom Protista, including forms having characteristics of both plants and animals. Protists include autotrophic algae, heterotrophic protozoans, and the fungi. Prior to the domain theory of life, the category of protists also included bacteria and viruses (when viruses are considered a form of life). In the Three Domains of Life theory put forth by Carl Woese in 1990, the kingdom Protista comprises the simplest organisms in the domain of Eukaryota. Most are unicellular, some are colonial, and others are simple multicellular organisms closely related to single protist cells. The name Protista was coined in New Latin by German biologist Ernst Haeckel in 1868, analogous to the term Primalia coined by Wilson and Cassin. Haeckel predated the theory of the three domains of life when he proposed Protista as the third kingdom of organized beings to include creatures of the simplest structure, not definitely classified as either animals or plants (e.g., Protozoa and Protophyta).
See also: ecology

proto-: [Greek *protos* = first.] Combining prefix meaning first formed, primitive, or original; as in protoclastic, Protozoa, and prototype.

Protochordata: [Greek *protos* = first + Latin *chorda* = cord; < Greek *khorde* = cord; < IE *gher* = gut, entrails.] n. A small marine animal belonging to one of the subphyla Hemichordata or Cephalochordata, which form a group considered to be related to ancestors of the vertebrates, and are characterized by a dorsal nerve cord, a notochord, and gill slits. The biological concept of 'chord' has evolved over time, beginning with the visceral meaning in Indo-European to that of the neural idea in New Latin.
Cited usage: All these modern protochordates are small and live in the sea [1933, L. A. Adams, Introd. Vertebrates i. 17].

prototheria: [Greek *protos* = first + *theria* = plural of *therion* = wild animal.] n. The lowest subclass of the class Mammalia, comprising the single order Monotremata, and considered to be the most primitive mammalian type.

protozoan: [Greek *protos* = first + *zoa* = plural of *zoon* = animal, living being.] n. Any organism of the subkingdom Protozoa, defined as single-celled, heterotrophic, eukaryotic organisms, such as: amoebas, paramecia, ciliates, flagellates, and sporozoans. Term coined by Goldfuss in 1818, which included higher forms of metazoans as well, such as: sponges, hydroids, corals, and crinoids. Traditionally, the animal kingdom has been divided into two subkingdoms, the Protozoa and Metazoa. In modern taxonomy, there are four kingdoms in the domain Eukaryota: Protista, Fungi, Plantae, and Animalia.

Cited usage: The name 'Protozoa' given by Goldfuss, meant the same as Oken's 'Urthiere.' It did not acquire its present significance until 1845, when von Siebold gave it a new meaning [1901, G. N. Calkins, Protozoa 28].
See also: Metazoa, Protista

proximal: [Latin *proximus* = nearest.] adj. In general usage, meaning in close physical position to. Said of a sedimentary deposit that forms closest to the provenance or source and is therefore usually the coarsest. In paleontology, nearest to the point of attachment.

psammite: [Greek *psammitos* = sandy < *psammos* = sand.] n. A clastic sediment or sedimentary rock composed of medium-sized sand particles having an average diameter of about 1mm. Psammite consists of clasts smaller than those of rudite and larger than those of pelite. Equivalent to Latin derived arenite. In 1921, Tyrrell defined psammite as the metamorphic derivative of arenite. This is probably the reason why many dictionaries define this general sedimentary term in the more specific sense "a species of micaceous sandstone."
Cited usage: Psammite, a term in common use among European geologists for fine-grained, fissile, clayey sandstones, in contradistinction to those which are more siliceous and gritty [1859, Page, Handbk. Geol. Terms].
See also: arenite, lutite, psephite, pelite, rudite

psephite: [Greek *psephios* = pebble, round stone.] n. A conglomerate or breccia composed of pebbles or cobbles. The term, by including breccias, has become strictly a size-related classification, as opposed to the original Greek etymology, which relates to both size and shape. The Latin-based equivalent term is rudite.
Synonyms: rudite
See also: arenite, lutite, rudite, psammite, pelite

pseudo-: [Greek *pseudes* = false < *pseudein* = to lie.] Combining prefix meaning false or imitation, used in such terms as: pseudomorph, pseudoconformity, and pseudobivalve.

pseudomorph: [Greek *pseudos* = false < *pseudein* = to lie + *morphos* = shape, form.] n. A mineral that has taken the external form of another mineral rather than the form normally characteristic of its own chemical composition, as with limonite pseudomorph after pyrite (limonite pseudomorphs pyrite). A pseudomorph is usually an atom by atom replacement of one mineral's chemistry for another. A pseudomorph is described as being 'after' the mineral that it has the crystal appearance of, as in malachite after azurite (malachite has pseudomorphed the form of azurite).
See also: paramorph

pseudotachylyte: [Greek *pseudes* = false < *pseudein* = to lie + German *Tachylyt* < Greek *takhu* = tachy- < *takhus* = swift + *lutos* = soluble < *luein* = to loosen; < IE *leu* = to loosen, divide.] n. Pseudotachylite is a subtype of impactite breccia that is found in large impact structures, such as meteorite and asteroid craters. These rocks are 'shock-faulted,' are often considered para-autochthonous, and are examples of palimpsest texture. Rocks whose texture is partially wiped out by a later geologic event are said to have palimpsest texture.
Cited usage: Pseudotachylytes have been found associated with several impact craters, on major tectonic faults, and have beengenerated experimentally by localized cataclasis and frictional melting, and by impact shock [2002, Pseudotachylytes that never melted, D. Rajmon, P. Copeland, and A. M. Reid, Lunar and Planetary Science XXXIII].
See also: impactite, breccia

psilo-: [Greek *psilos* = bare.] Combining prefix meaning bare or smooth, used in terms such as psilomelane or psilopsid.

psilomelane: [Greek *psilos* = bare + *melanos* < *melas* = black.] n. A steel-gray to iron-black, massive to fibrous to botryoidal mineral, barium manganese oxide hydroxide, $Ba,Mn_8O_{16}(OH)$ is used, Barium Manganese Oxide Hydroxide. Psilomelane, although not as common as pyrolusite, is still an important ore of manganese.

Psychozoic era: [Greek *psyko* = breath, life, soul + *zoikos* = pertaining to animals <*zoa* = plural of *zoon* = animal, living being.] adj. An obsolete term proposed by LeConte in 1877 to replace the term Quaternary period as the era of domination by humans, due to the fact that during this period the biosphere is largely controlled by living creatures having minds and souls. The prefix *psycho-* adopted the nuance of meaning relating to mind and things mental in the 17th century, whereas the proliferation of scientific terms using the prefix began in the late 19th century.
Cited usage: The Psychozoic era, or era of Mind. The Neolithic commences the Psychozoic era, or reign of man [1877, Le Conte, Elem. Geol. 269 & 561].
See also: Quaternary Period

psychro-: [Greek *psychro-* = cold.] Combining prefix meaning cold, used in such terms as: psychrometer, psychrotolerant, and psychrophile (organisms that develop at or very near the freezing point).

ptero-: [Greek *pteron* = feather, wing.] Combining prefix meaning winged, used in such terms as: pterosaur (winged lizard), pterodactyl (winged fingers), pterobranchia (winged fins) and pteropod (winged foot).

ptycho-: [Greek *ptyche* = fold.] Combining prefix meaning fold, used in such terms as ptychodont (folded teeth) and ptychopariid trilobites.

ptygmatic folds: [Greek *ptygma* = folded matter.] n. The convoluted and discordant bands that appear in the granitic dikes in some gneisses and migmatites. They are usually applied to thin convolute bands of intrusive igneous material that give the appearance of folds. Term coined in 1907 by Sederholm.
Cited usage: The term 'ptygmatic' was originally coined by Sederholm in 1907 to describe 'the primary folding caused by melting' in gneisses and migmatites. The word as defined, would embrace most of the contortions, many of which are now included in the term 'flow fold', commonly seen in migmatite zones the world over. The term was later restricted by Sederholm (1926) to those tortuous quartzo-felspathic veins, which occur in areas of granitization [1952, Geol. Mag. LXXXIX].

puddingstone: [British English *puddingstone* < Middle English *puddyng* = stuffed entrails + Old English *stan* = stone; < IE *stai* = stone.] n. A sedimentary rock consisting of rounded clasts cemented together by a siliceous matrix, and resembling pudding. In Great Britain, a term for silica-cemented conglomerate. The name comes from the lower Eocene Hertfordshire Puddingstone in England, whose rounded, brown, flint pebbles in white siliceous matrix resemble English plum pudding. The English have many terms using the word pudding as a descriptive adjective, such as: pudding-pie, pudding-ale, pudding-faced, pudding-headed, and pudding class.
Cited usage: Pudding stones differ from breccias, by being composed of rounded fragments, either of marble or hard stones [1839, The Civil Engineer & Architect's Journal II. 434].

pumice: [Latin *pumex* = alteration of *spuma* = foam.] n. A porous, spongelike, low-density, glassy volcanic rock full of cavities due to expanding gases that were liberated from solution in lava while it solidified. It is often light enough to float on water. Assimilated into Middle English as *pumis-stone*, then shortened by back-formation to pumice.
See also: holohyaline, obsidian

punctate: [Latin *punctatus* = pointed < *punctum* = point.] adj. Having minute, rounded and closely spaced pores or perforations, often serving as feeding holes or points of attachment in the shells of certain marine organisms, such as those found in brachiopod shells, pennate diatom tests, and sea-urchin shells.

punky: [English *ponk* = light ashes, dust, powder.] adj. Said of a weakly welded volcanic tuff or a semi-indurated limestone.

purga: [Karelian (dialect of Russian) *purgu* = fine snow blizzard.] n. A harsh, Arctic blizzard with fine snow.
Cited usage: Most of the high winds at Barnaul are associated with cyclonic storms during winter that may produce strong blizzards, the so-called buran or purga [1977, P. E. Lydolph, Geogr. U.S.S.R. 3rd ed. xvii. 377/2].

purl: [Old English *pirle* = a rill that whirls in agitation (perhaps reinforced by Norwegian *purla* = to gush, bubble out < *porla* = to gurgle).] n. A brook, stream, or rill that flows with whirling motion and moves swiftly around obstacles.
Synonyms: rill
See also: rill

puy: [French *puy* < Old French poi = hill, mount, hillock; < Latin *podium* = hill, peak, elevation, balcony; < Greek *podion* = base, diminutive of *pod-* = foot.] n. In volcanology, in general, a small, remnant volcanic cone; specifically, a cone located in the type locality of the region around Auvergne, France. Most of the puys of central France are small cinder-cones, with or without associated lava, whereas some of the others are domes of trachytic rock, like the Puy-de-Dome.
Cited usage: Clusters of small lateral cones or puys sprang up on their flank, like those on Mount Etna [1880, Dawkins, Early Man iv. 74].

pycno-: [Greek *pyknos* = dense.] Combining prefix meaning density; used in terms such as: pycnocline, Pycnopodia helianthoides, and pycnotheca. It often refers to an aspect of the density of the marine environment.

pygidium: [Greek *pygidion* = diminutive of *pug* = rump, buttocks.] n. The posterior body part or caudal segment in certain invertebrates, especially that part of the exoskeleton of a trilobite fossil.
Cited usage: (Of a trilobite fossil structure) The crust exhibits three regions.–1, a cephalic shield; 2, a variable number of movable 'body-rings' or thoracic segments; and 3, a caudal shield or pygidium [1872, Henry Alleyne Nicholson, Manual of Palaeontology 161].

pyralspite: [Composite mnemonic term from English *pyr*(ope) + *al*(mandine) +*sp*(essart)*ite*.] n. An informal subdivision of garnet including those that do not contain calcium, in particular: pyrope, almandine, and spessartine.
See also: ugrandite, garnet, pyrope

pyramid: [Latin *pyramides* < Greek *pyramis* < Egyptian *pimar* = pyramid < (possibly) *peremus* = a mathematical device using a papyrus reed to calculate the vertical height of a structure.] n. In general, a three-dimensional shape that has a polygonal base and triangular sides meeting at a single vertex not in the plane of the base. In crystallography, an open crystal form consisting of 3, 4, 6, 8, or 12 nonparallel faces that meet at a single point. The Egyptian pyramids have square bases with the vertex directly above the center point of the base.

pyrite: [Latin *pyrites* = flint, fire stone; < Greek *pyr* = fire; < IE *paew* = fire + Greek *-ites* < *-lite* = pertaining to rocks < *lithos* = rock, stone.] n. A brass-colored, isometric mineral, the native disulfide of iron, FeS_2. Pyrite commonly crystallizies as cubes and pyritohedrons. It is a

common iron ore and a source of sulfur dioxide. Pyrite is called firestone because it glitters and shines, as well as being able to produce sparks when struck. In mining regions, pyrite is often called fool's gold, although it really isn't terribly foolish to think of gold when finding pyrite in gold country, because, even though most pyrite-containing gravels don't contain gold, most auriferous gravels do contain pyrite. One of the earliest mentions of pyrite is by Theophrastus in the 4th century BCE, who, in his work "On Stones," mentions grains of pyrite embedded in lapis lazuli. Pyrite is also described by Pliny the Elder in his "Natural History" (77 CE), as well as by Agricola in *De Re Metallica* (*On Mining*) in 1556.
Cited usage: Pyrite is now only applied to the disulphide of iron which crystallizies in isometric forms [1896, Chester, Dict. Min. Names.]. The pyrite of most gold regions is auriferous [1868, James D. Dana, A System of Mineralogy 63].
Synonyms: firestone, fool's gold

pyro-: [Greek *pyr* or *pyro* = fire.] Combining prefix meaning fire, used in such term as: pyrite, pyroclastic, pyroxene, pyrrhotite, and pyromaniac. This suffix should not be confused with -*phyre*, meaning porphyry.
See also: porphyry

pyroclastic: [Greek *pyro* = fire + *klastos* = broken < *klaein* = to break.] adj. Composed chiefly of rock fragments of volcanic origin, broken by processes of eruption or extrusion.
Cited usage: I venture to suggest that we should distinguish between the three types of clastic rocks at present recognized by using the terms epiclastic, cataclastic, and pyroclastic [1887, J. J. H. Teall, Geol. Mag. Decade III. IV. 493].

pyrolusite: [German *pyrolusit* < Greek *pyro* = fire + *lousis* = a washing < *louein* = to wash; < IE *leu* = to wash.] n. A black to dark-gray, soft

tetragonal mineral, manganese oxide, MnO_2. Term coined in 1827 by German geologist Haidinger, due to its property of eliminating color from glass when heated, thus making it more transparent. IE derivatives include: deluge, alluvium, colluvium, and effluent.

pyrope: [Latin *pyropus* = gold-bronze, 'fiery-eyed'; < Greek *pyro* = fire + *op* = eye, face, see; < IE *ok* = to see.] n. A red-colored, isometric mineral of the garnet group, magnesium aluminum silicate, $Mg_3Al_2(SiO_4)_3$. Garnets are subdivided into two informal groups, ugrandite and pyralspite (of which pyrope is one). The members of this group form a solid solution series that do not contain calcium. Pyrope, along with most garnet types is considered a semiprecious gem. Term coined in 1803 by A.G. Werner, alluding to the Bohemian garnet or fire-garnet, a deep-red gem. IE derivatives include triceratops and phlogopite.
Cited usage: The pyrope, which has lately exfoliated from the class of garnets, has no difference but superior beauty [1804, Edin. Rev. III. 301]. It is likely that the mantle consists dominantly of magnesium olivine, together with smaller amounts of enstatite-diopside-pyrope garnet ($Mg_3Al_2Si_3O_{12}$) and perhaps some phlogopite mica [1977, A. Hallam, Planet Earth 34/3].
See also: ugrandite, pyralspite, andradite, uvarovite, almandine, grossularite, garnet

pyrophyllite: [Greek *pyro* = fire + *phyllon* = leaf.] n. A silvery-white to gray or brown, monoclinic mineral, hydroxy aluminum silicate, $Al_2Si_4O_{10}(OH)_2$. Pyrophyllite occurs naturally in soft, foliated, and compact masses. It is so named for its tendency to exfoliate in leaflike sheets when heated.

pyroxene: [Greek *pyro* = fire + *xenos* = stranger.] n. Any of a group of dark-colored rock-forming minerals, characterized by single chains of silicate tetrahedra, SiO_4. Pyroxenes exhibit prismatic cleavage at nearly 90°, have a general formula $XY(SiO_3)_2$, contain metallic oxides of iron, magnesium, calcium, sodium, or aluminum, and are common in both igneous and metamorphic rocks. Term coined in 1796 by René-Just Haüy, because the mineral group seemed atypical of those found in igneous rocks and therefore was a 'stranger to fire,' the heat needed to create igneous rocks. Another possible reason for the name stems from a mistaken belief that these silicate minerals were only accidentally assimilated into the lavas containing them.

pyrrhotite: [Greek *pyrrotes* = redness < *pyrros* = fiery < *pyro* = fire.] n. A brownish-bronze hexagonal mineral, iron sulfide, $Fe_{1-x}S$. It occurs most commonly with a massive habit, some specimens being magnetic. Pyrrhotite is used in the manufacture of sulfuric acid.
Synonyms: magnetic pyrites

qanat: [Arabic *qanat* = desert irrigation canal.] n. A human-constructed system for conveying water from mountain springs to the arable lands of the Middle Eastern and North African desert regions. Qanat are series of shafts dug down to the aquifer level located in the upper reaches of alluvial fans, which are then connected at the bottom by a gently sloping conveyance tunnel. *Cited usage*: (Qanat) systems are begun in the piedmont by excavating a vertical shaft down to the water table. At the point where water is reached a horizontal tunnel or gallery is then excavated away from the piedmont towards the gardens and villages requiring water [2003, Website of the United Arab Emirates Ministry of Information and Culture].
Alt. spelling: kanat
Synonyms: foggara
See also: foggara

quagmire: [Middle English *quag* < *quabbe* = marsh, bog; < Old English *owabba* = shake, tremble + Middle English *mire* = bog; < Old Norse *myrr* = bog.] n. A tract of wet and boggy ground too soft to sustain the weight of large animals, and sometimes floating over the substrate below, making it tremble or quake when depressed. Quagmire is a somewhat redundant term, because both *quag* and *mire* mean bog. First usage in English c. 1575.
Synonyms: quaking bog
Similar terms: fen, marsh

quantum: [Latin *quantus* = how much, how great.] n. In physics, the smallest packet of energy that an electron can give off or absorb when changing energy levels; this quantity is always transferred in whole number multiples of itself. In paleontology, quantum evolution is described as the important changes and biological adaptations that occur fairly rapidly in very small populations, leaving little fossil evidence before they spread and stabilize in large numbers. The term quantum evolution was coined in 1944 by George Gaylord Simpson in the book *Tempo and Mode in Evolution*. First usage in English c. 1615.

quaquaversal: [Latin *quaqua* = wherever + *versus* = towards.] adj. Said of a geologic structure that has beds dipping outward in all directions, such as a dome. Used in contrast to centroclinal, referring to a structure with beds dipping toward a common center. First usage in English c. 1725.
Cited usage: The slope and quaqua-versal dip of the beds [1830, Lyell, Principles of Geology I. 394].

quarry: [French *quarriere* = quarry < *quarre* = cut stone; < Latin *quadrum* = square; < IE *kwetwer* = four.] n. An open-pit excavation made by the extraction of rock that is useful for some structural application, such as a limestone quarry or a granite quarry. IE derivatives include: quaternary, fortnight, tetrahedral, and quarter. First usage in English c.1400.

quartz: [Uncertain origin, probably < Old German *twarc* < Slavic *tvrudu* = hard.] n. A variously colored and variously formed tectosilicate mineral, silicon dioxide, SiO_2. It occurs as macrocrystalline and cryptocrystalline varieties. Macrocrystalline varieties include: amethyst (purple), citrine (pale yellow), rock crystal (clear), rose quartz (pink), and smoky quartz (brown-black). The most common cryptocrystalline types include varieties of chalcedony, such as: agate, onyx, carnelian,

chrysoprase, flint, heliotrope, and jasper. The ancient Greeks believed that the water-clear variety of rock-crystal came from pure, frozen alpine water. They named it *kpbara*, meaning crystal. The term was given recognition by Georgius Agricola (Georg Bauer, 1494-1555) in his treatise on mineralogy, *Bermannus*. The German *quarz*, Dutch *kwarts*, and Italian *quarzo* are all cognates of the term quartz, and are probably derived from the same root.
Cited usage: White debas'd Crystal (which the Germans call Quartz) [1756, William Borlase, Observations on the Isles of Scilly 71]. I shall adopt this name of quartz in English as it has already gained access into other European languages [1772, translation of Cronstedt's Min. 57].

quasi-: [Latin *quasi* = as if, as it were, almost.] Combining prefix meaning similar or resembling, while having some quality, aspect, or source that makes it different from the base word.
Cited usage: In geologic writing the standards for contemporaneity and instantaneousness are more flexible. Perhaps, H. and G. Termier's (1956) expression quasi-instantaneous (quasi-instantané) could be recommended as a convenient term to indicate instantaneousness in the geologic sense [1958, Bull. Geol. Soc. Amer. LXIX. 111/2].

Quaternary: [Latin *quartum* = fourth in order; < IE *kwetver* = four.] adj. The second period of the Cenozoic era (1.81Ma-Present), characterized by several global glacial advances, and expansion of the territorial range of the hominids in the genus *Homo*. It is divided into the Pleistocene and Holocene epochs. The term Quaternary was coined by Desnoyers in 1826 to describe deposits overlying Tertiary strata in the Paris basin. It was subsequently redefined by Reboul in 1833 to include deposits whose fossils mainly represent extant organisms. The geologic eras were originally named Primary, Secondary, Tertiary, and Quaternary. The names Primary and Secondary have been

discarded, but Tertiary and Quaternary have been retained as periods within the Cenozoic Era.
Cited usage: The beginning of archaeology may be broadly held to follow on the last of the geological periods, the Quaternary [1910, Encycl. Britannica II. 344/2].

quebrada: [Spanish *quebrado* = broken < *quebrar* = to break.] n. Term used in South America and the U.S. Southwest desert regions for an episodic stream in a deep ravine, usually in a mountainous or plateau area. The name stems from the precipitous ravines creating a 'break' in the often flat, elevated topography.
Cited usage: Abrupt precipices occur in every part of the parent chain of the Andes near the equator, and diversify its appearance with the most horrid chasms, or rents, here called Quebradas, varying from 100 feet to 4 or 5,000 feet in depth [1845, Encycl. Metropolitana XIV. 565/1].
Similar terms: barranco, arroyo
See also: barranco, arroyo

radiolarian: [Latin *radiolus* = diminutive of *radius* = ray, spoke of a wheel, staff.] n. Any of a variety of marine protist zooplankton of the class Phaeodaria (formerly Radiolaria). Radiolarians are characterized by a siliceous skeleton (test) made up of variously shaped structural components. They are largely marine organisms occupying the photic zone, the earliest forms of which date back to the Cambrian Period. Originally named Polycystina by C. G. Ehrenberg in 1838, these marine sarcodina were renamed Radiolaria by Ernest Haeckel in 1862.
Cited usage: The order Radiolaria is defined as comprising those members of the Rhizopoda which possess a siliceous test or siliceous spicules [1872 Nicholson Palaeont. 66].
See also: ecology

radula: [Latin *radula* = scraper < *radere* = to scrape.] n. A raspy, phosphorus-based plate made of chitin that is present on the dorsal surface of most univalve molluscs, such as gastropods. It is covered with rows of tiny teeth and is used to help gather food into the mouth and grind it up for digestion. First usage in English c. 1750.
Cited usage: The snail combines the functions of teeth and tongue is a single organ: the radula [1975, Scientific American Feb. 106].

rahwacke: [German *rauh* = harsh + *wacke* = rock < *waggo* = boulder; < IE *wegh* = weigh.] n. A type of breccia usually derived from a porous cavernous carbonate rock, the pore spaces being filled with friable evaporite material that easily dissolves and falls out, leaving a rough surface.
Alt. spelling: rauhwacke
Synonyms: cargneule, cornieule
See also: cargneule

ram-: [Latin *ramus* = branch.] Combining prefix meaning branch or a branching out. It is used in the term ramus, a general term for a projecting branch or process that is found on both vertebrates and invertebrates and is composed of nearly any type of bodily tissue, such as: bone, nerve, or blood vessel in vertebrates, and chitinous exoskeletal material in invertebrates.
Alt. spelling: rami (pl.)

rambla: [Spanish *rambla* = dry ravine; < Arabic *ramlah* = sand.] n. A dry river bed, or the bed of an ephemeral stream. The term *arroyo* is the equivalent term in American Spanish.
Cited usage: The name Rambla is of Arab origin meaning 'bed of a seasonal river' [2001, Softguide to Spain].
Synonyms: arroyo

rampart: [French *remparer* = to fortify, defend against < *re* = back, against, again + *emparer* < *amparar* = to fortify, take possession of; < Latin *ante parare* = to prepare < *ante* = before + *parer* = to defend < *parare* = to prepare.] n. Any mound, wall, or barrier formed on the periphery of a geologically active structure, such as a volcano, plate boundary, or coral reef; in Canada, the steep bank of a river or gorge. In architecture, a fortification for defense made of a mound of raised earth. When Roman soldiers dug trenches to facilitate warfare, the resulting earth and stone piles they formed were called ramparts. First usage in English c. 1580.
Cited usage: Sometimes they (Mars impact craters) show more complex structure, like radial ridges, and double ramparts [2002, Ralph P. Harvey, Earth & Planets].

rand: [Toponymous back-formation after *Witwatersrand*, a region in South Africa; < Afrikaans *rand* = rocky ridge; < Old English *rand* = border, margin, bank, shield, margin.] n. A South African term for a long, low rocky ridge. The Witwatersrand in South Africa, popularly called the Rand, is a long ridge of Precambrian conglomerate (over 100km) lying west of Johannesburg. It forms the watershed between the Limpopo and the Orange Rivers. Since the late nineteenth century, the area has been the largest single producer of gold in the world. Because of this wealth, the rand has become the basic unit of currency in South Africa. The Krugerrand, a one-ounce gold coin, is named after Stephanus Johannes Paulos Kruger, the president of Transvaal, whose racist policies helped instigate the Boer War of 1899-1902.
Cited usage: The conflict that was inevitable from the moment that gold was discovered in the Rand...had to come [1899, George Bernard Shaw, Letters].

randkluft: [German *rand* = edge + *kluft* = crevice.] n. A crevasse or fissure that separates a moving glacier from its headwall rock. A randkluft is similar to a bergschrund, except that a bergschrund, instead of separating rock from ice, usually separates the downslope moving ice from the nonmoving upslope apron ice connected to the headwall.
Cited usage: Bergschrund, the crevasse which occurs at the head of a cirque or valley glacier and which separates the moving glacier ice from the rock wall and the ice apron attached to it. When the ice apron is absent the gap is known as a Randkluft [1958, Polar Record IX. 91].

rapakivi: [Finnish *rapakivi* = rotten stone, crumbly stone < *rapa* = mud + *kivi* = stone.] n. A rock or rock texture characterized by rounded crystals of potassium feldspar surrounded by mantles of finer-grained sodium feldspar, usually found in igneous rocks, but sometimes also in metamorphic rocks. Term coined in 1694 by Urban Hjarne for saprophytic outcrops that occur in southern Finland.

Cited usage: The stone is Rapakivi Granite, probably quarried from southeastern Finland. It formed during pre-Cambrian times around 1500 million years ago; packages of molten rock slowly cooled and the evidence can be seen for it in the Rapakivi Granite [2003, Anna Grayson et al., Essential Guide to Rocks, BBC Education Series].

rauk: [Toponymous term after Swedish *raukar*, a sea stack on Gotland Island.] n. A relict, limestone sea stack formed from biological reef sediments that remains standing after less resistant surrounding rock has eroded away. The type locality for rauken is the Baltic Island of Gotland, where carbonate reefs formed when the area was at a tropical latitude. The island's largest rauk is called the "Maiden."

ravine: [French *ravine* = violent rush < *raviner* = to hollow out; < Latin *rapere* = to seize; < IE *rep* = to snatch.] n. A steep-sided narrow gorge worn through elevated topography by rapidly running water. A ravine is smaller than a gorge but larger than a gully. The geologic sense of this term was perhaps influenced by the Latin *rapidus*, meaning rapid. Prior to its geomorphological sense, the 15th-century Middle English term *ravine*, meaning robbery, plunder, booty was more closely allied to the Latin and Indo-European roots, which give us the related terms, such as: raptor, rapacious, raven, rape, and surreptitious. First usage in English c. 1450.

razorback: [English *razorback* = a type of pig.] n. A narrow, sharp, steep-sided ridge. The use of razorback for a type of pig with a sharp ridgelike back dates from 1849. In geomorphology, the term first gained popularity in Australia and New Zealand.
Cited usage: Twice the way led along a real 'razor-back.' On both sides the mountain sloped precipitously [1911, Chambers's Jrnl. Dec. 30/1].

realgar: [Arabic *rehj al-ghar* = powder of the cave < *rehj* = powder + *al* = the + *ghar* = cave.] n. A soft, orange-red to bright-red, monoclinic mineral, arsenic sulfide, AsS. Realgar is an ore of arsenic and occurs in hot-spring deposits, volcanic sublimates, and certain karst limestones and dolomites, where it is commonly a cave deposit. Orpiment (As_2S_3), dimorphite (As_4S_3), alacrinite (As_8S_9), uzonite (As_4S_5), and duranusite (As_4S) are chemically similar minerals belonging to the realgar group.
Cited usage: Sulfur and arsenic readily unite by fusion and form a red vitreous semitransparent mass. The same substance is found native in different parts of Europe, and is called realgar [1812, Sir Humphry Davy, Chemical Philosophy 457].
Synonyms: red arsenic, red orpiment

recapitulation: [Latin *re* = back, again + *caput* = head.] n. The biological theory that, during development, an individual organism passes through the developmental stages of its ancestors, thus 'recapitulating' the phylogeny of its genetic lineage. Term coined in 1866 by Ernst Haeckel, who also coined the euphonic phrase "ontogeny recapitulates phylogeny." In 1977, Stephen Jay Gould resurrected the atheist Haeckel's ladder theory of evolution in his book *Ontogeny and Phylogeny.*
Synonyms: Haeckel's law
See also: phylogeny, ontogeny, ecology

Recent: [Latin *recens* = recent; < Greek *kainos* = recent; < IE *ken* = fresh, new, young.] adj. Used as a proper adjective in referring to the Recent epoch, synonymous with the Holocene epoch, and comprising the last 10,000 years of geologic history. IE derivatives include: Cenozoic, kainite, and the suffix *cene*, as in Oligocene.

recession: [Latin *re* = back, again + *cedere* = to go.] n. In general usage, the act of withdrawing or going back. In glaciology, a decrease in the length of a glacier, when wastage (ablation) exceeds accumulation (advance). In geomorphology, retreat of a cliff, escarpment, or headwall due to weathering and erosion. In marine geology, the landward retreat of a shoreline due to erosion, subsidence, sea level rise, or a combination of all three factors (known as retrogradation). In economics, a temporary decline or setback in economic activity or prosperity.
Similar terms: retrogression
See also: retrograde, progradation

reciprocal: [Latin *reciprocus* = alternating.] adj. In crystallography, referring to a lattice array of points formed by perpendiculars to each crystal plane intersecting at a common point of origin. In mathematics, an inverse relationship, as 2 is to ½, or n is to $1/n$. First usage in English c. 1565.

reconnaissance: [French *reconoissance* = recognition; < Latin *re* = back, again + *cognoscere* = to get to know.] n. An exploratory examination of any region or location, made as an initial survey in preparation for a more thorough investigation. Used as an adjective in terms, such as reconnaissance map and reconnaissance survey.

recrystallization: [Latin *re* = again, back + *crystallum* = crystal ice.] n. The formation of new crystal grains in a rock based on originally existing crystals. Recrystallized minerals are generally larger and are usually created in the solid state. Recrystallization is a fundamental process occurring during metamorphism. It is sometimes described as a progression of discrete phases: a) annihilation, b) rearrangement, c) nucleation, and d) aggregation.

recti-: [Latin *rectus* = straight, right.] Combining prefix meaning straight, as in: erect, rectify, direct, and coastal rectification; also meaning right, as in rectangle, rectilinear, and correct. This prefix can be misleading because *rectus* has a physical sense meaning straight and at right angles, as well as an ethical sense meaning correct. In ethics, straight can be synonymous with right and correct, as in 'straight shooter' or 'straight and narrow.' In a physical sense, straight and right have completely different meanings. But, as influenced by the ethical sense, the combining form *rect* in the physical sense can mean at right angles, as in rectangle, or in a straight line, as in erect.

recumbent: [Latin *recumbere* = to lie down < *re* = back, again + *cumbere* = to lie.] adj. An overturned fold whose hinge line and axial plane are nearly horizontal. In 1964, M. J. Fleuty suggested that the term recumbent fold be restricted to a fold whose axial plane does not dip more than 10°.

redox: [An acronym derived from English *red(uction)* + *ox(idation)*.] A coupled reaction process, often reversible, now broadly defined as either a loss of electrons (oxidation) or a gain of electrons (reduction). If the redox reaction involves ionic compounds, the electron transfer is complete, whereas, if the reaction occurs in covalent compounds, the electrons are shared. In the formation of water for example, hydrogen is oxidized and oxygen is reduced. Originally, oxidation was defined as a chemical reaction resulting in a net gain of oxygen, whereas reduction was a loss of oxygen. According to this definition, the formation of water could not be classified as a redox reaction.

reef: [Dutch *riffe* < Old Norse *rif* = ridge, rib.] n. A ridge of rock, shingle or sand that rises from the seafloor into shallow water, usually along coastal areas, and often breaking the surface at least at low tide. Reefs can be biogenic, such as those constructed by calcareous animals (corals, shellfish, and some algae); structural, such as those formed along folds or faults or near plate boundaries; or they can be remnant erosional features composed of country rock. In mining, a

vein of ore. Types of reefs include: patch reefs (small and circular in shape), pinnacle reefs (conical in form), barrier reefs (separated from the coast by a lagoon), fringing reefs (attached to a coast), and atolls (isolated reefs enclosing lagoons). Factors influencing reef growth include: water temperature (optimum 25°C), depth (must be less than 10 m), salinity (normal marine salinity is necessary), turbidity (coral growth requires clear water and an absence of terrigenous suspended sediment), wave action (intense wave action favors coral growth and is an agent of erosion), or structural deformity.

reflectance: [Latin *re* = back, again + *flectere* = to bend.] n. The proportion of energy reflected by a body compared to that which falls incident upon it.

reflux: [Latin *re* = back, again + *fluxus* = flow < *fluere* = to flow.] n. In general usage, a return flow, as in the backflow within a chemical reflux tube in a distillation apparatus. In petrology, the backflow of high-concentration brine across the barrier of an evaporite basin. The magnesium-enriched concentrate may contribute to dolomitization of carbonate rocks.

refraction: [Latin *re* = back, again + *frangere* = to break; < IE *bhreg* = to break.] n. A vector altering deflection of an energy wave due to a change in the medium it is traveling through, such as, a change in density of the rock though which a seismic wave is traveling, or a change in depth in the water through which a gravity wave is moving. IE derivatives include: breccia, breach, birefringent, and fractionation.

refugium: [Latin *re* = back, again + *fugere* = to flee.] n. A local region whose flora and fauna persisted through times of intense environmental change, such as: the driftless area in regions of glaciation, desert oases in regions of desiccation,

and cave populations in regions of excoriation.
Cited usage: The presence of endemic mammals
on the Island suggests that suitable survival areas
(refugia) existed there during the peaks of the last two
glaciations [1976, Islander (Victoria, B.C.) 16 May 7/1].

reg: [North African Arabic *reg* = desert blanket.]
n. An extensive area of desert flatland covered
with a mosaic of desert-varnished gravels, pebbles
and cobbles from which the sand has been
removed by wind and water; the stony blanket is
then cemented together by alkaline mineral
solutions to form a resistant pavement. The Arabic
reg and Greek *rhegos*, meaning blanket may have
the same root
Cited usage: Regs and serirs are planed areas
with a covering of boulders, which tumble from the
surface of the hamadas or from the plains below
them. The term reg is generally reserved for the
low plains used by caravans. Moreover, this term
is applied commonly to all bouldery ground which
has been subjected to deflation [1963, D. W. & E.
E. Humphries, translation of Termier's Erosion &
Sedimentation ii. 38].
Synonyms: hammada
See also: serir, hammada

regimen: [Latin *regere* = to rule, direct; < IE *reg*
= to move in a straight line.] n. In hydrology, the
behavioral and flow characteristics of a specific
drainage area with respect to time, ranging in size
from a single stream to an entire watershed.
Sometimes used incorrectly as a synonym for
regime, especially when applied to the volume of
water passing different cross-sections of a stream
within a specified time.
Cited usage: It will be useful to follow the
practice of engineers in reference to streams of
water, and refer to the system or activity of the
glacier as a whole, based on its meteorology,
economy, rate and possible type of flow, and
fluctuation, as the regimen of the glacier [1971, R.
F. Flint, Glacial & Quaternary Geol. iii. 47].

regolith: [Greek *rhegos* = blanket, rug + *lithos* =
rock, stone.] n. The general term for the layer of
loose rock and other unconsolidated material that
covers, in varying thicknesses, nearly the entire
bedrock of a planet. The regolith reaches its
greatest depth of over 60m in the tropics, where
weathering is most intense. Soil is the topmost
layer of the regolith where plant life abounds.
Although organisms such as large trees can
penetrate the lower extremes of the regolith for
water or stability, soil is where biological activity is
maximized. Because extraterrestrial planets have no
detectable biology, regolith is used as the preferred term for surficial unconsolidated material.
The term, coined by Merrill in 1897, gained favor
over the 'saprolith' of Becker and the 'sathrolith'
of Sederholm.
Cited usage: This entire mantle of unconsolidated material, whatever its nature or origin, it is
proposed to call the regolith, from the Greek words
rhegos, meaning a blanket, and lithos, a stone
[1897, G. P. Merrill, Treat. Rocks v. 299].

regur: [Hindi *regar* = black soil; < Telugu *regadi*
= clay.] n. In pedology, a dark and calcareous-rich
vertisol, high in montmorillonite clay, formed
mainly by the weathering of low-silica basaltic
rock, and common on the Deccan Plateau of Inda.
The English in India commonly called this soil
"black cotton soil" because they grew cotton on
it. There are four basic types of soils in India:
mountain, alluvial, regur, and red soils.
Cited usage: Regur, in its most characteristic
form, preserves the constant characters of being
highly argillaceous and somewhat calcareous, of
becoming highly adhesive when wetted, and of
expanding and contracting to an unusual extent
under the respective influences of moisture and
dryness [1879, Medlicott & Blanford, Manual
Geol. of India I. xviii. 429].

relief: [Italian *rilievo* = raised or embossed
works of Florentine art; < Latin *rilevare* = to raise.]

n. A descriptive term for the shape of the Earth's surface, with regard to differences in elevation (e.g., the relatively low relief of the plains compared to the high relief of the mountain regions). The topography of a region includes relief in the description of the configuration of a landscape.
See also: topography

remanent: [Latin *remanentem* = present participle of *remanare* = to remain.] adj. In geophysics, said of the magnetism that remains in rocks having iron-rich minerals that orient themselves to line up with the Earth's magnetic field. Upon lithification, this orientation is frozen in time and is used to quantify subsequent plate movement. In igneous rocks like oceanic basalt remanent magnetism is thermally acquired on cooling; in sedimentary rocks, it is depositionally acquired as individual clasts line up during compaction. The term is often called by the abbreviation NRM (natural remanent magnetization). Not to be confused with the term remnant.
Cited usage: The intensity of this remanent or permanent component of magnetization in basalts is invariably greater than that induced by the present Earth's field [1971, I. G. Gass et al., Understanding Earth xvi. 237/1].

rendzina: [Russian *rendzina* = calcareous soil; < Polish *redzina*.] n. A fertile soil rich in humus and calcium carbonate, typically developed under a cover crop and on top of soft carbonate bedrock. The A horizon is thin, dark, and humus-rich; there is no B horizon; and the C horizon is primarily calcareous bedrock. In the Russian soil classification system, rendzina soils are a subgroup of chernozem soils.
Cited usage: Recently under the influence of Russian soil workers the term 'rendzina' has been applied to all soils which have developed from the weathering of calcareous rocks [1928, C. L. Whittles, translation of E. Ramann's Evolution & Classification of Soils v. 91].
See also: chernozem, podzol

reniform: [Latin *renes* = kidneys.] adj. Shaped like a human kidney and used to describe a similarly shaped mineral habit.

reptile: [Latin *reptilis* = creeping < *repere* = to creep.] n. Any of the usually ectothérmic (cold-blooded) and egg-laying vertebrates belonging to the class Reptilia. Reptiles have lungs, are air-breathing in all developmental stages, and are covered by a layer of scales or plates. The class includes: lizards, snakes, turtles, and dinosaurs. The verb repent comes from the same root, which, according to the Bible, people had to do after Adam and Eve were exiled from the Garden of Eden for communing with a serpent.
Cited usage: Every neddre and every Snake and every Reptil which mai moeve [1390, John Gower, Confessio Amantis III].
See also: ectotherm, poikilotherm, endotherm, homoiotherm

reservoir: [French *reserver* = to keep, reserve; < Latin *reservare* = to keep back < *re* = back + *servare* = to keep.] n. A large natural or artificial receptacle, as in a lake or pond, used for the storage and regulation of water flow or consumption.

residue: [Latin *residuum* = that remaining < *residere* = to remain.] n. The remaining weathered erosional mineral and rock debris that covers the unweathered or only partially altered country rock after the soluble material has been dissolved away. As an adjective: in geomorphology, said of a feature, such as a rock, hill, or plateau that remains as a trace of a once larger mass; in ore deposits, said of a mineral deposit formed by mechanical concentration or by chemical alteration in the zone of weathering; in stratigraphy, said of a map that displays small-scale variations of a given stratigraphic unit; and in petroleum geology, said of oil that is left in the reservoir rock after the field has been depleted.

resonance: [Latin *resonantia* = echo < *resonare* = to resound < *re* = back, again + *sonare* = to sound.] n. The increase in amplitude in an oscillatory system that is exposed to a periodic force whose frequency is equal to or very close to the natural, undamped frequency of the system. This phenomenon causes augmented harmonic and synchronous vibration of the system, which can become catastrophic when the natural frequency of large-scale primary or secondary seismic waves are in harmonic resonance to the inherent vibrational period of surface structures, such as: buildings, bridges, pinnacles, and towers. *Cited usage*: When the frequency of the earthquake matches the natural frequency of a building, resonance occurs, and that building will shake more than others [1993, Janice VanCleave, Earthquakes, Wiley Pub.].

reticulate: [Latin *reticulum* = diminutive of *rete* = net.] adj. In paleontology and biology, the net-like form of some invertebrates, such as the pattern formed by the concentric rugae of a brachiopod shell intersecting with its radial costae. In petrology, secondary mineral formation that creates a grid around remnant primary minerals.

retrograde: [Latin *retrgradus* = to go back < *retro* = backward, behind + *gradus* = *walking* < *gradi* = to go; < IE *ghredh* = to walk, go.] adj. Used in a planetary context to denote a body moving in the opposite sense to that of most solar-system bodies; that is, clockwise as opposed to anticlockwise. Retrograde motion can be apparent, as in the apparent reversal of orbital motion between planets with different velocities, or actual, as in the absolute motion of the axial rotation of the planet Venus.

revetment: [Latin *re* = back, again + *vestire* = to clothe.] n. A protective facing of erosion resistant material, usually stone, built onto an embankment or shore structure. The term 'revet-crag' describes a series of narrow, pointed ridges of eroded strata inclined like a revetment against a mountain spur.

rhabdo-: [Greek *rhabdos* = rod.] Combining prefix meaning straight, rodlike, or tubelike. Used in such earth-related terms as: rhabdosome, a colony of stick-shaped marine Graptolites; rhabdolith, a minute, calcareous rodlike shell found in the ocean; and rhabdomancy, divination for ore or water by use of a stick.

rheid: [Greek *rhein* = to flow < *rheos* = stream.] n. A substance that deforms through viscous flow when subjected to stress at temperatures below its melting point. A rheid fold is a structural feature that has deformed as if it were fluid.

rheology: [Greek *rhein* = to flow < *rheos* = stream + *logos* = word, speech, discourse, reason < *legein* = to speak, to say; IE *leg* = to collect.] n. The study of the deformation and flow of matter, especially the non-Newtonian flow of liquids and the plastic flow of solids. Another term that may come from the same root is the name of the Greek goddess Rhea, wife of the Titan Cronos, and mother of Zeus and Poseidon.

rhodo-: [Greek *rhodon* = rose.] Combining prefix meaning red, used in mineralogy (rhodochrosite), marine biology (Rhodophyta), and petrology (rhodophyllite).

rhodochrosite: [Greek *rhodokhros* = rose-red colored < *rhodon* = rose + *khroma* = color.] n. A translucent, rose-colored, hexagonal mineral of the calcite group, manganese carbonate, $MnCO_3$. It usually has a pearly or vitreous luster.

rhombohedron: [Latin *rhombus* = flatfish, magician's circle, an equilateral parallelogram; < Greek *rhomb* = oblique equilateral parallelogram + *hedra* = side, face.] n. In crystallography, a rhombohedron is a prism with six faces, each a rhombus or an equilateral parallelogram. As the adjective rhombohedral, referring to a subdivision of the hexagonal system, containing both the rhombohedron and scalenohedron. First usage in English c. 1835.

rhourd: [Arabic *rhourd* = sand mountain.] n. A star-shaped dune formed at the site where two draa chains cross. When the wind blows continuously from one direction, it causes a crescent-shaped barchan dune to form. Bidirectional winds form snake-like dunes called siefs (from Arabic *seif*, meaning sword). Larger dunes or fields of dunes are called ergs. The highest of sand ridges called draas are over 300 meters high in the Sahara.

rhyolite: [Greek *rhux* = stream (of lava) < *rhein* = to flow + *lithos* = rock, stone.] n. A fine-grained, white to pinkish volcanic igneous rock, similar to granite in composition. Rhyolite lavas have high viscosity and typically exhibit flow banding. They commonly contain phenocrysts of quartz and alkali feldspar in a glassy to cryptocrystalline groundmass. The term was coined in 1860 by Baron von Richthofen (grandfather of the World War I aviator) for a variety of siliceous trachyte containing quartz that was found in Hungary. *Cited usage*: The name 'rhyolite' was proposed, early in 1860, for certain rocks frequently occurring on the southern slope of the Carpathians [1868, Mem. Calif. Acad. Sci. I. 50].
Synonyms: liparite
See also: liparite

rhythmites: [German *rhythmit* < Latin *rhythmus* = movement in time, especially with regard to rhymed poetry; < Greek *rhythmos* = measured flow < *rhein* = to flow.] n. Each member of a sequence of finely-textured, regularly repeating beds laid down in banded patterns of cyclic or rhythmic sedimentary sequences. Rhythmites, though most commonly associated with freshwater environ-

ments, can also be deposited by tidal action. Term coined by B. Sander in 1936. A. Knopf suggested that 'laminites' might be a better term than 'rhythmites' to avoid the implication of perfect periodicity in the recurrence of laminae.
Cited usage: The word 'rhythmite', used by Sander for the individual units of rhythmic beds, will be adopted here as a brief term to designate the couplet of distinct sedimentary types of rock, or the graded sequence of sediments, that form a unit bed or lamina in rhythmically bedded deposits [1946, Prof. Paper U.S.G.S. No. 212. 30/2].

ria: [Spanish *ria* = fjord-like inlet < *rio* = river.] n. A long narrow inlet of the sea formed by the partial submergence of an unglaciated river valley; generally, all types of subaerially carved troughs, including fjords, rias, dalmatians, and limians. *Cited usage*: The Spanish word, ria, introduced into German by Richthofen [Führer für Forschungsreisende (1886) ix. 309], and now widely adopted, means any broad or estuarine river mouth, and not necessarily an embayment produced by the partial submergence of an open valley in a mountainous coast, in the sense that Richthofen originally proposed [1915, Ann. Assoc. Amer. Geographers V. 82].
Synonyms: ria coast
See also: fjord

riebeckite: [Eponymous term after 19th-century German explorer and mineralogist Emil Riebeck.] n. A blue to black, monoclinic mineral of the amphibole group, hydroxy fluoro sodium iron silicate, $Na_2(Fe, Mg)_5(Si_4O_{11})_2(OH,F)_2$. It is isomorphous with magnesium-bearing glaucophane and occurs in alkali-rich rocks such as nepheline syenites and some banded iron formations. The fibrous asbestiform variety of riebeckite called crocidilite or "blue asbestos." First usage in English c. 1885.
Cited usage: This hornblende, named riebeckite, has the same composition as arfvedsonite, and is the analogue of aegerine of the augite series [1889, Jrnl. Chem. Soc. LVI. 109].

rill: [German *rille* (or Dutch *ril*) = running stream.] n. A small stream, brook, or rivulet of running water. In coastal geomorphology, a transient, thin, sheetlike channel on the beach carrying water back to the sea after a breaking wave (a runnel). This action leaves a 'rill mark' on the beach during a receding tide. In planetary geology, any long, narrow, sinuous valleys observed on the surface of the Moon. First usage in English c. 1535.
Cited usage: Tidal currents which occur during the retreat of the sea from a beach form a pattern of fine channels, particularly where the water is retarded by obstacles, pebbles or shells. These channels or rill-marks formed on the surface of moist, soft sand can be preserved by fossilization [1963, D. W. & E. E. Humphries, translation of Termier's Erosion & Sedimentation x. 211].
Synonyms: rundle, runnel
See also: runnel

rimaye: [Latin *rima* = fissure + *-aye* = collective suffix.] n. A crevasse or series of crevasses often found near the head of a mountain glacier.
Cited usage: At the foot of the snow-slope, just where it eases into the glacier, runs a long chasm in the ice, the bergschrund, randkluft, or rimaye [1920, H. Raeburn, Mountaineering Art viii. 107].
Synonyms: bergschrund
See also: bergschrund

rime: [Old English *hrim*, Dutch *rijm*, Icelandic *hrim*, Scandinavian *rim*, and Old French *rime* are likely all of Teutonic origin.] n. A coating of ice, as on grass and trees, formed when extremely cold water droplets freeze almost instantly on a cold surface. As an adjective, used in terms such as rime ice. Also used as a transitive verb, as in riming or rimed ice. Rime is generally distinguished from hoarfrost. First usage in English before 900.
Cited usage: Moonlight splendour of intensest rime, With which frost paints the pines in winter time [1820, Shelley, Witch Atl. Xliv].

Synonyms: hoarfrost
See also: hoarfrost

rincon: [Spanish *rincon* = corner.] n. A term used mainly in the U.S. Southwest for a square-cut recess in a cliff or mesa, a secluded valley, or a tight square bend in the course of a stream channel.

rip: [Flemish *rippen* = to strip away; < IE *reup* = to snatch.] n. A strong, narrow surface current that flows rapidly away from the beach, especially when the wave train approaches perpendicular to shore, or when water returning to sea is funneled through a break in an offshore bar. Also, a disturbed state of the sea, as at the interface where two opposing currents meet.
Cited usage: The name 'rip tide' is certainly not appropriate, since the current described has nothing to do with the tide. The name 'rip current' is suggested, since it is close to the other name and describes the way in which the current rips through the oncoming breakers [1936, F. P. Sherpard, Science 21 Aug. 181/2].

riparian: [Latin *ripa* = bank (of a river).] adj. The land that occurs adjacent to streams, lakes, ponds, wetlands, and other bodies of water. First usage in English c.1845.
Cited usage: Riparian is often used to refer to the vegetation at the sides of a river or creek, but in fact it refers to the location of that vegetation [2003, OSU Water Quality Program].

riparian: [Latin *ripa* = bank of a river + *arius* = pertaining to.] adj. Pertaining to or dwelling on the bank of a river. First usage in English c. 1845.

riprap: [Middle English *rappen* = a strike or blow.] n. A human-made deposit of large boulders or blocks thrown irregularly together, usually in a high-energy, near-shore environment for the purpose of preventing erosion. The term is an alliterative repetition, similar to an echoic term, intending to evoke the sound of waves hitting rocks. First usage in English c. 1575.

river: [Latin *ripa* = bank.] n. A general term for a relatively large freshwater stream, ultimately flowing downslope toward the sea, but often first flowing into a lake or another river. First usage in English c. 1275.

roche moutonnee: [French *roche* = rock + *moutonnee* = fleecy; < *mouton* = sheep; < IE *mel* = soft.] n. An asymmetric bedrock hill that has been covered by ice and shaped by glacial erosion. The glacial upstream (stoss) side is gently sloping, unjointed, and smoothed by abrasion, whereas the glacial downstream (lee) side is jointed, and is steeper due to plucking. Also used as a verb, as in 'roche-moutonneed.' Horace Benedict de Saussure coined the term in 1786, in reference to groupings of small rounded hillocks that are abundant in glacial terrain, and are usually covered with a thin veneer of vegetation. They suggested to him a fleece or a wig of a style termed moutonnée. De Saussure may also have been alluding to the similarity of a single one of the large rock formations looking like a grazing sheep, with its steep downslope side resembling the animal's head. He did not, however, associate them with glaciers. First usage in English c. 1840. *Cited usage*: Plus loin, derriere le village de Juviana ou Envionne on voit des rochers qui ont une forme que je nomme moutonnée [1786, H. B. De Saussure, Voyages dans Alpes II. xlviii. 512-3]. *Synonyms*: sheeprock, whaleback, embossed rock

rock: [Middle English *rokken* < Old English *rocc* = back formation of *stanrocc* = stone rock; also, French *roche* < Old French *roque* < Latin *rocca* = rock.] n. A naturally occurring, solid, and cohesive aggregate of one or more minerals, noncrystalline solids, or organic materials. Rocks are formed by one or more of the processes of pressure, heating, cooling, precipitation, evaporation, or chemical action. The term appears to have been used in Middle English and Old English principally to describe rock formations, as opposed to individual stones. First usage in English c. 1325.

rod: [Old English *rodd* = pole, offshoot, twig, stick.] n. A linear measure equal to 5.5 yards (16.5 feet). Also a square measure equal to 30.75 square yards; 160 square rods equals one acre. First used as a standard measure c. 1450. *Synonyms*: pole

rubble: [Middle English *robishe* = refuse, offal.] n. Loose angular rock fragments that overlie outcrops, or are found beneath alluvium or soil. Also, used for the loose, irregular, and artificailly broken rock fragments that are extracted from a stone quarry. First usage in English c. 1375. *Cited usage*: To this mass the provincial name of 'rubble' or 'brash' is given [1852, Lyell, Elements of Geology (ed. 4) vii. 81].

rubidium: [Latin *rubidus* = dark red < IE *reudh* = red, ruddy.] n. A highly reactive, silver-white, metallic element of the alkali group (symbol Rb). Discovered and named in 1861 by G. R. Kirchhoff and R. W. Bunsen (inventor of the *Bunsen* burner), upon witnessing previously unrecorded, distinctively dark-red spectral lines in the phyllosilicate mineral lepidolite from Saxony, thus the etymological reference to *rubidus*. IE derivative include: ruby, rutile, ruddy, rouge, and rubescent. *Cited usage*: Therefore we propose for this alkali metal, in respect to those two remarkable dark red lines, the name rubidium with the symbol Rb from rubidious which was used by the ancients to designate the deepest red [1861, G. R. Kirchhoff and R. W. Bunsen, University of Heidelberg].

ruby: [Latin *rubeus* = red; < IE *reudh* = red, ruddy.] n. A deep crimson to purple to pale rose-red gem, being a variety of corundum, Al_2O_3. Rubies are colored due to the presence of small amounts of chromium. IE derivatives include: rust, rubidium, rutile, and rouge. First usage in English c. 1300.
See also: corundum

rudist: [Modern Latin *Rudista* = bivalve mollusc family; < Latin *rudis* = unformed.] n. Any extinct bivalve mollusc belonging to the superfamily Rudistacea (subclass Heterodonta). The rudists inlcude several species of reef-building molluscs that are variously shaped in the form of horns, cones, or corals. They were usually attached to a substrate, and found with corals. Not to be confused with the term rudite.
Cited usage: The coral-like Rudists are important, forming reef environments in Tethys [1969, Bennison & Wright, The Geological History of the British Isles xiv. 332].
See also: rudite

rudite: [Latin *rudus* = broken stone, debris, rubble.] n. Any consolidated breccia or conglomerate consisting of particles larger than sand grains.
Cited usage: The consolidated rock whether conglomerate or breccia may be called a rudyte [1904, A. W. Grabau, Amer. Geologist XXXIII. 242]. Three textural types of rock may be recognized: (1) the rubble-rock or rubble-stone, or rudyte, which when the fragments are rounded is a conglomerate and when angular a breccia; (2) the sand-rock or sandstone or arenyte; and (3) the mud-rock or mud-stone or lutyte [1920, General Geol. Text, xviii. P. 569].
Alt. spelling: rudyte
Synonyms: psephite
See also: arenite, lutite, psephite, psammite, pelite

ruffle: [Middle English *ruffelen* = to roughen.] n. A disturbance on the surface of water. For-merly, a ripple mark left in sediment by such a disturbance. First usage in English c. 1375.

rugose: [Latin *ruga* = wrinkle.] adj. Said of a surface marked by rugae, being coarsely wrinkled, rough, and convoluted. Also, referring to extinct coral of the order Rugosa, which were solitary and not colonial, and were characterized by calcareous corallites. Also said of wrinkled pollen and spore sculptures. First usage in English c. 1700.

ruissellement: [French *ruisseler* = to stream, as with water or sweat; < Latin *rivus* = stream.] n. A pluvial catchment area that includes stream runoff from a watershed but not groundwater runoff. The term is largely restricted to French-speaking areas.

rumpfflache: [German *rump* = trunk, torso, body + *flachland* = flat land, plain < *abflachen* = to flatten.] n. A flat plain underlain by massive or undifferentiated rock, having no implication to the cycle of erosion. Term coined in 1953 by geomorphologist Walther Penck. He and William Morris Davis contributed greatly to the understanding and nomenclature of geomorphological cycles and provinces. He also coined the term endrumpf, which, along with rumpfflache, is closely related to Davis's 'peneplain.'
Synonyms: peneplain
See also: peneplain, endrumpf

runnel: [Middle English *rynel* < Old English = *rinnan* = to run; < IE *rei* = to flow, run + Old English *el* = diminutive suffix.] n. A small stream. Also, on an intertidal beach, a trough formed landward of the berm crest by the action of advancing and retreating tidal water. First usage in English c. 1575.
Cited usage: Herons stand in the little runnels which trickle over the flats [1883, C. Davies, Norfolk Broads xxvi. 198].
Synonyms: rundle, runlet, rindle
See also: rill, swale

rupicolous: [Latin *rupes* = rock, craig < + -*colous* = combining suffix meaning 'having a specific habitat' < *cola* = inhabitant, tiller < *colere* = to inhabit.] adj. Growing or living on or among rocks, as in a rupicolous plant or rupestrine organism. IE derivatives include: erupt, rupture, bankrupt, corrupt, and interrupt. First usage in English c. 1855.
Synonyms: saxigenous, rupestral, rupestrine
See also: saxigenous

rutile: [Latin *rutilus* = red; < IE *reudh* = red, ruddy.] n. A golden to reddish-brown to black tetragonal mineral, titanium dioxide, TiO_2. The colors are due to varying amounts of iron, as pure titanium oxide is white. Rutile is trimorphous with brookite and anatase. It commonly occurs as asicular needles in other minerals, especially quartz. Term coined by German mineralogist Abraham Gottlob Werner in 1803, alluding to its common reddish-brown color.

ruware: [Shona (native language of South Africa) *ruware* = rocky mound.] n. An area of bare plutonic rock with a slightly domed profile that locally outcrops where gently dipping exfoliation joints are exposed at the surface. It is considered the first stage in the erosional development of a dome or tor.
Similar terms: rock pavement

sabkha: [Arabic *sabkah* = a saline infiltration, salt flat.] n. A flat, salt-encrusted depression, on the coastal plain just above the high-tide level in arid to semi-arid climates. Periodic flooding and evaporation in the area causes accumulation of alternating layers of eolian clays and salts. Sabkhas are common in modern coastal environments such as the Red Sea, Persian Gulf, and Gulf of California. In Saudi Arabia, such features are most common in low-lying coastal plains, but older ones can be found in places as far inland as the edges of the Summan Plateau, some 125 km from the coast. The term has also been applied to any flat, coastal, or interior area where saline minerals crystallize at or near the surface, through deflation and evaporation.
Cited usage: The ground here is what the Arabs called 'sabkha', soft, crumbly salt marsh; sandy when it is dry and ready at a moment's notice to turn into a world of glutinous paste [1911, G. Bell, Let. 23 Feb. (1927) I. xii. 280].
Alt. spelling: sabkha, sabkhah, sibakh (pl.)
Similar terms: playa, dikakah, salt marsh
See also: dikakah, shatt, mamlahah

saccharoidal: [Latin *saccharon* < Greek *sakkhar* < Sanskrit *sarkara* = sugar.] adj. A texture of interlocking crystals resembling equigranular sugar, that is found in a variety of rocks, including: aplite, sandstone, quartzite, rock gypsum, and marble. Petrologically, the term is more specific and applies to the texture of equigranular igneous rocks with a grain size ranging from 0.05 to 10mm, and averaging 1mm. First usage in English c. 1830.
Cited usage: Ferruginous brown and red, coarse, friable sandstone, in some parts white and saccharoidal [1863, James D. Dana, Manual of Geology 383].

sackungen: [German *sackungen* = sag.] n. In structural geology, deep-seated rock creep, possibly triggered by earthquake shaking, and propagated by gravitational settling of a slablike mass into an adjacent valley. The top of the settled slab produces a ridge-top trench (sackung), that is aligned parallel to the crest line of the ridge.

saddle: [Middle English *sadel* < Old English *sadol* < IE *sed* = to sit.] n. In geomorphology, a low point in the crestline of a ridge, especially when a section of land, ice, or snow forming the ridge slopes sufficiently upward from a middle low point toward peaks on both sides, so that the feature resembles a ridding saddle in appearance. In structural geology, a depression along the surface axial trend of an anticline, concave in longitudinal section and convex in transverse section. IE derivatives include: sediment, sessile, and edaphic.
Cited usage: The northern Channel Islands and the Santa Monica Mountains are the crest of the same post-Miocene anticline, but are separated by a structural saddle exhibiting little evidence for late Quaternary folding [2000, Christopher Sorlen, Models for slip on blind thrust faults, Institute for Crustal Studies, Univ of California, Santa Barbara].
Similar terms: saddleback

sagger: [English corruption of *safeguard* < Middle English *savegarde* = protection.] n. A coarse fireclay found in association with the base layers of coal seams; used to form protective boxes for delicate ceramic objects in the firing process. Along with the term *seggard*, an obsolete

term meaning a protective article of dress, sagger has probably experienced parallel evolution from the word 'safeguard.'
Cited usage: A rich iron ore is packed, together with coke breeze (the reducing agent) and limestone, into clay containers called saggars (the term is taken from the pottery industry, and in fact the process is very similar to that used in making pottery [1960, Times Rev. Industry July 22/1].
Alt. spelling: sagre, saggar

salic: [A mnemonic term derived from *s*(ilica) + *al*(umina).] adj. Pertaining to the group of minerals that are light-colored and rich in silicon and aluminum, such as quartz, feldspar, and the feldspathoids. Also said of light-colored rocks containing one or more of these minerals in the norm of igneous rocks. As opposed to femic. Term coined in 1902 by Cross, Iddings, Pearson, and Washington.
Cited usage: To express concisely the two groups of standard minerals and their chemical characters in part, the words sal and fem have been adopted. The former mnemonically recalling the siliceous and aluminous character of its minerals [1902, W. Cross et al., Jrnl. Geol. X. 573].
See also: femic

salient: [Latin *salio* = leap < *salire* = to jump, leap.] adj. In general usage, said of material things that stand out beyond the general surface, or of immaterial things that are prominent and conspicuous in concept. In geomorphology, said of landforms which project, point, or rise out from the surrounding landscape. In structural geology, in fold-belt regions, said of the area where the axial traces are convex toward the perimeter of the fold belt. IE derivatives include: somersault, and salmon (the leaping fish).

salina: [Spanish *salina* < Latin *salinae* = salt pits, salt mine < *salinus* = of salt, salty < *sal* =

salt.] n. A salt lake, pond, well, spring, or marsh; also a salt pan, saltworks. The term is also used to describe a low, marshy area of land near the coast.

salinelle: [French *salin* = salty; < Latin *salinus* = salty < *sal* = salt.] n. A mud volcano that erupts saline material.
Similar terms: hervidero, macaluba
See also: hervidero, macaluba

salpausselkä: [Toponymous term after *Salpausselka Ridges*, three parallel ridges in Finland where dated Pleistocene-age recessional moraines exist.] n. A Finnish term for a steep recessional moraine. The term is named for the recessional moraine trending east-west across southern Finland that is regarded as marking the last advance of the ice sheet at the end of the Pleistocene.
Cited usage: The upper boundary of the Pleistocene is 'mirabile dictu' … it must lie somewhere between 10,000 and 10,500 years B.P., and may be defined as the time equivalents of sediments overlying the terminal moraines of the Valders (in North America) or the Salpausselkä (in Europe) [1968, R. W. Fairbridge, Encycl. of Geomorphology. 918].

salt: [Latin *sal* = salt; < Greek *hals* = salt, sea.] n. A chemical compound formed by replacing all or part of the hydrogen ions of an acid with metal ions or electropositive radicals. The most common natural salt is the mineral halite, NaCl. In Hebrew, the word for salt also means king. First usage in English before 900.

saltation: [Latin *saltare* = to leap (implying repetitive motion); < *salire* = to jump.] n. The dynamic transportation of denser particulate matter in a fluid medium by impact with the surface over which the fluid flows, the most common being sedimentary particles in air or water. Individual particles advance in episodic leaps

initiated by impact with another particle whose momentum is maintained laterally by fluid motion and downward by gravity. Individual particles may happen to bounce off the hard surface, hit another particle, or impart their momentum as a combined vector of both. First usage in English c. 1645.
Cited usage: Transportation may be regarded as the general movement of earth matter seaward by streams; it comprises carriage of material (a) in solution, (b) in suspension, and (c) in what may be denoted saltation [1908, W. H. McGee, Bull. Geol. Soc. Amer. XIX. 199].

saltpeter: [Medieval Latin *sal petrae* = salt of rock; < Latin *sal* = salt + *petra* = rock.] n. Name for either of the naturally occurring salts, niter, KNO_3, or nitratite, $NaNO_3$. So called because these salts commonly encrust rocks as a result of evaporation in arid regions. First usage in English c. 1300.
Synonyms: Chile saltpeter, soda niter, niter
See also: niter, caliche

sand: [Old English *sand* < Greek *psammos* = sand; < IE *samatha* = innumerability.] n. An unconsolidated rock fragment or detrital particle, smaller than a granule, but larger than a silt particle. Sand ranges in diameter from 0.062 to 2mm. Sand may be composed of any combination of mineral and rock fragments, but quartz is among the most common constituents. These water- and wind-worn particles are the material of beaches, riverbeds, and deserts. First usage in English before 900.
Cited usage: A packet of sand used as ballast for one of these balloons wound up in the hands of scientists at the U.S. Geological Survey's Military Geological Unit. Through painstaking research and microscopic analysis, the scientists pinpointed one of three balloon launch sites in Japan [1996, John McPhee, The Gravel Page (from Irons in the Fire)].
See also: psammite, arenite

sandur: [Icelandic *sandur* = sand.] n. Broad, gently sloping sheets of sand and gravel, composed of glacial outwash that accumulates at the foot of glaciers by deposition from meltwater streams. If unconstrained by topography, sandur eventually coalesce into progressively flatter outwash plains and fans.
Cited usage: The sandur (sandar is the Icelandic plural) landform is especially well developed along the south-central coast of Iceland, particularly in the 110 km of the coastal plain that lies between the eastern margin of Myrdalsjökull and Öraefajökull [1986, Richard S. Williams Jr., Glaciers and Glacial Landforms, in Geomorphology from Space, NASA publication edited by Nicholas M. Short, Sr. and Robert W. Blair, Jr.].
Alt. spelling: sandar (pl.)

sanidine: [German *sanidin* < Greek *sanis* = board.] n. A glassy, high-temperature monoclinic mineral, potassium aluminum silicate, $KAlSi_3O_8$. It is polymorphic with orthoclase and microcline, and forms a complete solid-solution series with high albite. It occurs commonly as phenocrysts in volcanic rocks, such as rhyolite and trachyte. The name alludes to its common tabular habit. Term coined in German as *sanidin* by K. W. Nose in 1808.
Cited usage: A name given to Glassy Felspar, on account of the tabular form of its crystals [1867, Brande & Cox, A Dictionary of Science, Literature and Art].

saponite: [German *seifenstein* = soapstone < *die seife* = soap + *der stein* = stone; also, influenced by Latin *sapon* = hair dye.] n. A magnesium-rich clay mineral of the montmorillonite (smectite) group, with the idealized formula $Mg_3(Al,Si)_4O_{10}(OH)_2$. Other hydrous phyllosilicates in the smectite group are aluminum-rich montmorillonite and iron-rich nontronite. Saponite is commonly found in veins and cavities in basalt and serpentinite, where it occurs as soft, soapy masses. Term coined in 1841 by Svanberg. First usage in English c. 1845.
Synonyms: bowlingite, mountain soap, piotine
See also: montmorillonite, beidellite, nontronite, smectite

sapphire: [Latin *sapphirus* = blue; < Greek sappheiros = lapis lazuli; < Hebrew *sappîr* = a precious stone.] n. A variouslycolored, transparent, hard, gem variety of corundum (other than ruby), Al_2O_3. Blue sapphires are particularly valued and are colored by impurities of titanium, cobalt, and iron. Sapphires may also be pink, purple, yellow, green, and orange. The chief sources of sapphires are: Sri Lanka, Kashmir, Burma, and Thailand. A rare, indigo to Berlin-blue, colored variety of quartz, colored by fibers of silicified crocidolite, is known as sapphire-quartz.
Cited usage: Bright azure rays from lively sapphyrs stream [1711, Alexander Pope, Temp. Fame 252].

sapro-: [Greek *sapros* = rotten, putrid.] Combining prefix meaning decomposed, rotten, or decayed; as in sapropel, saprophyte, saprogen or saprolite.
See also: sapropel, saprolite

saprolite: [Greek *sapros* = putrid, rotten + *lite* < *lithos* = rock.] n. Soft, usually clay-rich, thoroughly decomposed bedrock, formed in place by chemical weathering. Saprolite retains the fabric and structure of the parent rock. Term coined in 1895 by G. F. Becker.
Cited usage: Chemical weathering can produce a rotted rock-form known as a saprolite, which is the product of chemical changes which have taken place in situ [1977, A. Hallam, Planet Earth 48/1].
See also: grus

sapropel: [Greek *sapros* = rotten + *pelos* = mud.] n. An unconsolidated, gelatinous ooze of decomposed plant remains and algae, which has accumulated in an anaerobic environment on the floors of shallow swamps, lakes, and seas. An etymologically related term is pelite.
Cited usage: Aquatic ooze or sludge rich in organic matter, believed to be the source material for petroleum and natural gas [1982, Leet, L. Don, Physical Geology, 6th Edition].
See also: pelite

sard: [Toponymous term after Greek *Sardeis*, the ancient capital of Lydia, a wealthy country in Asia Minor c. 600 BCE.] n. A reddish, translucent variety of chalcedony, that is duller than carnelian. Carnelian and sard were the stones found near Sardeis that were most commonly used by the Greeks and Romans for carving and engraving (glyptic art). First usage in English c.1375.
Cited usage: The blood-red Sardian to its birthplace owes its name, to Sardis, whence it first arose [1860, Rev. C. W. King, Antique Gems 398].
Alt. spelling: sardius, sardine, sardian
See also: carnelian

sarsen: [Eponymous term after *Saracen* < Latin *Saracenus* = an Arab, Muslim, or "outlandish stranger"; < Greek *Sarakenos* = a member of a pre-Islamic nomadic people of the Syrian-Arabian deserts; < Arabic *sarq* = east, sunrise.] n. A large, residual mass of silicified sandstone or conglomerate (silcrete). Sarsen is found in the chalk downs of southeastern England, and includes the famous Hertfordshire puddingstone and Wiltshire sandstone. Remnant sarsen were given their anthropomorphic name after Muslim people in Old England, who, like the foreign conglomerates in chalkstone, were conspicuous among the country people. First usage in English c.1640.
Cited usage: The isolated blocks called Greywethers or Sarsen stones, scattered on the surface of the chalk downs [1888, J. Prestwich, Geol. II. 342].
Alt. spelling: sarsden, sarsdon, sarcen
Synonyms: Saracen stone, druid stone, greywether

sastrugi: [German *sastruga* < Russian *zastruga* = ridge, furrow, plane, smooth < *za* = beyond + *struga* = deep place into which one may fall; < IE *sreu-* = to flow.] n. Irregular ridges up to 5cm high that form by wind erosion of a level snow surface. They are oriented parallel to the prevailing wind and are prevalent on broad planar surfaces in polar

regions. The analogous Norwegian term is skavler. First usage in English c. 1835. IE derivatives include: rhyolite, diarrhea, and catarrh.
Alt. spelling: zastrugi, zastruga and sastruga (sing.)

sauconite: [Toponymous term after the Saucon Valley, near Bethlehem, Pennsylvania, the type locality of this mineral.] n. A zinc-rich clay mineral of the montmorillonite group, being the zinc analogue of Mg-rich saponite.

saussuritization: [Eponymous term after Horace Benedict de Saussure, Swiss geologist and mountaineer (1740-1799).] n. The hydrothermal or metamorphic alteration of calcic plagioclase to a grayish mineral aggregate containing varying amounts of albite, epidote, zoisite, calcite, sericite and zeolites. This process is especially common in mafic-rich basalts and gabbros. This suite of minerals was once thought to be a specific mineral, and was named saussurite. H. B. de Saussure, while professor at the University of Geneva (1762-1786), wrote *Regard sur la Terre* (*A Look at the Earth*) and *Voyages dans les Alpes* (*Journeys in the Alps*). He scaled Mt. Blanc in the French Alps and spent 11 days at 3,500m (11,500ft.) making measurements. He is thought to have popularized the term geology. First usage in English c. 1810.

savannah: [Spanish *savana* < Taino (Caribbean) *zavana* = grassland.] n. An open grassy plain with scattered drought-resistant trees and shrubs, usually in tropical or subtropical regions having alternate wet and dry seasons. Along the U.S. Southeast Atlantic Coast, the term is also applied to low, marshy alluvial flats with sporadic clumps of trees. Due to the flatness of a savannah, the likeness to the Spanish *sabana*, meaning sheet, has been suggested; however, the pronunciation of the two words is different. First usage in English c. 1850.
Cited usage: The treeless savana is called 'campo vero'; if the savannas are strewn with

clumps of low trees, they are 'serrados' [1920, M. E. Hardy, Geogr. of Plants iii. 142].
Alt. spelling: savanna

saxigenous: [Latin *saxum* = rock + *gen-us* = sprung from, begotten + *-ous* = abounding in, full of.] adj. Growing within and among rocks.
Synonyms: rupestral, rupicolous
See also: rupicolous

scabland: [Flemish *schab* = itch; < Latin *scabies* = contagious skin irritation caused by a parasitic mite < *scabere* = to scratch + Old English *land* = open land.] n. An elevated area of flat-lying basalt flows with patchy vegetation and thin soils. Scablands are characterized by deep dry channels scoured by glacial meltwater. The original sca-blands are those of the Columbia Plateau in eastern Washington State, U.S.A.. First usage in English c. 1920.
Cited usage: At peak stage Lake Missoula had a surface area of about 7,500 km^2 and contained an estimated 2,000 km^3 of water. All this water is thought to have discharged westward in a matter of a few days. This great flood moved boulders with diameters greater than 10 m and scoured a system of coulees across the Columbia Plateau. This great tract of flood-eroded topography is called the channeled scablands [1976, C. L. Matsch, N. Amer. & Great Ice Age vi. 74].

scallop: [Middle English *scalop* < Dutch *shelpe* < Old French *escalope* = shell, thin slice of meat.] n. In paleontology and biology, any various free-swimming marine bivalve molluscs of the family Pectinidae, having a rather flat, fan-shaped shell with a radiating fluted pattern. Scallops are able to swim by rapidly opening and closing their valves. First usage in English c.1375.
Cited usage: In the Scallops (Pecten) the edges of the mantle are studded with pearl-like points [1841, T. R. Jones, A General Outline of the Animal Kingdom. xxii. 391].

scaphopod: [Greek *skaphos* = boat + *pod* = foot.] n. Any benthic marine univalve mollusc belonging to the class Scaphopoda. Term coined in 1862 in modern Latin by Heinrich Georg Bronn, in his work *Klassen und Ordnung des Thier-Reichs* (*Classes and Orders of Thier-Reichs*). *Cited usage*: Scaphopods are small, marine, bilaterally symmetrical mollusks with an external, curved and tapering tubular shell open at each end… the scaphopod shell is composed of aragonite [1935, Twenhofel & Shrock, Invertebr. Paleontology ix. 360].

scar: [Gaelic *sgeir* = a rock in the sea; < Old Norse *sker* = a low reek in the sea; < IE *sker* = to cut.] n. In geomorphology, any isolated rock covered by the sea at high water or in stormy weather; a reef. Also, a steep cliff or craggy outcrop of a mountain. Certain British place names, such as Scarborough and The Skerries, are derived from this term. First usage in English c. 1325.
Alt. spelling: scaur
See also: skerry

scarp: [Back-formation from English *escarpment* = cliff; < French *escarper* = to cut steeply; probably < German *scharf* = sharp.] n. A cliff or steep slope produced by faulting, mass wasting, or erosion. Scarps are usually named according to their origin, as in: fault scarp, landslide scarp, or erosional scarp. Scarps generally are straight, and of considerable linear extent, as in the erosional scarps along the edge of a planteau or mesa. First usage in English c. 1585.
Cited usage: The scarps of the hills face indiscriminately all points of the compass [1802, Playfair, Illustr. Huttonian Theory 410].
Synonyms: escarpment, escarp, cuesta (Gt. Britain)
See also: escarpment

schist: [French *schiste* < Latin *schistos* = fissile (stone), a kind of iron ore; Greek *skhistos* = split, divisible; < *skhizein* = to split.] n. A strongly foliated metamorphic rock produced by regional metamorphism, in which the mineral grains are easily visible and are aligned in planar fashion. The granular texture and banded structure of schist is usually composed of laminated, flaky layers of chiefly micaceous minerals with alignment at right angles to the direction of stress, leaving no trace of the original bedding. This planar (parallel to subparallel) display of such minerals, as platy micas and elongate amphiboles, is called 'schistosity', and allows the rock to be easily split into slabs. Schist is coarser-grained than phyllite, and lacks the segregation into dark and light bands that is characteristic of gneiss. When a basic igneous rock is metamorphosed it forms a hornblende-schist (amphibolite), a greenschist if it contains a planar fabric, or a greenstone if no fabric is present. Thus, in this latter context 'schist' refers to the fabric component and not to the overall rock type. First usage in English c. 1780.

schizo-: [Greek *schizein* = to split.] Combining prefix meaning to split, used in such terms as: schizohaline, schizomycete, schizophrenic, schizocarp and schizodont.
See also: schist

schlieren: [German *schliere* = streaks < *slier* = mud, slime.] n. Elongate lensoid or tabular concentrations of minerals in igneous intrusions that appear as either dark or light streaks, have gradational boundaries toward the host plutonic rock, and differ in composition from the principal mass. Their origin is yet unknown, and has been variously ascribed to flow banding, crystal segregation, and sequential stages of crystal nucleation during cooling. In physics, schlieren is used to describe the visible streaks in a transparent medium where each streak is a region of different density and refractive index. First usage in English c. 1885.

schorl: [German *schorl* = black tourmaline < *schrul* = mining term for small black stones.] n. A

term for the common, opaque, black, iron-rich variety of tourmaline. Schorl occurs in pegmatites, greisens, and as an accessory mineral in some granites. It can occur as acicular, radiating, needle-like crystals. An obsolete use of schorl applied the term to other minerals, such as: white schorl for albite, green schorl for epidote or actinolite, blue schorl for hauyne, and volcanic schorl for augite. *Cited usage*: Tourmaline forms in a variety of geologic settings. It occurs most often in granite pegmatites [1998, Tourmaline: Mineral Information Page, Gem & Mineral Miners, Inc.].
Alt. spelling: shorl
See also: tourmaline

schuppen: [German *schuppe* = scale < *schuppen* = to scale, descale.] n. A structural fabric resulting from the stacking of rock fragments, particles, or tectonic units. Also, used as an adjective for imbricate, as in pebble beds showing an imbricate structure, the pebbles leaning in the direction of the current.
Similar terms: imbricate stucture, schuppenstruktur
See also: imbricate

science: [Middle English *science* = knowledge, learning; < Latin *scientia* = knowledge < *scire* = to know; < IE *skei* = to cut, split.] n. The systematic accumulation of knowledge through observation, identification, description, and experimental investigation of phenomena which leads to a connected body of demonstrated truths and observed facts that can be systematically classified and colligated into general laws. This knowledge includes reliable methods for the confirmation of old laws and discovery of new truths within a specific realm of science. First usage in English c. 1325.

scintillation: [Latin *scintillare* = to sparkle < *scintilla* = spark.] n. A spark or flash of light. In physics, a flash of light produced by a radioactive

particle or other ionizing agent when it impacts a phosphor. In gemology, the sparkling display of light reflected from the polished facets of a gemstone. First usage in English c.1620.

sclera-: [Greek *skleros* = hard.] Combining prefix meaning hard, used in such terms as: scleractinian coral, sclerite, scleroderm, sclerotize, arteriosclerosis, and scleracoma (hard parts of a radiolarian).

scoria: [Latin *scoria* = dross; < Greek *skoria* = slag < *skor* = excrement, dung, refuse; < IE *sker* = dung, excrement.] n. A highly vesicular aphanitic or porphyritic volcanic rock, generally of basaltic composition. Scoria makes up the surface of basaltic lava flows, and is the dominant rock in such pyroclasts as cinders and bombs formed by fountaining eruptions of basalt. Accumulations of scoria around a volcanic vent form cinder cones that may accumulate to heights of about 300m, especially around Strombolian vents. Scoria is surface-cooled on exposure to air, and distended by expanding gases. Scoria is the singular noun, but the false singular scorium is sometimes used. IE derivatives include scatology and skatole (a white crystalline organic compound with a strong fecal odor.
Cited usage: The ejected scoriae of volcanoes are receptacles in which mineral products previously unknown are constantly discovered [1830, Herschel, Studies in Nat. Phil. 294].
Alt. spelling: scoriae (pl.)

scree: [Old English *scridan* = to glide; < Old Norse *skridha* = landslide < *skridha* = to slide.] n. As used in the U.S., the rock fragments making up a talus. In Europe, often used as a loose equivalent for talus. First usage in English c. 1780.
Cited usage: Skirl, or screes, small stones or pebbles [1781, James Hutton, Tout to Caves 2nd ed.].

scum: [Dutch *schum* = foam, scum; < IE *skeu* = to cover.] n. A film of varied composition, such as those made up of decomposed organic matter, that floats on a relatively stagnant body of water. In metallurgy, slag or dross appearing on the top surface of ingots during pouring. First usage in English c. 1225.

sea: [Old English *sae* = sea; < Dutch *zee*; < Old Norse *sjar* = sea.] n. The term refers both to the areas of saltwater covering much of the Earth's surface (the seven seas), and, in a more restrictive sense, to bodies of saltwater that are more or less confined by land boundaries, but are connected to the great oceans. Examples of seas in the latter sense are: the Mediterranean, the Aegean, and the Caribbean Seas connected with the Atlantic Ocean; the Arabian Sea, a division of the Indian Ocean; and the China and Japan Seas connected to the Pacific Ocean. Subdivisions of great seas, such as the Adriatic is integral to the Mediterranean, can also be called seas. Some large bodies of salt water that are entirely land-locked are also called seas, such as: the Caspian Sea, the Sea of Aral, and the Dead Sea. The ultimate source of this term is uncertain, but it is used extensively in northern and central Europe as part of large water-body names, such as: the Zuider Zee (North Sea) and Gosausee in Austria (named after the location, Gosau).
Cited usage: Ofer sæ side [c.1100. Beowulf, line 2394]. This precious stone, set in the siluer sea [1593, Shakespeare, Richard II, Act II. scene i. 46. God moves in a mysterious way, His wonders to perform; He plants his footsteps in the sea, And rides upon the storm [1779, Cowper, Olney Hymns III. xv. 3].

Secchi disk: [Eponymous term in honor of the Italian papal scientific advisor and astrophysicist Pietro Angelo Secchi (1818-78).] n. A checkered white-and-black disc used in oceanography and limnology to determine the transparency of water, and, by inference, the amount of suspended material in the water column. The disc is allowed to sink to determine the depth at which it ceases to be visible from the surface. The first Secchi disk was lowered from the papal steam yacht,

l'Immacolata Concezione, in the Mediterranean Sea on April, 20, 1865. Peitro Angelo Secchi (1818-1878), the scientific advisor to the pope, initiated a series of seven experiments over a six-week period at the request of Commander Cialdi, head of the Papal Navy. In absolutely pure water, the theoretical maximum value would be between 70-80 meters (230-262 ft). The deepest published Secchi disk reading was 66 meters (217 ft) in the Sargasso Sea. The smallest Secchi depth recording was 1-2 cm (0.4-0.8 in) in Spirit Lake, Oregon, after the eruption of Mt. St. Helens.

secretion: [Latin *secernere* = to separate, set aside, to sever < *se* = apart + *cernere* = to separate; < IE *krei* = to sieve, discriminate, distinguish.] n. In paleontology and biology, the process by which plants and animals release substances, often to form hard parts such as exoskeletons or chrysalii. Also, an obsolete term in geology for precipitation or deposition of mineral matter from solution inside a rock cavity, forming a geode or amygdule. IE derivatives include: excrete, excrement, and discriminate.
Cited usage: In a true concretion, the material at the centre has been deposited first, and has increased by additions from without. Where, on the other hand, cavities have been filled up by the deposition of materials on their walls, and gradual growth inward, the result is known as a secretion [1882, Geikie, Archibald, Text-Book of Geology 4th ed., p. 95]
See also: concretion

sectile: [Latin *sectus* = cuttable < *secare* = to cut; < IE *sek* = to cut.] adj. Said of a mineral that can be cut smoothly with a knife. IE derivatives include: section, sector, segment, dissect, insect, intersect, and transect.
Cited usage: Sectile are those fossils whose integrant particles are coherent, but not perfectly immoveable one among another. Sectile is a medium between brittle and malleable [1805, Weaver, translation of 1774 Abraham Gottlob

Werner, *Von den Āᵍusserlichen Kennzeichen der Fossilien* (*A Treatise on the External Characters of Fossils*)196].

section: [Latin *sectus* = cuttable < *secare* = to cut; < IE *sek* = to cut.] n. In general geologic usage, the successive rock units or geologic structure represented by a graphic display of either an exposed rock surface, or the rock surface as it would appear if cut through by an intersecting plane. In surveying, one of the 36 square and equal subdivisions of a township in the U.S. Public Land Survey system, usually being a tract of land one square mile in area. In mineralogy and petrology, a thin section is a slice of a mineral or rock glued to a glass slide and ground to 0.03mm thickness so that light can transmit and optical properties can be determined. First usage in English c.1555.
Cited usage: A number of groups or stages similarly related constitute a series, section (Abtheilung) or formation, and a number of series, sections, or formations may be united into a system [1882, Geikie, Textbook Geol. vi. 635].

secule: [Latin *seculum* = age, generation.] n. A rarely used term for a unit of geological time corresponding to a biostratigraphic zone. Term coined in1903 by A. J. Jukes-Browne.
Cited usage: The term hemera may, however, be occasionally convenient to signify the duration of a subzone, as age signifies the duration of a stage, but if we want to avoid confusion we must not speak of the hemera of a zone. For this another word should be coined I would suggest that the Latin word seculum will furnish us with 'secule' which finds an actual French equivalent in siècle [1903, A. J. Jukes-Browne, Geol. Mag. X. 37].
Similar terms: moment

sediment: [Latin *sedimentum* = act of settling, sinking down < *sedere* = to sit, settle.] n. Uncon-solidated particulate material that has been transported some distance and then deposited in layers from water, air, or ice. Sediment originates by several possible mechanisms: a) breakdown of pre-existing rock, b) secretion by organisms, c) settling of dead biologic material, d) chemical precipitation from solution, and e) settling of cosmic debris. Examples of sediment include: water-transported alluvium, wind-blown loess, and glacial till. First usage in English c. 1545.

seep: [Dutch *sijp* < Frisian *sipe* = a ditch, channel.] n. Generally, a small area where fluid, such as water or oil, migrates slowly to the land surface. In general usage, the verb seep means to pass slowly through the interstices of a porous medium. First usage in English c. 1785.
Cited usage: Almost without exception, seeps are at topographically low spots where water has accumulated. Oil rises to the surface of the water, covering it with an iridescent film [1966, McGraw-Hill Encycl. Sci. & Technology X. 60/2].

seiche: [French *seiche* = exposed lake bottom < *sec* = dry; < Latin *siccus* = dry.] n. An oscillatory wave formed in enclosed or semi-enclosed basins, such as those occurring in lakes, bays, fjords, and gulfs. Seiches may be caused by such disturbances as: earthquakes, large changes in barometric pressure, and wind. Once the body of water is disturbed, gravity acts to restore the horizontal surface, and the resulting wave travels the length of the basin at a velocity dependent on the water depth. The wave is then reflected back from the basin's end, generating interference. Repeated reflections generate standing waves having one or more locations within the basin, called nodes, that experience no vertical motion. The length of the lake is an exact multiple of the distance between nodes. The period of the seiche is a function of the basin's horizontal dimensions, and ranges in duration from seconds to hours. First usage in English c. 1835.

seif: [Arabic *sayf* = sword; < Aremaic *sayp* < Greek *xiphos* = sword.] n. A sharp-crested, tapering set of longitudinal sand ridges or dunes. Seif are composed of loose, fine-to-medium grained, well-sorted particles. They are common in the Sahara Desert, and range up to 300 m in height and 300 km in length. The flat-topped varieties are usually called sand ridges, whereas the sinuous saw-toothed varieties are more often called seifs. Strong but infrequent crosswinds can widen the dunes and create steep slip faces on both dune flanks. Surfaces on the lower flanks of both sides of the ridge are wind compacted, whereas loose sand will be found on the ridge top.
Cited usage: In form it is a typical sif (seif) dune, a long straight ridge of sand with a single longitudinal chain of crests rising to billowy pyramids set at regular intervals, in silhouette something like huge saw-teeth. [1931, Geogr. Jrnl. LXXVIII. 16].
Alt. spelling: sif, saif
Synonyms: sword dune, alab, uruq, slouk, sigmoidal dune

seismology: [Greek *seismos* = earthquake < *seiein* = to shake, quake + *logos* = word, speech, discourse, reason < *legein* = to speak, to say; IE *leg* = to collect.] n. The study of earthquakes, as well as the study of the structure and properties of the Earth by using natural and human-induced seismic waves as a source of information. Deep earthquakes provide information about the structure of the Earth's mantle, whereas shallow earthquakes provide detailed information about the structure of the crust.
Synonyms: tectonics
See also: tectonics

selenite: [Latin *selenites* = gypsum crystals; < Greek *lithos selenites* = moonstone < *Selene* = goddess of the moon < *selas* = light, brightness + -*ites* < -*lite* = pertaining to rocks < *lithos* = rock, stone.] n. The clear, colorless, platy variety of the mineral gypsum, hydrated calcium sulfate,

$CaSO_4 \cdot 2H_2O$. As described by ancient writers, the Greek *lithos selenites* (moonstone) had properties distinct from clear crystals of gypsum. A related term is selenium.
See also: -ite

selenium: [Latin *selenites* = gypsum crystals; < Greek *lithos selenites* = moonstone < *Selene* = goddess of the moon < *selas* = light, brightness.] n. An element resembling sulfur and telluriumt, selenium was formerly classed among the metals, but is now considered a nonmetal (symbol Se). Selenium has several allotropic forms, one being a dark grey to black solid with a metallic lustre, and sometimes referred to as metallic selenium.. An important property of selenium is that its electrical resistance is greatly decreased by exposure to light: hence its use in the photophone or radiophone of A. Graham Bell. Term coined in 1818 by Jöns Jacob Berzelius, alluding to the resemblance of the properties of the new element to those of tellurium.
Cited usage: The analogy of tellurium has induced me to give it the name of selenium [1818, Jöns Jacob Berzelius, Annual of Philos. XI. 292].

selvage: [Dutch *zelfegge* < *selfegghe* = self edge.] n. In petrology, a marginal zone at the border of an intrusion or lava flow, such as the chilled margin of a dike or basaltic pillow. In ore deposits, a thin alteration zone at the borders of a metalliferous vein, often being a layer of clayey or earthy material. In textiles, selvage represents a surplus edge of a flat piece of cloth or other material not intended for final use. First usage in English c. 1450.
Cited usage: Selvages of hydrothermal alteration of quartz monzonite along the contacts of some larger aplite-pegmatite bodies [1958, Economic Geology LIII. 292]. The majority of M(id-)O(cean)R(idge)B(asalt) lavas have a pillow form and glass selvages–characteristics of rapid cooling in seawater [1981,

Cambridge Encycl. of Earth Sci. 211].
Alt. spelling: selvedge
Synonyms: salband

semblance: [French *sembler* = to resemble; <
Latin *simulare* = to simulate < *similis* = like.] n. In
seismic surveying, a multi-trace geometric seismic
attribute that is a measure of coherence. The
semblance is used to differentiate similar from
dissimilar seismic samples (dissimilar or incoherent
seismic data are caused by stratigraphic or
structural changes, such as at the edge of a bed).

semi-: [Latin *semi* = half; < IE *semi* = half.]
Combining prefix meaning half or partly, used in
terms such as: semidiurnal, semi-arid, and
semiconductor. The Greek-derived prefix *hemi-*,
also used in modern English, comes from the
same IE root.

senile: [Latin *senescentem* = aged < *senescere* =
to grow old.] adj. In limnology, said of a lake that
is approaching extinction due to sediment filling.
In biology, referring to cells of an organism that
show a diminished power of growth after each
successive replication. In geomorphology, a nearly
obsolete term referring to a landscape or topographic feature that is approaching a base-level
plain or the end of an erosional cycle. Geologists
have generally abandoned use of anthropomorphic terms for describing geologic processes, such
as: birth, growth, maturity, old age, senescence,
senility, and death.
Cited usage: Senile topography, the configuration of land which prolonged degradation has
reduced nearly to a base-level plain. [1902,
Webster Suppl.].
Synonyms: senility

sepiolite: [Greek *sepion* = cuttlebone < *sepia* =
cuttlefish < *sepein* (possibly) = to make rotten.] n.
A white to light yellow, orthorhombic clay mineral,
hydrated hydroxy magnesium silicate,

$Mg_4Si_6O_{15}(OH)_2 \cdot 6H_2O$. Sepiolite usually occurs in
compact, lightweight, claylike masses, called
meerschaum and is found in hydrothermal veins
and weathered serpentinite. It is plentiful in the
Mediterranean area, chiefly Asia Minor, and is
used to make smoking pipes, cigarette holders,
and small carvings..
Synonyms: meerschaum

septum: [Latin *saeptum* = partition < *saepire* =
to enclose < *saepes* = fence, hedge.] n. In structural geology, a relatively thin zone of generally
older metamorphic rock that separates two
igneous intrusions. In paleontology, the calcareous partition between chambers of a cephalopod.
In biology, a thin partition or membrane that
divides two cavities or soft masses of tissue in an
organism, such as the nasal septum or the atrial
septum of the heart.
Cited usage: When these thin septa are converted
into iron pyrites, their edges appear like golden
filigrane work, meandering amid the pellucid spar
[1836, Buckland, Geol. & Min. xv. sec.4 I. 347].
Alt. spelling: septa (pl.)

serac: [French *serac* = cottage cheese; < Latin
seraceum < *serum* = whey.] n. In glaciology, a
large pointed pinnacle of glacial ice formed
between, and isolated by, intersecting crevasses.
Seracs are often found at a sudden increase in the
slope of the glacier and result from tensional
failure in the more rigid upper portion of glacial ice.
This is due to the stretching that occurs when a
glacier moves over a convex slope, spreads out
over a plain, or passes around a bend in a valley.
The etymology was inspired by the pinnacles'
form and color resembling cottage cheese curds.
First usage in English c. 1855.
Cited usage: These ridges are often cleft by
fissures thus forming detached towers of ice ... to
such towers the name séracs is applied [1860,
Tyndall, Glac. I. vii. 51].
Alt. spelling: sérac

sericite: [Latin *sericus* = silken.] n. A fine-grained variety of muscovite or similar potassium-rich mica, usually formed by hydrothermal alteration of feldspars or other aluminosilicates, often with the formula $K_2Al_4(Si_3AlO_{10})_2(OH)_2$. It is common wherever hydrothermal alteration has occurred, as near ore deposits, fault gouge, and in schists and phyllite. The presence of sericite gives feldspar a cloudy appearance when viewed in thin section. Term coined in 1852 by K. List.
Cited usage: Sericite of K. List, is regarded by him as near Damourite [1854, Dana, Syst. Min. (ed. 4) II. 223]. Sericite-Schist.This is a schistose rock closely allied to the porphyroids, and consists of sericite, fragments of quartz, etc. [1879, Rutley, Study of Rocks 296].

series: [Latin *serere* = to join.] n. The chronostratigraphic equivalent of an epoch. A series is a subdivision of a system, and can be subdivided into stages. It denotes the layers of strata or the body of rock formed during one epoch. Originally, when age-dating was solely a relative process highly dependant on correlating fossil assemblages, a series was defined as a set of successive deposits or group of successive formations having certain common fossil or mineral features. The term is also more loosely used for any assemblage of usually conformable, successive strata without regard to the hierarchical rank of the assemblage. At the meeting of the International Geological Congress in 1881, a scheme of nomenclature was adopted in which the stratigraphical terms: group, system, series, and stage correspond to the units of time: era, period, epoch, and age. Use of 'group' as a chronostratigraphic term is diminishing, therefore, the term 'system' and its subdivisions are now regarded as the primary time-stratigraphical terms. When used formally the initial letter of the term is capitalized, as in Lower Cretaceous Series.
Cited usage: As equivalents of Series, the terms Section or Abtheilung may be used. According to this scheme, we would speak of the Palaeozoic Group or Era, the Silurian System or Period, the Ludlow Series or Epoch, and the Aymestry Stage or Age [1881, Geol. Mag. Decade II. VIII. 558].

serir: [Arabic *serir* = dry.] n. A desert plain from which the sand has been removed by wind erosion, leaving a ground-covering layer of rounded pebbles. Over time, the surface veneer gets more tightly compacted, weathered, and carbonate-cemented to form a fragile yet weather-resistant armor. Serir are extensive in the Saharan plains, specifically those of Libya and Egypt.
Cited usage: Evaporation of capillary water may cause the precipitation of calcium carbonate, gypsum, and other salts that cement the pebbles together to form a desert conglomerate. In the western Sahara such a surface is known as a 'reg', whereas in the eastern Sahara it is called a 'serir' [1974, Encycl. Britannica Micropaedia III. 486/2].
Synonyms: reg, desert pavement

serpentine: [Latin *serpentinus* = serpentlike < *serpens* = serpent < *serpere* = to creep.] n. A mineral group that comprises the rock serpentinite, consisting mainly of polymorphs of hydroxy magnesium silicate, $Mg_3Si_2O_5(OH)_4$, including antigorite, chrysotile, and lizardite. The serpentine group of minerals forms primarily by retrograde metamorphism of ultramafic rocks containing olivine and enstatite (mainly ocean crustal material). The term alludes to the appearance of serpentinite rocks that resemble legless reptiles, with its distinct dull green color, greasy luster, slippery feel, sinuous bedding, and mottled metamorphic effect. The serpentine mineral group should not be confused with serpentinite, the rock, formed by the hydrothermal alteration of ultramafic parent rocks. First usage in English c. 1375.

serrate: [Latin *serra* = saw.] adj. Said of topographic features, mineral interactions, and biological organisms that possess a row of small projections resembling the teeth of a saw, such as: a jagged mountain profile, the serrate texture of megacrysts in igneous rocks, a Stegosaurus dinosaur, or a serrated plant leaf. First usage in English c. 1595.

sesquioxide: [Latin *sesqui* = contraction of *semis-que* = a half in addition (to the whole) < *semis* = one half + *que* = in addition + Greek *oxus* = sharp, acid.] n. An oxide containing three atoms of oxygen and two of another element or radical, therefore, there is 1.5 times more oxygen than the accompanying cation. The mineral hematite (Fe_2O_3) and corundum (Al_2O_3) are sesquioxides. First usage in English c. 1830.

sessile: [Latin *sessilis* = sitting down < *sedere* = to sit.] adj. Said of those plants or animals permanently attached to a substrate at their base, and therefore, not free to move about. Examples of sessile aquatic creatures, include: subtidal and intertidal anemones, echinoderms, and algae. Of benthic creatures, the opposite of sessile is vagile, which comes from the Latin *vagus*, meaning wandering. First usage in English c. 1720.
See also: vagus, isopod

seston: [Greek *sestos* = that which is filtered < *sethein* = to filter, strain.] n. The general term for all fine particulate material suspended in water, both living (bioseston) and nonliving (abioseston or tripton). Seston is that portion that is retained by fine-meshed sieves. Term coined in 1912 by R. Kolkwitz, alluding to filtration by use of sieves. *Cited usage*: The seston consists of bioseston, or plankton and nekton, which latter is ordinarily quantitatively negligible, and of abioseston or tripton. The entire mass of suspended matter in a volume of free water is called seston, the nonliving part, tripton [1967, G. E. Hutchinson, A Treatise on Limnology. vol. II. Introduction to Lake Biology and Limnoplankton, xix pp 235-243].
See also: tripton

seta: [Latin *saeta* = bristle.] n. A slender and usually rigid protruding bristles, such as those found on cat or seal whiskers, the quills of a porcupine, or the whorled branches of the horsetail plant (genus Equisetum). First usage in English c. 1790.
Alt. spelling: setae (pl)

shaft: [German *schacht* = cylindrical shaft; < Greek *skaptein* = to dig.] n. A nearly vertical excavation into the Earth with a relatively small cross-section, made for access to mine workings, tunnelling, or the like. A mineshaft is a vertical entry to a mine, as opposed to an adit, which is a horizontal entry. First usage in English before 1000.
See also: adit

shale: [Old English *scealu* = scale, shell; < Old Norse *skal* = scale; < IE *skel* = to cut.] n. A fine-grained, fissile, argillaceous sedimentary rock composed of clay-sized and silt-sized particles of various minerals deposited in thin laminae. Due to the alignment of clay particles relative to the bedding planes, shale splits easily along closely-spaced laminae. Shales are characteristic of low-energy environments, such as: lake and lagoonal deposits distal portions of river deltas, flood-plains, and deepwater marine environments. Etymologically related cognates of shale include: a) Old English *stanscalu*, meaning soilless rocky stratum (referring to easily loosened material from a mine or quarry), and b) German *schale*, meaning thin layer of ore or stone. First usage in English c. 1745.

shard: [Old English *sceard* = cut, notch; < IE *skel* = to cut.] n. In general usuage, a fragment. In volcanology, a vitric fragment formed during or after volcanic eruptions usually produced by disintegration of pumice. Shards have a character-istically curved surface, usually consisting of fragments of bubble walls. In archeology, broken

fragments of earthenware pottery are called potsherds. Pyroclastic shards can have a similarly curved surface compared to the pottery shards often found in archeological digs. First usage in English before 1000.
Alt. spelling: sherd

shelf: [German *schelf* < IE *skel* = to cut.] n. In marine geology, generally referring to the continental shelf, the area of shallow seafloor contiguous with exposed landmass, which dips gently away from the shoreline to about 200 meters depth. Beyond the shelf edge, the continental slope begins, extending into deeper water. In general: a) a projecting ledge of rock, whether exposed or buried, b) an aquatic reef or sandbar, or c) bedrock itself. A special category of shelves are marine carbonate platforms. If they have steep, sharply delineated, horizontal shelf areas in carbonate-dominated sedimentary environments, they are referred to as platforms. If they are more uniformly sloping shallow marine carbonate areas, they are termed ramps. First usage in English c. 1375.

shergottite: [Toponymous term after the city of Sherghati, India (English = Shergotty).] n. A basaltic, achondrite type of stony meteorite containing abundant augite. Also, the mineral complex found in these meteorites. The Shergotty stone was found in 1865. Two other stony meteorite finds, one at El-Nakhla, Egypt, in 1911 and another at Chassigny, France, in 1815, were found to be similar to the Shergotty stone, but different from all other meteorites and Earth rocks. This group of igneous achondrites acquired the abbreviated name SNC meteorites, which stands for the shergottite-nakhlite-chassignite group of meteorites, and all are considered to have originated from the planet Mars.
See also: nakhlite

shingle: [Middle English *shyngle* < Norwegian *singl* or *singling* = coarse gravel, small round stones; (possibly influenced by Latin *scindula* = a split or cut piece < *scindere* = to split).] n. In coastal geology, loose, water-worn beach gravel and pebbles consisting of relatively large clasts (0.75-7.5cm), smoothed and rounded by abrasion, and lacking in admixed finer material. Shingle consists of erosion resistant minerals, such as flint which is the chief constituent of the shingle beaches in southeastern England (e.g., Chesil Beach in Dorset, England). The term may have echoic origins, perhaps from chink. The modern *sh* spelling evolved from a *ch* spelling, such as in shiver from chiver. The term is more widely used in the British Isles than in the U.S.
Cited usage: The Sea of Faith / Was once, too, at the full, and round earth's shore... / Of the night-wind, down the vast edges drear / And naked shingles of the world [1867, Matthew Arnold, Dover Beach].

shoal: [Middle English *shold* = shallow(s) < Old English *sceald* = shallow.] n. In general, a relatively shallow area in a body of water. In particular, a submerged bar or ridge covered by unconsolidated sediment that rises significantly above the seafloor. The term can be used as several parts of speech: a) as a noun, as in 'the protective coastal shoal,' b) as a verb, as in 'the sandbar shoals abruptly below the point', or c) as an adjective, as in 'the entire length of shoal coast is good for tidepooling'. The etymology is not to be confused with a homograph of 'shoal,' from Old English *scolu*, meaning troop, or division of an army. The modern meaning of this term is: a large number of aquatic animals swimming together.

shor: [Russian *shor* = in Turkestan, a salt lake; < Turkish *shor* = an elongated saline depression located in Turkestan; < Urdu *shor* = to become saline, salty.] n. In Turkestan, a salt lake, a long saline valley, or an impression left in a dry lake

bed, similar to a playa. Turkestan is the present area of the Turcoman Desert.
Alt. spelling: sor

shore: [Old English *scora* = shore; < IE *sker* = to cut.] n. The narrow strip of land bordering any body of water, the upper boundary being the landward limit of wave action, whereas the seaward limit is the low-tide line. The equivalent landform on either side of a river is more often called the bank. Some combining forms are: shoreface, shoreline, shoreland, backshore, foreshore, onshore, offshore, longshore, and nearshore.

shott: [Arabic *satt* = bank, coast < *satta* = to exceed, deviate; < Semitic *stt* = tear apart, split.] n. A shallow brackish lake or marsh in Northern Africa, specifically southern Tunisia and northern Algeria. Shott are usually dry in the summer and covered with saline deposits. The term is now more generally used for large playa lakes, often tens of kilometers in diameter.
Cited usage: Some parts of the Sahara are below the level of the sea, and here are formed what are called chotts or sebkhas, open depressions without any outlets, inundated by torrents from the southern slopes of the Atlas in winter and covered with a saline efflorescence in summer [1891, Baron Lyon Playfair in Report to the British Assoc. for the Advancement of Sci. 1890 876].
Alt. spelling: shatt, chott
Synonyms: playa, cavir
See also: sabkah, playa

shuga: [Kwakiutl *shuga* = a form of sea ice; < Chinook *shuga* = sugar.] n. An accumulation of small, spongy, white lumps of sea ice, usually less than 10cm in the longest dimension. Shuga forms from grease ice (coagulated frazil crystals), frazil ice, or surface slush (water-saturated snow). If wind action is strong, the shuga or frazil ice may be thoroughly distributed through the thermally mixed upper layer of seawater, before the next stage of ice formation occurs. In 1982, the National Climatic Data Center created a scale (from 0 - 9) classifying sea ice by age. Shuga is the youngest sea ice at level 0. A folk etymology states that the term *shuga* was influenced by its granular resemblance to sugar.
Cited usage: Grease ice is called shuga when ice accumulations reach several centimeters in size [1999, Parkinson et al., Variability of the Arctic Sea Ice Cover 1978-1996, J. Geophys. Res., 104, 20, 837-20, 856].
Synonyms: frazil ice, grease ice, slush
See also: frazil

sial: [A portmanteau term from *si*(lica) + *al*(uminum).] n. A term, used loosely, for the continental crust of the Earth. The term alludes to the fact that the continental crust is rich in silica (SiO_2) and alumina (Al_2O_3). Sial is used in contrast to sima, the equivalent term for the oceanic crust, which has higher amounts of magnesia (MgO).
Synonyms: continental crust
See also: sima, sifema

siderite: [Latin *sideritis* < Greek *sideros* = iron.] n. A gray to brown mineral, iron carbonate, $FeCO_3$. Siderite occurs either massive or as rhombohedral crystals, which effervesce slightly in warm hydrochloric acid. It occurs in hydrothermal veins, and in sedimentary deposits, where it often forms concretions. Siderite also occurs as a gangue mineral in hydrothermal veins together with other metallic ore. It is also used as a synonym for iron-nickel meteorites (although common in museum collections, iron meteorites form only a few per cent of observed meteorite falls). Other uses of the term siderite, now obsolete, include: a name for lodestone in the 16th century; a name for a phosphate of iron used by Bergman in 1790 when he thought he'd discovered a new mineral; and as a blue variety of quartz named by Moll in 1797. In mineralology, the term siderate was applied to iron carbonate in 1844 by Haidinger.

Cited usage: The great division of meteorites into iron masses or siderites, mixed masses or siderolites, and aërolites or stony meteorites, seems to be a sufficiently logical division [1875, Nature XII. 521/1]. Siderite, or spathic ore, so called from its sparry or glassy crystals, is a combination of iron with carbonic acid (Fe OCO$_2$) [1894, Harper's Mag. Jan. 410].
Synonyms: chalybite, spathose iron

siderophile: [Greek *sidero* < *sideros* = iron + *philos* = loving.] n. An element that dissolves easily in molten iron. Also, an element concentrated in the metallic rather than the silicate or sulfide phases of meteorites. Siderophile elements are preferentially concentrated in the Earth's core according to V. M. Goldschmidt's threefold system for dividing the elements in the solid Earth. Goldschmidt published his tripartite division of Earth's elements, along with his explanation of the fundamental laws of geochemistry, in a series of monographs titled '*Geochemische Verteilungsgesetze der Elemente*' (*Geochemical Laws of Distribution of the Elements*).
See also: chalcophile, lithophile

siderose: [Latin *sideritis* < Greek *sideros* = iron.] adj. n. An obsolete term referring to a substance containing or resembling iron. Term was originally proposed to replace ferruginous for forms of iron other than the oxides, for example: in iron carbonates, such as siderite and ankerite. Now sometimes incorrectly used as a synonym for siderite.
Cited usage: If the mineral is siderite you should use sideritic not siderose. The former refers to the mineral siderite while the latter refers to it having an iron content and is a synonym of ferruginous [2003, James A. Dockal, Sedimentary Petrology Laboratory Manual, Univ. of North Carolina].
See also: siderite

sierra: [Spanish *sierra* = saw; < Latin *serra* = saw.] n. A rugged range of mountains having an irregular or jagged profile, rising in peaks which suggest the teeth of a saw. Sierra is often used in names for mountain ranges, such as the Sierra Nevada of California, and the Sierra Madre Occidental and the Sierra Madre Oriental of Mexico.
Alt. spelling: serra

sifema: [Acronym from English *si*(ilica) + *fe*(rric) + *ma*(gnesium).] n. An obsolete term for the ultrabasic subcrustal layer underlying the sima. Term coined by Van Bemmelen in 1949 to describe the lower, theoretical layer of the sima, referred to by some authors as the ultrasima.
See also: sial, sima

sigmoidal: [Greek *sigma* = 18th letter of the Greek alphabet; < Hebrew *samekh* < Semitic *smk* = to support.] adj. Having the shape of a crescent or the letter S. In structural geology, used in reference to recumbent folds. In geomorphology, used to describe steep-sided, S-shaped sand dunes formed by opposing winds of nearly equal intensity.

sike: [Old English *sic* = small stream.] n. Name used in the north of England for a runnel, rill, streamlet, or trickle of water, often dry in the summer and running through flat marshy ground.
Alt. spelling: syke
Synonyms: sitch (British)

sikussak: [Inuit *sikussak* = thick old sea ice.] n. Very thick and old arctic sea ice. Sikussak forms in sheltered areas such as fjords, blocking the entrance for a time before eventually fragmenting during warm periods. Because it is formed partly from snow, sikussak resembles glacier ice.
Cited usage: During cold climatic events, multi-year shorefast sea ice (sikussak) formed in these fjords and trapped the icebergs [2000, J.A. Dowdeswell et al., Sedimentology, v. 47, p. 557-576].
See also: nunatak

silcrete: [Portmanteau word from English *sil*(ica) < Latin *silex* = flint + English (con)*crete* < Latin *concrescere* = united, or connected by growth.] n. A firmly indurated mixture of detrital clasts of any size that are bound together by siliceous cement to form concretionary rock, such as silica-cemented breccia, conglomerate, or sandstone. Where the cementing agents are iron or calcium compounds, analogous terms can be used, such as ferricrete and calcrete.
See also: calcrete, silica

silex: [Latin *silex* = flint, hard stone.] n. A term used in France for flint (cryptocrystalline quartz). The term is also used in France as a synonym for chert. As the related term silexite, Pliny the Elder used the term silex in reference to various forms of quartz.
Cited usage: Silex is the name of a genus of semi-pellucid stones [1753, Chambers' Cyclopaedia Supplement].

silica: [Latin *silex* = flint, silica.] abbr. The naturally occurring dioxide of silicon, SiO_2. Silica occurs in five crystalline polymorphs: quartz, crystobalite, tridymite, coesite, and stishovite. As a cryptocrystalline substance, silica forms chalcedony. As the gem opal, silica is in the amorphous hydrated form, $SiO_2 \cdot nH_2O$.

silicate: [Latin *silicis* < *silex* = flint, hard stone.] n. As used in J. D. Dana's original classification system, the class of mineral containing silica (SiO_2). The silicates make up about 95% of the Earth's crust and upper mantle, and comprise 75% of all known minerals. The different arrangements of silicon-oxygen tetrahedra have been used to define subclasses of the silicates, which are: nesosilicates, sorosilicates, cyclosilicates, inosilicates, phyllosilicates, and tectosilicates.

silicon: [Latin *silex* = flint, silica.] n. An opaque, silver-white, semi-metallic element (symbol Si). After oxygen, it is the most abundant element in the Earth's crust, and most commonly occurs in the silicate class of minerals. Industrially, silicon is used to make semiconductors and silicones, whereas the silicates are used in refractories to make brick, pottery, and ceramics. Term coined by Scottish chemist Thomas Thomson in 1817 in place of Sir Humphry Davy's silicum, and as an analogous term to boron and carbon, because of the chemical resemblance of these three elements.
Cited usage: The base of silica has been usually considered as a metal, and called silicium. But as it bears a close resemblance to boron and carbon, it is better to class it along with these bodies, and to give it the name of silicon [1817, T. Thomson, Syst. Chem. 5th ed. I. 252].

sill: [Old English *syll* = threshold.] n. A concordant, tabular, igneous intrusion emplaced along bedding planes in thicknesses up to 300 meters. Most sills are hypabyssal, medium-grained, mafic intrusions, commonly composed of diabase and dolerite of Pennsylvanian age. The Great Whin Sill in the Pennine Mountains is the largest exposed hypabyssal intrusion in Britain.
Cited usage: Today the edge of this sill (the Palisades Sill) is exposed along the Hudson River, where the hard igneous rock resists erosion and forms the Palisades Cliffs [2004, J. Bret Bennington, Physical Geology: Intrusive and Plutonic Igneous Rocks, Hofstra University, N.Y.].

sillar: [Peruvian Spanish *sillar* = altered ignimbrite.] n. A local Peruvian name for ignimbrites that have been pervasively altered by an aqueous vapor. Also, an ignimbrite or volcanic tuff that has not become indurated by welding.
Cited usage: There has been a tendency to call all such deposits 'welded tuffs.' For those in which induration is primarily the result of recrystallization, and for those in which the fragments have little cohesion, another term is desirable. The local term 'sillar' commonly used in the Arequipa region [in Peru], has been

applied in the present paper [1948, C.N. Fenner, 1948, Bull. Geol. Soc. Amer. LIX. 883].

sillimanite: [Eponymous term after American mineralogist Benjamin Silliman (1779-1864).] n. A white, gray, or brown orthorhombic mineral, aluminum silicate, Al_2SiO_5. It is trimorphous with andalusite and kyanite and forms at high temperatures and pressures in regionally metamorphosed terrains. Sillimanite occurs as long, slender, prismatic crystals, or as fibrous and felted aggregate masses, particularly in metamorphosed aluminous rocks. The name was coined in 1824 by G. T. Bowen, in honor of Benjamin Silliman, the first professor of mineralogy at Yale University.
See also: kyanite, andalusite

silt: [Middle English *cylte* = silt; < Swedish *sila* = to strain, filter; < Anglo-Saxon *sihan* = to filter.] n. A sedimentary material consisting of very fine particles (0.002 - 0.05mm) intermediate in size between sand and clay. Silt is commonly characterized by a high content of clay minerals, and tends to be deposited in estuaries, lagoons, and other sheltered, low-energy marine and nonmarine environments.
See also: sand, clay, gravel

Silurian: [Eponymous term after Latin *Silures*, a name given by the Romans to a group of people once occupying southeastern Wales c. 50 CE.] adj. Referring to the third period of the Paleozoic Era, from 443.7-416Ma. The name was given to the system of Paleozoic rocks lying immediately below the Devonian or Old Red Sandstone in Britain. The Latin name for the Silurian people is possibly based on one of two deities: Solis, the Roman sun god, or Sulis Minerva, goddess of healing who presided over the hot spring beside the river Avon near the ancient city of Bath. The Romans created Sulis Minerva by combining Sul, the Celtic goddess of arcane prophecy who guarded the entrance to the underworld and Minerva, the Roman goddess of arts and science. The territory of the Silures included such places in Wales as: the Brecon Beacons National Park, and the South Welsh Valleys. The Silures were difficult for the Romans to conquer and control. They were leaders in the opposition to the Roman westward advance. Finally overcome, their bitter resistance may explain why they were not granted self-governing *civitas* status until the early 2nd century. Term coined by R. I. Murchison.

silver: [Old English *seolflor* = silver; < Akkadian *sarpu* = refined silver < *sarapu* = to smelt, refine.] n. A soft, lustrous, white to gray element (symbol Ag). It occurs in the native state as the cubic mineral silver. It is characterized by its lustrous white color, high malleability, ductility, and superior electrical conductivity. Silver also occurs combined with other elements, in such minerals as: acanthite, Ag_2S; and pyrargyrite, Ag_3SbS_3. Various forms of the metal, pure or mixed with other substances, are distinguished by descriptive defining terms, such as: hornsilver, redsilver, rubysilver, capillary silver, or shell silver. There are a variety of metallic substances resembling silver, such as: cat silver, German silver, and mock silver. The chemical symbol Ag comes from Latin *argentum*, meaning silver, metal, or white money (from IE *arg*, meaning white, or to shine). First usage in English c. 825.

sima: [Acronym English *si(lica)* + *ma(gnesium)*.] n. The mafic-rich basaltic to peridotite-type rocks that make up the oceanic crust as well as the underlayment of the continental crust. The sima is denser, more malleable, and more easily magnetized than the sial.
Cited usage: The earth's crust consists of two quite different materials: sima, a silicon-magnesium rock, usually basalt, (characteristic of oceanic crust) and sial, a silicon-aluminum rock, (typically granitic and characteristic of continental crust). Wegener thought that the sialic continental plates

sail across the simatic ocean crust like icebergs in the ocean [2003, Earth in Upheaval Website].
See also: sial, sifema

sinistral: [Middle English *sinistre* = unfavorable; < Latin *sinister* = on the left, unlucky.] adj. In general, relating to the left-hand side. In paleontology and biology, pertaining to the direction of coiling in gastropods as viewed from the apex (apical view). In structural geology, pertaining to a strike-slip fault in which the motion of the block on the opposing side of the fault from an observer is towards the left (left-lateral). One assumption for the term assuming a pejorative connotation relates to the simple fact that left-handed (right-brained) people are a minority group. Left-handedness is also associated with being clumsy due to the Latin *sinistritis*, meaning awkwardness.
Cited usage: Basalts occurring around the Tanlu fault zone, a sinistral, translithospheric fault [Alfred T. Anderson, Jr., Journal of Geology, Vol. 107, No.3].
See also: dextral

sinter: [Old English *sinder* = scoria, slag of metal; < German *sinter* = dross, metal slag.] n. In general, to form a coherent mass by heating without melting. In geochemistry, a hard, incrusting chemical sediment or crust, as of porous silica, deposited by precipitation of minerals usually facilitated by heat, as in the siliceous sinter around geysers and hot springs. The English term sinter has been equated with ashes, but it has no etymological connection with the French *cendre* or Latin *cinerem*, meaning ashes. The notion that sinter is connected to ashes has given rise to its alternate spelling of cinder.
Cited usage: The hot springs precipitate vast quantities of siliceous sinter [1830, Lyell, Principles of Geology I. 213]. Iron sinter is a yellowish or brownish hydrous arsenate of the peroxyd of iron [1857, James D. Dana, Manual of Mineralogy p. 249].
Alt. spelling: cinder
Synonyms: geyserite

sirocco: [Italian *scirocco* = hot wind; < Arabic *sharaqa* = the sun rise < *sharq* = east; < Semitic *srq* = to become red (as the rising sun).] n. A regional term for a type of warm, southeasterly wind generated in the Sahara desert of North Africa. It moves ahead of an eastward-travelling depression bringing hot, dry, dusty conditions to portions of North Africa and the Eastern Mediterranean Levant. It then picks up moisture as it crosses the Mediterranean bringing hot and humid southeasterly winds to southern Italy, Sicily, and the Mediterranean islands. The Italian spelling *scirocco* is now rarely used in English.
Cited usage: A khamseen was blowing; this wind, which is an exaggerated scirocco, brings clouds of hot sand from the desert [1860, Mrs. Harvey, Cruise Claymore vii. 134].
Alt. spelling: scirocco (Italian)
Synonyms: khamseen
See also: khamseen, calina

Skargaard: [Norwegian *skargaard* = archipelago.] n. Norwegian term for group of islands in the sea.

skarn: [Swedish *skarn* = dung, filth.] n. Originally, a Swedish mining term applied to the silicate gangue of certain Archean iron-ore and sulfide mineral deposits, especially where these occur in limestone or dolomite. Now, more generally, any lime-bearing siliceous contact metamorphic rock composed of calcium, magnesium, and iron silicates (with or without iron, copper, and manganese sulfides and oxides), which has been derived from limestone or dolomite by the metasomatic introduction of large amounts of silicon, aluminium, iron, and magnesium from a nearby igneous intrusion, usually a granite. The term skarn was originally coined in Sweden for deposits of iron found in rocks near an igneous pluton. Today, it more broadly refers to any metamorphic zone that develops around igneous intrusive contact areas. Tactite skarn, and later simply tactite, was the term then applied to the more specifically iron-rich variety. Until recently skarn deposits have been considered not worth mining. In addition to their being a contact metamorphic layer on the surface of a plutonic outcrop, this idea may have led Swedish miners to make scatological comparisons.

Cited usage: I have of late years adopted a very convenient Scandinavian word, skarn, for the zone of altered rock, usually calcareous, that contains such minerals as epidote, schorl, idocrase, at times augite, hornblende, mica, etc., and I speak of these altered rocks as epidote skarn, etc [1901, H. Louis, Trans. Inst. Mining & Metallurgy X. 49].

Synonyms: tactite, pyrometasomatic rocks

skauk: [Kwakiutl *skauk* = a form of deity.] n. Term coined by Taylor in 1951 to describe an extensive network of crevasses in a glacier.

skerry: [Scottish *skerry* = diminutive form of sker; < Swedish and Old Norse *skar* = an isolated rock in the sea.] n. A rugged insulated sea rock or stretch of rocks, covered by the sea at high water or in stormy weather. Seal skerry (a mainly Scottish word meaning 'small rocky island' or 'reef') was named after the seals that use it as a haul-out point

Synonyms: skar

skew: [Middle English *skewen* = to escape, run sideways; < Old North French *eskiuer* = awry, obliquely, askew.] n. An oblique course or direction. In statistics, the amount of asymmetry relative to the average shown in a frequency distribution of a minimum number of data points. A frequency distribution has a positive skew if the mean is larger than the median.

slag: [German *schlacke* = fragments of metal struck off by hammering < *slahan* = to strike, slay.] n. In volcanology, a clinkerlike lump of lava. In ore deposits, a by-product of metal refining, mostly from the reduction of iron in a blast furnace. It

forms from silica-based impurities that react with calcium oxide produced by thermal decomposition of limestone added to the furnace. This high-temperature reaction produces the slag calcium silicate ($CaO + SiO_2 = CaSiO_3$). The molten slag is less dense than molten iron, so it floats on the surface from where it can be skimmed off. The volcanic term almost certainly derived from the mining term because the *Oxford English Dictionary* reports the earliest geologic use late in the 18th century, whereas European miners were using the word slag in the late 16th century.

Cited usage: At the first melting of the after Gripple … was mad therof 288lbs. of lead besids the slaggs and stones (sic) [1552, P. H. Hore, Wexford II. 236].

slake: [Middle English *slaken* = to abate; < Old English *slaec* = sluggish; < Old Dutch *slaken* = to make slack, relax, diminish.] v. To combine chemically with water molecules in a true chemical combination, used especially in reference to the slaking of calcium oxide to calcium hydroxide. The etymology is reflected in the chemical meaning because the reactivity of anhydrous lime (CaO) diminishes with increased exposure to moisture, thereby slaking it to $Ca(OH)_2$.

slate: [Middle English *sclate* = slate; < French *esclat* = splinter < *esclal* = small piece of wood used as a tile < *esciater* = to break into pieces.] n. A fissile, fine-grained, low-grade metamorphic rock formed mainly by the metamorphosis of argilla-ceous mudstone or shale, in which platy minerals (phyllosilicates) enter parallel alignment causing the rock to cleave or split into thin plates or slabs of considerable durability and tensile strength. The etymology may also be influenced by the German *schieissen*, meaning to split. Slaty cleav-age is a superinduced structure, meaning that splits generally do not occur along bedding planes, but along planes of cleavage, which intersect the bedding at some often high angle.

Cited usage: Many other rocks are improperly called slate, if they are thin bedded and can be used for roofing…. One of the best known of these is the "Stonesfield slate", which is a Jurassic limestone occurring near Oxford, England and famous for its fossils [Encyclopedia Brittanica, 1911].

slickenside: [English *slicken* = dialectical alteration of slick; < Middle English *slike* = silken, sleek; < Old English *slician* = freshly smoothed; < IE *lei* = slimy + English *side*.] n. A lineated and polished, usually striated rock surface caused by frictional movement of one rock mass sliding over another on a fault plane or bedding surface. The plane may itself be mineralized by a coating, usually quartz or calcite, which has striations showing the direction of movement. In mineralogy, a specular variety of galena found in Derbyshire. In pedology, crack surfaces created by expansion and contraction of soils that have a high percentage of swelling clays.

slieve: [Gaelic *sliabh* = mountain.] n. Term used in Ireland for a mountain. Slieve Gullion and Slieve Donard are mountain ranges in Northern Ireland that reach elevations over 500m and 800m, respectively.

slikke: [Dutch *slikke* = mudflat.] n. A tidal flat or mudflat, especially one rich in decaying organic material.

slime: [Old English *slim* < Swedish *slem* = expectoration, mucus, phlegm; < Latin *limus* < IE *slei* = slime, slimy.] n. Soft, glutinous mud or alluvial ooze. In biology, a mucous substance secreted by various organisms, such as catfishes, slugs, and many seaweeds. IE derivatives include: slip, slick, lime, loam, liniment, and schlep.

slough: [Old English *slogh* = a hollow place; < German *schlucken* = to swallow; < Celtic *sloc* = pool, ditch.] n. A sluggish to stagnant marshy area located in a swale or slow-draining lowland.

In North America, a popular term along the Mississippi River for a side channel that accommodates floodwater. In mining, rock fragments that fall into a borehole and must be washed out. *Cited usage*: Opposite the gravel island is a slough aligned with a grassed depression. Both features undoubtedly carry water during flood flow [1970, G. H. Dury, Rivers & River Terraces, Leopold & Wolman, vii. 199].
Alt. spelling: slew, slue, sleugh

sluice: [Latin *excludere* = to shut out.] n. A constructed dam or embankment for impounding the water of a river, canal, or other small waterway. A sluice is provided with an adjustable gate for controlling the volume of water passing through. *Cited usage*: All eight sluice-gates of Pak Mool Dam were opened yesterday after the Electricity Generating Authority of Thailand (Egat) agreed to compensate protestors with 5 million bhat [2001, The Nation Magazine, June 3].

sluit: [Dutch *sloot* = ditch; < *spruit* = shoot, spring < *spruiten* = to spring up.] n. A term used in Southern Africa for a usually dry gully produced by erosion in a natural fissure enhanced by heavy rain.

slump: [German *schlumpen* = to come about, happen by accident.] n. A landslide in which an exposed portion the landmass slides a short distance with some degree of cohesion and usually slight backward rotation owing to the concavity of the surface of separation from the parent mass.

slurry: [Middle English *slori* < *sloor* = mud.] n. A relatively low-viscosity fluid composed of water and solid fragments. Slurries are often made by people, and are created for ease of transport, such as a coal slurry for transport through a pipeline. Demand for slurry pumps in the African mining industry is generated by the production of gold, base metals, diamonds, coal, and less valuable ores as well.

smectite: [Latin *smecticus* < Greek *smektikos* = Fuller's Earth (montmorillonite clay), < *smectein* = to wipe, cleanse + *-ites* < *-lite* = pertaining to rocks < *lithos* = rock, stone.] n. A common use meaning for clay that includes the dioctahedral (montmorillonite) and trioctahedral (saponite) types. In 1932, it was proposed in the publication *American Mineralogist* that the terms smectite and montmorillonite were synonymous to the extent that smectite should be dropped from the lexicon of earth science.

smog: [Portmanteau word from English *sm(oke)* + *(f)og*.] n. Fog that has become mixed and polluted with smoke. Specifically, a form of air pollution produced by the photochemical reaction of sunlight with hydrocarbons and nitrogen oxides that have been released into the atmosphere, especially by automotive emissions. The word smog is first recorded in 1905 in a newspaper report of a meeting of the Public Health Congress. During that meeting, Dr. H. A. des Vœux gave a paper entitled "Fog and Smoke," in which, in the words of the *Daily Graphic* of July 26, "he said it required no science to see that there was something produced in great cities which was not found in the country, and that was smoky fog, or what was known as 'smog.'" The next day the *Globe* remarked that "Dr. des Vœux did a public service in coining a new word for the London fog."

smut: [German *smutz* = fat, grease < *schmutz* = dirt, filth.] n. In coal-mining parlance, smut is bad, soft, and earthy coal.
Cited usage: The uppermost seam of coal is commonly soft and dusty, it is vulgarly called smut [1799, Geological Essays 292].

sodium: [English *soda* = sodium hydroxide (NaOH); < Latin *sodanum* = headache remedy + -

ium = suffix denoting a metal; symbol *Na* < Latin *nitron* < Greek *nitros* < Hebrew *neter* = soda + Latin *-ium* = suffix for a metal.] n. A soft, light, malleable, silver-white metallic element of the alkali series (symbol Na). Like most of the alkali metals, sodium is common in nature, but only in the form of bound compounds. First isolated by Sir Humphry Davy in 1807 by electrolysis of caustic soda (NaOH), hence the name sodium (*soda* is a back-formation from Neo-Latin *sodanum*, meaning headache remedy). The chemical symbol *Na* comes from Berzelius's proposal to call the element *natrium*, a shortened form of *natronium*.
Cited usage: I discovered sodium a few days after I discovered potassium in the year 1807 [1812, Sir Humphry Davy, Chemistry and Philosophy 331].
See also: natron

sofar: [Acronym from English *so(und)* + *far* = fixing and ranging.] adj. The Sofar Channel is a deepwater zone in the oceanic water column at a depth of about 1500 m where the velocity of sound is at a minimum, but due to minimal loss of intensity may be transmitted over long distances. Sound passing laterally through the Sofar Channel is refracted in a sinusoidal manner with little loss of energy, causing sound energy to be trapped in a zone of well-defined depth. The now famous Sofar Channel is named after a wartime system of monitoring sound pulses in water created by explosives detonated for the purpose of rescuing aviators downed at sea. It is a memorable term because sound waves travel so far within the channel.
Cited usage: The SOFAR channel may be used for the transmission of sound over long distances, exceeding 28,000 km, and can be used to track free-drifting, subsurface, neutrally buoyant floats [Oxford Dictionary of Science].

soil: [Old English *soyl* = piece of ground, place; < Latin *solium* = seat; < IE *sed* = to sit.] n. The earth or ground; the face or surface of the earth. It is counterintuitive that the English term soil is derived from the Latin *solium*, meaning seat rather than *solum*, meaning soil, but there is evidence for cross influence by both Latin words. IE derivatives include: subsidence, sessile, and sediment.

solano: [Spanish *solano* < *sol* = sun.] n. A southeasterly wind that brings hot and humid weather to the southern coast of Spain via the Straits of Gibralter. It is generated by high pressure over western Europe during periods of low pressure southwest of Gibralter in the Atlantic east to Morocco. This phenomenon occurs more frequently during hurricane season (July-October).
Cited usage: The Solanus or Easterne winde is commonly hote and troublesome in Spaine [1604, E. G[rimstone] D'Acosta's Hist. Indies iii. ii. 120].
Synonyms: levanter

solfatara: [Toponymous term after Italian *Solfatera*, a dormant volcano west of Naples near Pozzuolli, Italy.] n. A type of fumarole or vent associated with volcanoes that give off primarily steam and hydrogen sulfide gas. Solfatara, the most active volcanic area of Campi Flegrei, is one of the most famous tourist sites of Italy.
Alt. spelling: solfaterra
See also: fumarole

solifluction: [Latin *solum* = soil, ground, earth + *fluctin* = a flowing < *fluere* = to flow.] n. The gradual movement of waterlogged soil or other surface material down a slope, especially where the subsoil is frozen and acts as a barrier to the percolation of surface water.
Synonyms: congelifluction

solstice: [Latin *solstitium* = point at which the sun seems to stand still < *sol* = sun; < IE *sawel* = sun + Latin *stitium* = a stoppage < *sistere* = to come to a stop; < IE *sta* = to stand.] n. One of two times each year that the sun reaches its apparently furthest distance from the Earth's equator, and

appears to halt before starting to come closer again. This is due to the 23.5° angle that Earth's axis of rotation is tilted relative to the plane of its orbit around the sun. A graphic sinusoidal plot of the sun's position relative to the equator shows that the peaks mark the summer solstice, the valleys mark the winter solstice, and the midpoints crossing the x-axis are the equinoxes. IE derivatives of *sta* include: staurolite, stamen, static, stem, and prostitute.

solum: [German *solumhorizont* = solum horizon; < Latin *solum* = base, ground.] n. A set of soil horizons that are related through the same cycle of pedogenic processes. The solum of a contemporary soil presently at the surface consists of all horizons now forming; they are the A, E, and B horizons, the transitional horizons, and some O horizons. Introduced by Benjamin Frosterus in 1922 at the international "Conference Extraordinaire Agropedologiqu".
Cited usage: Solum and soils are not synonymous. Some soils include layers that are not affected by soil formation. These layers are not part of the solum [1993, Soil Survey Manual, U.S.D.A. Handbook No. 18, U.S. Government Printing Office, October].
Alt. spelling: sola (pl)

solvent: [Latin *solvere* = to loosen; < IE *leu* = to loosen or divide.] n. A substance, usually a liquid, that can dissolve another substance, or in which another substance will dissolve. Water is a particularly strong and versatile solvent. IE derivatives include loess and tachylyte.
See also: loess, tachylyte

somma: [Toponymous term named after Italian *Somma-Vesuvius*, volcanic mountain complex in Italy..] n. A somma volcano is an ancient caldera having steep inner walls, being partially filled by a new central cone. Examples of other modern somma volcanoes are the Kolokol Group in the

Kurile Islands and Hakone in Japan. Mt. Vesuvius is a typical example of a volcano withn a volcano made by an outer broken cone, Mt. Somma.

sorosilicate: [Greek *soros* = coffin, heap + Latin *silex* or *silic* = flint, hard stone.] n. The structural group of silicates characterized by isolated pairs of binary SiO_4 silica tetrahedra that share an oxygen atom at a common apex.
Cited usage: In the sorosilicates (two isolated tetrahedra which share one oxygen atom) calcium is present mainly in the epidote group [1972, R.W. Fairbridge, Encycl. Geochemistry & Environmental Sci. 101/1].
See also: cyclosilicate, inosilicate, nesosilicate, phyllosilicate, tectosilicate, silicate

sound: [Old English *sund* = water, sea, swimming; < Old Norse *sund* = strait, ferry, swimbladder.] n. An inlet of the sea, often a long and wide body of water connecting larger bodies of water. A sound is generally larger than a strait or a channel. In biology, the air bladder of a fish.

spa: [Toponymous term after Belgian *Spa*, town in Belgium famous for therapeutic hot springs.] n. A mineral springs that acts as a watering place, often attributed with life-affirming health and medicinal qualities.

spall: [German (possibly) *spellen* = to split.] n. A relatively thin flat stone chip that flakes off a rock surface, usually in reference to exfoliation fragments from the surface of an exposed igneous intrusion. In archaeology, a thin elongate chip of stone flaked from a stone tool. Also, used as the verb 'to spall,' meaning to chip or to fragment. Exfoliation causes lenticular sheets of various sizes to spall, or peel off a mountain.

spar: [Old English *spaerstan* = gypsum; < German *spar* = crystalline stone.] n. A general term for any nonmetallic crystalline mineral that is translucent or transparent, light-colored, and lustrous, usually of easy cleavage. The term spar is synonymous with the obsolete term spath, but spath does persist in combined forms, such as: feldspathoid, spathic, spathose, and spathization. There are many varieties of spar that are distinguished by giving them a unique modifier, such as: bitter-, Bolognian-, calcareous-, Derbyshire-, Icelandic-, pearl-, and tabular-spar.
Synonyms: spath (obs.)
See also: spathic, feldspar

spathic: [Old English *spaerstan* = gypsum; < German *spar* or *spat* = crystalline stone.] adj. An obsolete term for a mineral that resembles or has the properties of spar, especially with regard to cleavage.
Synonyms: spathose
Similar terms: feldspathic
See also: spar, feldspar

species: [Latin *specie* = a seeing or appearance, < the verb *specere* = to look < IE *spek* = to observe.] n. A grouping of organisms whose offspring can also procreate. By this definition donkeys and horses are distinct species that can mate with each other and produce mules, but the mule is not a species because it cannot procreate. A species is now defined as the most discrete grouping in the Linnaean classification system. IE derivatives are: specimen, aspect, and special.
Cited usage: In his publications, Linnaeus provided a concise, usable survey of all the world's plants and animals as then known, about 7,700 species of plants and 4,400 species of animals [The Linnean Society of London, Burlington House, Piccadilly, London].

specimen: [Latin *specimina* = things examined < *specere* = to look at.] n. An individual item or organism that is representative of a larger group, such as a genus, family, class, or phylum. Also, a

sample from an individual taken for examination.
Synonyms: example
See also: species

Spectre of the Brocken: [French *spectre* = an apparition, phantom, or ghost, especially one of a terrifying nature or aspect; < Latin *spectrum* < *specere* = to look, see + German *Brocken* = toponymous term after the highest peak (1,143m) in the Harz Mountains of Central Germany.] n. An observer's shadow cast upon a cloud; this phenomenon usually occurs when the observer is on a mountain top or ridge and a low sun casts the shadow onto a fog or cloud in the valley below. Although the shadow is essentially the same size as the person, the observer sometimes gains the impression that it is gigantic. This is likely the result of a comparison between the nearby shadow and distant objects glimpsed through the cloud. It was named after early observations made by climbers on the Brocken, a peak in the Harz Mountains of Germany.
See also: Brocken

speleology: [French *speleologie* < Latin *speleum* = cave; < Greek *spelaion* = cave + Greek *logos* = word, speech, discourse, reason < *legein* = to speak, to say; IE *leg* = to collect.] n. The exploration and study of caves, whether scientific, casual, or for sport. Spelunking is a derived term coined to describe the act of exploring in caves.
Synonyms: caving, potholing

speleothem: [Greek *stalaktos* = dripping < *stalassein* = to drip + *thema* = that which is laid down, deposit.] n. Any structure which is formed in a cave by the deposition of minerals from water, (e.g., a stalactite or stalagmite). Any of the secondary mineral cave deposits formed by chemical deposition, most commonly of calcite. Term coined by G. W. Moore in 1952.
Cited usage: In an effort to relieve the ambiguities of 'formation', the term speleothem is

proposed. It is suggested that the word be used as a general term for secondary mineral deposits formed from water in caves, such as stalactites, helictites, rimstone terraces [1952, G. W. Moore, National Speleological Society News June 2/1].

spessartine: [Toponymous term after German *Spessart*, a hilly area in northwestern Bavaria, West Germany.] n. A red to brownish manganese aluminum mineral of the garnet group, $Mn_3Al_2(SiO_4)_3$. It occurs in granite and pegmatites, often as as small dodecahedral porphyroblasts. This nesosilicate of the isometric crystalline system is relatively rare due to replacement of calcium in grossular by manganese. Term coined by Francois Sulpice Beudant (1787-1850) in his 1832 *Traité Elémentaire de Minéralogie*. Not to be confused with the rock spessartite.
See also: ugrandite, pyralspite, pyrope, uvarovite, almandine, grossularite, garnet

spessartite: [Toponymous term after German *Spessart*, a hilly area in northwestern Bavaria, West Germany.] n. A lamprophyre rock with mafic phenocrysts of hornblende and pyroxene in a sodic plagioclase matrix. The term enters English from the French *spessartine*, meaning the mineral spessartine. Term coined by Rosenbusch in 1896 in the same hills where F. S. Beudant discovered spessartine 60 years prior.
See also: spessartine, grossular, garnet

sphalerite: [Greek *sphaleros* = slippery, deceitful < *sphallein* = to trip.] n. A zinc-iron sulfide mineral, (Zn,Fe)S. Sphalerite is in the isometric crystal system, is the primary ore of zinc, and occurs in yellow-brown or brownish-black crystals or cleavage masses. Term coined by E. F. Glocker in 1847 because it is easily mistaken for galena (PbS). This mineral was named blende by G. Agricola in 1546 and is from the German *blenden*, meaning to blind or deceive, because the mineral resembles galena in appearance but has no lead. Because of this miners often rejected it as worthless, and gave it such names as: mock ore, false lead, and blackjack.

sphene: [French *sphene* < Greek *sphen* = wedge.] n. A titanium accessory mineral ($CaTiSiO_5$) found in granitic and calcium-rich metamorphic rocks, sometimes used as a gemstone. Term coined by René-Just Haüy in 1801 for the wedge shape of its crystals. The name has been replaced by titanite.
Cited usage: A greyish, compact and impure variety of sphene, known as leucoxene frequently occurs in basic igneous rocks as an alteration product of ilmenite and rutile [Encycl. Brittanica, 1911].
Alt. spelling: sphen
Synonyms: titanite

spherite: [German *spharit* < *spharisch* = spherical; < Greek *phaira* = ball, sphere.] n. In petrology, a sedimentary rock that emulates a conglomerate or breccia because it is composed of gravel-sized particles, but it is not a clastic rock, rather a constructional aggregate. In mineralogy, a semiprecious, hydroxy aluminum phosphate mineral occurring in globular concretions; used as a substitute for turquoise. The term can also be used to describe a single spherical grain in a sedimentary rock, such as an oolith or spherulite. The petrologic meaning of the term was introduced in 1911 by Grabau (1870-1946).
Cited usage: It appears that sphaerite is wholly identical with variscite [1950, Amer. Mineralogist, XXXV, p. 1059].
Alt. spelling: spheryte, sphaerite
Synonyms: variscite
See also: rudite, clastic

spherulite: [Latin *sphaerula* = diminutive of *sphaera* = sphere; < Greek *phaira* = ball, sphere.] n. A usually acicular, radial, and coarsely crystal-line aggregate, often formed in situ by devitrification of volcanic glass. Spherulites vary in size from microscopic to macroscopic. Also, spherulites are defined as small spherical concretions less than 5mm in diameter, and having a relatively dense calcareous shell around a sparry calcite core. This type of spherulite most commonly originates by recrystallization or by biologic processes. First usage in English c. 1820.
Alt. spelling: spherolite, sphaerolite

spicule: [Latin *spiculum* = sharp point, sting, dart < *spica* = ear of corn.] n. A small silicate or carbonate needlelike structure acting as support for the soft tissue of an invertebrate, such as those found in sponges, radiolarians, chitons. An etymologically related term is Spica, the brightest star in the constellation Virgo. First usage in English c. 1780.

spinel: [Italian *spinella* = little thorn; < Latin *spina* = thorn.] n. A variably-colored, isometric mineral, magnesium aluminum silicate, $MgAl_2O_4$. Spinel is found in pegmatites and in rocks formed by contact metamorphism of dolomite. Some spinel crystals are of gem quality and resemble the ruby. Spinel is the great imposter of gemstone history: many famous rubies in crown jewels around the world are actually spinel. The most famous is the Black Prince's Ruby, a magnificent 170-carat red spinel that currently adorns the Imperial State Crown in the British Crown Jewels. The Timur Ruby is actually a 352-carat red spinel owned by the Royal Family of Britain; it has the names of some of the Moghul emperors engraved on its face. Spinel is so named because of its sharply pointed, spine-shaped, octahedral crystals of the isometric system. First usage in English c. 1525.
Alt. spelling: spynel, spinell, espinal, and spinelle

spit: [Old English *spitu* = pointed stick; < French *epois* = points of a deer's horn.] n. A small point or

tongue of land connected to the mainland and running into the sea. Also, a long narrow shoal extending from the shore into the sea, usually composed of sand and gravel deposited and maintained by longshore drift. First usage in English before 1000.
Cited usage: The Tomales Point Trail, which winds its way for close to five miles along a spit of land between Tomales Bay and the Pacific Ocean, is the best place for spotting Tule elk [2002, Janice Cooke Newman, Hiking the Tomales Point Trail].

spod-: [Greek *spodos* = ashes, dross, dust.] Combining prefix for things decomposed, leached, or highly oxidized, such as: spodosol and podzol soils, spodiosite, and spode.

spodumene: [Greek *spodousthai* = to be burned to ashes < *spodos* = wood ashes.] n. A white to green to pink monoclinic mineral, lithium aluminum silicate, $(LiAlSi_2O_6)$. Spodumene is an ore of litium, and occurs in granitic pegmatites, often as crystals up to 10 meters long. Term coined by B. J. d'Andrada in 1800 because spodumene becomes ash gray when exposed to air. Varieties of spodumene are based on color; kunzite is a pink to lilac variety, and hiddenite is a green variety. Both varieties have eponymous etymologies, the first after George Kunz and the second after A. E. Hidden.
Cited usage: Spodumene was so called because it assumes a form like ashes before the blowpipe [1837, James D. Dana, A System of Mineralogy p. 305].

sporbo: [Acronym from English *s*(mooth) *po*(lished) *r*(ound) *b*(lack) *o*(bject).] n. A synonym for oolite. Term coined in the San Juaquin Valley, California by geologists in the petroleum industry. Sporbo is sometimes blue or brown instead of black, but the acronym remains intact.

Cited usage: Some pellets are found in California oil shale of the Miocene age. The common name for these pellets is Sporbo [2001, James W. Farr, The Fecal Pellets of Polychaete Worms].
Similar terms: oolite
See also: oolite

spore: [Greek *spora* = seed; < IE *sper* = to strew.] n. A minute reproductive body characteristic of spore-bearing flowerless plants. IE derivatives include diaspora and sperm. First usage in English c.1875.

spruit: [Afrikaans *spruit* = intermittent stream; < Dutch *spruten* = to sprout.] n. An intermittent stream, usually dry except during heavy rains or flash floods.
Synonyms: arroyo

squall: [An echoic term from English *squeal* < Scandinavian *sqvala* = various screaming sounds.] n. A strong and violent sudden burst of wind, or a short-lived local thunderstorm often accompanied by rain. First usage in English c. 1695.

stadia: [Middle English *stadia* < Greek *spadion* = racetrack < *span* = to pull (possibly) altered by *stadios* = firm.] n. In surveying, a measuring instrument made of a rod placed at one end of the distance to be measured and a pair of horizontal lines, usually made of parallel-placed hairs, on the diaphragm of a telescope placed at the other end. An ancient Greek and Roman linear measure, roughly equal to one-eighth of a Roman mile, (185m or 600ft). In the English Bible, the stadia is called a furlong, the unit distance in modern horse racing.
Cited usage: The idea to measure the distances by a scale and the micrometer of a telescope was proposed by an Italian engineer about 45 years ago, and the name of Stadia (scale) was given by him to that kind of measure [1865, Mayer, Franklin Inst. Jrnl.].

stagnation: [Latin *stagnare* = to stagnate, be overflowed < *stagnum* = pool.] n. A nearly motionless condition of water, unstirred by current or waves, usually for a period long enough for it to become foul or stale from standing. In glacial geology, the state of a glacier that has ceased to flow. Etymologically related terms include the verb 'stanch,' meaning to make stagnant, and the noun 'stank,' meaning water retained by an embankment. First usage in English c. 1665.

stalactite: [Greek *stalaktos* = dripping < *stalassein* = to drip + *-ites* < *-lite* = pertaining to rocks < *lithos* = rock, stone.] n. An elongate, conical or cylindrical, tubular mineral formation created by repetitive dripping of mineral-rich water from overlying semi-soluble rock into a hollow earthen cavity below. These projections resemble icicles and are usually composed of calcite or aragonite, but may form from other minerals as well. Stalactites are speleothems that grow downward from above,while stalagmites grow up from the bottom, fed by still mineral-rich water dripping from the stalactites directly overhead. Olaus Wormius coined both stalagmite and stalactite c.1675.
See also: speleothem, stalagmite

stalagmite: [Greek *stalagma* = a dropping.] n. A tubular incrustation or deposit, like a fatter form of an inverted stalactite, which develops on the base of a cavern or hollow. It is formed by the same dripping solution that deposits stalactites above. Olaus Wormius coined both stalagmite and stalactite c. 1675.
See also: stalactite

stamukha: [Russian *stamukha* = grounded hummock ice.] n. A form of ice that was grounded on tidelands and repeatedly covered by tidewater or storm surge before floating free. Stamukhas mark the seaward boundary of 'fast ice' along the broad ice shelves of the Arctic Ocean. Although used frequently by sea ice specialists, the term is not defined in the World Meteorological Organization Sea-Ice Nomenclature.
Cited usage: Ice that is grounded on tidelands and repeatedly covered by tidewater before floating free can be far more massive and dangerous to shipping than ambient level ice and offshore ridges. These ice blocks, or "stamukhas", carry heavy sediment loads and therefore are more massive than sea ice ridges of the same volume [2000, Orson Smith, Formation and Decay of Samukhas, Cook Inlet Alaska, University of Alaska, Anchorage].

staurolite: [Greek *stauros* = cross + *lithos* = rock, stone.] n. A brown to black monoclinic mineral, hydroxy iron magnesium, aluminum silicate, $(FeMg)_2Al_9Si_4O_{23}(OH)$. It is found frequently in cruciform twins and is sometimes used as a gem. Term coined in 1792 by Delamétherie.
Cited usage: Staurolite occurs chiefly in the crystalline schists [1888, Rutley Rock-forming Min. 163].

steno-: [Greek *stenos* = narrow.] Combining prefix meaning narrow, or having a narrow range. Occurs in terms, such as: stenotopic, stenohaline, stenothermal, and stenographic.

stentorg: [Swedish *sten* = stone + *torg* = marketplace.] n. A well-defined felsenmeer, or block field, found on the crest and flanks of an esker, often highlighted by paleoshoreline stripes.
See also: felsenmeer, esker

steppe: [Russian *step* = steppe.] n. A vast semiarid grass-covered plain, such as those found in southeastern Europe, Siberia, Mongolia, Patagonia, and central North America. First usage in English c. 1670.

sternum: [Greek *sternon* = breast, breastbone.] n. A long flat bone in most vertebrates that is situated along the ventral midline of the thorax, and articulates with the ribs. First usage in English c. 1665.

stibnite: [Middle English *stibium* = antimony; < Latin *stibium* = antimony; < Greek *stibi* = a grey cosmetic used to darken the eyelids.] n. A lead-gray, orthorhombic mineral, antimony sulfide, Sb_2S_3. Stibnite has perfect cleavage and is the most common ore of antimony. Term coined in 1832 by François Sulpice Beudant (1787-1850). The mineral was mentioned by the Greek writer Dioscorides and the Roman encyclopedist Pliny the Elder. In addition to stimmi and stibi, they also called it platyophthalmon.
Synonyms: antimonite

stoichiometry: [Greek *stoikheion* = element + *metron* = to measure; < IE *me* measure.] n. Calculation of the quantities of reactants and products in a chemical reaction; the quantitative relationship between the substances in a reaction or compound based on molecular weight. Term coined by J. B. Richter in 1792 in his *Anfangsgründe der Stöchiometrie* (*Initial Explanations of Stoichiometry*), to quantify the relative amounts in which acids and bases neutralize each other.
Cited usage: Stoichiometry, the weight relations in chemical reactions, must be kept in mind in the study of rates of reaction [1975, R. F. Brown, Organic Chem. viii. 192].

stone: [Old English *stan* = stone; < IE *stai* = stone.] n. A synonym for rock, especially when used in the combined form, e.g., sandstone, siltstone, dolostone, or millstone. Also, often used as the diminutive of rock when used as a stand-alone term. The word stone was adopted early into the Teutonic languages, which evolved into the Russian *stena*, meaning wall. IE derivatives include tungsten and the surname suffix -*stein*, as in Goldstein or Bernstein. First usage in English before 900.

stonewort: [Greek *stia* = pebble + Old English *wyrt* = root, plant.] n. Any of various submerged aquatic algae of the phylum Chlorophyta that are frequently encrusted with calcium carbonate deposits and have nodes with whorled, filamentlike branches. Possibly named by Vaillant c. 1720 after a now uncertain Roman plant species of the family of aquatic carbonate-rich plants Characeae. First usage in English c. 1580.
Synonyms: Chara, brittlewort

stope: [Old English *stoepe* = step; < German *stope* = step.] n. In mining, a form of excavation in the form of steps made by cutting into steeply inclined or vertical veins. In petrology, used as a verb for the process in which a magma makes its way through country rock by quarrying. First usage in English c. 1745.
Cited usage: If the specific gravity is lower than the corresponding solid rock, a magma could stope its way into rocks of similar chemical composition [1962, W. T. Huang, Petrology iv. 104].

strandline: [Old Norse *strond* = border, edge, coast.] n. The land bordering a sea, lake, or river. In a more restricted sense, that part of a shore that lies between high and low tide levels. First usage in English c. 1900.

strath: [Scottish < Gaelic *strath* = valley.] n. In geomorphology, a broad, undissected remnant of a flat valley floor that undergoes post-uplift erosional dissection. In marine geology, a long, broad, steep depression on the continental shelf, usually formed by a glacier. First usage in English c. 1535.
See also: glen

strato-: [Latin *stratum* = something spread or laid down < *sternere* = to extend, stretch, spread out; < IE *ster* = to spread.] Combining form meaning laid down flat. In meteorology, used to form names for mixed types of cloud structure, such as stratocumulus and cirrostratus. In geology, used in such terms as: stratigraphy, stratovolcano, and stratum.

stratosphere: [Latin *stratus* = a spreading out < *sternere* = to stretch; < IE *ster* = to spread + Latin *sphaera* = sphere.] n. The region of the atmosphere above the troposphere and below the mesosphere where clouds rarely form, extending from about 11 kilometers (7 miles) to 50 kilometers (31 miles) above the Earth's surface and in which temperature increases gradually to about 0°C (32° Fahrenheit). IE derivatives include: stromatolite, structure, sternum, strew, and perestroika. First usage in English c. 1905.

stratum: [Latin *stratum* = something spread or laid down < *sternere* = to extend, stretch, spread out; < IE *ster* = to spread.] n. A tabular or sheetlike body or layer of sedimentary rock, visually distinct from adjacent strata. In stratigraphy, it has been defined as a unit that may be composed of a number of beds, or as a layered part of a bed greater than 1cm thick. First usage in English c. 1595.
Alt. spelling: strata (pl.)

stratus: [Latin *stratum* = something spread or laid down < *sternere* = to extend, stretch, spread out; < IE *ster* = to spread.] n. Along with cirrus and cumulus, stratus is one of the three basic cloud forms. Usually existing at low elevations, they appear as broad sheets of nearly uniform thickness that develop upward from flat bases. Thick and gray in color, they rarely form at elevations higher than 1,500m and, though rarely producing heavy precipitation, will often touch ground to form fog. Most of the names given to clouds (cirrus, cumulus, stratus, nimbus, and their combinations) were coined in 1803 by the English meteorologist Luke Howard. He immediately gained international fame, becoming a favorite among writers and painters. Goethe, Constable, and Coleridge revered him, and in so doing they helped legitimize the young science of meteorology.
Cited usage: This application of the Latin word stratus is a little forced. But the substantive stratum, did not agree in its termination with the other two [cirrus, cumulus], and is besides already used in a different sense even on this subject, e.g. a stratum of clouds; yet it was desirable to keep the derivation from the verb sterno, as its significations agree so well with the circumstances of this cloud [1803, Luke Howard, On the Modification of Clouds].
Alt. spelling: strati (pl.)

strontium: [Toponymous term after *Strontian*, a village in Scotland.] n. A soft, silver-white metallic chemical element of the alkaline-earth group (symbol Sr). It is chemically similar to calcium and occurs, combined with other elements, in various minerals, such as strontianite and celestite. Strontium is easily oxidized, igniting spontaneously in air when finely divided. Strontian, Scotland, is the type locality where strontium was first discovered. First usage in English c. 1805.

suevite: [Toponymous term after German *Suevit* or *Schwabia* < Latin *Subia* or *Suabia* = an area of Germany.] n. A type of welded breccia found associated with impact craters, resembling a tuff breccia or pumiceous tuff but of nonvolcanic origin. They show signs of shock and aerodynamic shaping, evidence of impact metamorphism. Originally a rock from the Ries impact crater near Nördlingen in West Germany that fit the criteria, now used generally for this type of impactite. The type locality Swabia was a former German duchy occupying a region now covered by the state of Baden-Württemberg and part of Bavaria.
Cited usage: The so-called 'suevite' rocks of the Ries are almost identical to some of the surface samples from the fragmented lunar 'regolith' [1970, New Scientist, 23 July 174].
See also: fladen

sulfur: [Middle English *sulphur* = brimstone; < Latin *sulpur* = sulfur.] n. A pale yellow, nonmetallic, naturally occurring element (symbol S). Sulfur usually occurs as yellow crystals, but is also found in several other allotropic forms, one of which is red. Sulfur occurs as the native element in yellow crystals at hot springs and fumaroles, and combined with other elements in minerals such as: galena, gypsum, epsomite, and barite.It is chemically similar to oxygen, but is less reactive and more acidic. Sulfur is used in numerous industrial applications, such as: black gunpowder, rubber vulcanization, insecticides, pharmaceuticals, and the preparation of sulfuric acid. It is recovered commercially from salt domes, such as those along the Gulf of Mexico. Jupiter's moon Io owes its colors to various forms of sulfur. The spelling 'sulfur' predominates in U.S. technical usage, whereas both 'sulfur' and 'sulphur' are common in general usage. British usage tends to favor the use of 'sulphur' for all applications. The International Union of Pure and Applied Chemistry (IUPAC) has apparently decided that it has jurisdiction over both British and American written English as well as over chemical nomenclature, because people in Britain are now expected to use the 'f' word. First usage in English c.1325.
Synonyms: sulphur

supergene: [Greek *super* = over + *genesis* = origin, creation < *genesthai* = to be born, come into being; < IE *gen* = to give birth, beget.] n. In ore deposits, referring to mineral deposits formed near the Earth's surface at the water table, usually by descending meteoric solutions. First usage in English c. 1945.

swale: [Old Norse *svalr* = cool.] n. A low, generally level tract of land, usually moist or marshy, and formed in undulating ground moraine due to uneven glacial deposition. Also, a long, narrow, usually shallow trough between ridges on a beach, running parallel to the coastline. First usage in English c. 1425.
Synonyms: runnel
See also: runnel

syn-: [Latin (from Greek) *syn* = similar, together, alike; < IE *ster* = to spread.] Combining prefix meaning of similar form, structure, appearance, or the like; used in such terms as: syncline, synform, synonym, and syntactic.

synantexis: [Greek *syn* = together, alike + *tekein* = to melt.] n. An obsolete term for the formation of a primary mineral by the reaction of two other minerals. Not to be confused with syntexis, which has to do with formation of magma.
See also: syntexis, anatexis

syncline: [Latin *syn* < Greek *syn* = similar, together, alike + *clinare* = to incline.] n. A concave upward fold in which the stratigraphically younger layers of rocks are located in the core, having a strike direction in line with the fold axis, and a dip generally trending away from the axial plane. In repeated formation, synclines alternate with anticlines in a sinusoidal pattern. First usage in English c. 1870.
Similar terms: synform

synclinorium: [Latin *syn* < Greek *syn* = similar, together, alike + *clinare* = to incline + *oros* = mountain.] n. A mountain range exhibiting alternating folds of synclines and anticlines. First usage in English c. 1875.
Cited usage: The mountain range, begun in a geosynclinal, and ending in a catastrophe of displacement and upturning, is appropriately named a synclinorium [1880, James D. Dana, Manual of Geology 3rd ed. 821].
See also: anticline

syngenesis: [Greek *syn* = similar, together, alike + *genesis* = origin, creation < *genesthai* = to be

born, come into being; < IE *gen* = to give birth, beget.] n. In sedimentology, the formation or stage of accumulation of unconsolidated sediments in situ, including changes to particles in suspension within the depositional basin. In biology, the formation of the germ in sexual reproduction by fusion of the male and female elements, so that the substance of the embryo is derived from both parents. Term coined by Alexander E. Fersman in 1922. Although the parameters of this term are still in dispute in some parts of the world, in the U.S. it is equivalent to early diagenesis.
Similar terms: epigenesis

synoptic: [Greek *synopsis* = general view < *syn* = similar, together, alike + *opsis* = view; < IE *ok* = to see.] adj. In meteorology, pertaining to simultaneously existing weather conditions over a large area at the same point in time, which, when analyzed, help predict and give a description of the weather. Also said of a weather map showing these conditions. IE derivatives include: phlogopite, triceratops, autopsy, dropsy, walleye, daisy, and pinochle. First usage in English c.1760.

syntaxis: [Greek *syn* = similar, together, alike + *tassein* = to arrange < *tak*.] n. An arrangement of fold axes or trust faults in mountain ranges showing convergence towards a common point, associated with a sharp bend in the orogenic belt. It is a megascopic structure in which a bundle of folds or faults is pinched in the middle to create the image of a bow-tie-shaped map pattern. This term is a translation by Sollas of the German term *schaarung*, coined by Eduard Suess in 1901.
Cited usage: In the direction of the syntaxis, i.e. towards the west, their strike bends back in an arc [1909, W. J. Sollas, translation of E. Suess's Face of Earth IV. 289].
See also: virgation

syntexis: [Greek *syntekein* = to fuse together < *syn* = together + *texis* = to melt.] n. In petrology, any kind of reaction between a rising body of magma and the surrounding country rock. Also, the formation of magma from the melting of two or more rock types. Or, alteration of an existing magma by the melting or assimilation of country rock. Term coined by F. Loewinson-Lessing in 1911.
Cited usage: When the re-melted portion of the crust is composed of different rocks, eruptive, or sedimentary, the process is rather a 'syntexis', as I have called it, an assimilation which is followed by liquation and differentiation [1911, F. Loewinson-Lessing, Geol. Mag. VIII. 295].
See also: anatexis, palingenesis, synantexis

synthetic fault: [Greek *syn* < *sun* = together, similarly, with, united + *tithenai* = to place, put; < IE *dhe* = to set, put.] n. Faults that have the same dip direction as the rocks displaced by the fault, and that rotate fault-bounded blocks so that the net slip on each fault is less than it would be without rotation. Often said of minor faults that are subsidiary to and formed in the same stress regime as a major fault, and are oriented at low angles to the major fault. Term coined c.1925 by H. Cloos. IE derivatives include thesis and facies.
Synonyms: homothetic fault
See also: homothetic

system: [Latin *systema* < Greek *synistanai* = to combine < *syn* = together, with, combined, united + *histanai* = set up, establish.] n. A major lithostratigraphic division, composed of a number of rock series and corresponding to a geologic period in time; hence, all the rocks deposited during a specific geologic period.
Cited usage: I venture to suggest the term 'Silurian system' should be adopted as expressive of the deposits which lie between the old red sandstone and the slaty rocks of Wales [R. I. Murchison, philosophical Magazine, VII. 48].

tachy-: [Greek *takhys* = swift.] Combining prefix meaning rapid, accelerated, or condensed. For example, tachylite is rapidly cooled volcanic glass, while tachygenesis is condensed development of primitive phylogenetic stages in the early development of an organism.

tachylyte: [German *tachylyt* < Greek *takhys* = swift + *lutos* = soluble; < *luein* = to loosen; < IE *leu* = to loosen, divide.] n. A black to brown or green volcanic glass formed by rapid cooling of molten basalt. It is commonly found on the chilled margins of lava flows, dikes, and sills. An abundance of crystallites, of varying composition in the glass, contribute to the color of tachylyte. Named in reference to its easy fusibility. First usage in English c. 1865. IE derivatives include: loess, analysis, dissolve, and soluble.
Alt. spelling: tachylite
See also: loess, solvent, pseudotachylyte

taconite: [Toponymous term after the Taconic Mountains in New York State.] n. A term used initially in the Lake Superior region for generally low-grade, hard, siliceous banded or disseminated iron formation. Taconite contains about 30% iron, and consists of ferruginous chert (often jaspery), interbedded with layers of iron minerals, such as magnetite, hematite, siderite, or other hydrous iron silicates. The term is now used more generally for low-grade,

siliceous iron formation, which can be concentrated by fine grinding and magnetic separation. *Cited usage*: In the Mesabi district the local name 'taconite' is applied to the ferruginous chert [1905, Economic Geology I. 48].

tactite: [Unknown origin.] n. A coarse-grained, iron-rich rock formed by contact metamorphism and metasomatism of carbonate rocks. Tactite is generally considered a type of iron-rich skarn, and characteristically contains iron-bearing garnet (andradite) and pyroxine (hedenbergite), epidote, and wollastonite. Not to be confused with tektite.
Synonyms: tactite skarn
Similar terms: skarn
See also: skarn

tafoni: [Italian tafoni = plural form of tafone = window, perforation.] n. A honeycomb texture in rocky outcrops that often hollows out to small, shallow caves. The morphology of these cavernous rock domes is characterized by a helmet-shaped outer roof and an arched-shaped cavern, often with a partially overhanging visor. Tafoni occur in many rock types, but are most common in calcite-cemented sandstones. Heavy winter rains seep into the sandstone along cracks and planes of soft rock. The rain water contains carbon dioxide from the air which dissolves the calcium carbonate that holds the sandstone grains together. The dry season allows the rock to dry out, and the calcium carbonate is then drawn to the surface by the capillary action of water. As the water evaporates, the calcium carbonate is left within a few feet of the surface to form a hard shield that resists erosion. The interior of the rock, however, is left without much cement. It therefore easily crumbles away, and is removed by water, wind, and animal activity once the crust is broken.. The honeycomb effect is enhanced when the calcite cement is of variable extent and strength within the rock. Tafoni has also been observed in other types of sedimentary rock, as well as in some

igneous rocks, usually granite. These forms have been exposed to solutions that are rich in silica and iron, or in halite. In granitic tafoni, salt or silica/iron-rich solutions appear to differentially penetrate the crystal matrix, resulting in the characteristic honeycomb texture. Tafoni formation is most common in regions having a moderately dry climate, along with a prolonged dry season, but they also occur in coastal settings in a variety of climates. Tafoni were first described around 1900 on the Mediterranean islands of Corsica and Sardinia, the most pronounced being the now famous "Fungo" rock. The ancient Romans turned this tafone into a resting point for horses on the road to Olbia. Some other locations where tafoni are found include: Manciano Mine in the Grosseto Province of Tuscany, Italy; Ayers Rock in Uluru National Park in Central Australia; Murphy's Haystacks near Streaky Bay, and Granite Island in Southern Australia; Ucontitchie Hill near Mt. Wudinna on the Eyre Peninsula, Australia; the El Chorro region of Andalucia, Spain; Castle Rock State Park, El Corte de Madera, and Long Ridge Open Space Preserve, all in the Santa Cruz Mountains of California. Tafoni can also be found at Salt Point and Bean Hollow, along the Northern California Coast.
Cited usage: A phased model of tafone evolution is proposed whereby the features pass through four phases of development: initiation, enlargement, amalgamation and degradation; much of the evidence suggests that the tafoni are actively developing under current environmental conditions [1997, A. Mellor, J. Short, and S. J. Kirkby, Tafoni in the El Chorro area, Andalucia, southern Spain, 22, 817-833].
Alt. spelling: tafone (sing.)

taiga: [Russian *taiga* < Altai *tayga* = forest-covered mountain; < Athabaskan *taiga* = land of little sticks.] n. The swampy coniferous forest area of Siberia. Also, the zone of temperate coniferous forest stretching across Europe and North America. First usage in English c. 1885.

talc: [Latin *talcum* = shiny mineral; < Arabic *talq* = talc, mica, selenite.] n. An extremely soft, light-green to gray monoclinic mineral, hydroxy magnesium silicate, $Mg_3Si_4O_{10}(OH)_2$. Talc is most commonly formed by hydration and metamorphism of magnesium silicates, such as olivine and enstatite in mafic and ultramafic rocks. It is also formed by contact metasomatism of dolomite. It usually occurs in foliated, fibrous, or granular masses, and has a greasy feel. In addition to talc itself, Arabic and medieval writers applied the term to various soft and shiny minerals, such as mica and selenite. First usage in English c. 1600.

talus: [French *talu* = slope; < Latin *talus* = ankle.] n. A sloping accumulation of angular rock fragments at the base of a cliff or steep slope, formed chiefly by rock falls and slides. First usage in English c. 1700.
Cited usage: It is only at a few points that the grassy covering of the sloping talus marks a temporary relaxation of the erosive action of the sea [1830, Lyell, Principles Geology I. ii. xx. 266].
Alt. spelling: taluses (pl.)

tangent: [Latin *tangere* = to touch.] n. A line, curve, or surface that meets and touches a curved line or surface at a single point. Although tangent surfaces touch, they do not intersect or cross over each other. In trigonometry, a tangent or tangent function is the ratio of the length of the side opposite an angle to the length adjacent to that angle in a right triangle. It is referred to as a tangent function, because, when graphed, the function theoretically touches the asymptote at a single point at infinity. First usage in English c. 1585.

tantalum: [Eponymous term after the mythic Greek character *King Tantalos*, who was condemned to Hades.] n. A gray, hard, and rare metallic element (symbol Ta). It occurs in the minerals columbite and tantalite. Anders Gustaf Ekeberg discovered tantalum in 1802, and named it

after the mythic king who was forced to stand in a pool of water that receded when he reached to drink from it. The name alludes to the high resistance of tantalum to acid, even when immersed in it.
Cited usage: This new recruit among the metals I call Tantalum, partly following the custom which favours names from Mythology, partly in allusion to its incapacity, when immersed in acid, to absorb any and be saturated [1802, Ekeberg, Kongl. Vetenskaps Acad. Handl. XXIII. 80].

taphrogeny: [Greek *taphros* = pit, trench + *genesis* = origin, creation < *genesthai* = to be born, come into being; < IE *gen* = to give birth, beget.] n. The process of formation of rift zones, characterized by high-angle normal faults and subsidence, as the result of tensional forces in the crust. Taphrogeny is often considered the first stage of continental rifting and plate separation. Term coined in 1922 by German geologist Erich Krenkel, in reference to the East African Rift Zone.
Cited usage: Taphrogenesis (taphrogeny) was coined for the East African rifts and is largely descriptive. In general, however, it implies tensional forces as opposed to horizontal compression for orogeny and differential vertical movements for epeirogeny [1963, E. S. Hills, Elem. Structural Geol. xi. 315].

tarn: [Icelandic *tjarn* = a small land-locked mountain lake having no significant tributaries.] n. In glacial geology, a lake occupying an ice-gouged, horseshoe-shaped basin or cirque formed at the source of an alpine glacier. Also, a term used in Michigan for any lake or pool that occurs in a tract of swamp, bog, and muskeg. The term has been expanded in general use to mean any small lake. First usage in English c.1325.
Cited usage: The largest river in the world takes its most remote origin among the Andean Highlands, in a little inky tarn [1880, Haughton, Physical Geog. v. 235].
Synonyms: cirque-lake

tautonym: [Greek *tauto* = the same + *-onym* < *onoma* = name.] n. In biology and paleontology, taxonomic nomenclature in which the genus and species both have the same name, the genus name being derived from that of the species (e.g., *Gorilla gorilla*, meaning western mountain gorilla, and *Vulpes vulpes*, meaning red fox). This construction is common in zoology, but is now forbidden in botany under the International Code of Botanical Nomenclature. For example, under the new code the plant name Armoracia armoracia (common horseradish) had to be changed to *Armoracia rusticana*. First usage in English c. 1895.

taxonomy: [Greek *taxis* = arrangement < *tassein* = to arrange + *nomia* = related to < *nomos* = law < *nemein* = to distribute.] n. The science of classifying plants, animals, fungi, and microorganisms into categories based on shared features. As originally developed by Carolus Linnaeus (1707-1778), organisms were grouped by visible physical characteristics, but, recently, gene sequencing techniques have allowed greater refinements of taxonomic classification. In 1813, Swiss botanist Augustin Pyrame de Candolle coined the term taxonomy in the scheme of plant classification that he published in *Theorie elementaire de la botanique* (*Elementary Theories of Botany*).

tectofacies: [Greek *tekton* = builder; < IE *teks-* = to weave + Latin *facia* < *facies* = form, figure, appearance < (perhaps) *facere* = to make, or < *fa-* = to appear, shine.] n. A group of strata having different tectonic aspects when compared to laterally equivalent stratigraphic units. Term coined by Laurence L. Sloss in 1949. The term is of limited practical value because the nature of a tectofacies is often only understood after the area has been outlined based on other criteria. Not to be confused with *tectonic facies*.
See also: tectonic facies

tectonic facies: [Greek *tektonikos* = pertaining to building < *tekton* = builder; < IE *teks-* = to weave + Latin *facia* < *facies* = form, figure, appearance < (perhaps) *facere* = to make, or < *fa-* = to appear, shine.] n. A group of rocks having characteristics resulting mainly from tectonic activity, such as mylonites and fault breccias. Term coined by B. Sander in 1912. Not to be confused with tectofacies.
See also: tectofacies

tectonics: [Greek *tektonikos* = pertaining to building; < *tekton* = builder; < IE *teks-* = to weave.] n. The branch of geology dealing with the major architectural and structural features of the Earth's lithosphere, along with the processes by which they originated and evolved. Tectonics is similar to structural geology, but deals with larger scale features. The modern global tectonic theory is plate tectonics. First usage in English c. 1630. IE derivatives include: architect, technology, and badger (the builder).

tectonite: [Greek *tektonikos* = pertaining to building < *tekton* = builder; < IE *teks-* = to weave.] n. Any rock whose structural fabric reflects its deformational history, especially one that shows continuous solid flow during formation. Term coined by B. Sander in 1930.
Cited usage: Rocks that owe their present characters to the integration of differential movements, Sander (1930) calls tectonites in distinction to non-tectonites, which are formed under conditions involving no differential movement [1933, Amer. Jrnl. Sci. CCXXV. 433].
Alt. spelling: tektonite
See also: palimpsest

tectosilicate: [Greek *tekton* = builder; < IE *teks-* = to weave + Latin *silex* or *silic* = flint, hard stone.] n. The silicate subclass in which the four oxygen atoms of each SiO_4 tetrahedron are shared with four neighboring tetrahedra in a three-dimensional framework. The ratio of silicon to oxygen is 1:2. The etymological reference to building has to do with the framework structure of the linked SiO_4 tetrahedra.
Cited usage: All the tektosilicates are colorless, white, or pale gray when free from inclusions [1959, Berry & Mason, Mineralogy xv. 471].
See also: cyclosilicate, inosilicate, nesosilicate, phyllosilicate, sorosilicate, silicate

teilzone: [German *teil* = part + Greek *zone* = girdle, celestial region < *zonunai* = to gird.] n. The time span during which a species occupied a particular region. Term coined in 1914 by J. F. Pompeckj. The synonym *teilchron* was coined by Arkell in 1933.
Cited usage: Abundance Zone (also called Peak Zone, Acme Zone): Subset of teilzone where index species reaches some higher level of abundance [1998, Donald R. Prothero, Bringing Fossils to Life: An Introduction to Paleobiology].
Alt. spelling: teil-zone
Synonyms: teilchron

tejon: [Spanish *tejon* = round gold ingot.] n. In geomorphology, a solitary, prominent, disk-shaped feature, separated by erosion from the feature to which it was originally connected. Term used primarily in the U.S. Southwest, and also in place names, such as Tejon, New Mexico, and Tejon Pass, California.

tektite: [Greek *tektos* = molten < *tekein* = to melt.] n. A black, greenish, or yellowish, pitted and rounded body of high-silica glass of nonvolcanic origin. Tektites are spheroidal, discoid or dumbbell shaped, and usually walnut sized. They range in weight from a few grams to 13kg. Most tektites are found in groups over widely separated geographic areas. Their composition is generally unrelated to the local geology. They are made of a glass called lechatlierite, having a composition more like shale than obsidian. Prior to the Moon

landings, they were considered by some to be of lunar origin, but now, tektites are generally believed to result from large hypervelocity impacts on the Earth, combined with accompanying shock-melting of primarily target sediments. Term coined in 1900 by Eduard Suess, who thought they were meteorites that had undergone melting.
Cited usage: There are eight main areas in the world generally accepted to be genuine tektite-fields [1968, R. A. Lyttleton Mysteries Solar Syst. vi. 183].

tellurium: [Latin *tellus* = Earth, land < *terra* = earth, land + *ium* = suffix for a metal.] n. A tin-white, lustrous, brittle, relatively rare, semimetallic element (symbol Te). By visual characteristics it was originally classified as a metal, but in chemical properties it is akin to sulfur, and selenium. It is isomorphous with anitmony, arsenic, and bismuth, and it occasionally occurs as native rhombohedral crystals. Tellurium was originally discovered in gold ore from Transylvania by Franz J. Muller von Reichenstein, chief inspector of mines and salt works. The ore was variously known as *Faczebajer weisses blattriges Golderz*, meaning white, leafy gold ore from Faczebaja or *antimonalischer goldkies*, meaning antimonic gold pyrite. The term tellurium was coined by M. H. Klaproth in 1798, in intentional contrast to the element uranium, from the Greek *Oranos*, meaning heaven, which Klaproth discovered in 1789.

temblor: [Spanish *temblor* = a trembling, earthquake < *temblar* = to shake; < Latin *tremulare* < *tremulus* = shaking < *tremere* = to tremble.] n. An earthquake. First usage in English c. 1895.
Cited usage: One freshet of one Ohio River a dozen years ago took more lives than all the temblores in California in a century and a half have taken [1896, Land of Sunshine, July].
Synonyms: earthquake

tenor: [Middle English *tenor* = uninterrupted course; < Latin *tenor* = course, continuity, tone; < *tenorem* = contents, holding on; < *tenere* = to hold; < IE *ten* = to stretch.] n. In ore deposits, a term designating the grade of ore, meaning its element or mineral content in percent, grams per ton, or parts per million. In music, a term for the adult male voice, intermediate between the bass and the alto, which carries the sustained melody. The etymology alludes to the Latin root *tenorem*, meaning contents. Etymologically related terms include: attenuate, tendon, and tenure. First usage in English c. 1275.

tension: [Latin *tendere* < IE *ten* = to stretch.] n. In structural geology, a state of stress that tends to pull a body apart by the action of stretching or extension, such as occurs at a divergent plate boundary. Etymologically related terms include: hypotenuse, attenuate, and tendon. First usage in English c. 1530.
Cited usage: Sigmoidal tension gashes form along zones of ductile shear. En echelon tension fractures (gashes) may indicate the direction of displacement by two features [1992, Eldridge M. Moores, Structural Geology].

tepetate: [Mexican Spanish *tepetate* < Nahautl (Aztec) *tepetatl* = stone matting.] n. A term synonymous with caliche for a calcareous crust coating solid rocks at or near the surface in regions of arid or semiarid climate. In Mexico, the term is used to describe a hardened land surface with various degrees of infertility; this hardened surface layer is formed from pyroclastic materials, usually tuff.

tephra: [Greek *tephra* = ash; < IE *dhegh* = to burn.] n. A synonym for pyroclastic material that is solid matter of any size, shape, or origin. It is explosively ejected by an erupting volcano. First usage in English c. 1940.

Cited usage: The author suggests (volcanic) ash or (better) tephra as a collective term for all clastic ejectamenta [1944, Geografiska Annaler XXVI. 210].

tepui: [Pemón (South American indigenous language) = mountain.] n. Remnants of giant sandstone plateaus that are deeply eroded along fractures caused by crustal movement. The type locality occurs in the table mountains of the Guiana Highlands, composed of Precambrian sandstone that towers above the surrounding savanna, often with sheer cliffs over 1,000m high. Some of the most outstanding tepuis are Autana, Neblina, Auyantepui, and Roraima. Due to their isolated environments resembling islands, tepuis are host to unique suites of flora and fauna. In addition, Angel Falls in Venezuela, the world's tallest waterfall, plunges off the top of Auyantepui. Sir Walter Ralegh, in 1595, was the first European to descibe tepuis.
Alt. spelling: tepuy

terra cotta: [Italian *terra* < Latin *terra* = earth; < IE *ters* = to dry + Italian *cottal* = baked, cooked; < Latin *cocta* < *coquere* = to cook.] n. A hard, low-fired, unglazed clay of a characteristically brownish-red color; used in pottery, ornaments, and building construction. Also, the brownish-red color these clay products display. First usage in English c.1720. Etymologically related terms include: terrigenous, terrain, Mediterranean, subterranean, terrestrial, and territory.

terrain: [Latin *terrenum* = of the earth < *terra* = earth; < IE *ters* = to dry.] n. In general, a particular geographic area or region including the surficial topographic features of the area. In geology, a name for a connected series, group, or system of rocks or formations; a stratigraphical subdivision. Not to be confused with terrane. Etymologically related terms include: terrigenous, terrain, Mediterranean, subterranean, terrestrial, and territory. First

usage in English c.1725. IE derivatives include: terrane, torrid, and thirst.
Cited usage: The union of several formations constitutes a geological series or a district (terrain); but the terms rocks, formations, and terrains, are used as synonymous in many works on geognosy [1823, translation of Humboldt's Geognostic essay on the trend of rocks in two hemispheres].

terrane: [Alteration of *terrain* < Latin *terrenum* = of the earth < *terra* = earth; < IE *ters* = to dry.] n. A coherent body of rock of regional extent that is fault-bounded and is considered a distinct tectonostratigraphic unit, having a different geologic history from that of contiguous terranes. A terrane is a discrete allochthonous fragment of oceanic or continental crust that was accreted to the craton at an active plate boundary. Although the term was used in the 1800s for a contiguous rock series, the modern usage was introduced in 1972 by W. Porter Irwin, in reference to an accreted fault-bounded unit. Not to be confused with terrain. IE derivatives include: terrain, terrane, torrid, and thirst.
Cited usage: Terrane...is used for any single rock or continuous series of rocks of a region, whether the formation be stratified or not [1864, James D. Dana, Manual of Geology 81]. The word terrane proposed by Prof. Gilbert to be used for a stratigraphical subdivision of any magnitude [1889, Q. Jrnl. Geol. Soc. XLV. 63].
See also: terrain

Tertiary: [Latin *tertius* = third < Greek *trias* = three < IE *trei* = three + Latin *rius* = of or pertaining to.] n. The initial period of the Cenozoic era (65.5-1.81Ma). The Tertiary period is commonly divided into the Paleogene period (65.5-23.8Ma) and the Neogene period (23.8-1.81Ma). The term Tertiary originated with the first attempts to develop a geologic time scale, in which the fundamental subdivisions were: Primary, Secondary

Tertiary, and Quaternary. Primary and Secondary were abandoned, but Tertiary and Quaternary were retained as periods of the Cenozoic Era. IE derivatives include: trio, trinity, -troika (Russian), and trikaidekaphobia (the abnormal fear of the number thirteen).
Cited usage: Arduino, in his memoirs on the mountains of Padua, Vicenza, and Verona, first recognized the distinction between primary, secondary, and tertiary rocks [1830, Lyell, Principles of Geology I. 49].
See also: Cenozoic, Paleogene, Neogene, Quaternary

Tethys: [Latin < Greek *Tethys* = Titaness, sea-goddess who was both sister and wife of Oceanus.] n. The name of a large sea that occupied the region between the proto-Eurasian and proto-African plates, in the general position of the Apline-Himalayan orogenic belt during the late Mesozoic and early Cenozoic eras. The Tethys Sea finally closed during the Eocene and Miocene epochs when Africa collided with Eurasia. The Tethys Sea was named in 1893 by Austrian geologist Eduard Suess. In paleogeographic reconstructions, three Tethys Seas have been recognized. In Permean times, the Paleo-Tethys was a large embayment on the eastern side of the supercontinent Pangea. The Mesozoic Tethys Sea expanded westward, splitting Pangea into the supercontinents of Gondwanaland (south) and Laurasia (north). Finally, the Tethys Seaway was formed during Jurassic times when a shallow arm opened between Central Asia and Southern Europe. This last manifestation of the Tethys Sea closed during formation of the Alpine-Himalayan Mountain belt. The Mediterranean Sea is a small remnant of the Tethys Sea.
Cited usage: Modern geology permits us to follow the first outlines of the history of a great ocean which once stretched across part of Eurasia. The folded and crumpled deposits of this ocean stand forth to heaven in Thibet, Himalaya, and the Alps. This ocean we designate by the name 'Tethys', after the sister and consort of Oceanus [1893, E. Suess, in Nat. Sci. II. 183].

teton: [French *teton* = breast; also, Teton (dialect of Dakotan Sioux) *thi' huwa* = dwellers on the prairie.] n. One of the five dialects of Dakotan Sioux, and the group of prairie dwelling, nomadic buffalo hunters who spoke Teton. In geography, the name of a mountain range in the North American Rockies. The Sioux called the mountains *Teewinot*, but due to their mammary-shaped appearance and the similarity in pronunciation to the Dakota Sioux, the French-Canadian trappers of the 1830's called them the Tetons.
Cited usage: This trade, as small as it may appear, has been sufficient to render the Tetones independent of the trade of the Missouri [1806, correspondence between President Thomas Jefferson and explorers Lewis and Clark].

tetragonal: [Greek *tetragonos* = quadrangle < *tetra* = four + *gonos* = angled.] adj. In crystallography, referring to one of the six crystal systems, having three perpendicular axes and fourfold rotation, the vertical axis being a different length than the two equal length horizontal axes. First usage in English c.1570.
Similar terms: quadrangular

tetrahedron: [Greek *tetraedros* = four-faced < *tetra* = four + *hedra* = face (of a geometric solid); < IE *sed* = to sit.] n. In crystallography and geometry, a four-faced polyhedron, each face being a triangle, three of which meet at a common apex; also described as a triangular pyramid. First usage in English c. 1565.

thalassic: [French *thalassique* < Greek *thalassa* = sea.] adj. A little-used term pertaining to seas and oceans. First usage in English c. 1855.

Cited usage: Thalassicus, applied by Brongniart to the strata of superior sediment, i.e. those found from the surface of the earth to the limestone exclusively: thalassic [1860, Mayne Exposition Lexicon].

thalweg: [German *thal* = valley + *weg* = way; < IE *wegh* = to go.] n. In hydrology and geomorphology, the line connecting the deepest points of a stream bed or valley, whether occupied by water or dry. First usage in English c. 1860.
Cited usage: Thalweg is a German geographical term, employed in the records of the congress of Berlin, which designates the line of lowest level formed by the two opposite slopes of a valley [1881, Harper's Mag. LXIV. 275]. The opposite of the talweg itself is a divide, i.e., the lines joining all high points in topography [1968, R. W. Fairbridge, Encycl. Geomorphology 1149/1].
Alt. spelling: talweg

theca: [Greek *theca* = case, receptacle; < IE *dhe* = to set, put.] n. In biology and paleontology, a case, covering, or sheath, such as the calcareous plates of an echinoderm, the protective cell covering of algae, an individual graptolite tube, the calcareous wall of a coral polyp, the covering of the pupa of certain insects, or the spore case of a moss. First usage in English c. 1660. IE derivatives include: faculty, thesis, synthesis, and bodega.

thermo-: [Greek *thermos* = warm, hot; < IE *gher* = to heat, to warm.] Combining prefix relating to temperature, usually a warming above the ambient temperature, for example thermocline, thermohaline, and thermophilic.

thermocline: [Greek *thermos* = warm, hot + *klinein* < IE *klei* = to lean.] n. In limnology and oceanography, a temperature gradient that is characteristic of the horizontal layer of water, where temperature decreases most rapidly with depth. It is characterized by having the greatest temperature change per unit change in depth, with a correspondingly large change in density. The thermocline lies directly below the surficial mixed layer. First usage in English c. 1895.

thixotropic: [Greek *thixis* < *thinganein* = to touch; < IE *dheigh* = to form, build + Greek *tropos* = changeable < *trepein* = to turn.] adj. The property of certain gels to become fluid when agitated and to revert back to a gel when left to stand. First usage in English c. 1925. IE derivatives include dough, and fiction.

tholeiitic: [Toponymous term after *Tholey*, a village in Saarland, Germany.] adj. Referring to a silica-oversaturated basaltic rock that is based on occurrence of quartz in the norm. Tholeites are rich in aluminum, and low in potassium and are typically formed at oceanic ridges. The term silica-oversaturated indicates the presence of quartz in the normative composition, where the norm is based on chemistry, and the mode on visual analysis. Term coined by Steininger in 1840, originally applying it to a sill of altered andesite. In its present meaning, the term was first used in 1933 by Kennedy.
Cited usage: Geochemists have known for some time that [ocean] ridge basalts tend to be tholeiitic whereas alkali basalts are more likely to be found away from ridges [1970, Nature 12 Dec. 1030/2].
Synonyms: subalkaline basalt

thorium: [Eponymous term after Norse god of thunder and war; < Old Norse *Thorr* = Thor; < IE *sten* = to thunder.] n. A soft, ductile, radioactive metallic element of the actinide series (symbol Th). Due to a protective oxide coating, thorium retains its silvery luster for months. It is believed that thorium is three times as abundant as uranium within the Earth and is therefore responsible for a large fraction of Earth's interior heat. Thorite ($ThSiO_4$), the mineral from which thorium was first

derived, was discovered by H. M. Esmark in 1828 near Brevik, Norway, but Jons Jakob Berzelius isolated the element, and named it thorium in 1829. The metal was named thorinum in England for many years. Etymologically related terms include: thunder, tornado, and dunderhead.
Cited usage: Thorium...gives no fewer than seven radio-active products, in the following order: mesothorium, radiothorium, thorium X, thorium emanation, and thorium A, B, and C [1907, Athenæum 31 Aug. 244/2].

thufa: [Icelandic *thufa* = tussock.] n. A usually grass-covered earth hummock, about the size of a soccer ball, which arises when water in the soil freezes; the resulting expanded icy mounds create a condition difficult for farming or even walking. The peaks on rims of certain volcanic mountain craters in Iceland are sometimes called thufa, e.g., Vesturthufa, Midthufa, and Nordthufa glaciers. Icelanders have a special name for navigating over thufa fields called *thufnagangar*, meaning thufa-walking. It is said that after thufa-walking a considerable distance, it is similar to first stepping off a small boat after a long cruise. Not to be confused with tufa.
Cited usage: The nature and geomophological relationships of earth hummocks (thufa) in Iceland [1992, John Gerrard, Geomorph. Suppl.].
Alt. spelling: thufur (pl.)

tide: [Middle English < Old English *tid* = division of time; < IE *da* = to divide.] n. A periodic diurnal to semidiurnal rise and fall of the world ocean surface level caused by gravitational interaction between the Earth, the Moon, and the Sun. Because of its mass, the Sun has the strongest tidal pull on the Earth. But, due to its closer proximity to the Earth, the Moon has a greater effect on the variation of the Earth's tides (spring tides vs. neap tides). The three celestial bodies cycle between a linear and orthogonal relationship, creating a fortnightly cycle (every two weeks) ranging from weak (neap) to strong (spring) tides. First usage in English before 900. IE derivatives include: geodesy, demography, and time.
See also: fortnight, neap, spring

till: [Old English *tilian* = cultivate, tend, work at.] n. In glacial geology, sediment composed of unsorted and unconsolidated heterogeneous material ranging in size from clay to boulders, deposited directly by a glacier without subsequent reworking by meltwater. The term alludes to the observation that glacial deposits are crudely similar, by their heterogenous and unconsolidated nature, to a field that has been tilled with a plow.
See also: moraine, drift, glacier

tin: [German *zinn* < Middle English *tinnen* < Cornish *sten* = tin; probably from an uncertain Indo-European source.] n. A malleable, silvery metallic element (symbol Sn). There are 26 known isotopes of tin, of which 5 are radioactive. The elemental symbol *Sn* is derived from the Latin *stannum*, meaning tin, or an alloy of silver and lead, which stems from the Cornish *sten*, meaning tin. Due to evidence that tin was imported from Cornwall into Italy after invasion of Britain by Julius Caesar, the Latin *stannum* is derived from the same pre-Germanic root via Cornish. The main ore of tin is cassiterite (SnO_2), which is named after the Greek *Kassiterides*, meaning 'tin islands'. The Greek poet Homer referred to tin as "the metal from the land of the Kassi." First usage in English before 900.

tindar: [Icelandic *tindur* = pinnacle, summit; < Norwegian *tind* = glacial horn.] n. A glacial horn that is detached from the main mountain range by lateral recession of cirques cutting though an upland spur between two glacial troughs.

titanium: [Eponymous term after *Titananem* one of the children of the Titans in Roman mythology.] n. A strong, low-density, highly corrosion-resistant, lustrous white metallic element (symbol Ti). It is not found naturally as the native element, but occurs widely in igneous rocks as the minerals rutile (TiO_2) and ilmenite ($FeTiO_3$). Because it is as strong as steel yet only 40% as dense, titanium is highly desirable in applications where weight must be minimized, such as in aviation and cycling. In addition, its corrosion resistance makes it a popular metal for use in joint replacements, implants, and body piercings. In 1791, William Gregor of Cornwall discovered the mineral ilmenite, and named it menachanite. He regarded it as the oxide of a new metallic element that he called menachin (or menacha), after the Menachan Valley near Cornwall, England. No recognition was given to this discovery however, until the famous chemist Martin H. Klaproth isolated titanium from the mineral rutile in 1795 and named the element after the Titans. In Greek and Roman mythology, the Titans, known as the elder gods, were a race of giants who personified the forces of nature. *Cited usage*: I think it best to chose such a denomination as means nothing of itself, and thus can give no rise to any erroneous ideas. In consequence of this, as I did in the case of Uranium, I shall borrow the name for this metallic substance (titanium) from mythology [1795, Klaproth].

tonalite: [Toponymous term after *Passo del Tonale*, a pass in the Italian Alps.] n. In igneous petrology, originally a particular variety of quartz-diorite containing nearly equal proportions of hornblende and biotite, with plagioclase as the main feldspar, and alkali feldspar is less than 8% of total feldspars. Now defined in IUGS classification as an igneous plutonic rock in which quartz (Q) isbetween 20-60, and P/(A+P) is greater than 90. Term coined in 1864 by G. vom Rath, after finding the rock near Tonale Pass on Monte Adamello in the Eastern Alps of Italy.

Cited usage: Lindgren has defined tonalite (or quartz diorite), as containing less than 8 per cent of alkali feldspar, granodiorite as containing 8-20 per cent of alkali feldspar [1913, Jrnl. Geol. XXI. 213].

topaz: [Middle English *topace* < Latin *topazus* < Greek *topazos* = topaz; < Sanskrit (possibly) *tapas* = fire, heat.] n. A colorless to blue, yellow, brown, or pink orthorhombic aluminum silicate, $Al_2SiO_4(F,OH)_2$. Topaz is found in association with highly siliceous igneous rocks and tin-bearing ore deposits. It is used as a gemstone and has a hardness of 8 on the Mohs scale. According to Pliny, topaz was named after the island Topazos south of Greece in the Red or Arabian Sea, but the stone obtained there was likely the mineral chrysolite or peridot, a gem variety of olivine. Due to the common golden-brown color of much gem topaz, the term may be derived from Sanskrit *tapas*, meaning heat, fire.

topo-: [Greek *topos* = place.] Combining prefix meaning place, used in such terms as: topography, topology, and toponym (a name given to a person or thing in recognition of its place of origin).

topography: [Greek *topos* = place + Latin *graphicus* < Greek *graphikos* < *graphe* = writing < *graphein* = to write; < IE *gerbh* = to scratch.] n. In general, the makeup, distribution, and arrangement of any surface, especially with regard to the relief and position of any feature on that surface. In cartography, the graphic representation on a map of the surface features of a region, both natural and of human construction. The terms relief and topography are not synonymous. First usage in English c.1425. *See also*: relief

topology: [Greek *topos* = place + *logos* = word, speech, discourse, reason < *legein* = to speak, to say; IE *leg* = to collect.] n. In general, the scientific study of a particular place, location, structure, or shape. In crystallography, the connectivity of a crystal structure in terms of its molecular bonds. Structures with the same topology can be made geometrically identical by distortions that do not break these bonds. In mathematics, the study of properties of geometric shapes that are not changed by stretching or bending. In this context, hula hoops and window frames are topologically identical. First usage in English c.1655, referring to an aspect of botany concerned with the various locations where plants are found. First defined as a branch of mathematics in 1847 by J.B. Listing (1808-1882) in his book *Vorstudien zur Topologie* (*Studies on Topology*).

tor: [Old English *torr* = tower, rock; < Gaelic *torr* = hill, mound; < Welsh *twr* = heap, pile; < Latin (possibly) *turris* = high structure.] n. An isolated crag or rocly peak of a hill or mountain, specifically one exposed by weathering and erosion of the surrounding rock. Tors are often granitic in composition, heavily jointed, and unusual in shape. The term is frequently used in place names, such as Mynydd Twrr (Holyhead Mountain) and Glastonbury Tor in Somerset (important location in King Arthur's Legends). First usage in English before 900.

tornado: [Spanish *tronada* = thunderstorm; < *tronar* < Latin *tonare* < IE *ten* = thunder + Spanish *tornar* = to turn.] n. A localized, violently destructive windstorm occurring over land and characterized by a rapidly rotating funnel-shaped column of air that can vary greatly in width and intensity, from a few yards to more than a mile in diameter. Tornadoes are usually accompanied by a funnel-shaped downward extension from a cumulonimbus cloud. Classic examples of fully developed tornadoes are found most frequently on the west coast of Africa and in the U.S. Midwest. Tornado and its variations ternado, tronada, and tournado, appear to have been coined by 16th-century

Spanish explorers. It is now thought that tornado was a poor adaptation of the Spanish *tronada*, meaning thunderstorm. There was then an attempt to modify the term by incorporating the word *tornar*, meaning to turn. It has, in effect, become a back-formed portmanteau word from *tor*(nar) and (tro)*nada*. IE derivatives relating to thunder include: Thursday, Thor, dunderhead and blunderbuss.
Synonyms: twister, waterspout (over water)

torrent: [Latin *torrens* = burning, roaring, boiling < *torrere* = to dry by heat, to burn; < IE *ters* = to dry.] n. A rapidly rushing stream, often flowing with destructive force and usually applied to the flow of water. IE derivatives include: terrain, thirst, terrigenous, torrid, and mediterranean.

tourmaline: [Sinhalese *tòramalli* = tourmaline, carnelian.] n. Tourmaline is a group name for many closely related hydroxy fluoro sodium aluminum borosilicates with the general formula, $Na(Mg,Fe,Li,Al,Mn)_3Al_6(BO_3)_3Si_6O_{18}(OH,F)_4$. The tourmaline group occurs in 3-, 6-, or 9-sided prisms, which are often vertically striated. It is commonly found as an accessory mineral in granitic pegmatites, silicic igneous rocks, and metamorphic rocks. The four most common varieties of tourmaline can be distinguished by formula, color, and transparency. Elbaite is the most common gemstone variety, appears in a wide variety of colors, and is transparent to translucent. Schorl is black, opaque, abundant, and is a common accessory mineral in igneous and metamorphic rocks. Dravite forms translucent brown crystals that can grow to a large size. Uvite is a translucent green to opaque calcium-sodium-rich variety. The name tourmaline, from the Sinhalese *tòramalli*, was originally applied to an assortment of colored stones, mainly zircons. Dravite was named after the Drave region in Carinthia. Elbaite was named after the isle of Elba in Italy. Schorl is an old German mining term for gangue minerals. Uvite is named for the Uva district in Sri Lanka. Tourmaline has many scientific and technological uses due to the fact that an electrical charge can be induced in some tourmaline crystals simply by applying pressure to the crystal in the direction of the vertical crystal axis. This effect is known as piezoelectricity, and has many uses in pressure measuring equipment and other scientific applications. Some tourmalines also show pyroelectricity, which occurs when the crystal is heated yielding a positive charge at one end of the crystal and a negative charge at the other.

trachy-: [Greek *trachys* = rough.] Combining prefix meaning rough, used in such terms as: trachyte, trachycarpous, trachytoid, and trachea.

trachyte: [Greek *trachys* = rough + *ites* < *lithos* = rock, stone.] n. A fine-grained, generally porphyritic volcanic rock, with alkali feldspar and minor mafic minerals as the chief phenocrysts. It is the volcanic equivalent of the plutonic rock syenite, and grades into latite when alkali feldspar decreases, and into rhyolite when quartz increases. Term coined by René-Just Haüy c. 1800 for volcanic rocks of this type, which commonly have a rough-feeling surface, and are plentiful in the Auvergne region of France.
Cited usage: These isles are formed of brown trachyte....full of crystals of glassy felspar [1830, Lyell, Principles of Geology I. 386].

trade wind: [Back-formation from English *to blow trade* = to blow steadily from same direction; < German *trate* = footstep, trace, track, way, passage.] n. Major tropical winds that move from subtropical highs toward equatorial low-pressure zones and are influenced in direction by the Coriolis effect. They blow from the northeast in the Northern Hemisphere and from the southeast in the Southern Hemisphere, and are strongest over the long stretches of ocean in the Pacific and

Atlantic. The term 'trade winds' has a common folk etymology. With all the merchant vessels that used the trade winds to travel westward in the ever-growing world trade market of the 17th and 18th centuries, it is natural to think that the term derived from the trading vessels using these winds to transport goods. But, the term's derivation has nothing to do with trade in the sense of commerce, though the importance of those winds to navigation led 18th-century sailors and lexicographers to understand the term in that way. The actual etymology refers to winds that "blow trade", meaning in a constant course or manner, such as a wind that blows steadily in the same direction. Apparently the word was introduced into English in the 14th century from Hanseatic German, perhaps directly from mariners' jargon for the course or track of a ship, and later applied to the constant winds of the subtropics. First usage in English c. 1630.

trans-: [Latin *trans* = across, beyond, through; < IE *ter* = to cross over.] Combining prefix meaning across, beyond, over, or through; used in such terms as: transitional, translation, translucent, transfluvial, and transmission. *Trans-* is often shortened to *tra-*, as in trajectory and tradition. The English prefix *trans-* is derived from the Old French prefix *tres-*, which survives in some English words, such as trespass, and in borrowed phrases, such as *tres gauche*. IE derivatives include: transient, transom, and avatar (< Sanskrit *tirati*, meaning he crosses over).

transgression: [Latin *trans* = across, beyond, through + *gradi* = to step, to go; < IE *gredh* = to walk, to go.] n. The covering over of land areas by the sea. Transgressions may be local or global in extent. When global, they are due to eustatic changes in sea level, with or without concomitant subsidence of land masses. In stratigraphy, evidence for marine transgressions is found in the rock record as marine sediments deposited

uncomformably over older rocks. First usage in English c. 1425. An IE derivative is the contrasting term retrograde.
Cited usage: In a section 'Upon Abrasion and Transgression', the author insists upon the paramount influence of the sea as an agent in planing down the surface of the land [1882, Geikie, Nature 13 July 242/2].
See also: retrograde

translational: [Latin *trans* = across, beyond, through + *latus* = brought, carried, borne.] adj. Said of uniform movement in one direction, without rotation. In structural geology, said of fault-block displacement in which features remain parallel because the blocks have not rotated relative to one another. First usage in English c. 1325.

translucent: [Latin *translucere* = to shine through < *trans* = across, beyond, through + *lucere* = to shine; < IE *leuk* = light, brightness.] adj. Capable of transmitting light, but causing sufficient diffusion to prevent perception of distinct images. In mineralogy, said of a mineral through which light, but not image, is transmitted. In contrast to opaque and transparent. Etymologically related terms include: luster, lunar, pellucid, and Lucifer. First usage in English c. 1595.
Synonyms: pellucid
See also: opaque, transparent

transparent: [Latin *trans* = across, beyond, through + *parere* = to show.] adj. Capable of transmitting light so that objects can be seen as if there were no intervening material. In mineralogy, said of a mineral through which both light and image are transmitted. Etymologically related terms include: appearance, apparition, and parent. First usage in English c. 1400.
See also: opaque, translucent

travertine: [Toponymous term after the ancient Roman town of Tibur (presently Tivoli) near Rome,

Italy where there are large deposits of travertine (Italian *travertino* < *tivertino* < Latin *tiburtinus* = stone of Tibur).] n. A usually dense, finely crystalline, white or light-colored massive or concretionary limestone, deposited by chemical precipitation from calcium carbonate-bearing groundwater. Travertine is commonly formed in the conduit and around the mouth of hot springs, as speleotherms in caverns, as vein fillings, and in soil crusts. Bagni di Tivoli, lying 20km east-northeast of Rome, is a famous travertine location, where deposits over 90m thick have been quarried since ancient times. Combined with pozzolana cement, the ancient Romans used blocks of *lapis Tiburtinos* ("stone of Tibur", or travertine) to construct the Colosseum, the Trevi Fountain, the aqueducts, and parts of the Christian Basilica built under Emperor Constantine in 324 CE, predating St. Peter's Basilica. First usage in English c. 1550.
See also: pozzollana, calcite, speleothem

tremolite: [Toponymous term after *Valle Tremola*, a valley in the Swiss Alps.] n. A white to dark-gray monoclinic mineral of the amphibole group, calcium magnesium silicate, $Ca_2Mg_5Si_8O_{22}(OH)_2$. It occurs in fibrous masses, thin-bladed crystals, or compact aggregates, generally in metamorphic rocks, such as dolomitic marble and talc schists. Term coined in 1796 for the discovery locality near the Tremola Valley at Campolongo in the St Gotthard region of Switzerland, site of the *Val Tremola*, one of the highest passes in the Alps. First usage in English c. 1795.
Cited usage: Not far from St. Gothard, it is found mixed with tremolite, and stratified [1799, Kirwan, Geological Essays 219].

tremor: [Middle English *tremor* = terror; < Old French < Latin *tremor* = a trembling < *tremere* = to tremble.] n. A shaking or vibrating movement. In seismology, a minor earthquake, especially a foreshock or an aftershock. A tremor is often associated with involuntary movement in animals, with symptoms of shaking, quivering, and agitation, and usually resulting from sickness, trauma, or intense emotion. First usage in English c. 1350.

trench: [French *tranche* < Old French *trenche* = an act of cutting, a gash, a cut < *trenchier* = to cut; < Latin *truncus* = trunk; < IE *ter* = to cross over, to pass through.] n. In geomorphology, any long, narrow depression, produced naturally in the Earth's surface by erosion or tectonism. An example is a straight, commonly U-shaped valley between two mountain ranges, such as the Tintina Trench in British Columbia. In marine geology, a narrow, elongate depression of the deep-sea floor, associated with a subduction zone, and oriented parallel to a volcanic arc. A trench is commonly parallel to the continental edge. Such trenches can be located at the junction between two oceanic plates or between an oceanic and continental plate. They can be several kilometers deeper than the adjacent seafloor, and many thousands of kilometers long. First usage in English c. 1375.

trend: [Middle English *trenden* < Old English *trendan* = to revolve < *trind* = ball, round lump.] n. In general English usage, the direction, course, or prevailing tendency of a thing, an action, or a concept. In paleontology, the evolution of a specific structure or morphologic characteristic within a group of organisms. In structural geology, the compass direction of a linear feature. In stratigraphy, a trend map is one that displays systematic large-scale features of a given unit, such as its compositional or structural change. In ore deposits and petroleum geology, a mineralized belt, or an oil and gas reservoir. In statistics, the direction or rate of increase or decrease in magnitude of the individual members of a time series of data. The etymological connection to a sphere lies in the fact that as you traverse the circumference, your direction is always changing at the same rate, therefore creating a trend. First usage in English before 1000.

Cited usage: In Louisiana, the 'trend' (main potential gas~producing formation) lies four miles beneath the green bayous and sugar-cane fields [1977, Time 5 Dec. 59/1].

Triassic: [Latin *trias* = three < Greek *treis* = three < IE *trei* = three.] adj. The first period of the Mesozoic era, from 251-199.6Ma. Term coined in 1834 by Frerich August von Alberti, alluding to the rock strata being divisible into three groups, originally named in German as: a) the Keuper, b) the Muschelkalk, and c) the Bunter (brown or 'bunt' colored) Sandstein. The middle unit, from German *muschel*, meaning bivalve, and *kalk*, meaning limestone, is a slight misnomer because the preponderance of marine mollusks occurring in these Triassic limestones were actually brachiopods and cephalopods, not bivalves. Etymologically related terms are: trinity, trio, and tritium. *Cited usage*: The Triassic and Permian formations show a time of 'great physical disturbance, volcanic eruptions discharging vast beds of lava and layers of volcanic ash' [1889, Science-Gossip XXV. 122/2].

triboluminescence: [Greek *tribos* = rubbing + Latin *lumen* = light + *-esere* = to assume a certain state.] n. Emission of light produced by mechanical energy, such as friction or pressure, usually involving the scratching, rubbing, or crushing of certain crystals. Examples of substances exhibiting triboluminescence include the minerals: quartz, fluorite, sphalerite, and wintergreen LifeSavers. There are two types of triboluminescence. The first type results from the storage of electrons that have been ejected by penetrating gamma radiation. A small mechanical shock is then sufficient for these electrons to overcome their energy barrier and cascade down to ground state. The second type is observed most frequently in sugars, and results from the breaking of certain bonds. This breakage creates free bonds, which immediately absorbs and ionizes nitrogen from the

atmosphere, producing a characteristic green or blue-green flash. Maple syrup sucrose produces an especially strong effect. Term coined in 1889 by E. Wiedemann.
Cited usage: According to the mode of excitation I distinguish Photo-, Electro-, Chemi-, and Tribo~luminescence [1889, E. Wiedemann, Philos. Mag. Sept. 151].

triclinic: [Greek *treis* = three + *klinos* = to incline, lean, slope.] adj. Of the six crystal systems, the one characterized by crystalline forms in which the three axes are of unequal length, and are obliquely inclined to one another. Triclinic crystals have a center of symmetry, with equivalent faces opposite one another. First usage in English c. 1850.
Cited usage: In the Triclinic System, the three axes are unequal, and all the intersections are oblique [1854, James D. Dana, A System of Mineralogy Introduction 4th ed. 29].
Similar terms: anorthic, tetartoprismatic, doubly oblique

tridymite: [Greek *tridymos* = threefold < *treis* = three + *didymos* = twin.] n. A high-temperature, hexagonal polymorph of silica, SiO_2. It occurs as small hexagonal plates or scales, and usually occurs in cavities of volcanic rocks, such as in rhyolite, trachyte, and andesite. One of the best known localities is Cerro San Cristobal in Hidalgo, Mexico, where another high-temperature polymorph of silicic cristobalite is found. Term coined in 1866 by Gerhard vom Rath, alluding to the small, thin, pseudo-hexagonal plates, which are usually twinned together in groups of three.
See also: cristobalite

trilobite: [Greek *trilobos* = three-lobed < *treis* = three + *lobos* = lobe.] n. Any extinct, marine arthropod of the class Trilobita, restricted to the Paleozoic era. They are characterized by a flattened, oval-shaped body divided into three distinct lobes (cephalon, thorax, pygidium) by two deep furrows. There are many factors that make trilobites excellent index fossils for dating rocks of the Paleozoic era, such as: a) worldwide distribution, b) more than 1,500 identified genera, c) 300 million-year duration, and d) easily preserved exoskeletal remains. In addition, according to a 1997 study, trilobites molted their lobes individually, making an enormous contribution to the fossil record.

tripoli: [Toponymous term after Tripoli, capital city of Libya in North Africa.] n. A light-colored, porous, friable, siliceous sedimentary rock, occurring in powdery or fine-earth masses, usually resulting from weathering of chalcedonic chert, siliceous limestone, or lithified diatomaceous ooze. It is widely used in industry as a polishing powder.
Synonyms: infusorial earth, rotten-stone

tritium: [Greek *tritos* = third < *treis* = three.] n. A radioactive isotope of the element hydrogen (symbol H^3). Tritium has two neutrons as well as a proton in the nucleus, and a half-life of 12.5 years. Tritium constitutes one part in 10^{18} of all naturally occurring hydrogen. It is produced for use in fusion reactors, as a constituent of hydrogen bombs, and as an isotopic label or tracer. Before it was discovered, Urey and Murphy recommended the name tritium for this isotope of hydrogen. Its eventual discovery in 1934 was the result of a group effort by Lord Rutherford, Luis Alvarez, Ernest Lawrence, Sir John Cockroft, and Williard Libby.
Cited usage: If the H^3 isotope is discovered, we would recommend to the discoverer the consideration of the name tritium for it [1933, Urey & Murphy, Jrnl. Chemistry & Physics I. 513].
See also: deuterium, hydrogen

trona: [Swedish *trona* < Arabic *tron* < *natron* = soda, hydrous sodium carbonate.] n. A white or

yellow-white, monoclinic mineral, hydrated sodium carbonate sodium bicarbonate, $Na_2CO_3 \cdot NaHCO_3 \cdot 2H_2O$. It occurs as fibrous or columnar layers in evaporite beds, particularly resulting from evaporation of playa lakes. Large deposits are found in various places in North Africa and America. The term *trona* was formed by the linguistic process of apocope, the loss of one or more sounds from either end of a word, in this case from natron. Near Death Valley, California the Kerr-McGee Company created the town of Trona in an area of ancient salt-rich playa lake beds. First usage in English c. 1795.
See also: apocope

trondhjemite: [Eponymous term after the city of *Trondhjem* (now *Trondheim*), in central Norway.] n. A white to light-colored plutonic rock, composed primarily of plagioclase (especially oligoclase) and quartz, with some biotite, but little or no alkali feldspar. Term coined in 1916 by Viktor M. Goldschmidt. The type locality is at Follstad, about 50 kilometers south of the city of Trondheim.

tropics: [Greek *tropikos* = pertaining to the 'turning' of the sun at the solstice *trope* = a turning; < IE *trep* = to turn.] n. In astronomy, the two solstitial lines of latitude on the Earth, being the parallel lines of latitude at 23°27' north or south of the equator, properly named Tropic of Cancer (north) and Tropic of Capricorn (south). They represent the angle of tilt of the Earth's axis of rotation from the perpendicular to the plane of its orbit around the Sun. They are the points furthest north and south where the Sun can appear directly overhead. In meteorology and geography, the climatic 'torrid zone' between the latitudes 23°28' north and south, where the climate is characterized by high temperature and humidity, abundant rainfall, and potentially rapid biological growth. As the adjective 'tropical', used to describe this type of climate, as well as in reference to specific climatic phenomena, such as:

tropical cyclone, tropical desert, and tropical lake. The Tropics of Cancer and Capricorn are so named because the Sun appears in these astronomical constellations at noon on the solstices. Etymologically related terms include entropy and troposphere. First usage in English c. 1375.

troposphere: [Greek *trope* = a turning; < IE *trep* = to turn + *sphaira* = sphere.] n. The lowest region of the atmosphere, starting at the Earth's surface and extending up to the tropopause (the boundary between the troposphere and stratosphere), ranging in height from approximately 8 kilometers (5 miles) at the poles to approximately 18 kilometers (11 miles) at the equator. It is characterized by rapid decrease of temperature with altitude, and strong convection currents, such as the Hadley, Ferrel, and Polar gyres. The boundary between the troposphere and the stratosphere varies between about 15 kilometres (9 miles) at the Equator and 8 kilometers (5 miles) at the North and South Poles. In the troposphere, temperature decreases linearly with altitude at a rate of about 1°C per 100m (4°F per 1000ft). This is called the lapse rate. Pressure, on the other hand, decreases exponentially with altitude. First usage in English c. 1905.
Cited usage: M. Teisserenc de Bort discovered that the atmosphere is divided into two parts, the troposphere, which extends from the surface to about 7 miles, and the stratosphere, which lies above [1914, Q. Jrnl. R. Meteorol. Soc. XL. 108].

tsunami: [Japanese *tsu* = harbour, port + *nami* = waves.] n. A brief series of long-wavelength, low-amplitude, undulations on the surface of the sea, due to displacement of the seafloor caused by an earthquake or other disturbance causing displacement, such as a submarine landslide or volcanic eruption. Tsunami travel at great speed, have long wave lengths, and low wave heights in deep water, but, as they enter shallow water near a land mass, they convert their high velocity and force into

surges capable of attaining great elevations above sea level, with associated devastation of the coast. First translated into English around 1900, the term tsunami was adopted for general use in 1963 at an international scientific conference. Tsunami are often called 'tidal waves,' which is a misnomer because they are unrelated to tides, and are not caused by gravitational interactions of celestial bodies.
Cited usage: 'Tsunami!' shrieked the people; and then all shrieks and all sounds and all power to hear sounds were annihilated by a nameless shock as the colossal swell smote the shore with a weight that sent a shudder through the hills [1897, Lafcadio Hearn Gleanings, Buddha-Fields i. 24].

tsuyu: [*tsu* = plum + *uu* = rain.] n. The tsuyu is a cloudy and rainy period in early summer in Japan; it is the same as the Chinese *Meiyu*. All areas except the northern island of Hokkaido have this rainy season that lasts from early June to late July. In the southern Nansei islands the tsuyu starts and ends about one month earlier. The term relates to plums because the tsuyu occurs during the plum-ripening season. Historically, planting of the year's rice crop is completed just before tsuyu begins.
Synonyms: baiu
See also: baiu

tufa: [Italian *tufo* = a soft stone for building; < Latin *tofus* = porous, loose stones of various kinds.] n. Tufa is a porous, chemical, sedimentary rock composed of calcium carbonate. It forms at springs, seeps, along streams, and in lakes. Tufa usually forms by direct precipitation or evaporation. It is often stratified, and its formation is often aided by algae or bacteria. Tufa may form towers up to several meters in height. It can also form mounds over 100m high. Not to be confused with tuff. This confusion likely originated in Italy, because the term 'tufa' was used there to describe certain pyroclastic rocks, especially

from Mount Vesuvius. The term 'calcareous tufa' was used there when referring to the calcareous sedimentary rock.
See also: tuff

tuff: [Italian *tufo* < Latin *tofus* = porous, loose stones.] n. Lithified volcanic ash. Tuff and tufa are a tough pair of terms to distinguish if you are an etymologist, but pretty easy to distinguish if you are a geologist. Both words come from the Italian root *tufo*, defined by Florio as "a kind of soft, crumbling, or mouldring stone to build withal." Numerous citations in science and literature indicate that the two terms were considered synonymous prior to the 20th century. For example; in the Ansted Glossary of Geology c. 1850, tufa and tuff are treated equally as, "an Italian name for a variety of volcanic rock of earthy texture made up of fragments of volcanic ashes." Conversely, in 1881, the Raymond Mining Glossary reports: "tuff or tufa, a soft sandstone or calcareous deposit." In 1867, in the Ansted Dictionary of Science, an early distinction between the volcanic and sedimentary rocks were made by use of the adjectival modifiers volcanic and calcareous, used with the interchangeable tuff or tufa. Ansted's examples are: "Tufa is a name applied in Italy to certain porous loose rocks... Volcanic Tufa is the material under which Pompeii was buried... Calcareous Tufa when consolidated passes into Travertine." To date, the exact time in lexicography when the terms tuff and tufa developed separate meanings remains unclear. But someone realized that they could do away with the modifiers 'volcanic' and 'calcareous' if they just separated the terms. This, apparently, appealed to everyone's tendency toward precision. Thus, the terms became unique, but still close enough in sound and spelling to cause confusion.
Cited usage: Tufa, Tuff, an Italian name for a variety of volcanic rock of earthy texture, made up of fragments of volcanic ashes [1850,

Ansted, Elementary Geology Glossary] Tuff or
Tufa, a soft sandstone or calcareous deposit
[Raymond Mining Glossary, 1881]
See also: tufa

tumescence: [Latin *tumescere* = to begin to
swell < *tumescere* = to swell; < IE *teu* = to swell.]
n. The swelling of a volcanic edifice due to an
increase in pressure caused by accumulation of
magma in the reservoir. IE derivatives include:
thigh, tumor, butter, tomb, and thousand
(swollen hundred).
Cited usage: The tumescence of the Hawaiian
volcanoes, however, is slight, only one metre or so
at the summit [1976, P. Francis, Volcanoes x. 310].

tungsten: [Swedish *tungsten* = heavy stone.]
n. A hard, brittle, corrosion-resistant, gray to
white metallic element (symbol W). It occurs in
such minerals as wolframite, and scheelite.
Among the metals, tungsten has the highest
melting point (3422 degrees C), and lowest
vapor pressure of any metal. At high temperatures it also has the highest tensile strength. In
1556 Georgius Agricola, author of De Re
Metallica (*On Mining*), described a mineral he
called *lupi spuma*, meanig wolf's foam, which
translates to the German *wolf rahm*, meaning
the mineral wolframite (Fe,Mn)WO$_4$. He coined
the name because during extraction the reactant
element tin was "eaten, like the wolf eats the
sheep." In 1781 Carl W. Scheele analyzed the
mineral, then called *tungsten* (now *scheelite*)
and found that it contained a new element
which he called *Lapis ponderosus*, meaning
heavy stone, which in Swedish translates to
tungsten.
Cited usage: Tungsten...which has also been
called Scheelium and Wolframium, was first
obtained by Messrs. de Luyart (in 1783), from
the tungstic acid previously discovered by
Scheele, in 1781 [1841, Brande, Man. Chem.
5th ed. 921].

turbidite: [Latin *turbidus* = full of confusion,
disordered, muddy, perplexed, violent < *turba* =
crowd, disturbance.] n. Sedimentary deposits
formed by turbidity currents in relatively deep
water compared to the point of origin. They can
occur in lakes, but are a more common feature at
the base of the continental slope outward to the
abyssal plain. Turbidites commonly show predict-
able changes in bedding from coarse layers at the
bottom to finer laminations at the top that result
from different settling velocities of the variously
sized particles (known as the Bouma sequence).
Term coined by Philiip Kuenen in 1957.
Cited usage: The term 'turbidite' for all deposits
of turbidity currents is more appropriate, and the
writer accepts this verbal suggestion of C. P. M.
Frijlinck [1957, P. H. Kuenen, Jrnl. Geol. LXV].
See also: turbidity current, Bouma sequence

turbidity current: [Latin *turbidus* = full of
confusion, disordered, muddy, perplexed, violent <
turba = crowd, disturbance.] n. A bottom-flowing
density current, usually in water, carrying suspended sediment moving by gravitational flow
downslope until the sediment drops out of
suspension due to reduced energy (caused by
friction, reduced slope, and spreading of the
density current). Turbidity currents originate and
are triggered by one or more of the following
factors: an oversupply of sediment, tectonic
forces, meteorological disturbances, or remotely
generated tsunami, or seiche. Turbidity currents
are characteristic of trench slopes of convergent
plate margins and continental slopes of passive
margins. Term coined by Johnson in 1939 (after
evidence by Daly in 1936) for the agent causing
submarine canyons.
See also: turbidite

turquoise: [Toponymous term after Turkestan,
where the mineral was found in ancient times.] n.
A blue to blue-green, opaque, triclinic mineral,
hydrated hydroxy copper aluminum phosphate,

$CuAl_6(PO_4)_4(OH)_8 \cdot 5H_2O$. Turquoise commonly displays a botryoidal habit, and occurs in the supergene oxide zone of copper deposits. It is isomorphous with chalcosiderite. Turquoise is considered among the most valuable of the non-transparent precious stones. The name comes from the French *pierre turqueise*, meaning Turkish stone, alluding to Turkestan. The mineral is named after Turkestan, where it was first discovered; however, true oriental turqoise is found in Persia. Turquoise found in Central America was called *chalchihuitl* in Nahuatl and was used in jewelry by the Aztecs. First usage in English c. 1375.

typhoon: [Arabic *tufan* = delude; < Greek *tuphon* = father of the winds, whirlwind; < Chinese *taifung* < *tai* = great, big + *feng* = wind.] n. A tropical cyclone, equivalent to the Caribbean *hurricane*, but occurring in the western Pacific or Indian Oceans, chiefly during the period from July to October. The history of the term typhoon is an example of linguistic convergence. It arose as a meteorological term independently, both in Greece and in China, before assuming its current form in English. The Greek word *tuphon* passed through Arabic into English in 1588 as *tufan*, originally referring specifically to a severe storm in India. The modern English meaning of typhoon was influenced by a borrowing from the Chinese *taifung*, meaning great wind. The terms were coincidentally similar, and therefore changed into the English *tuffoon* in 1699. The spelling *typhoon* first appeared in 1819 in Shelley's "Prometheus Unbound."

ubac: [French *ubac* = north-facing (shady) slope; < Latin *opacus* = sombre, dark.] n. The shady northern side of an alpine mountain, characterized by a lower timberline and lower snow line than the sunny southern side (adret slope). Conditions tend to be cooler and moister in ubac regions. Plants that live on adret slopes, such as those of the chaparral community, must be adapted to more arid conditions than those found on ubac terrain. It is preferable to build dwellings located in the High Alps (Hautes-Alpes) of France and Switzerland facing south (à l'adroit or adret).
See also: adret

ugrandite: [Composite mnemonic term from English *u*(varovite) + *gr*(ossular) +*an*(dra)*dite*.] n. An informal subdivision of garnet including those that contain calcium, in particular: uvarovite, grossularite, and andradite.
See also: pyralspite, garnet, uvarovite

ulexite: [Eponymous term after German chemist Georg Ludwig Ulex (1811–1883).] n. A white, triclinic mineral, hydrated hydroxy sodium calcium borate, $NaCaB_5O_9 \cdot 8H_2O$. Ulexite is usually found with the mineral borax and is directly deposited as fine acicular crystals in reniform masses (cottonball ulexite) and saline crusts in arid regions, resulting from evaporation of boron-rich water in ephemeral alkali lakes called playa lakes. Ulexite is also found in veinlike beds composed of closely packed fibrous crystals which display unique fiber-optic transmission properties, leading to the name TV rock. First usage in English c. 1865.
Cited usage: Ulexite, a name given to native borate of lime (Hayesine), after Ulex, by whom it was analysed [1867, Brande & Cox Dict. Sci.].

ultra-: [Latin *ultra* < IE *al* = beyond.] Combining prefix meaning spatially beyond, as in ultraviolet or ultra-stellar; meaning on the other side of, as in ultramontane; or meaning extremely, as in: ultramafic, ultrabasic, ultramylonite, ultraplankton, ultracold, and ultramarine. The term also stands alone as a noun, meaning an extremist person, especially with regard to politics. It is first recorded in this sense in 1817, used as a back-formation of *ultra-royaliste*, meaning an extreme royalist loyal to the crown. IE derivatives include: alien, parallax, allegory, alias, altruism, and adultery.

umbo: [Latin *umbo* = knob, projection, shield boss; < IE *ombh* = navel.] n. A projection of a round or conical form; a knob. In biology and paleontology: a) the humped part of a bivalve shell, b) the apical part of a brachiopod or crustacean valve, or c) a central, rounded, and elevated protuberance on discoidal foraminifera tests. The term is often mistakenly used as a synonym for beak, but both terms are frequently needed to describe different features on the same organism. First usage in English c. 1720.
Cited usage: Like the umbo of the Romans Which fiercest foes could break by no means [1721, Jonathan Swift, George-nim-Dan-Dean's Answer to Sheridan, 33].

unaka: [Toponymous term after the Unaka Mountains in the U.S. Southwest; < Cherokee *unaka* = white.] n. In geomorphology, a term coined by Hayes in 1899 for an erosional remnant or residual mass that is topographically higher than the surrounding peneplain. A unaka is sometimes capped by the remains of an older peneplain. In 1939, Lobeck extended the term to refer to a group of monadnocks perched at the head of a stream valley where erosion has not yet reduced the entire area to a level peneplain. The type locality lies in the Unaka Mountains in the U.S. Southeast from Clingman's Dome, Tennessee's highest point (2,025m) to northern Georgia, comprising segments of the Blue Ridge, Great Smoky, and Appalachian Mountains. The Unakas are sometimes referred to as the Unicoi Mountains.

unconformity: [Middle English < Old English *un* < IE *ne* = not + Latin *conformare* = to shape after < *con-* < *cum* = with, together + *formare* = to shape < *forma* = shape.] n. A buried surface of erosion and/or nondeposition marking a break or gap in the geologic record and representing a missing interval of time. Unconformities are produced by depositional, erosional, and tectonic changes which cause an interruption in the continuity in the rock record. Although recognized and described by James Hutton in the late 1700's, the term was coined in 1805 by Robert Jameson.
See also: nonconformity, disconformity, paraconformity

unda: [Latin *unda* = wave.] n. Combining prefix or stand-alone term applied to the environment of sedimentation that lies in the zone of wave action, or swash zone. The term was coined by John L. Rich in 1951. He also coined the terms *undaform*, referring to the topographic feature, and *undathem*, referring to the rocks. These terms, since the advent of sequence stratigraphy, have come into use.
Cited usage: The topographic surface assumed a form resembling in cross section the classic delta, with topset (unda), foreset (clino) , and bottomset (fondo) beds. The writer uses the parenthetical terms, plus the suffix *-form*, to designate the particular topographic element; and *-them* similarly

for the corresponding rocks, as proposed by the late John L. Rich [1964, D. C. Van Siclin, Symposium on cyclic sedimentation: Kansas Geological Survey, Bulletin 169, pp. 533-539].

uni-: [Latin *unus* = one, single.] Combining prefix meaning to have, as in unipolar; provided with, as in unidimensional; composed or consisting of, as in unilateral; characterized by one thing that is specified or connoted by the second element in the term, such as: uniform, unicline, and unicycle; or to make, form into, or cause to become one, as in unify. The Greek equivalent prefix is *mono-*.

uranium: [Eponymous term after *Uranus* (Greek *Ouranos*), Roman god who personified the heavens and sired the Titans.] n. A heavy, radioactive metallic element of the actinide series (symbol U). In 1789, Martin H. Klaproth isolated uranium oxide and deduced the existence of elemental uranium in pitchblende ores from Johannorgenstadt in the Erzegebirge (Metalliferous Mountains) of Saxony. He resisted the public temptation to name the element *Klaprothium*, after himself, instead naming it after *Uranus*, the farthest planet from the sun then known. The planet was discovered in 1781 by Sir William Herschel, who initially named it 'Georgium sidus'. It was finally named Uranus by Bode, in keeping with the tradition of naming the planets after ancient mythology. When elements 93 and 94 were created many years later, the names *neptunium* and *plutonium* continued the allusion to the outer planets of the solar system.

urstromtal: [German *urstromtal* = ancient river valley.] n. A wide, shallow valley shaped like a trench, and excavated by a temporary meltwater stream that runs parallel to the front margin of a continental ice sheet. Once formed, the urstromtal serves as a path for collecting distal meltwater and groundwater that have permeated into deeper layers of the water table.

uvala: [Serbo-Croatian *uvala* = hollow, depression.] n. A karst valley, being a sizeable depression in the ground surface occurring in karst regions. They are generally complex, closed topographic basins composed of multiple smaller depressions called dolines.
Cited usage: Dr. Cvijic's researches have led him to consider the uvala (Karstmulde) as an intermediate form between doline and polye. The uvala is a large, broad sinking in the karst with uneven floor, formed by the breaking down of the wall between a series of dolines [1902, Geogr. Jrnl. XX. 429].
Alt. spelling: ouvala
See also: polje, karst, doline

uvarovite: [Eponymous term after Count Sergei Semenovitch Uvarov.] n. An emerald-green, vitreous, transparent to translucent, isometric mineral of the garnet group, calcium chromium silicate, $Ca_3Cr_2(SiO_4)_3$. Named in 1832 by the German academician Ivanovich Gess, after Count S. S. Uvarov (1785–1855), president of St. Petersburg Academy. It is reported that I. Gess was actually G. H. Hess. This discrepancy stems from the fact that Gess changed his name from Germain Henri Hess to Germain Ivanovich Gess after moving from Switzerland to Russia. The type locality is the Saranovski Mine in the Central Ural Mountains, of Russia.
Alt. spelling: uwarowite, ouvarovite
See also: ugrandite, pyralspite, pyrope, grossularite, andradite, almandine, garnet

vadose: [Latin *vadosus* = shallow < *vadum* = a shallow body of water.] adj. Of, relating to, or being water that is located in the zone of aeration above the water table. Vadose is used in such terms as: vadose water, vadose solution, and vadose zone. First usage in English c.1895.

vagile: [Latin *vagus* = wandering, straying.] adj. Said of a plant or animal that is free to move about or be dispersed within an environment, as opposed to sessile. Not to be confused with motile, from Latin *motus*, meaning motion, and defined as having the power to move spontaneously.

valley: [Middle English *valey* < Old French *valee* < Latin *vallis* = valley; < IE *wel* = to turn, roll.] n. A long depression or hollow lying between hills or stretches of high ground and usually having a river or stream flowing along its bottom to an outlet. Valleys are usually formed by stream erosion, but can also be formed by tectonic activity. In modern use a valley is distinguished from a vale by having steeper sides; a vale often forms the broad and flatter part of a valley. First usage in English c. 1275.

vallis: [Middle English *valey* < Old French *valee* < Latin *vallis* = valley; < IE *wel* = to turn, roll.] n. A valley on another planet or on the Moon. Vallis is usually used as a proper name, such as the singular Vallis Shroteri on the Moon, or the plural Valles Marineris on Mars.

vapor: [Middle English *vapour* < Old French *vapeur* < Latin *vapor* = steam.] n. A gaseous substance that can coexist with a corresponding liquid, but is not normally a gas under standard conditions at the Earth's surface. In contrast, the term *gas* is used to describe a substance that is a gas at standard temperature and pressure. First usage in English c. 1350.
Cited usage: As man, brid, best, fisshe, herbe, and greene tree The feele in tymes with vapour eterne [c.1374, Chaucer, Troylus iii. 11]. The aeriform state of liquids is known by the name of *vapour*, while gases are bodies which, under ordinary temperature and pressure, remain in the aeriform state [1863, E. Atkinson, trans. of Ganot's *Elements do physique* (*Elementary Treatise of Physics, Elementary and Applied*) IV].

vara: [Spanish and Portuguese *vara* = rod, yardstick; < Latin *vara* = forked pole, trestle < *varus* = bent, bow-legged.] n. A traditional unit of linear measure used in Spanish- and Portuguese-speaking countries, primarily in Latin America. The length of the Spanish vara is roughly 84cm, whereas the Portuguese vara is about 110cm. The standard vara was the vara of Castile, (about 0.8359 meter, subdivided into 3 pies or 4 palmos). In 1536, the Viceroy Don Antonio Mendoza instituted an ordinance that established the vara as the fundamental unit of length in New Spain. Terms still in use in northern Mexico and Texas that are based on the vara include: *millionada*, meaning a million square varas, and *ayuntamiento*, meaning a township equal to 20,000 square vara.
Alt. spelling: bara

variole: [Medieval Latin *variola* = pustule, pox; < Latin *varius* = speckled, variegated.] n. A pea-size spherule, usually composed of radiating

crystals of plagioclase or pyroxene in mafic igneous rocks. The term was first used in geology in 1890, in reference to the spotted appearance of the weathered surfaces of fine-grained to aphanitic rocks containing spherulites. It was first used in English in 1826, alluding to any spot on the body resembling a smallpox marking. An etymologically related term, *variolite*, was coined by Johannsen in 1938 for such rocks.
Cited usage: The spherulites or 'varioles' (of the variolite-diabase) are grouped or drawn out in bands parallel to the surface [1890, Quart. Jrnl. Geol. Soc. XLVI. 312].

varve: [Swedish *varv* = layer, a complete turn < *varva* < Old Norse *hverfa* = to bend.] n. A pair or couplet of thin, usually fine-grained sedimentary layers, that contrast in color and texture, and represent the deposit of a single year in still water. This commonly occurs in lakes, fjords, and other protected bodies of water in colder climates, showing seasonal variation when freezing conditions of winter give way to sediment-laden meltwater of spring and summer. Varves have been used to establish the chronology of glacial advances and retreats. First usage in English c. 1910.
Cited usage: The Swedish word varv, (old spelling: hvarf), means as well a circle as a periodical iteration of layers. An international term for the last sense being wanted it seems suitable to use the transcription varve [1912, G. De Geer, Compt. Rend. XI Session Congrès Géol. Internat. 253].
Alt. spelling: hvarf

varzea: [Brazilian Portuguese *varzea* = flooded forest, meadow, plain.] n. Alluvial flood plains that stretch across low, seasonally flooded river drainages. The term was named for large drainage basins of major rivers in Brazil, such as: the Central and Lower Amazon, the Madeira River, the Purus River, and their tributaries. The annual floods of sediment-laden water may rise 6-12m (20-40ft), and

inundate the flood plain for up to 8 months. The boundaries of the broad varzea are defined by stretches of savanna near the forest's edge.

vascular: [Latin *vas* = vessel.] adj. Characterized by, or containing vessels that carry or circulate fluids, such as blood, lymph, or sap, through the body of an animal or plant. An etymologically related term is vase. First usage in English c. 1670.

vauclusian: [Toponymous term after *Fontaine-de-Vaucluse*, a medieval village in a closed valley at the southwestern corner of the mountainous *Plateau de Vaucluse*, near Avignon, France.] adj. Applied to a type of artesian spring, often large, occurring in karst regions.
Cited usage: Cheddar Gorge....and the spectacular Wookey Hole with its 'Vauclusian spring', show true karstic features [1937, Wooldridge & Morgan, Physical Basis Geogr. xix. 294].
Synonyms: gushing spring

vein: [Middle English *veine* < Old French *vaine* < Latin *vena* = vein.] n. In petrology, a hydrothermal intrusion formed by deposition of minerals in the open space along a fracture, such as a joint or fault. In coal mining, the term vein has also been used for a coal bed. In botany, a vascular bundle of a leaf that provides a framework for the conducting tissues. In physiology, the vessels that conduct blood to the heart. In water-witching or dowsing, a term for the purported narrow channels through which water supposedly flows, as revealed by practitioners using questionable methods of detection. First usage in English c. 1275.

veld: [South African English *veld* = open country < Dutch *velt* = field.] n. In South Africa, large regions of open country or open pasture-land. *Veld* is frequently used in conjunction with modifiers that help define a locality's character,

such as: bush-, grass-, high, low, sour, and sweet veld. First usage in English c. 1800.
Cited usage: Here for the first time we bivouacked in what is called the Veld [1835, A. Steedman, Wanderings and Adventures in the Interior of South Africa I. ii. i. 92].
Alt. spelling: veldt, velt

ventifact: [Latin *ventus* = wind + *factus* = past participle of *facere* = to make.] n. A faceted stone shaped or altered by wind-blown sand. First usage in English c. 1910.
Cited usage: If a general expression be required for any wind-shaped stone, we might speak of a 'ventifact', on the analogy of artifact [1911, J. W. Evans, Geol. Mag. Decade V. VIII. 335].

ventral: [Latin *venter* = abdomen.] adj. Of or belonging to the anterior or lower surface. In biology, relating to the front part of the body, especially that part near the belly. In botany, situated on the upper or inner side of a plant organ that is facing the axis, as on a leaf. First usage in English c. 1735.
Cited usage: The longitudinal gangliated chain of articulated animals is often distinguished as the ventral cord [1874, Carpenter, Physiology].
See also: dorsal

vertebrate: [Latin *vertebratus* = jointed, articulated < *vertebra* = joint, of the spine < *vertere* = to turn; < IE *wer* = to turn.] n. Any member of the subphylum Vertebrata, being a major division of the phylum Chordata. Vertebrates are characterized by a segmented protective spinal column which contains a long integral part of the central nervous system; a distinct and relatively large head; a muscular system comprised of mostly paired masses; and an internal skeleton composed of bone and/or cartilage. Primary groups include: fish, amphibians, reptiles, birds, and mammals. Jean Babtiste Lamarck (1744-1829) coined the terms vertebrate and invertebrate and published them in his *Philosophie Zoologique* in 1809.

vesicle: [Old French *vesicule* < Latin *vesicula* = small blister < *vesica* = bladder, blister.] n. In petrology, a small cavity of variable shape formed in volcanic rock by entrapment of gas bubbles during solidification. As the adjective vesicular, used to refer to the texture of a rock, especially lava, having a high percentage of gas-formed cavities, such as vesicular basalt. In biology and paleontology, a membrane-enclosed cavity in a plant or animal, such as a blood, seminal, or umbilical vesicle. First usage in English c. 1575.

vestigial: [Latin *vestigium* = footstep, footprint, trace, mark.] adj. In biology, a surviving trace of some larger or more important body part formerly existing in all or most members of a species (e.g., a vestigial organ). In geomorphology, a trace, small portion, or erosional remnant of some pre-existing larger structure, a vestigial valley, for example. First usage in English c. 1880.

Vesuvian: [Toponymous term after *Monte Vesuvio* (Mount Vesuvius), an active volcano near Naples, Italy; < Latin *Vesouuios* = son of Ves (alternate name of Hercules) < *Ves* = persona of Zeus, god of rains and dews.] adj. Referring to an explosive type of volcanic eruption characterized by blocky pyroclastic fragments and great quantities of ash-laden gas violently discharged to form a cauliflower-shaped cloud high above the volcano. Although Mount Vesuvius is still an active volcano, a Vesuvian-type eruption refers to its cataclysmic eruption on August 24, 79 CE. This epic eruption buried the towns of Pompeii and Herculaneum, and resulted in the death of Pliny the Elder, who was sailing northward with a fleet of ships and stopped to witness the event and to rescue survivors. First usage in English c. 1670.
Cited usage: There is a tendency in almost all the Vesuvian dikes to divide into horizontal prisms [1833, Lyell, Principles of Geology III. 125].
Synonyms: Vulcanian

vesuvianite: [Toponymous term after *Monte Vesuvio* (Mount Vesuvius), an active volcano near Naples, Italy; < Latin *Vesouuios* = son of Ves (alternate name of Hercules) < *Ves* = persona of Zeus, god of rains and dews.] n. A brown to yellow or green, vitreous, tetragonal mineral, hydroxy calcium magnesium aluminum silicate $Ca_{10}Mg_2Al_4(SiO_4)_5(Si_2O_7)_2(OH)_4$. It is commonly found in skarns formed by contact-metasomatism of limestones. The term vesuvianite replaced the now obsolete term idocrase. Term coined in 1795 by Abraham Gottlob Werner, for the fine crystals he found in the caldera of Monte Somma within which Vesuvius was constructed.
Cited usage: Vesuvianite was first found among the ancient ejections of Vesuvius and the dolomitic blocks of Monte Somma [1892, E. S. & James D. Dana, A System of Mineralogy 6th ed. 480].
Synonyms: idocrase (obs.)
See also: idocrase

vicinal: [Latin *vicinus* = neighboring.] adj. In mineralogy, a crystal face that approximates, resembles, or takes the place of a fundamental crystalline form or face, which it closely approximates in angle. In general English usage, belonging to neighbors or neighborhood. First usage in English c. 1620.

virga: [Latin *virga* = rod, stick, twig.] n. Long, streaky wisplike trails of precipitation that fall from water-saturated clouds, but evaporate before reaching the ground. Virga appear more distinctive when viewed with backlight from the sun. Term coined in meteorology in 1938 because the etymology is descriptive of the phenomenon, and because VIRGA was coined as an aviation forecasting acronym standing for Variable Intensity Rainfall Gradient Aloft.
Alt. spelling: virgae (pl.)
Synonyms: fall-stripes, fallstreaks

virgation: [Latin *virga* = rod, stick, twig.] n. In structural geology, a system of divergent faults that branch out like twigs from a larger limb of a tree. In geomorphology, a series of mountain ranges diverging from a common center.
Cited usage: The Western Balkans form in their southern part six ranges, the orographical expression of a geological 'virgation' [1897, Geogr. Jrnl. (R.G.S.) IX. 87].
See also: syntaxis

viscosity: [Late Latin *viscosus* = viscous, full of birdlime; < Latin *viscum* = mistletoe, birdlime made from mistletoe berries; < Greek *ixos* = mistletoe.] n. The tendency of a fluid to resist molecular motion by internal friction, thus increased viscosity means increased resistance to flow or change of shape. The temperature of a fluid is generally inversely proportional to its viscosity. The etymological connection between a physical property and a botanical substance stems from the properties of birdlime being a thick, sticky substance prepared from mistletoe or holly bark and used to catch small birds. The C.G.S. unit of viscosity measurement is called the poise, an eponymous term after French physician Jean Poiseuille (1797-1869). First usage in English c. 1400.

vitalism: [French *vitalisme* < Latin *vitalis* < *vita* = life.] n. A term stemming from the teachings of Plato, being a spiritual or metaphysical idea that an immaterial and supernatural force, distinct from physical or chemical forces, is present in organisms. This force supposedly determines the direction of the energies, desires, and evolution of the creature within which it resides. The idea of vitalism was integral to Plato's philosophical teachings of his theory regarding the fundamental elements of the universe: Earth, Air, Wind, and Fire, which were animated by a fifth element, or vital force called the Ether. First usage in English c. 1820.

vitrain: [Portmanteau word from vitr(eous) + (fus)ain; < Latin *vitreus* = of glass, glassy, bright < *vitrum* = glass + French *fusain* = spindle tree, charcoal from spindle tree; < Latin *fusago* = spindle < *fusus* = spread out, wide, copious.] n. A lithotype of coal visually typified by vitreous luster, dark to black color, and conchoidal fracture. Microscopically, containing at least 95% humic-rich, vitrinite macerals. Term coined in 1919 by M. C. Stopes.
Cited usage: These four distinguishable ingredients, all of which are to be found in most ordinary bituminous coals, I name provisionally as follows: (i) Fusain. (ii) Durain. (iii) Clarain. (iv) Vitrain [1919, M. C. Stopes, Proceedings of Royal Society Bull. XC. 472].
See also: clarain, durain, fusain, maceral

vitro-: [Latin *vitro* < *vitrum* = glass.] Combining prefix meaning glass, used in terms such as: vitric tuff, vitrain, and vitriol. The term in vitro, coined in New Latin, literally means 'in glass', and stands for biological experiments performed outside a living body, or under artificial conditions (in contrast to in vivo, meaning 'in a living body').

vitrophyre: [Latin *vitrum* = glass + Greek *(por)phyr* = porphyry.] n. Any porphyritic volcanic rock having a glassy groundmass.
Cited usage: Vogelsang has proposed to classify this type [Porphyritic] in three divisions: 1st, Granophyre, 2nd, Felsophyre, 3rd, Vitrophyre, where the ground-mass is a glassy magna [1882, Geikie, Textbook Geol. ii.].

vlei: [Dutch *vlei* = shallow lake, marsh < *vallei* = valley.] n. In South Africa, a shallow lake or pool of water, especially an intermittent one in low-lying ground covered with water during the rainy season. In the U.S. Middle Atlantic states, a swamp or marsh.
Cited usage: The Hottentots look anxiously around for the well known 'vlei' [1849, E. E. Napier, Excurs. S. Africa II. 179] (Ed. note: Hottentot is a pejorative term originally used by Dutch settlers in South Africa, in reference to native Africans of the region, who actually call themselves *Khoikhoi*, meaning 'genuine people' or 'men of men').
Alt. spelling: vly, vley

voe: [Norwegian *vaag* < Icelandic *vagr* = bay, inlet.] n. A bay, creek, or inlet, especially as used in the Scottish Orkney and Shetland Islands. Possibly related to Icelandic *ver*, meaning sea.

volcano: [Italian < Spanish *volcán* < Latin *volcanus, vulcanus* = fire, flames < *Vulcan* = Roman god of fire and forge.] n. A usually conical hill or mountain built up around a vent through which, during episodic periods of activity, lava, pyroclastic material, and gases erupt. In general, the term can be applied to any eruption of any material, such as mud, salt, or sand, that mimics the classic conical shape of an igneous volcano. In Roman mythology, Vulcan was the god of fire, and his forge lay inside Mt. Etna in Sicily. When he was busy forging tools and weapons for the gods, fiery eruptions emanated from the top of Mt. Etna. Unlike the term volcano, volcanology, being the study of volcanoes, can also be spelled with a *u*, as in the British variant vulcanology. First usage in English c. 1610.

vug: [Cornish *vug* < *vooga* = cavern.] n. Originally used in mining for a cavity in a rock or vein, usually lined with crystals of a different mineral composition from the enclosing rock. In pedology, a relatively large, distinct, and irregular void in soil material. First usage in English c. 1815.
Cited usage: The sound which the miner hears, may reasonably be accounted for by presuming him to be at work in the immediate neighbourhood of a cavity, or as he terms it, a voog (vug) [1818, W. Phillips, Geol. 207].
Alt. spelling: vugg, vugh, voog

wadi: [Arabic *wadi* = valley, ravine, river bed; < West Semitic *wdy* = to put, place, cast < *awada* = to cut off, kill.] n. In desert regions of Southwest Asia and certain Arabic-speaking countries, a dry, steep-sided ravine, streambed, valley, or channel which becomes a watercourse in the rainy season, often with an associated oasis in a closed basin at its termination. Also said of the stream or torrent running through such a landform.
Cited usage: A stair of rock brought us into a wâdy (Sidri), enclosed between red granite mountains...I cannot too often repeat, that these wâdys are exactly like rivers, except in having no water [1856, Arthur Penrhyn Stanley, Sinai & Palestine in Connection with their History i. ii. 70].
Alt. spelling: wady, ouady

wallow: [Old English *wealwian* = to turn, roll about in, wallow.] n. A depression filled with water or mud, and used by animals as a watering hole. First usage in English before 900.
Cited usage: Let the dog turne to the vomit, and the swine to the walow [1591, H. Smith, Serm., Jacob's Ladder (1601) 545].

water: [Middle English < Old English *waeter* = water; < IE *wed* = water, wet.] n. A clear, colorless, odorless, and tasteless polar binary compound, H_2O. At conditions occurring on the Earth's surface, it is found in all three phases; solid, liquid, and gas. It is the most widely used solvent (the 'universal solvent'), and is essen-tial for most plant and animal life. Many scales of measurement are based on properties of water, such as the Celsius scale for temperature, specific gravity for density, and the liter for volume. A variant of the Indo-European root led to the Greek word *hydor*, meaning water. Other IE derivatives include otter, vodka, and whiskey. First usage in English before 900.
Cited usage: Water, water, everywhere, Nor any drop to drink. Water, water, everywhere, And yet the boards did shrink [1798, Samuel Taylor Coleridge, Rime of the Ancient Mariner].
See also: hydro

wave: [Middle English *waven* < Old English *wafian* = wave; < IE *webh* = to weave, move quickly.] n. Waves are disturbances that move, or propagate through a medium, and exhibit the following properties: a) exhibit no net transport of material, b) transport energy, c) have characteristic waveforms and periodicity, d) propagate at uniform waveform-independent speed, and e) have speeds dependent on the medium. Waves are categorized according to certain unique aspects, such as light waves, gravity waves, mechanical waves, transverse waves, and longitudinal waves. In oceanography, used in such terms as: wind wave, capillary wave, and tidal wave (not to be confused with tsunami). In seismology, used in such terms as: seismic wave, P-wave, and S-wave. In Old English, the term *waw* was a synonym for wave, but lost popularity and eventually became obsolete. As a verb meaning to move to and fro or up and down, wave was first used c. 1350. As a noun, wave was first used to mean a moving billow of water in 1526. It wasn't applied to physical phenomena, such as heat and light, until 1832. IE derivatives include: weave, waffle, waver, and wobble.

wax: [Old English *weax* = wax.] n. Any of various natural, solid, noncrystalline, greasy, organic substances derived from petroleum (mineral waxes), or other natural sources (animal and vegetable waxes). Examples of wax include: paraffin, ozocerite, carnauba, and beeswax. First usage in English before 900.

weather: [Middle English *wether* < Old English *weder* < IE *we* = to blow.] n. The state or condition of the atmosphere at a given place and time with respect to variables, such as: temperature, moisture, wind direction and velocity, and barometric pressure. Also with respect to quantity of sunshine, cloudiness, and the presence or absence of rain, hail, snow, thunder, or fog. Weather is generally described as a temporary condition, in contrast to climate which is the characteristic weather conditions that prevail in a particular region. IE derivatives include: wind, window, vent, wing, and nirvana. First usage in English before 900. *See also*: climate

weir: [Middle English *were* < Old English *wer* = weir < IE *wer* = to cover.] n. A barrier or dam in a stream, designed to restrain water for the purpose of raising or diverting it and passing it through a desired channel for the purpose of driving a mill wheel, irrigation, power generation, or simple regulating its flow. IE derivatives include: aperture, operculum, and warren. First usage in English before 900.

wind: [Old English *wind* = wind; < IE *we* = to blow.] n. Naturally moving air that moves roughly parallel to the Earth's surface. First usage in English before 900.

windkanter: [German *wind* = wind + *kant* = corner or angle or a polygon, a tilted position.] n. A ventifact that is bounded by smooth faces or facets cut by wind-born abrasive sediment. The faces intersect at sharp angles, and are usually cut at different times, either when the wind had changed direction for a long time or when the object is blown over onto its flattened face.

xeno-: [Greek *xenos* = stranger, foreigner, guest.] Combining prefix meaning foreign, alien, or different; used in such terms as: xenocryst, xenomorph, xenolith, xenon, xenophobia, and xenology (study of extraterrestrial phenomena).

xenoblast: [Greek *xenos* = stranger, foreigner, guest + *blastos* = sprout, shoot, germ.] n. A type of crystalloblast characterized by low formation energy, being a mineral that grows during metamorphism, but lacks development of characteristic crystal faces. Term coined by German petrologist Friedrich Johann Karl Becke in 1903.

xenocryst: [Greek *xenos* = stranger, foreigner, guest + Latin *crystallum* = crystal ice; < Greek *krystallos* = clear ice, rock crystal.] n. A crystal that is foreign to the igneous rock in which it occurs. It represents a crystal from some outside source that was engulfed by the magma, and is retained as an inclusion in the resulting rock.

xenon: [Greek *xenos* = stranger, foreigner, guest.] n. A colorless, odorless, heavy, inert, gaseous element (symbol Xe). In a discipline called xenology, it is used as the basis for dating early events in the chronology of the planetary system. The term alludes to the occurrence of the gas in only trace amounts in the Earth's atmosphere. Xenon was discovered in 1898 by English chemists Sir William Ramsay and Morris M. Travers, shortly after their discovery of the elements krypton and neon.
Cited usage: In September, 1898, the discovery of another gas was announced; it was separated from krypton by fractionation, and possessed a still higher boiling point. We named it *xenon* or the *stranger* [1904, Sir William Ramsay, Nobel Prize acceptance speech].

xeric: [Greek *xeros* = dry.] adj. Of, characterized by, or adapted to an extremely dry habitat. In comparison to mesic and hydric. First usage in English c. 1926.
See also: mesic, hydric

xylo-: [Greek *xylon* = wood.] Combining prefix meaning wood, as in xylem, xylinite, xylopal (wood opal), xylostroma, and xylophone.

yardang: [Turkish *yar* = steep bank, precipice.] n. An elongate ridge of soft but coherent sediment that has been grooved, fluted, pitted and streamlined by wind erosion. Yardangs can be meters to kilometers in length, and develop in arid deserts having minimal plant cover or soil development, and where strong, unidirectional winds occur during most of the year. The alternating ridges and furrows run essentially parallel to the dominant wind direction. Exploring the Taklimakan Desert in northwest China in the 1890s, Swedish explorer Sven Hedin encountered a strange landform of parallel ridges separated by gullies, which his

guides called *jardang* (the ridges). Massive yardangs are common on the surface of Mars. In 1975, the Egyptian geologist Farouk El-Baz, based on images from space, theorized that the statue of the Egyptian Sphinx was carved on a yardang. Not to be confused with the term *yarding*, meaning herding up in forested lowland areas called *yards*.
Cited usage: At intervals furrows or trenches in the clay sub-soil, called jardangs, traced between long elevations or ridges, crop up amongst the dunes [1904, S. Hedin, Scientific Results Journey in Central Asia I. xxvii. 439].
Alt Spelling: yarding; jardang

yazoo: [Toponymous term after Yazoo River in Mississippi; < *Yazoo* = name given to a Native American tribe.] adj. Referring to a tributary that flows parallel to the main stream, especially a tributary forced to flow on the overbank side of a natural levee of the main stream. The Yazoo River, after flowing roughly 300km (190mi) joins the Mississippi at Vicksburg. The river was named in 1682 by French explorer Sieur de La Salle (1643-1687), in reference to the people living near its mouth.

zastrugi: [Russian *zastrúga* = small ridge, furrow < *zastrugát* = to plane, smooth < *strug* = plane.] n. One of a series of irregular ridges formed on relatively level snow surfaces by wind erosion and deposition, aligned parallel to the direction of the prevailing wind.
Cited usage: We were guided by the wave-like stripes of snow (sastrugi) which are formed, either on the plains on land or on the level ice of the sea, by any wind of long continuance [1840, E. J.

Sabine, translation of Admiral von Wrangell's Narrative of an Expedition to the Polar Sea 1820, 21, amp 23 vii. 146].
Alt Spelling: sastrugi

zenith: [Middle English *senith* < Old French *cenith* < Latin *semita* = path; < IE *mei* = to change, go, move.] n. The point in the sky that is directly above the observer. The opposite of the nadir point. An etymologically related term is azimuth. First usage in English c. 1375.
Cited usage: The point that is rycht abufe our hede is callit zenyth,....ande as oft as ve change fra place to place, as oft ve sal hef ane vthir zenytht [1549, probably Robert Wedderburn, The Complaynt of Scotland vi. 50].

zeolite: [Greek *zein* = to boil; < IE *yes* = to foam, bubble + Greek *-ites* < *-lite* = pertaining to rocks < *lithos* = rock, stone.] n. Any member of a large group of hydrated tectosilicates that can hold a highly variable number of water molecules in the generally large voids present in their framework structure The zeolite structure resembles a cage, and the water molecules, along with dissolved cations (such as Na, Ca, K) have considerable freedom of movement. This feature leads to a common descriptive term calling zeolites 'molecular sieves'. A member in the family called 'molecular sieves' composed of crystalline hydrated aluminum silicate minerals displaying a cage structure, which encloses cavities occupied by cations and water molecules, both of which have considerable freedom of movement. Zeolites are low-temperature minerals that crystallize from late-stage hydrothermal solutions, and are mostly found in the filled cavities of amygdaloidal basalt. They display reversible dehydration. Examples of zeolites include: natrolite, stilbite, analcime, chabazite, thomsonite, and heulandite. The name zeolite alludes to the fact that many species of zeolites fuse readily upon heating, and intumesce or swell up due to the escape of water. The term

was coined in 1756 by Swedish mineralogist Axel Fredrick Cronstedt, who observed, upon rapidly heating a natural zeolite, that it began to dance about as the water was expelled, hence the Greek root meaning "stone that boils." More than 150 zeolites have been synthesized, and 40 naturally occurring zeolites are known. IE derivatives include yeast and eczema.
Cited usage: This name is given by Mr. Cronstedt to a stone described by him in the Transactions of the Academy of Sciences at Stockholm for the year 1756, the peculiar properties of which have induced that mineralogist to consider it as forming a distinct order of earths, called zeolites [1777, Dict. Chem. III. X8].

zeugen: [German *zeuge* = a witness, someone who gives evidence in a court of law < *bezeugen* = to witness.] n. Tabular masses of resistant rock, ranging in height from 2 to 50 meters, that are left standing on a pedestal of softer rocks. Zeugen result from differential erosion by the scouring effect of wind-blown sand in a desert region being driven into the base by the wind. The term may allude to the human shape.
Alt Spelling: zeugen (pl.)
Synonyms: witness rock
Similar Terms: mushroom rock

zinc: [German *das Zink* < *zinke* = spike, prong, point; < Pre-Germanic *tindja* = tine, prong.] n. A blue to white, lustrous, metallic element (symbol Zn). It occurs most commonly in the mineral sphalerite, ZnS, and is exceedingly rare in the native state. It is used to form alloys such as brass and solder. The element was named in 1526 by Paracelsus (1493-1541), based on the sharp pointed appearance of its crystals after smelting. (Born Theophrastus Bombastus von Hohenheim, later in life Paracelsus took the pseudonym meaning 'superior to Celsus', the early Roman physician.) Around the same time, Georgius Agricola, author of De Re Metallica (*On Min-*

ing), reported on a white metal he called "contrefey" because it was used to imitate gold. This metal, scraped from the walls of smelting furnaces in the Harz Mountains, was actually zinc.

zircon: [German *zirkon* < Arabic *siriqun* < Greek *surikon* < Persian *zargon* = fire color; < Avestan *zar* = gold; < IE *ater* = fire + Persian *gun* = color.] n. A variously colored, dense, adamantine tetragonal mineral, zirconium silicate, $ZrSiO_4$. Zircon occurs in pegmatites and as a common accessory mineral in siliceous igneous rocks, schists, gneisses, and in sedimentary rocks derived from these sources. It also occurs in beach and river placer deposits. It is the chief ore of the element zirconium, and the colorless variety can be cut into hard, brilliant gemstones resembling diamonds. First usage in English c. 1790.
Similar Terms: jacinth, zirconite, hyacinth, jargon

zirconium: [German *zirkon* < Arabic *siriqun* < Greek *surikon* < Persian *zargon* = fire color; < Avestan *zar* = gold; < IE *ater* = fire + Persian *gun* = color.] n. A lustrous, grayish-white, ductile metallic element (symbol Zr). It is obtained primarily from zircon, and is used in the manufacture of corrosion-resistant alloys for use in extreme conditions, such as in nuclear reactors. Term coined as *zirconia* in 1789 by Martin Heinrich Klaproth, but the element was first isolated in 1824 by Swedish chemist Jons Jakob Berzelius as an iron-grey powder made by heating potassium zirconofluoride with metallic potassium.

zoisite: [Eponymous term after Baron Sigismund Zois von Edelstein of Slovenia, discoverer of the mineral.] n. A white, gray, brown, green, or rose colored orthorhombic mineral of the epidote group, hydroxy calcium aluminum silicate, $Ca_2Al_3O(SiO_4)(Si_2O_7)(OH)$. It is dimorphic with clinozoisite and occurs in skarns and calcareous schists. Originally named *saualpite* by Zois in

1805 when he discovered it in the Sau-Alp Mountains, it was later renamed after him by Abraham Gottlob Werner in 1811.

zoo-: [Greek *zoo* < *zoion* = animal, living being; < IE *gwei* = to live.] Combining prefix meaning animal, used in such terms as: zoology, zootrophic, zooid, zooxanthellae, and zoolith. Sometimes now also used in biology and botany to denote having the power of spontaneous movement or motility, as in zoospores that occur in certain fungi, algae, and protozoans. IE derivatives include: biota, amphibian, symbiosis, zodiac, hygiene, victual, survive, and vivid.

zooplankton: [Greek *zoo-* = animal + *planktos* = to wander, roam or drift.] n. Free-floating (planktonic) aquatic heterotrophs, including radiolaria, foraminifera, copepods, euphausiids, and rotifers. Also used for various meroplankton in the juvenile state, including corals, anemones, and jellies (cnidarians and ctenophores). First usage in English c. 1900.

zygote: [Greek *zygotos* = yoked < *zygoun* = to yoke.] n. A cellular body of living protoplasm formed by the fusion (conjugation) of two reproductive cells or gametes. The zygote is a temporary state lasting from fertilization until the first cell cleavage (mitosis). First usage in English c. 1885.
Synonyms: zygocyte

Appendix I
Linguistics

aphesis: [Greek *aphesis* = a letting go < *ap* = off, away + *ienai* = to send, let go.] n. The loss of one or more sounds from the beginning of a word, for example *squire* < *esquire*. Term coined by original editor of the Oxford English Dictionary Sir James Augustus Henry Murray in 1880.
Cited usage: The Editor can think of nothing better than to call the phenomenon aphesis [1880, J.A. H. Murray, Trans. Philol. Soc. 175].
See also: apocope, syncope

apocope: [Greek *apokoptein* = to cut off < *apo* = away from; < IE *ap* = off, away + *koptein* = to cut.] n. The loss of one or more sounds from the end of a word. Examples in Modern English include: *mos'* < *most, sing* < Middle English *singen*, and in Britain *pud* from *pudding*. IE derivatives include: ablation, off, ebb, and compost.
See also: aphesis, syncope

assimilation: [Latin *assimilare* = to liken < *ad* = to, toward + *similis* = like.] n. In linguistics, the process by which a sound is modified so that it becomes similar or identical to an adjacent or nearby sound, such as the prefix *in* changing to *im* in the English *impossible* to make both sounds labial.
See also: assimilation (in body of dictionary)

back-formation: [German *ruckbildung* = back-formation.] n. A new word created by removing an affix from an already existing word, as *vacuum clean* from *vacuum cleaner*, or by removing what is mistakenly thought to be an affix, as in the modern word pea from the earlier English singular and plural *pease*. A back-formed term usually looks like the root-word, whereas the word that it derives from looks like the derivative itself (but it is not). There are two types of the linguistic process called back-formation. The first type is based on misunderstanding, as in the case of the English word *pea*. In Middle English the ancestor of pea was *pease*, a form that functioned as both singular and plural. In other words, the *s* was part of the word, not a plural ending. But around the beginning of the 17th century people began to interpret the sound represented by *s* as a plural ending, and a new singular, spelled pea in Modern English, was developed. In the second type of back-formation, as seen in the case of *baby-sit*, first recorded in 1947, and *babysitter*, first recorded in 1937, no misunderstanding is involved. The agent noun *babysitter* with its –er suffix could have been derived from the verb *baby-sit*, as diver was from dive, but the evidence shows that the pattern was reversed, and the agent noun preceded the verb from which it would normally have been derived.
Cited usage: Scavenger, the noun, is the origin, in English, from which to scavenge is a back-formation, the normal verb being to scavenger [1926, Fowler, Modern English Usage].

cf: [Latin *confer* = to compare, bring together for joint examination < *con-* < *cum* = together + *ferre* = intensive meaning to bear or bring.] abbr. An abbreviation used in lexicography telling the reader to see a related word or entry (the phrase 'See also' is used in preference to cf in this dictionary).

dissimilation: [Latin *dis* = apart, away + *assimilare* = to liken < *similis* = like, similar.] n. In linguistics, the differentiation of two identical

sounds occurring near each other in a word, by change of one of them, such as the change from *r* to *l* in English *marble* from French *marbre*. This process is in opposition to assimilation (*in* becomes *im* in English *impossible*, making both the *m* and *p* labial sounds). Although seldom used in geology, this term is similar to fractional crystallization. The opposite term, assimilation, however, is used in geology.
See also: assimilation

frequentative: [Latin *frequentem* = crowded, frequent.] n. In linguistics and grammar, said of a verb or verbal form that serves to express the frequent repetition of an action. The Latin root is cognate with *farcire*, meaning to stuff.

homonym: [Greek *homos* = same, similar + *onyma* = name.] n. In general usage, one of two or more words that have the same sound and often the same spelling but differ in meaning. Examples are: wave and wave, or bear and bare, but not wind and wind. In paleontology and biology, any one of two or more identical names that have been duplicated for a taxon of the same rank and thus has been rejected. Homonyms, homophones, and homographs are linguistic terms for groups of two or more words that always have different meanings, but may or may not be spelled the same or sound the same. Homophones always sound the same, but may be spelled the same or differently, such as: bank and bank, or led and lead. Homographs are always spelled the same, but may or may not sound the same, such as: fair and fair, or wind and wind. Thus, homonyms and homophones are essentially synonymous, whereas homographs are only sometimes synonymous with the other two. First usage in English c.1640.

hybrid: [Latin *hibrida* = a cross-bred animal (mongrel), offspring of a tame sow and a wild boar.] n. In lexicography, used as a noun for a word composed of elements drawn from different languages. Examples of etymological hybrids include: paleosol < Greek *paleo* + Latin *solum*; sesquioxide < Latin *sesqui* + Greek *oxus*; haplogranite < Greek *haplo* + Latin *granum*; and sociology < Latin *socius* + Greek *logos*. First usage in English c.1600.

metanalysis: [Greek *meta* = beside + *analysis* = a dissolving < *analuein* = to undo < *ana* = throughout + *luein* = loosen.] n. A linguistic term relating to the transposition of a letter between a noun and its indefinite article. Examples of metanalysis are: 'a newt' < 'an ewt,' or the reciprocal 'an umpire' < 'a numpire.' Term coined by Danish linguist Otto Jespersen in 1914.

metathesis: [Greek *metatithenai* = to transpose < *meta* = beside + *tithenai* = to place; < IE *dhe* = to set, put.] n. The process by which a new word is formed by transposing the letters, sounds, or syllables of an existing word. Examples of metathesis involving letter transposition include: 'bird' < 'brid' and 'dirt' < 'drit.' An example of metathetic sound transposition can be heard in the pronunciation of the letter *r* in the word comfortable, colloquially pronounced 'comfterble.'

parasynthesis: [Greek *para* = beside, along side + *synthesis* = composition < *synthema* = collection, connection, token < *syntithemai* = to put together < *syn* = similar, together, alike + *tithenai* = to place.] n. The formation of words by a combination of compounding and adding an affix, as in downhearted, formed from down plus heart plus -ed, not down plus hearted.

portmanteau word: [French *portmanteau* = a hinged suitcase with two separate compartments < *porte-* < *porter* = to carry + *manteau* < Old French *mantel* < Latin *mantellum* = cloak.] n. A word created by juxtaposing parts (phonemes or morphemes) of two different words, thereby making a new-sounding word from the sounds of

both. The meaning of the new word is a merging of the meanings of both original words. The term was originally conceived in this sense in 1872 by Lewis Carroll in *Through the Looking Glass*. He used the term in describing to the character Alice the meanings of several strange new words contained in the poem '*The Jabberwocky*'. In addition to 'slithy', he explained that 'chortle' comes from chuckle and snort, and 'mimsy' comes from flimsy and miserable. Other examples include: 'momentaneous' from momentary and instantaneous, 'editating' from editing and annotating, 'bash' from bang and smash, 'blurt' from blow and spurt, 'glimmer' from glint and shimmer, 'scuzzy' from scum and lousy, 'prissy' from prim and sissy, and 'snazzy' from snappy and jazzy. Many modern terms indicating new aspects of popular culture that have arisen from amalgams of two other forms are now familiar words, such as: cineplex, docudrama, rockumentary, and televangelist.
Cited usage: Well, slithy means lithe and slimy. . . You see it's like a portmanteau there are two meanings packed up into one word [1872, Lewis Carroll, Through the Looking Glass].

sic: [Latin *sic* = so, thus.] conj. Used to indicate a quoted passage that has been retained in its original form, especially one containing an error or unconventional spelling, whether written intentionally or unwittingly.

syncope: [Middle English *sincopis* < Latin *sycope* < Greek *synkoptein* = to cut short < *syn* = similar, together, alike + *koptein* = to strike.] n. The loss of one or more sounds from the middle of a word, for example *fortnight* < *fourteen-night*, *foc's'le* < *forecastle*, *bos'n* < *boatswain*, and *shepherd* < *sheep herd*.
Cited usage: For the record, when bits are nicked off the front end of words it's called aphesis, when off the back it's called apocope, and when from the middle it's syncope [1990, Bill Bryson, Mother

Tongue: English & How It Got That Way].
See also: aphesis, apocope

tautology: [Greek *tautologos* = redundant < *tauto* = the same + *logos* = saying, discourse.] n. In general, a needless repetition of a word, phrase, or statement that usually amounts to a redundancy as a fault of style. In logic, a statement composed of simpler statements that are based on the same fundamental idea, but are stated in different words. The effect is to fashion a statement that is logically true, whether or not the simpler statements are factually true or false. Being a statement that is true by its own definition, a tautology is essentially uninformative. Examples of tautology are: "All crows are either black, or they are not black; or, "Tomorrow, either it will rain or it will not rain."

toponym: [from Greek *topos* = place + *onoma* = name.] n. A name derived from a place or region. Used primarily in this book as the adjective, toponymous, referring to a term named after a place. The term is a back-formation after *toponymy*, meaning the study of place names or a region or language. In antiquity, people were frequently associated by name to the place, or tribal locale, from which they were known to come, for example, Jesus the Nazarene and Atilla the Hun.
Cited usage: The Zapotec people used extensive toponyms, glyphic place signs for important landmarks, mountains in particular [Scientific American, Feb. 48/1].
Similar Terms: toponymy

Appendix II
Abbreviations

Amer. : [America] Generally referring to the United States of America.

Amer. Jrnl. Sci. : [American Journal of Science]

amu : [atomic mass unit]

anc. : [ancient] Contemporary with or prior to Greco-Roman times.

ann. : [annual] Published or occurring once each year.

arch. : [architecture]

assoc. : [association]

BCE: [Before the Common Era] BCE is a non-denominational alternative to B.C. (Before Christ). Along with the abbreviation A.D., it refers to the Christian dating scheme based on the birth of Jesus Christ being time zero.

bif: [banded iron formation]
See also: bif

bk: [book]

Brit. : [Britain, British] In reference to England, Scotland, and Wales, but not Ireland.

bull. : [bulletin] Referring to a newsletter published by some group, association, or agency, often at irregular intervals.

c. : [Latin *circa* = approximately] Usually used in reference to an approximate year when something occurred.

c.g.s. : [centimeter/gram/second]

Cambr. : [Cambrian]
See also: Cambrian

CE: [Common Era] CE is a non-denominational alternative to A.D. (Anno Domini), meaning 'in the year of our lord,' a Christian reference to the birth of Jesus Christ, which is now almost universally accepted as time zero for counting years.

cent. : [central]

cf. : [Latin *confer* = compare, see] See the item referenced after cf.

chem. : [chemistry, or chemical]

com. : [communication]

comm. : [commission]

Cty. : [county]

dept. : [department]

dict. : [dictionary]

div. : [division]

e.g. : [Latin *exempli gratia* = for example]

ecol. : [ecology]

ed. : [edition, editor]

elem. : [elementary]

encyc. : [encyclopedia]
Alt Spelling: encyclopaedia (Brit.)

et al: [Latin *et alli* = and others] Used in reference to additional contributors to a publication or project who aren't specifically named.

etc. : [Latin *et cetera* = and so on] Used in reference to things that are closely related to the specifically mentioned item, but are not actually cited.

expl. : [exploration]

First Int. Cong. of Soil Sci. {First International Congress of Soil Science]

fund. : [fundamental(s)]

Ga: [Latin *giga annum* = one billion years] Used in reference to billions of years ago.
See also: Ga

gen. : [general]

geochem. : [geochemistry]

geog. : [geography]

geol. : [geology, geological]

geomorph. : [geomorphology]

geophys. : [geophysics]

glac. : [glacial, glaciology]

gloss. : [glossary]

glossogr. : [glossography < Greek *glossa* = tongue + *graphia* = writing] The writing of comments, explanations, or glosses (words inserted in text to explain foreign or difficult words). Also, by extension, the explanation or definition of a term given in a glossary or dictionary.

Gr. : [Greek]

gram. : [grammatical]

Gt. : [Great] Used in place names, such as Gt. Britain and the Gt. Barrier Reef.

hist. : [history]

husb. : [husbandry]

i.e. : [Latin *id est* = that is] *Id est* was formerly used in works written in Latin as a brief way to say 'that is to say' when explaining a word or phrase; retained in English in the same sense, but usually only used as an abbreviation.

I.U.P.A.C. : [International Union of Pure and Applied Chemistry]

ibid. : [Latin *ibidem* = in the same place] Used in footnotes and bibliographies to show that the cited book, chapter, article, or page is the same as in the previous reference.

IE: [Indo-European] In reference to the language that was the root for Sanskrit and most languages now spoken in India and Europe.

illustr. : [illustration]

impr. : [impression]

induct. : [induction, inductive]

Inst. : [Institute] Used in place names, generally those of distinction for being an organization of people dedicated to preserving and advancing aspects of culture, research, art, etc., such as: The Massachusetts Institute of Technology (MIT), The National Institute of Standards and Technology, and The Art Institute of Chicago.

intell. : [intelligence]

inter. : [international]

introd. : [introduction]

invert. : [invertebrate]
See also: invertebrate

IR: [infrared] The range of invisible radiation wavelengths from about 750 nanometers, just longer than red in the visible spectrum, to 1 millimeter, on the border of the microwave region.

itin. : [itinerary < Latin *iter* = journey] An account or record of a journey.

IUGS: [International Union of Geological Sciences]

jrnl. : [journal]

kingd. : [kingdom]

let. : [letter(s)]

lex. : [lexicon, lexicographic] Referring to dictionaries or glossaries, or to the making of them.

Linn. : [Linnaean]
See also: Linnaean

Lond. : [London, England]

Ma: [Latin *mega-annum* = one million years] A time measurement used in reference to millions of years ago
See also: Ma

man. : [manual]

manip. : [manipulation]

mem. : [memoir(s) < Old French *memoire* = memory; < Latin *memor* = mindful] An autobiographical sketch, a personal account of one's experiences, or the report of the proceedings of a learned society.

Metrop. : [Metropolitan < Latin *metropolis* = mother-city; < Greek *meter* = mother + *polis* = city]

microgr. : [micrograph] A drawing or photographic reproduction of an object as viewed through a microscope.

microsc. : [microscope, microscopy]

min. : [mineral, mineralogy]
See also: mineral, mineraloid

modif. : [modification]

monogr. : [monograph] A scholarly piece of written work on a specific topic, can be a succinct essay or a lengthy book.

Mt. : [Mount, Mountain]

n. : [noun < Latin *nomen* < Greek *onoma* = name, noun] The part of speech that is used to name a person, place, thing, quality, or action. A noun function as the subject or object of a verb, or as an object of a preposition.

n.b. : [Latin *nota bene* = note well] Used to direct attention to something particularly important.

narr. : [narrative] Consisting of or characterized by the telling of a story. Some narrative styles are: chronicle, saga, ballad, epic, fable, history, or tale.

NASA: [National Aeronautic and Space Administration]

nat. : [national]

observ. : [observation]

orig. : [origin, original]

OUCC: [Oxford University Caving Club]

outl. : [outline]

p. : [page]

paleontol. : [paleontology]
Alt Spelling: palaeontology (Brit.)
See also: paleontology

petrogr. : [petrography]
See also: petrography

petrol. : [petrology]
See also: petrology

pf. : [perfect]

Pg. : [Portuguese]

philos. : [philosophy; < Greek *philosophos* = lover of wisdom < *philos* = lover + *sophos* = wisdom] In ancient times, a system of thought based on the love and pursuit of knowledge by intellectual prowess, physical pursuit, and moral self-discipline. In modern times, extended

to the investigation of the nature, causes, or principles of reality, knowledge, or values, based on logical reasoning rather than empirical methods.

phys. : [physical, physics]

physiogr. : [physiography] Physical geography, or the study of the natural features of the Earth's surface, especially in its current aspects, including land formation, climate, currents, and distribution of flora and fauna.

pl. : [plural]

pop. : [population]

pp. : [pages]

pract. : [practical]

preh. : [prehistoric] Relating to the time before history was recorded by humans.

primit. : [primitive < Latin *primus* = first] Not derived from something else. Referring to an early or original state. In anthropology, referring to a nonindustrial, often tribal culture, especially one that is characterized by a low level of economic complexity. In linguistics, serving as the basis for derived or inflected forms; also, a proto-language.

princ. : [principles; < Latin *princeps* = leader, emperor] A basic truth, law, or assumption, as compared to 'principal', meaning first, highest, or foremost in importance. Both words have the same Latin root.

prof. : [professional]

prov. : [province; < Latin *provincia* = official duty] An administrative division of a country, state, kingdom, or empire. Applied to the North

American Colonies of Great Britain, now provinces of Canada. In Roman History, a territory outside Italy that fell under Roman dominion, and was administered by a governor sent from Rome.

prov. : [proverb; < Latin *pro* = forward + *verbum* = word] A concise sentence, often metaphorical, poetical, or alliterative in form, that expresses some truth ascertained by experience or observation, especially one that is familiar to all. An adage.

P-T-X: [pressure-temperatrure-composition] Used in describing fundamental physical conditions affecting a system, often including time as a factor, then written P-T-X-T

publ. : [publication; < Latin *publicus* = public]

rep. : [report]

rev. : [review; < Latin *re* = again + *videre* = to view] To write or give a critical report, or a chronology of past experiences or observations.

rom. : [romance]

Rom. : [Roman]

s.v. : [Latin *sub verbo* = under verb] Used in a bibliographical reference to tell what word or phrase to use referencing a topic in another work; for example: (BGloss. Geol 4, s.v. "oxbow lake") tells the reader to see the entry "oxbow lake" in Bates and Jackson's 4th edition Glossary of Geology.

sci. : [science]

Scot. : [Scotland]

sec. : [section]

sec. : [second; < Latin *secundus* = second, following] The time needed for a cesium-133 atom to perform 9,192,631,770 complete oscillations.

secr. : [secrets]

seismol. : [seismology]
See also: seismology

serm. : [sermon; < Latin *sermon* = discourse] A discourse, usually religious in nature, delivered as part of a service. Sermons often have a tone of reproof, reproach, or exhortation, and are often noted for their length and tediousness.

Silur. : [Silurian]
See also: Silurian

sk. : [sketches]

SNC: [shergottite, nakhlite, and cassignite]
See also: shergottite, nakhlite

Soc. : [Society; < Latin *socius* = companion]

spec. : [species]

std. : [standard]

stp: [standard temperature and pressure] In physics and chemistry, 0°C at 1 atmosphere of pressure. In aviation, 15°C at 1 atmosphere and 0% humidity.

strat. : [stratigraphy]
See also: stratigraphy

struc. : [structure; < Latin *structus* < *struere* = to construct]

stud. : [studies; < Latin *studere* = to study]

subterr. : [subterranean; Latin *sub* = below, beneath, under + *terra* = earth] Situated beneath the Earth's surface, and operating underground.

supp. : [supplement; < Latin *supplere* = to complete] A section added to a piece of writing designed to give further information, or to correct errors. Also, a separate section devoted to a special subject inserted into a publication.

surg. : [surgical, surgery; < Greek *kheirourgos* = working by hand < *kheir* = hand + *ergon* = work]

surv. : [survey; < Latin *supervidere* < *super* = over + *videre* = look] To examine or look at comprehensively. In geography, to determine the boundaries, area, or elevations of the land or a structure by means of measuring angles and distances.

syst. : [system; < Greek *syn* = together, with, beside + *histanai* = to set up, establish] A group of interacting, interrelated, or interdependent elements forming a complex whole.

T: [temperature; Latin *temperatura* = due measure < *temperatus* < *temperare* = to mix < *tempus* = time, season]

teach. : [teachings]

tech. : [technology; Greek *tekhnologia* = systematic treatment of an art or craft < *tekhne* = skill + *logos* = discourse, word] The application of science, especially to industrial or commercial objectives.

tr. : [translation; Latin *translatus* = transference < *trans* = cross, over + *litus* = brought]

trav. : [travels; Middle English *travailen* = to toil]

treat. : [treatise; < Latin *tractare* = to drag about, deal with] A systematic, usually extensive written discourse on a subject.

txtbk. : [textbook; < Latin *texere* = to weave + Old English *boc* = charter, beech tree]

U.S. : [United States of America]

U.S.D.A. : [United States Department of Agriculture]

U.S.G.S. : [United States Geological Survey]

Univ. : [University; Latin *universitas* = the whole, a corporate body < *universus* = whole < *unus* = one + *vertere* = to turn]

UV: [ultraviolet; < *ultra* = extreme, beyond + *viola* = purplish flower, violet] The range of invisible radiation wavelengths from about 4 nanometers, on the border of the x-ray region, to about 380 nanometers, just beyond the violet in the visible spectrum.

v. : [verb; < Latin *verbum* = word] The part of speech that expresses existence, action, or occurrence.

vasc. : [vascular; < Latin *vasculum* < *vas* = vessel] Characterized by vessels that carry or circulate fluids, such as blood, lymph, or sap, through the body of an animal or plant.

vol. : [volume; < Latin *volumen* = roll of writing < *volvere* = to roll] A collection of written or printed sheets bound together.

volc. : [volcano]
See also: volcano

voy. : [voyage; < Latin *viaticus* = of a journey < *via* = road]

wks. : [works; < Old English *weorc* = work; < IE *werg* = to do] The output of a writer, artist, or musician, usually considered as a whole (e.g., the works of Shakespeare).

y.b.p. : [years before present]